ELLIPTIC, HYPERBOLIC AND MIXED COMPLEX EQUATIONS WITH PARABOLIC DEGENERACY

Including Tricomi–Bers and Tricomi–Frankl–Rassias Problems

PEKING UNIVERSITY SERIES IN MATHEMATICS

Series Editor: Kung-Ching Chang *(Peking University, China)*

Vol. 1: An Introduction to Finsler Geometry
by Xiaohuan Mo (Peking University, China)

Vol. 2: Numerical Methods for Exterior Problems
by Lung-An Ying (Peking University & Xiamen University, China)

Vol. 3: Approaches to the Qualitative Theory of Ordinary Differential Equations: Dynamical Systems and Nonlinear Oscillations
by Tongren Ding (Peking University, China)

Vol. 4: Elliptic, Hyperbolic and Mixed Complex Equations with Parabolic Degeneracy: Including Tricomi–Bers and Tricomi–Frankl–Rassias Problems
by Guo Chun Wen (Peking University, China)

Peking University Series in Mathematics — Vol. 4

ELLIPTIC, HYPERBOLIC AND MIXED COMPLEX EQUATIONS WITH PARABOLIC DEGENERACY

Including Tricomi–Bers and Tricomi–Frankl–Rassias Problems

Guo Chun Wen

Peking University, China

World Scientific

NEW JERSEY • LONDON • SINGAPORE • BEIJING • SHANGHAI • HONG KONG • TAIPEI • CHENNAI

Published by

World Scientific Publishing Co. Pte. Ltd.
5 Toh Tuck Link, Singapore 596224
USA office: 27 Warren Street, Suite 401-402, Hackensack, NJ 07601
UK office: 57 Shelton Street, Covent Garden, London WC2H 9HE

British Library Cataloguing-in-Publication Data
A catalogue record for this book is available from the British Library.

Peking University Series in Mathematics — Vol. 4
ELLIPTIC, HYPERBOLIC AND MIXED COMPLEX EQUATIONS WITH PARABOLIC DEGENERACY
Including Tricomi–Bers and Tricomi–Frankl–Rassias Problems

ISBN-13 978-981-277-942-7
ISBN-10 981-277-942-6

Printed in Singapore by World Scientific Printers (S) Pte Ltd

Preface

This book is a continuation and development of the author's book (see [86]33)), and mainly deals with various boundary value problems for equations and systems of mixed (elliptic-hyperbolic) type with parabolic degeneracy. For this the corresponding boundary value problems for elliptic and hyperbolic complex equations of first and second orders are firstly considered, in which some representations and a priori estimates of solutions for the above problems are given, and the existence and uniqueness of solutions of these problems are proved. In the whole book we apply the complex analytic method in details.

In Chapters I and II, by using some new methods, we mainly investigate some discontinuous boundary value problems for some classes of elliptic complex equations of first and second orders with smooth and nonsmooth parabolic degenerate lines, which include the discontinuous Riemann-Hilbert boundary value problem, the mixed boundary value problem and discontinuous oblique derivative boundary value problem. In which we first reduce the degenerate elliptic equations or systems to some complex equations with singular coefficients, after this it is not very difficult to obtain a priori estimates of solutions for the above boundary value problems, and then the existence and uniqueness of solutions of these problems can be proved. As an application of the above results, we discuss a boundary value problem in axisymmetric filtration with homogeneous medium.

In Chapter III, on the basis of notations of hyperbolic numbers and hyperbolic complex functions, the hyperbolic systems of first order equations and hyperbolic equations of second order with some conditions can be reduced to complex forms. Moreover, several boundary value problems, mainly the Riemann-Hilbert boundary value problem, oblique derivative boundary value problem for first and second orders hyperbolic complex equations with parabolic degeneracy are discussed, which includes the Dirichlet boundary value problem as a special case. In addition, we discuss the Cauchy problem for second order hyperbolic equations with degenerate rank 0.

In Chapter IV, by using the notations of complex numbers in elliptic domains and hyperbolic numbers in hyperbolic domains, we mainly introduce the Riemann-Hilbert boundary value problem for first order linear and quasilinear complex equations of mixed type in special domains and general domains, the results obtained in which are the preparation for latter chapters.

For the classical gas dynamical equation of mixed type due to S. A. Chaplygin [17], the first really deep results were published by F. G. Tricomi [78]1). In Chapters V and VI, on the basis of the results obtained in Chapters I–IV, we consider the Tricomi boundary value problem, oblique derivative problem for second order linear and quasilinear complex equations of mixed type with parabolic degeneracy in several domains including general domains and multiply connected domains. We mention that in the books [12]1),3) and [74], the authors investigated the Tricomi problem for the Chaplygin equation: $K(y)u_{xx}+u_{yy}=0$ and special second order equations of mixed type with parabolic degeneracy in some standard domains and general domains by using the methods of integral equations and energy integral, but the methods are not simple. In the present book, we apply the uniqueness and existence of solutions of discontinuous boundary value problems for elliptic, hyperbolic complex equations and other methods to obtain the solvability results of several discontinuous oblique derivative problems for second order equations of mixed type, which include the Tricomi problem as a special case. Besides, we also discuss the Frankl problem and the exterior Tricomi problem for general equations of mixed type with parabolic degeneracy.

There are two characteristics of this book: one is that elliptic, hyperbolic and mixed complex equations are included in several forms and the quasilinear case, and boundary value conditions are almost considered in the discontinuous Riemann-Hilbert problem and oblique derivative problem, especially multiply connected domains are considered. Another one is that several complex methods are used to investigate various problems about complex equations of elliptic, hyperbolic and mixed type, for example the complex functions in elliptic domains and the hyperbolic complex functions in hyperbolic domains are used, and in general we first discuss the corresponding problems for first order complex equations, and then the problems for second order equations can be solved. The above method is different from the methods used by other authors. We mention that some boundary value problems in gas dynamics can be handled by using the results as stated in this book.

The great majority of the contents originates in investigations of the author and cooperative colleagues, and many results are published here for the first time, in which many open problems are solved, for instance the Tricomi problem for second order equations of mixed type in special and general multiply connected domains posed by L. Bers in [9]2), and the existence, regularity of solutions of some boundary value problems for second order equations of mixed type with smooth and nonsmooth degenerate lines posed by J. M. Rassias in [71]2). After reading the volume, it can be seen that many questions remain for further investigations.

In order to conveniently understand the purpose of the book, we give a sketchy introduction about the idea of the book. It is known that the general second order linear equation of mixed type is as follows

$$K(y)u_{xx}+u_{yy}+a(x,y)u_x+b(x,y)u_y+c(x,y)u+d(x,y)=0 \text{ in } D, \tag{1}$$

especially

$$K(y)u_{xx}+u_{yy}=0 \text{ in } D \tag{2}$$

is the so-called famous Chaplygin equation in gas dynamics, where $K(y)$ possesses the derivative $K'(y)$ and $yK(y)>0$ on $y\neq 0$, $K(0)=0$, and D is a bounded domain including the domain D^+ in the upper-half plane, the domain D^- in the lower-half plane and the line γ on the real axis, obviously equation (1) in D^+ is the elliptic type and in D^- is the hyperbolic type, and γ is a parabolic degenerate line. From [9]2) and [71]2), we can see the mechanical background of equations (1) and (2). The so-called Tricomi problem is to find a solution $u(z)$ of equation (1) or (2) satisfying the boundary conditions

$$u(z)=\phi(z) \text{ on } \Gamma,\ u(z)=\psi(z) \text{ on } L_1 \text{ or } L_2, \tag{3}$$

where Γ is the partial boundary of D^+ in the upper-half plane and L_1, L_2 are two characteristics in the lower-half plane, here $L_1\cup L_2\cup\gamma$ is the boundary of D^-, and $z_0=x_0+jy_0$ is the intersection point of L_1 and L_2. The above Tricomi problem can be divided into two boundary value problems, namely the mixed boundary value problem

$$u(z)=\phi(z) \text{ on } \Gamma,\ u_y=r(x) \text{ on } \gamma, \tag{4}$$

of equation (1) in D^+, and the mixed boundary value problem

$$u(z)=\psi(z) \text{ on } L_1 \text{ or } L_2,\ u_y=r(x) \text{ on } \gamma \tag{5}$$

of equation (1) in D^-, where $r(x)$ on γ is an undetermined function, which can be determined by the boundary condition of Problem T of the mixed equations.

According to the mathematical view, the centrical subject of the Tricomi problem is to prove the existence and uniqueness of solutions for the Tricomi problem, and verify some regularity of solutions of the above problem. In recent half century, many mathematical authors investigated the problem about several equations of mixed type and obtained many interesting results (see [9]2), [12]1),3), [71], [74], [86]33) and so on). However the Tricomi problem of equation (1) has not been completely solved, and the obtained results almost require some unnecessary conditions, besides for instance the Tricomi problem for second order equations of mixed type in multiply connected domains posed by L. Bers in the book [9]2), i.e. Tricomi-Bers problem, and the existence, regularity of solutions of above problems for mixed equations with nonsmooth degenerate line in several domains posed by J. M. Rassias in the book [71]1), i.e. Tricomi-Frankl-Rassias problems have not been solved. The purpose of the book is just to introduce some new results about the above problems. As stated before we need first introduce the related results about degenerate elliptic equations and degenerate hyperbolic equations, and then the problems about equations of mixed type are discussed (see Sections 3 and 4, Chapter VI).

Our method of handling mixed equations is the complex analytic method, which is different to the methods of other authors, i.e. through the transformation of functions: $W(z) = [H(y)u_x - iu_y]/2$ in D^+ and $W(z) = [H(y)u_x - ju_y]/2$ in D^-, here $H(y) = \sqrt{|K(y)|}$, $Y = G(y) = \int_0^y H(t)dt$, i and j are the imaginary unit and hyperbolic imaginary unit with the conditions $i^2 = -1$ and $j^2 = 1$ respectively, and denote $Z = x + iY$ in $\overline{D^+}$ and $Z = x + jY$ in $\overline{D^-}$, the equation (1) can be reduced to the complex equation of first order with singular coefficients

$$W_{\overline{Z}} = A_1(Z)W(Z) + A_2(Z)\overline{W(Z)} + A_3(Z)u(Z) + A_4(Z) \text{ in } D_Z, \quad (6)$$

where D_Z is the image domain of D with respect to the mapping $Z = Z(z) = x+iY = x+iG(y)$ in D^+ and $Z = Z(z) = x+jY = x+jG(y)$ in D^-. Moreover we find the directional derivatives of boundary condition (3) or (4), (5) according to the parameter of arc length of the boundaries Γ and L_1 or L_2, then (4), (5) can be written in the complex form

$$\operatorname{Re}[\overline{\lambda(Z)}W(Z)] = R(Z) \text{ on } \Gamma \cup L_1 \cup \gamma \text{ or } \Gamma \cup L_2 \cup \gamma, \quad (7)$$

in which $\lambda(Z) = i$ or $-j$, $R(Z) = -r(z)/2$ on γ, this is a boundary condition of Riemann-Hilbert type, we can use the similar method as stated in [86]11),33) and [87]1) to solve the problem of equation (6), but we must give some important modifications, and then the Tricomi problem for equation (1) can be solved.

In this book, we mainly consider three classes of second order equations of mixed type with parabolic degeneracy, i.e. the equation (1),

$$K_1(y)u_{xx}+|K_2(x)|u_{yy}+a(x,y)u_x+b(x,y)u_y+c(x,y)u+d(x,y)=0 \text{ in } D, \quad (8)$$

and

$$K_1(y)u_{xx}+|K_2(y)|u_{yy}+a(x,y)u_x+b(x,y)u_y+c(x,y)u+d(x,y)=0 \text{ in } D, \quad (9)$$

in which $K_1(y)$, $K_2(y)$ satisfy the condition similar to that of $K(y)$ in (1), and $K_2(x)$ possesses the derivative $K_2'(x)$ and $xK_2(x)>0$ on $x\neq 0$, $K_2(0)=0$. It is easy to see that in (8), the real axis and the imaginary axis are parabolic degenerate lines, hence the parabolic degenerate curve is not smooth near the origin point. Except the above linear equations, we also discuss some quasilinear or nonlinear elliptic, hyperbolic and mixed equations with parabolic degeneracy.

The more general boundary value problem is the oblique derivative boundary value problem, which is to find a continuous solution $u(z)$ of (1) in $\overline{D}$ satisfying the boundary conditions

$$\begin{aligned}&\frac{1}{2}\frac{\partial u}{\partial l}=\frac{1}{H(y)}\mathrm{Re}[\overline{\lambda(z)}u_{\tilde{z}}]=\mathrm{Re}[\overline{\Lambda(z)}u_z]=r(z) \text{ on } \Gamma\cup L_1,\\ &\frac{1}{H(y)}\mathrm{Im}[\overline{\lambda(z)}u_{\tilde{z}}]|_{z=z_0}=b_0,\, u(z_1)=b_1,\, u(z_2)=b_2,\end{aligned} \quad (10)$$

in which l is a given vector at every point $z\in\Gamma\cup L_1$, $W(z)=u_{\tilde{z}}=[H(y)u_x-iu_y]/2$, $\Lambda(z)=\cos(l,x)-i\cos(l,y), \lambda(z)=\mathrm{Re}\lambda(z)+i\mathrm{Im}\lambda(z)$, $\cos(l,n)\geq 0$ for $z\in\Gamma$, n is the outward normal vector of Γ, and $W(z)=u_{\tilde{z}}=[H(y)u_x-ju_y]/2$, $\lambda(z)=\mathrm{Re}\lambda(z)+j\mathrm{Im}\lambda(z)$ for $z\in L_1$, z_1, z_2 are two end points of γ, b_0,b_1,b_2 are real constants. We can give the well-posed formulation according to the index of boundary conditions in the elliptic domain. It is clear that the oblique derivative problem includes the Tricomi problem as a special case. In the book, we mainly handle the oblique derivative problem about elliptic equations, hyperbolic equations and mixed equations with parabolic degeneracy, which include the above three classes of equations. Except the above boundary value problems, we also introduce the discontinuous oblique derivative problem, exterior Tricomi-Rassias problem and the Frankl problem in Sections 3-5 of Chapter V. Besides, the Tricomi problem for second order equations of mixed type with parabolic degeneracy in general domains and multiply connected domains are discussed. This book has not been related to the Tricomi problem for equations in higher dimensional domains (see [71]2),6)), some problems on this hand remain to be further investigated.

Finally the author would like to acknowledge the support of NSFC (No. 10671207), and Prof. J. M. Rassias, for his many valuable opinions for improving the manuscript of the book.

Beijing
June, 2007

Guo Chun Wen
Peking University

Contents

CHAPTER I
ELLIPTIC COMPLEX EQUATIONS OF FIRST ORDER

In this chapter, we mainly discuss the discontinuous Riemann-Hilbert boundary value problem for some degenerate elliptic systems of first order equations. Firstly we reduce the above systems to a class of complex equations with singular coefficients, give the representations and a priori estimates of solutions of the boundary value problem for the class of degenerate elliptic complex equations, and then prove the existence and uniqueness of solutions for the boundary value problem.

1 The Discontinuous Riemann-Hilbert Problem for Nonlinear Uniformly Elliptic Complex Equations of First Order

First of all, we reduce general uniformly elliptic systems of first order equations with certain conditions to the complex equations, and then give estimates of solutions of the discontinuous Riemann-Hilbert problem for the complex equations, finally we verify the solvability of the boundary value problem.

1.1 Reduction of general uniformly elliptic systems of first order equations to standard complex form

Let D be a bounded simply connected domain in $\mathbf{R}^2$ with the boundary ∂D. Without loss of generality we can assume that ∂D is a smooth closed curve, because the requirement can be realized through a conformal mapping. We first consider the linear uniformly elliptic system of first order equations

$$\begin{aligned} a_{11}u_x + a_{12}u_y + b_{11}v_x + b_{12}v_y &= a_1u + b_1v + c_1, \\ a_{21}u_x + a_{22}u_y + b_{21}v_x + b_{22}v_y &= a_2u + b_2v + c_2, \end{aligned} \tag{1.1}$$

where the coefficients $a_{jk}, b_{jk}, a_j, b_j, c_j (j, k=1, 2)$ are known real bounded measurable functions of $(x, y) \in D$. The uniform ellipticity condition in D

is as follows

$$\begin{aligned} J &= 4K_1K_4 - (K_2+K_3)^2 \\ &= 4K_5K_6 - (K_2-K_3)^2 \geq J_0 > 0,\ K_1 > 0 \text{ in } D, \end{aligned} \tag{1.2}$$

in which J_0 is a positive constant and

$$K_1 = \begin{vmatrix} a_{11} & b_{11} \\ a_{21} & b_{21} \end{vmatrix},\ K_2 = \begin{vmatrix} a_{11} & b_{12} \\ a_{21} & b_{22} \end{vmatrix},\ K_3 = \begin{vmatrix} a_{12} & b_{11} \\ a_{22} & b_{21} \end{vmatrix},$$

$$K_4 = \begin{vmatrix} a_{12} & b_{12} \\ a_{22} & b_{22} \end{vmatrix},\ K_5 = \begin{vmatrix} a_{11} & a_{12} \\ a_{21} & a_{22} \end{vmatrix},\ K_6 = \begin{vmatrix} b_{11} & b_{12} \\ b_{21} & b_{22} \end{vmatrix}.$$

From $J > 0$ it follows that

$$K_1K_6 > 0, \text{ or } K_1K_6 < 0, \text{ i.e. } K_1 > 0,\ K_6 \neq 0.$$

We can assume that $K_6 > 0$. Hence from the elliptic system (1.1), we can solve v_x, v_y and obtain the system of equations

$$\begin{aligned} v_y &= au_x + bu_y + a_0u + b_0v + f_0, \\ -v_x &= du_x + cu_y + c_0u + d_0v + g_0, \end{aligned} \tag{1.3}$$

where $a = K_1/K_6$, $b = K_3/K_6$, $c = K_4/K_6$, $d = K_2/K_6$, and the uniform ellipticity condition (1.2) is transformed into the condition

$$\Delta = \frac{J}{4K_6^2} = ac - \frac{1}{4}(b+d)^2 \geq \Delta_0 > 0,\ a > 0, \tag{1.4}$$

here Δ_0 is a positive constant and a, b, c, d are bounded for almost every point in D. Noting that

$$\begin{aligned} &z = x+iy,\ w = u+iv,\ w_z = \frac{1}{2}(w_x - iw_y),\ w_{\bar z} = \frac{1}{2}(w_x + iw_y), \\ &u_x = \frac{1}{2}(w_z + \overline{w}_{\bar z} + w_{\bar z} + \overline{w}_z),\ u_y = \frac{i}{2}(w_z - \overline{w}_{\bar z} - w_{\bar z} + \overline{w}_z), \\ &v_x = \frac{i}{2}(-w_z + \overline{w}_{\bar z} - w_{\bar z} + \overline{w}_z),\ v_y = \frac{1}{2}(w_z + \overline{w}_{\bar z} - w_{\bar z} - \overline{w}_z), \end{aligned}$$

the system (1.3) can be written in the complex form

$$w_{\bar z} = Q_1(z)w_z + Q_2(z)\overline{w}_{\bar z} + A_1(z)w + A_2(z)\overline{w} + A_3(z), \tag{1.5}$$

where

$$Q_1(z)=\frac{-2q_2}{|q_1+1|^2-|q_2|^2},\ Q_2(z)=\frac{|q_2|^2-(q_1-1)(\overline{q_1}+1)}{|q_1+1|^2-|q_2|^2},$$

$$q_1(z)=\frac{1}{2}[a+c+i(d-b)],\ q_2(z)=\frac{1}{2}[a-c+i(d+b)].$$

On the basis of

$$|q_1+1|^2-|q_2|^2=\frac{1}{4}[(2+a+c)^2+(d-b)^2]$$
$$-\frac{1}{4}[(a-c)^2+(d+b)^2]=1+a+c+\left(\frac{d-b}{2}\right)^2+\Delta\geq 1+\Delta,$$

the uniform ellipticity condition (1.4) can be written in the complex form

$$|Q_1(z)|+|Q_2(z)|\leq q_0<1, \tag{1.6}$$

in which q_0 is a non-negative constant. If the coefficients $a_{jk},b_{jk}\in W_p^1(D)$, $p>2$, $j,k=1,2$, then the following function $\eta(z)$ can be extended in $D_R=\{|z|\leq R\}$ $(\supset D,\ 0<R<\infty)$, such that $\eta(z)\in W_p^1(D_R)$, thus the Beltrami equation

$$\zeta_{\bar z}-\eta(z)\zeta_z=0,$$
$$\eta(z)=\frac{2Q_1(z)}{1+|Q_1|^2-|Q_2|^2+\sqrt{(1+|Q_1|^2-|Q_2(z)|^2)^2-4|Q_1|^2}} \tag{1.7}$$

has a homeomorphic solution $\zeta(z)\,(\in W_{p_0}^2(D_R))$, and its inverse function $z(\zeta)\in W_{p_0}^2(G_R)$, herein $G_R=\zeta(D_R)$ and $p_0\,(2<p_0\leq p)$ is a positive constant. Setting $w=w[z(\zeta)]$, the complex equation (1.5) is reduced to the complex equation

$$w_{\bar\zeta}=Q(\zeta)\bar w_{\bar\zeta}+B_1(\zeta)w+B_2(\zeta)\bar w+B_3(\zeta), \tag{1.8}$$

in which

$$Q(\zeta)=\frac{Q_2[z(\zeta)]}{1-\eta[z(\zeta)]\overline{Q_1[z(\zeta)]}},$$

$$B_1(\zeta)=\{A_1[z(\zeta)]+\overline{A_2[z(\zeta)]}Q(\zeta)\eta[z(\zeta)]\}\bar z_{\bar\zeta},$$

$$B_2(\zeta)=\{A_2[z(\zeta)]+\overline{A_1[z(\zeta)]}Q(\zeta)\eta[z(\zeta)]\}\bar z_{\bar\zeta},$$

$$B_3(\zeta)=\{A_3[z(\zeta)]+\overline{A_3[z(\zeta)]}Q(\zeta)\eta[z(\zeta)]\}\bar z_{\bar\zeta}.$$

Setting $W(\zeta)=w(\zeta)-Q(\zeta)\overline{w(\zeta)}$, the complex equation (1.8) can be transformed into the complex equation

$$W_{\bar\zeta}=C_1(\zeta)W+C_2(\zeta)\overline{W}+C_3(\zeta), \tag{1.9}$$

in which

$$C_1(\zeta)=\frac{B_1+(B_2-Q_{\bar\zeta})\overline{Q}}{1-|Q|^2},\ C_2(\zeta)=\frac{B_1Q+B_2-Q_{\bar\zeta}}{1-|Q|^2},\ C_3(\zeta)=B_3,$$

(see [86]9), [87]1)). This is a standard complex form of the uniformly elliptic system (1.1), which is called the nonhomogeneous generalized Cauchy-Riemann system, and the solution of homogeneous generalized Cauchy-Riemann system in D is called the pseudoanalytic function (see [9]1)) or the generalized analytic function (see [81]1)).

For the nonlinear uniformly elliptic system of first order equations

$$F_j(x,y,u,v,u_x,v_x,u_y,v_y)=0 \text{ in } D,\ j=1,2, \tag{1.10}$$

under certain conditions, we can transform the system into the complex form

$$w_{\bar z}=F(z,w,w_z),\ F=Q_1w_z+Q_2\overline{w}_{\bar z}+A_1w+A_2\bar w+A_3,\ z\in D, \tag{1.11}$$

where $Q_j=Q_j(z,w,w_z), j=1,2,\ A_j=A_j(z,w),\ j=1,2,3$ (see [86]9), [87]1)). We assume that equation (1.11) satisfy the following conditions.

Condition C :

1) $Q_j(z,w,U)\,(j=1,2)$, $A_j(z,w)\,(j=1,2,3)$ are measurable in $z\in D$ for all continuous functions $w(z)$ in $D^*=\bar D\backslash Z$ and all measurable functions $U(z)\in L_{p_0}(D^*)$, and satisfy

$$L_p[A_j,\overline{D}]\le k_0,\ j=1,2,\ L_p[A_3,\overline{D}]\le k_1, \tag{1.12}$$

where $Z=\{z_1,...,z_m\}$, $z_1,...,z_m$ are different points on the boundary ∂D arranged according to the positive direction successively, $U(z)\in L_{p_0}(D^*)$ means $U(z)\in L_{p_0}(\tilde D^*)$, $\tilde D^*$ is any closed subset in D^*, and $p_0,p\,(2<p_0\le p), k_0,k_1$ are non-negative constants.

2) The above functions are continuous in $w\in\mathbf{C}$ for almost every point $z\in D, U\in\mathbf{C}$, and $Q_j=0\,(j=1,2)$, $A_j=0\,(j=1,2,3)$ for $z\notin D$.

3) The complex equation (1.11) satisfies the uniform ellipticity condition

$$|F(z,w,U_1)-F(z,w,U_2)|\le q_0|U_1-U_2|, \tag{1.13}$$

for almost every point $z\in D$, in which $w,U_1,U_2\in\mathbf{C}$ and $q_0\,(<1)$ is a non-negative constant.

1.2 Representation of solutions of discontinuous Riemann-Hilbert problem for elliptic complex equations

Let D be a bounded domain in $\mathbf{C}$ with the smooth boundary $\partial D = \Gamma$. Now we formulate the discontinuous Riemann-Hilbert problem for equation (1.11).

Problem A The discontinuous Riemann-Hilbert boundary value problem for (1.11) is to find a continuous solution $w(z)$ in D^* satisfying the boundary condition:

$$\mathrm{Re}[\overline{\lambda(z)}w(z)] = r(z),\ z \in \Gamma^* = \partial D\backslash Z, \tag{1.14}$$

where $\lambda(z), r(z)$ satisfy the conditions

$$\begin{gathered} C_\alpha[\lambda(z), \Gamma_j] = \sup_{\Gamma_j} |\lambda(z)| + \sup_{z_1 \neq z_2} \frac{|\lambda(z_1) - \lambda(z_2)|}{|z_1 - z_2|^\alpha} \leq k_0 \\ C_\alpha[R_j(z)r(z), \Gamma_j] \leq k_2,\ j = 1, ..., m, \end{gathered} \tag{1.15}$$

in which $\lambda(z) = a(z) + ib(z), |\lambda(z)| = 1$ on ∂D, and $Z = \{z_1, ..., z_m\}$ are the first kind of discontinuous points of $\lambda(z)$ on ∂D, Γ_j is an arc from the point z_{j-1} to z_j on ∂D, and does not include the end points z_{j-1}, z_j $(j = 1, 2, ..., m)$, herein $z_0 = z_m$, $R_j(z) = |z - z_{j-1}|^{\beta_{j-1}}|z - z_j|^{\beta_j}$, $\alpha\,(1/2 < \alpha \leq 1), k_0, k_2, \beta = \min(\alpha, 1 - 2/p_0), \beta_j (0 < \beta_j < 1), \gamma_j$ are non-negative constants and satisfy the conditions

$$\beta_j + \gamma_j < \beta,\ j = 1, ..., m, \tag{1.16}$$

where $\gamma_j (j = 1, ..., m)$ are as stated in (1.17) below. Problem A with $A_3(z) = 0$ in D, $r(z) = 0$ on Γ^* is called Problem A_0.

Denote by $\lambda(z_j - 0)$ and $\lambda(z_j + 0)$ the left limit and right limit of $\lambda(z)$ as $z \to z_j$ $(j = 1, 2, ..., m)$ on ∂D, and

$$\begin{gathered} e^{i\phi_j} = \frac{\lambda(z_j - 0)}{\lambda(z_j + 0)},\ \gamma_j = \frac{1}{\pi i}\ln\frac{\lambda(z_j - 0)}{\lambda(z_j + 0)} = \frac{\phi_j}{\pi} - K_j, \\ K_j = \left[\frac{\phi_j}{\pi}\right] + J_j,\quad J_j = 0 \text{ or } 1,\quad j = 1, ..., m, \end{gathered} \tag{1.17}$$

in which $0 \leq \gamma_j < 1$ when $J_j = 0$, and $-1 < \gamma_j < 0$ when $J_j = 1$, $j = 1, ..., m$. The index K of Problems A and A_0 is defined as follows:

$$K = \frac{1}{2}(K_1 + \cdots + K_m) = \sum_{j=1}^{m}\left[\frac{\phi_j}{2\pi} - \frac{\gamma_j}{2}\right]. \tag{1.18}$$

If $\lambda(x)$ on Γ is continuous, then $K = \Delta_\Gamma \arg \lambda(x)/2\pi$ is a unique integer. Now the function $\lambda(x)$ on Γ is not continuous, we can choose $J_j = 0$ or 1, hence the index K is not unique. If we choose $K = -1/2$, then the solution of Problem A is unique.

In order to prove the solvability of Problem A for the complex equation (1.11), we need to give a representation theorem for Problem A.

Theorem 1.1 *Suppose that the complex equation* (1.11) *satisfies Condition C, and* $w(z)$ *is a solution of Problem A for* (1.11). *Then* $w(z)$ *is representable by*

$$w(z) = \Phi[\zeta(z)]e^{\phi(z)} + \psi(z), \tag{1.19}$$

where $\zeta(z)$ *is a homeomorphism in* $\bar{D}$, *which quasiconformally maps* D *onto the unit disk* $G = \{|\zeta| < 1\}$ *with boundary* $L = \{|\zeta| = 1\}$, *such that three points on* Γ *are mapped onto three points on* L *respectively,* $\Phi(\zeta)$ *is an analytic function in* G, $\psi(z), \phi(z), \zeta(z)$ *and its inverse function* $z(\zeta)$ *satisfy the estimates*

$$C_\beta[\psi, \bar{D}] \le k_3, C_\beta[\phi, \bar{D}] \le k_3, C_\beta[\zeta(z), \bar{D}] \le k_3, C_\beta[z(\zeta), \bar{G}] \le k_3, \tag{1.20}$$

$$L_{p_0}[|\psi_{\bar{z}}| + |\psi_z|, \bar{D}] \le k_3,\ L_{p_0}[|\phi_{\bar{z}}| + |\phi_z|, \bar{D}] \le k_3, \tag{1.21}$$

$$C_\beta[z(\zeta), \bar{G}] \le k_3,\ L_{p_0}[|\chi_{\bar{z}}| + |\chi_z|, \bar{D}] \le k_4, \tag{1.22}$$

in which $\chi(z)$ *is as stated in* (1.27) *below,* $\beta = \min(\alpha, 1-2/p_0)$, $p_0\,(2 < p_0 \le p)$, $k_j = k_j(q_0, p_0, \beta, k_0, k_1, D)\,(j = 3, 4)$ *are non-negative constants dependent on* $q_0, p_0, \beta, k_0, k_1, D$. *Moreover, if the coefficients* $Q_j(z) = 0\,(j = 1, 2)$ *of the complex equation* (1.11) *in* D, *then the representation* (1.19) *becomes the form*

$$w(z) = \Phi(z)e^{\phi(z)} + \psi(z), \tag{1.23}$$

and when $K < 0$, $\Phi(z)$ *satisfies the estimate*

$$C_\delta[X(z)\Phi(z), \bar{D}] \le M_1 = M_1(p_0, \delta, k, D) < \infty, \tag{1.24}$$

in which

$$X(z) = \prod_{j=1}^{m} |z - z_j|^{\eta_j},\ \eta_j = \begin{cases} |\gamma_j| + \tau,\ \gamma_j < 0,\ \beta_j \le |\gamma_j|, \\ |\beta_j| + \tau,\ \textit{for other case}, \end{cases} \tag{1.25}$$

here $\gamma_j\,(j = 1, ..., m)$ *are real constants as stated in* (1.17) *and* $\tau, \delta\,(0 < \delta < \min(\beta, \tau))$ *are sufficiently small positive constants,* $k = (k_0, k_1, k_2)$, *and* M_1 *is a non-negative constant dependent on* p_0, δ, k, D.

Proof We substitute the solution $w(z)$ of Problem A into the coefficients of equation (1.11) and consider the following system

$$\psi_{\bar z}=Q\psi_z+A_1\psi+A_2\bar\psi+A_3,\ Q=\begin{cases} Q_1+Q_2\dfrac{\overline{w_z}}{w_z} \text{ for } w_z\neq 0,\\ 0 \text{ for } w_z=0 \text{ or } z\notin D,\end{cases}$$

$$\phi_{\bar z}=Q\phi_z+A,\ A=\begin{cases} A_1+A_2\dfrac{\overline{w}-\overline{\psi}}{w-\psi} \text{ for } w(z)-\psi(z)\neq 0,\\ 0 \text{ for } w(z)-\psi(z)=0 \text{ or } z\notin D,\end{cases} \tag{1.26}$$

$$W_{\bar z}=QW_z,\quad W(z)=\Phi[\zeta(z)].$$

By using the continuity method and the principle of contracting mapping, we can find the solution

$$\psi(z)=Tf=-\frac{1}{\pi}\iint_D\frac{f(\zeta)}{\zeta-z}d\sigma_\zeta,$$
$$\phi(z)=Tg,\ \zeta(z)=\Psi[\chi(z)],\ \chi(z)=z+Th \tag{1.27}$$

of (1.26), where $f(z),g(z),h(z)\in L_{p_0}(\overline{D})$, $2<p_0\leq p$, $\chi(z)$ is a homeomorphism in $\bar D$, $\Psi(\chi)$ is a univalent analytic function, which conformally maps $E=\chi(D)$ onto the unit disk G (see [81]1)), and $\Phi(\zeta)$ is an analytic function in G. We can verify that $\psi(z),\phi(z),\zeta(z)$ satisfy the estimates (1.20) and (1.21). It remains to prove that $z=z(\zeta)$ satisfies the estimate (1.22). In fact, we can find a homeomorphic solution of the last equation in (1.26) in the form $\chi(z)=z+Th$ such that $[\chi(z)]_z,[\chi(z)]_{\bar z}\in L_{p_0}(\bar D)$ (see [87]1)). Next, we find a univalent analytic function $\zeta=\Psi(\chi)$, which maps $\chi(D)$ onto G, hence $\zeta=\zeta(z)=\Psi[\chi(z)]$. By the result on conformal mappings, applying the method of Lemma 2.1, Chapter II in [87]1), we can prove that (1.22) is true. When $Q_j(z)=0$ in D, $j=1,2$, then we can choose $\chi(z)=z$ in (1.27), in this case $\Phi[\zeta(z)]$ can be replaced by the analytic function $\Phi(z)$, herein $\Psi(z),\zeta(z)$ are as stated in (1.27), it is clear that the representation (1.19) becomes the form (1.23). Thus the analytic function $\Phi(z)$ satisfies the boundary conditions

$$\mathrm{Re}[\overline{\lambda(z)}e^{\phi(z)}\Phi(z)]=r(z)-\mathrm{Re}[\overline{\lambda(z)}\psi(z)],\ z\in\Gamma^*. \tag{1.28}$$

On the basis of the estimate (1.20), by using the methods in the proof of Theorems 1.1 and 1.8, Chapter IV in [87]1), we can prove that $\Phi(z)$ satisfies the estimate (1.24).

1.3 Existence of solutions of discontinuous Riemann-Hilbert problem for nonlinear complex equations in upper half-unit disk

We first consider a special domain, i.e. D is an upper half-unit disk with the boundary $\Gamma'=\Gamma\cup\gamma$, where $\Gamma=\{|z|=1, \text{Im}z>0\}$ and $\gamma=\{-1<x<1, y=0\}$.

Theorem 1.2 *Under the same conditions as in Theorem* 1.1 *for the above domain D, the following statements hold.*

(1) *If the index $K\geq 0$, then Problem A for* (1.11) *is solvable, and the general solution includes $2K+1$ arbitrary real constants.*

(2) *If $K<0$, then Problem A has $-2K-1$ solvability conditions.*

Proof Let us introduce a closed, convex and bounded subset B_1 in the Banach space $B=L_{p_0}(\bar{D})\times L_{p_0}(\bar{D})\times L_{p_0}(\bar{D})\,(2<p_0\leq p)$, whose elements are systems of functions $q=[Q(z), f(z), g(z)]$ with the norm $\|q\|=L_{p_0}(Q,\bar{D})+L_{p_0}(f,\bar{D})+L_{p_0}(g,\bar{D})$, which satisfy the conditions

$$|Q(z)|\leq q_0<1\,(z\in D),\ L_{p_0}[f(z),\bar{D}]\leq k_3,\ L_{p_0}[g(z),\bar{D}]\leq k_3, \tag{1.29}$$

where q_0, k_3 are non-negative constants as stated in (1.13) and (1.21). Moreover introduce a closed and bounded subset B_2 in B, the elements of which are systems of functions $\omega=[f(z), g(z), h(z)]$ satisfying the condition

$$L_{p_0}[f(z),\bar{D}]\leq k_4,\ L_{p_0}[g(z),\bar{D}]\leq k_4,\ |h(z)|\leq q_0|1+\Pi h|, \tag{1.30}$$

where $\Pi h=-\frac{1}{\pi}\iint_D[h(\zeta)/(\zeta-z)^2]d\sigma_\zeta$.

We arbitrarily select $q=[Q(z), f(z), g(z)]\in B_1$, and using the principle of contracting mapping, a unique solution $h(z)\in L_{p_0}(\bar{D})$ of the integral equation

$$h(z)=Q(z)[1+\Pi h] \tag{1.31}$$

can be found, which satisfies the third inequality in (1.30). Moreover, $\chi(z)=z+Th$ is a homeomorphism in $\bar{D}$. Now, we find a univalent analytic function $\zeta=\Psi(\chi)$, which maps $\chi(D)$ onto the unit disk G as stated in Theorem 1.1. Moreover, we find an analytic function $\Phi(\zeta)$ in G satisfying the boundary condition in the form

$$\text{Re}[\overline{\Lambda(\zeta)}\Phi(\zeta)]=R(\zeta),\ \zeta\in L^*=\zeta(\Gamma^*), \tag{1.32}$$

in which $\zeta(z)=\Psi[\chi(z)]$, $z(\zeta)$ is its inverse function, $\psi(z)=Tf$, $\phi(z)=Tg$, $\Lambda(\zeta)=\lambda[z(\zeta)]\exp[\phi(z(\zeta))]$, $R(\zeta)=r[z(\zeta)]-\text{Re}[\overline{\lambda(z(\zeta))}\psi(z(\zeta))]$, where

$\Lambda(\zeta)$, $R(\zeta)$ on L^* satisfy the conditions similar to those of $\lambda(z)$, $r(z)$ in (1.15) and the index of $\Lambda(\zeta)$ on L^* is K. In the following, we first consider the case $K \geq 0$. By using Theorem 1.1, we can find the analytic function $\Phi(\zeta)$ in the form (1.73), Chapter I, [87]1), here $2K+1$ arbitrary real constants can be chosen. Thus the function $w(z) = \Phi[\zeta(z)]e^{\phi(z)} + \psi(z)$ is determined. Afterwards, we find out the solution $[f^*(z), g^*(z), h^*(z), Q^*(z)]$ of the system of integral equations

$$f^*(z) = F(z,w,\Pi f^*) - F(z,w,0) + A_1(z,w)Tf^* + A_2(z,w)\overline{Tf^*} + A_3(z,w), \tag{1.33}$$

$$Wg^*(z) = F(z,w,W\Pi g^* + \Pi f^*) - F(z,w,\Pi f^*) + A_1(z,w)W + A_2(z,w)\overline{W}, \tag{1.34}$$

$$\begin{aligned} S'(\chi)h^*(z)e^{\phi(z)} = {} & F[z,w,S'(\chi)(1+\Pi h^*)e^{\phi(z)} + W\Pi g^* + \Pi f^*] \\ & - F(z,w,W\Pi g^* + \Pi f^*), \end{aligned} \tag{1.35}$$

$$Q^*(z) = h^*(z)/[1+\Pi h^*],\ S'(\chi) = [\Phi(\Psi(\chi))]_\chi, \tag{1.36}$$

and denote by $q^* = E(q)$ the mapping from $q = (Q,f,g)$ to $q^* = (Q^*, f^*, g^*)$. According to Lemma 5.5, Chapter III, [87]1), we can prove that $q^* = E(q)$ continuously maps B_1 onto a compact subset in B_1. By means of the Schauder fixed-point theorem, there exists a system $q = (Q,f,g) \in B_1$, such that $q = E(q)$. Applying the above method, from $q = (Q,f,g)$, we can construct a function $w(z) = \Phi[\zeta(z)]e^{\phi(z)} + \psi(z)$, which is just a solution of Problem A for (1.11). As for the case $K < 0$, it can be similarly discussed, but we first permit that the function $\Phi(\zeta)$ satisfying the boundary condition (1.32) has a pole of order $\|[K+1]\|$ at $\zeta = 0$, if $-2K$ is an even integer, then we need to add a point condition: $\mathrm{Im}[\overline{\lambda(z_0')}w(z_0')] = b_0$, z_0' is a fixed point on $\Gamma\backslash Z$, b_0 is a real constant, and then find the solution of the nonlinear complex equation (1.11) in this case. From the representation $w(z) = \Phi[\zeta(z)]e^{\phi(z)} + \psi(z)$, we can derive the $-2K-1$ solvability conditions of Problem A for (1.11).

Besides, we can discuss the solvability of the discontinuous Riemann-Hilbert boundary value problem for the complex equation (1.11) in the upper half-plane and the zone domain. For some problems in nonlinear mechanics as stated in [61]2),[91], it can be solved by the results in Theorem 1.2.

1.4 The discontinuous Riemann-Hilbert problem for nonlinear complex equations in general domains

In this subsection, let D' be a general simply connected domain with the boundary $\Gamma' = \Gamma_1' \cup \Gamma_2'$, herein $\Gamma_1', \Gamma_2' \in C^1_\mu$ $(0<\mu<1)$ and their intersection points z', z'' with the inner angles $\alpha_1\pi, \alpha_2\pi\,(0<\alpha_1,\alpha_2<1)$ respectively. We discuss the nonlinear uniformly elliptic complex equation

$$w_{\bar z}=F(z,w,w_z),\ F=Q_1w_z+Q_2\overline{w}_{\bar z}+A_1w+A_2\bar w+A_3,\ z\in D', \tag{1.37}$$

in which $F(z,w,U)$ satisfies Condition C in D'. There exist m point $Z=\{z_1=z',...,z_n=z'',...,z_m=z_0\}$ on Γ' arranged according to the positive direction successively. Denote by Γ_j the curve on Γ' from z_{j-1} to $z_j\,(j=1,2,...,m)$, and Γ_j does not include the end points $z_{j-1}\,(j=1,...,m)$.

Problem A' The discontinuous Riemann-Hilbert boundary value problem for (1.37) is to find a continuous solution $w(z)$ in $D^*=\overline{D'}\backslash Z$ satisfying the boundary condition:

$$\begin{aligned}&\mathrm{Re}[\overline{\lambda(z)}w(z)]=r(z),\ x\in\Gamma^*=\Gamma'\backslash Z,\\&\mathrm{Im}[\overline{\lambda(z_j')}w(z_j')]=b_j,\ j=1,...,2K+1,\end{aligned} \tag{1.38}$$

where $z_1',...,z_{2K+1}'\,(\notin Z)$ are distinct points on Γ' and $b_j\,(j=1,...,2K+1)$ are real constants, and $\lambda(z)$, $r(z)$, $b_j\,(j=1,...,2K+1)$ are given functions satisfying

$$\begin{aligned}&C_\alpha[\lambda(z),\Gamma_j]\le k_0,\ C_\alpha[R_j(z)r(z),\Gamma_j]\le k_2,\ j=1,...,m,\\&|b_j|\le k_2,\ j=1,...,2K+1,\end{aligned} \tag{1.39}$$

in which $\alpha\,(1/2<\alpha<1)$, k_0, k_2 are non-negative constants, $R_j(z)=|z-z_{j-1}|^{\beta_{j-1}}|z-z_j|^{\beta_j}$, and assume that $\beta_j+\gamma_j<\beta=\alpha_0\min(\alpha,1-2/p_0)$, $\gamma_j,\beta_j(j=1,...,m)$ are similar to those in (1.16) and (1.17), $\alpha_0=\min(\alpha_1,\alpha_2)$, and $K(\ge -1/2)$ is the index of $\lambda(z)$ on Γ', which is defined as in (1.18).

In order to give the uniqueness result of solutions of Problem A' for equation (1.37), we need to add one condition: For any complex functions $w_j(z)\in C(D^*), U_j(z)\in L_{p_0}(D^*)\,(2<p_0\le p, j=1,2)$, the following equality holds:

$$F(z,w_1,U_1)-F(z,w_1,U_2)=Q(U_1-U_2)+A(w_1-w_2)\ \text{in}\ D', \tag{1.40}$$

in which $|Q(z,w_1,w_2,U_1,U_2)|\le q_0$, $A(z,w_1,w_2)\in L_{p_0}(\overline{D'})$. Especially, if (1.37) is a linear equation, then the condition (1.40) obviously holds.

Applying a similar method as stated in the proof of Theorem 1.1, we can prove the following theorem.

Theorem 1.3 *If the complex equation* (1.37) *in* D' *satisfies Condition C, then Problem* A' *for* (1.37) *is solvable. If Condition C and the condition* (1.40) *hold, then the solution of Problem* A' *is unique. Moreover the solution* $w(z)$ *can be expressed as* (1.19) *satisfying the estimates* (1.20) – (1.22), *in which* $\beta = \alpha_0 \min(\alpha, 1 - 2/p_0)$. *If* $Q_j(z) = 0\,(j = 1, 2)$ *in* D' *in* (1.37), *then the representation* (1.19) *becomes the form*

$$w(z) = \Phi(z)e^{\phi(z)} + \psi(z), \tag{1.41}$$

and $w(z)$ *satisfies the estimate*

$$C_\delta[X(z)w(z), \overline{D'}] \le M_2 = M_2(p_0, \delta, k, D') < \infty, \tag{1.42}$$

in which

$$\begin{gathered} X(z) = \prod_{j=1, j\neq 1, n}^{m} |z - z_j|^{\eta_j} |z - z_1|^{\eta_1/\alpha_1} |z - z_n|^{\eta_n/\alpha_2}, \\ \eta_j = \begin{cases} |\gamma_j| + \tau, \text{ if } \gamma_j < 0,\ \beta_j \le |\gamma_j|, \\ |\beta_j| + \tau, \text{ if } \gamma_j \ge 0, \text{ and } \gamma_j < 0, \beta_j > |\gamma_j|, \end{cases} \end{gathered} \tag{1.43}$$

here $\gamma_j(j = 1, ..., m)$ *are real constants as stated in* (1.17), $\tau, \delta\,(0 < \delta < \min(\beta, \tau))$ *are sufficiently small positive constants, and* $M_2 = M_2(p_0, \delta, k, D')$ *is a non-negative constant dependent on* p_0, δ, k, D' (see [86]33),[92]6)).

2 The Riemann-Hilbert Problem for Linear Degenerate Elliptic Complex Equations of First Order

In this section we discuss the Riemann-Hilbert Problem for linear degenerate elliptic systems of first order equations in a simply connected domain. We first give the representation of solutions of the boundary value problem for the systems, and then prove the uniqueness and existence of solutions for the problem.

2.1 Formulation of the Riemann-Hilbert problem for degenerate elliptic complex equations

Let D be a domain in the upper half-plane with the boundary ∂D, which consists of $\gamma = \{-1 < x < 1, y = 0\}$ and a curve $\Gamma(\in C^1_\mu, 0 < \mu < 1)$

with the end points $-1, 1$ in the upper half-plane. We consider the linear degenerate elliptic equation of first order

$$\begin{cases} H(y)u_x - v_y = a_1u + b_1v + c_1 \\ H(y)v_x + u_y = a_2u + b_2v + c_2 \end{cases} \text{ in } D, \tag{2.1}$$

where $H(y) = \sqrt{K(y)}$, $G(y) = \int_0^y H(t)dt$, $G'(y) = H(y)$, $K(y) = y^m h(y)$ is continuous in $\overline{D}$, here m is a positive number and $h(y)$ is a continuously differentiable positive function in $\overline{D}$, and $a_j, b_j, c_j\ (j = 1, 2)$ are functions of $z\,(\in D)$. The following degenerate elliptic system is a special case of system (2.1) with $H(y) = y^{m/2}$:

$$\begin{cases} y^{m/2}u_x - v_y = a_1u + b_1v + c_1 \\ y^{m/2}v_x + u_y = a_2u + b_2v + c_2 \end{cases} \text{ in } D. \tag{2.2}$$

For convenience, we mainly discuss equation (2.2), and equation (2.1) can be similarly discussed. From the ellipticity condition in (1.2), namely

$$J = 4K_1K_4 - (K_2 + K_3)^2 = 4H^2(y) > 0 \text{ in } \overline{D}\backslash\gamma \tag{2.3}$$

and $J = 0$ on $\gamma = \{-1 < x < 1, y = 0\}$, hence system (2.1) or (2.2) is elliptic system of first order equations in D with the parabolic degenerate line $\gamma = (-1, 1)$ on the x-axis. Setting $Y = G(y) = \int_0^y H(t)dt$, $Z = x + iY$ in $\overline{D}$, if $H(y) = y^{m/2}$, $Y = \int_0^y H(t)dt = 2y^{(m+2)/2}/(m+2)$, then its inverse function is $y = [(m+2)Y/2]^{2/(m+2)} = JY^{2/(m+2)}$. Denote

$$\begin{aligned} w(z) &= u + iv,\ w_{\bar{z}} = \frac{1}{2}[H(y)w_x + iw_y] \\ &= \frac{H(y)}{2}[w_x + iw_Y] = H(y)w_{x-iY} = H(y)w_{\overline{Z}}, \end{aligned} \tag{2.4}$$

then the system (2.1) can be written in the complex form

$$\begin{gathered} w_{\bar{z}} = H(y)w_{\overline{Z}} = A_1(z)w + A_2(z)\overline{w} + A_3(z) = g(Z) \text{ in } D_Z, \\ A_1 = \frac{1}{4}[a_1 + ia_2 - ib_1 + b_2],\ A_2 = \frac{1}{4}[a_1 + ia_2 + ib_1 - b_2],\ A_3 = \frac{1}{2}[c_1 + ic_2], \end{gathered} \tag{2.5}$$

in which D_Z is the image domain of D with respect to the mapping $Z = Z(z) = x + iY = x + iG(y)$ in D. If the slopes of the Γ at $z = \mp 1$ are satisfied the conditions $-\infty < \partial y/\partial x \le 0$, $0 \le \partial y/\partial x < \infty$ respectively, then $\partial Y/\partial x = (\partial Y/\partial y)(\partial y/\partial x) = H(y)\partial y/\partial x = 0$ at $z = \mp 1$ respectively, i.e.

the inner angles of ∂D_Z are equal to π in D_Z at $Z=\mp 1$; if the slopes of the Γ at $z=\mp 1$ are satisfied the conditions $0\le \partial y/\partial x<\infty$, $-\infty<\partial y/\partial x\le 0$ respectively, then $\partial Y/\partial x=(\partial Y/\partial y)(\partial y/\partial x)=H(y)\partial y/\partial x=0$ at $z=\mp 1$ respectively, i.e. the inner angles of ∂D_Z are equal to 0 in D_Z at $Z=\mp 1$. If the boundary $\partial D\backslash\gamma(\in C^1_\mu)$ is a curve with the form $x=G(y)/\alpha_1-1\,(\alpha_1\ne\pm 1)$ and $x=1-G(y)/\alpha_2\,(\alpha_2\ne\pm 1)$ near the points $z=-1,1$ respectively, then the inner angles of the boundary ∂D_Z in $Z-$plane at $Z=-1,1$ are equal to $\tan^{-1}\alpha_1(\alpha_1\ge 0),\pi-\tan^{-1}(-\alpha_1)(\alpha_1\le 0)$ and $\tan^{-1}(-\alpha_2)(\alpha_2\le 0),\pi-\tan^{-1}\alpha_2(\alpha_2\ge 0)$ respectively, especially if $\alpha_1=1,\alpha_2=-1$, then the inner angles are equal to $\pi/4$. If $Y_x=Y_yy_x=H(y)/x_y=\pm\infty$ at $Z=\pm 1$, which include $x_y=0$ and $\pm H^2(y)$ at $Z=\pm 1$, in this case the inner angles of the curve $\tilde\Gamma=Z(\Gamma)$ and $\tilde\gamma=Z(\gamma)$ in $Z=x+iY-$plane at $Z=\pm 1$ are equal to $\pi/2$. For equations (2.5), we can give a conformal mapping $\zeta=\zeta(Z)$, which maps the domain D_Z onto D_ζ, such that line segment $\gamma=(-1,1)$ and boundary points $-1,1$ are mapped onto themselves respectively, and the boundary $\partial D_\zeta\backslash\gamma(\in C^1_\mu)$ is a curve with the form $\mathrm{Re}\,\zeta=G(\mathrm{Im}\zeta)-1$ and $\mathrm{Re}\,\zeta=1-G(\mathrm{Im}\zeta)$ near the points $\zeta=-1,1$ respectively. Denote by $Z=Z(\zeta)$ the inverse function of $\zeta=\zeta(Z)$, thus equation (2.5) is reduced to

$$\begin{gathered} w_{\overline{\zeta}}=g[Z(\zeta)]\overline{Z'(\zeta)}/H(y),\ \text{i.e.}\\ w_{\overline{\zeta}}=[A_1(z)w+A_2(z)\overline{w}+A_3(z)]\overline{Z'(\zeta)}/H(y)\ \text{ in } \overline{D_\zeta}. \end{gathered}\tag{2.6}$$

In this section, there is no harm in assuming that the boundary Γ is a curve with the form $x=G(y)-1$ and $x=1-G(y)$ near the points $z=-1,1$ respectively.

Suppose that equation (2.5) satisfies the following conditions: **Condition** C

The coefficients $A_j[z(Z)]\,(j=1,2,3)$ in (2.5) satisfy

$$L_\infty[A_j(z(Z)),\overline{D_Z}]\le k_0,\ j=1,2,\ L_\infty[A_3(z(Z)),\overline{D_Z}]\le k_1,\tag{2.7}$$

where $z(Z)$ is the inverse function of $Z(z)$, and k_0, k_1 are non-negative constants.

Now we formulate the Riemann-Hilbert boundary value problem as follows:

Problem A Find a solution $w(z)$ of (2.5) in D, which is continuous in $D^*=\overline{D}\backslash\{-1,1\}$ and satisfies the boundary conditions

$$\mathrm{Re}[\overline{\lambda(z)}w(z)]=r(z)\ \text{ on } \partial D^*=\partial D\backslash\{-1,1\},\ \mathrm{Im}[\overline{\lambda(z_0)}w(z_0)]=b_0,\tag{2.8}$$

where $\lambda(z) = a(x) + ib(x)\,(|\lambda(z)| = 1)$, b_0 is a real constants, $z_0\,(\in \Gamma\backslash \{-1,1\})$ is a point, and $\lambda(z)\,r(z)$, b_0 satisfy the conditions

$$\begin{gathered} C_\alpha[\lambda(z),\Gamma] \le k_0,\ \ C_\alpha[\lambda(z),\gamma] \le k_0, \\ C_\alpha[r(z),\Gamma]\le k_2,\ C_\alpha[r(z),\gamma]\le k_2,\ |b_0|\le k_2, \end{gathered} \tag{2.9}$$

in which $\alpha\,(0 < \alpha < 1), k_0, k_2$ are non-negative constants. In particular, if $\lambda(z) = a(x) + ib(x) = 1$, then Problem A is the Dirichlet boundary value problem, which will be called Problem D. Denote by $\lambda(z_j-0)$ and $\lambda(z_j+0)$ the left limit and right limit of $\lambda(z)$ as $z \to z_j (j = 1, 2)$ on ∂D^*, and

$$\begin{gathered} e^{i\phi_j} = \frac{\lambda(z_j-0)}{\lambda(z_j+0)}, \gamma_j = \frac{1}{\pi i} \ln \left[\frac{\lambda(z_j-0)}{\lambda(z_j+0)}\right] = \frac{\phi_j}{\pi} - K_j, \\ K_j = \left[\frac{\phi_j}{\pi}\right] + J_j,\ J_j = 0 \ \text{ or } \ 1,\ j = 1,2, \end{gathered} \tag{2.10}$$

in which $z_1 = -1, z_2 = 1$, $0 \le \gamma_j < 1$ when $J_j = 0$, and $-1 < \gamma_j < 0$ when $J_j = 1$, $1 \le j \le 2$, and

$$K = \frac{1}{2}(K_1 + K_2) = \frac{1}{2}\sum_{j=1}^{2}[\frac{\phi_j}{\pi} - \gamma_j]$$

is called the index of Problem A. If $\lambda(z)$ on ∂D is continuous, then $K = \Delta_\Gamma \arg \lambda(z)/2\pi$ is a unique integer. If the function $\lambda(z)$ on ∂D is not continuous, we can choose $J_j = 0$ or 1, hence the index K is not unique. We shall only discuss the case $K = 0$ later on, and the other cases for instance $K = -1/2$, the last point condition in (2.8) should be cancelled, we can similarly discussed.

2.2 Representations and estimates of solutions of Riemann-Hilbert problem for elliptic complex equations

It is clear that the complex equation

$$w_{\overline{Z}} = 0 \text{ in } \overline{D_Z} \tag{2.11}$$

is a special case of equation (2.5). On the basis of Theorem 1.3, we can find a unique solution of Problem A for equation (2.11) in $\overline{D_Z}$.

Now we consider the function $g(Z) \in L_\infty(D_Z)$, and first extend the function $g(Z)$ to the exterior of $\overline{D_Z}$ in $\mathbf{C}$, i.e. set $g(Z)=0$ in $\mathbf{C}\backslash\overline{D_Z}$, hence we can only discuss the domain $D_0=\{|x|<R_0\}\cap\{\text{Im}Y\geq 0\}\supset\overline{D_Z}$, here $Z=x+iY$, R_0 is a positive number. In the following we shall verify that the integral

$$\begin{gathered}\Psi(Z)=Tg/H=-\frac{1}{\pi}\iint_{D_0}\frac{g(t)/H(\text{Im}t)}{t-Z}d\sigma_t \text{ in } D_0,\\ L_\infty[g(Z),D_0]\leq k_3,\end{gathered}\tag{2.12}$$

satisfies the estimate (2.13) below, where $H(y)=y^{m/2}$, m ia a positive number. It is clear that the function $g(Z)/H(y)$ belongs to the space $L_1(D_0)$ and in general is not belonging to the space $L_p(D_0)\,(p>2, m\geq 2)$, and the integral $\Psi(Z_0)$ is definite when $\text{Im}Z_0>0$. If $Z_0\in D_0$ and $\text{Im}Z_0=0$, we can define the integral $\Psi(Z_0)$ as the limit of the corresponding integral over $D_0\cap\{|\text{Re}t-\text{Re}Z_0|\geq\varepsilon\}\cap\{|\text{Im}t-\text{Im}Z_0|\geq\varepsilon\}$ as $\varepsilon\to 0$, where ε is a sufficiently small positive number. The Hölder continuity of the integral will be proved by the following method.

Lemma 2.1 *If the function $g(Z)$ in D_Z satisfies the condition in* (2.12), *and $H(y)=y^{m/2}$, where m is a positive number, then the integral in* (2.12) *satisfies the estimate*

$$C_\beta[\Psi(Z),\overline{D_Z}]\leq M_1,\tag{2.13}$$

where $\beta=2/(m+2)-\delta$, δ is a sufficiently small positive constant, and $M_1=M_1(\beta,\,k_3,H,D_Z)$ is a positive constant.

Proof We first verify the boundedness of the integral in (2.12), as stated before, if $H(y)=y^{m/2}$, then $H(y)=J^{m/2}Y^{m/(m+2)}$. For any two points $Z_0=x_0\in\gamma=(-1,1)$ on $x-$axis and $Z_1=x_1+iY_1(Y_1>0)\in D_0$ satisfying the condition $2\text{Im}Z_1/\sqrt{3}\leq|Z_1-Z_0|\leq 2\text{Im}Z_1$, this means that the inner angle at Z_0 of the triangle $Z_0Z_1Z_2\,(Z_2=x_0+iY_1\in D_0)$ is not less than $\pi/6$ and not greater than $\pi/3$, choose a sufficiently large positive number q, from the Hölder inequality, we have $L_1[\Psi(Z),D_0]\leq L_q[g(Z),D_0]L_p[1/H(\text{Im}t)(t-Z),D_0]$, where $p=q/(q-1)\,(>1)$ is close to 1. In fact we can derive as follows

$$\begin{aligned}&|\Psi(Z_0)|\leq|\frac{1}{\pi}\iint_{D_0}\frac{g(t)/H(\text{Im}t)}{t-Z_0}d\sigma_t|\leq\frac{1}{J^{m/2}\pi}L_q[g(Z),D_0]\\ &\times[\iint_{D_0}|\frac{1}{t^{m/(m+2)}(t-Z_0)}|^p d\sigma_t]^{1/p}=\frac{1}{J^{m/2}\pi}L_q[g(Z),D_0]\,J_1^{1/p},\end{aligned}\tag{2.14}$$

where

$$
\begin{aligned}
J_1 &= \iint_{D_0} \left|\frac{1}{t^{m/(m+2)}(t-Z_0)}\right|^p d\sigma_t \\
&\le \iint_{D_0} \frac{1}{|t|^{pm/(m+2)}|\mathrm{Im}(t-Z_0)|^{p\beta_0}|\mathrm{Re}(t-Z_0)|^{p(1-\beta_0)}} d\sigma_t \\
&\le \left|\int_0^{d_0} \frac{1}{Y^{pm/(m+2)}|Y-Y_0|^{p\beta_0}} dY \int_{d_1}^{d_2} \frac{1}{|x-x_0|^{p(1-\beta_0)}} dx\right| \le k_4,
\end{aligned}
$$

in which $d_0 = \max_{Z\in\overline{D_0}} \mathrm{Im}Z$, $d_1 = \min_{Z\in\overline{D_0}} \mathrm{Re}Z$, $d_2 = \max_{Z\in\overline{D_0}} \mathrm{Re}Z$, $\beta_0 = 2/(m+2) - \varepsilon$, $\varepsilon\,(< 1/p - m/(m+2))$ is a sufficiently small positive constant, we can choose $\varepsilon = 2(p-1)/p\,(< 2/(m+2))$ such that $p(1-\beta_0/2) < 1$ and $p[m/(m+2)+\beta_0] < 1$, and $k_4 = k_4(\beta, k_3, H, D_0)$ is a non-negative constant.

Next we estimate the Hölder continuity of the integral $\Psi(Z)$ in $\overline{D_0}$, i.e.

$$
\begin{aligned}
|\Psi(Z_1) - \Psi(Z_0)| &\le \frac{|Z_1 - Z_0|}{\pi}\left|\iint_{D_0} \frac{g(t)/H(\mathrm{Im}t)}{(t-Z_0)(t-Z_1)} d\sigma_t\right| \\
&\le \frac{|Z_1-Z_0|}{J^{m/2}\pi} L_q[g(Z), D_0]\left[\iint_{D_0}\left|\frac{1}{t^{m/(m+2)}(t-Z_0)(t-Z_1)}\right|^p d\sigma_t\right]^{1/p},
\end{aligned}
\tag{2.15}
$$

and

$$
\begin{aligned}
J_2 &= \iint_{D_0}\left|\frac{1}{t^{m/(m+2)}(t-Z_0)(t-Z_1)}\right|^p d\sigma_t \\
&\le \iint_{D_0} \frac{|\mathrm{Re}(t-Z_0)|^{p(\beta_0/2-1)}|\mathrm{Re}(t-Z_1)|^{p(\beta_0/2-1)}}{|t|^{pm/(m+2)}|\mathrm{Im}(t-Z_0)|^{p\beta_0/2}|\mathrm{Im}(t-Z_1)|^{p\beta_0/2}} d\sigma_t \\
&\le \int_0^{d_0} \frac{1}{Y^{pm/(m+2)}|\mathrm{Im}(Y-Z_0)|^{p\beta_0/2}|\mathrm{Im}(Y-Z_1)|^{p\beta_0/2}} dY \\
&\times \int_{d_1}^{d_2} \frac{1}{|\mathrm{Re}(t-Z_0)|^{p(1-\beta_0/2)}|\mathrm{Re}(t-Z_1)|^{p(1-\beta_0/2)}} d\mathrm{Re}t \\
&\le k_5 \int_{d_1}^{d_2} \frac{1}{|x-x_0|^{p(1-\beta_0/2)}|x-x_1|^{p(1-\beta_0/2)}} dx,
\end{aligned}
$$

where $\beta_0 = 2/(m+2) - \varepsilon$ is chosen as before and

$$
k_5 = \max_{Z_0, Z_1\in D_0} \int_0^{d_0} [Y^{pm/(m+2)}|\mathrm{Im}(Y-Z_0)|^{p\beta_0/2}|\mathrm{Im}(Y-Z_1)|^{p\beta_0/2}]^{-1} dY.
$$

Denote $\rho_0 = |\mathrm{Re}(Z_1 - Z_0)| = |x_1 - x_0|$, $L_1 = D_0 \cap \{|x - x_0| \le 2\rho_0, Y = Y_0\}$ and $L_2 = D_0 \cap \{2\rho_0 < |x - x_0| \le 2\rho_1 < \infty, Y = Y_0\} \supset [d_1, d_2]\backslash L_1$, where ρ_1 is a sufficiently large positive number, we can derive

$$\begin{aligned}
&J_2 \le k_5[\int_{L_1} \frac{1}{|x - x_0|^{p(1-\beta_0/2)}|x - x_1|^{p(1-\beta_0/2)}} dx \\
&+ \int_{L_2} \frac{1}{|x - x_0|^{p(1-\beta_0/2)}|x - x_1|^{p(1-\beta_0/2)}} dx] \\
&\le k_5[|x_1 - x_0|^{1-2p+p\beta_0} \int_{|\xi|\le 2} \frac{1}{|\xi|^{p(1-\beta_0/2)}|\xi \pm 1|^{p(1-\beta_0/2)}} d\xi \\
&+ k_6|\int_{2\rho_0}^{2\rho_1} \rho^{p\beta_0 - 2p} d\rho|] \le k_7|x_1 - x_0|^{1-p(2-\beta_0)} = \\
&= k_7|x_1 - x_0|^{p(2/(m+2)-\varepsilon+1/p-2)},
\end{aligned}$$

in which we use $|x - x_0| = \xi|x_1 - x_0|$, $|x - x_1| = |x - x_0 - (x_1 - x_0)| = |\xi \pm 1||x_1 - x_0|$ if $x \in L_1$, $|x - x_0| = \rho \le 2|x - x_1|$ if $x \in L_2$, choose that $p(> 1)$ is close to 1 such that $1 - p(2 - \beta_0) < 0$, and $k_j = k_j(\beta, k_3, H, D_0)$ $(j = 6, 7)$ are non-negative constants. Thus we get

$$|\Psi(Z_1) - \Psi(Z_0)| \le k_7|Z_1 - Z_0||x_1 - x_0|^{2/(m+2)-\varepsilon+1/p-2} \le k_8|Z_1 - Z_0|^{\beta}, \tag{2.16}$$

in which we use that the inner angle at Z_0 of the triangle $Z_0Z_1Z_2$ $(Z_2 = x_0 + iY_1 \in D_0)$ is not less than $\pi/6$ and not greater than $\pi/3$, and choose $\varepsilon = 2(p-1)/p$, $\beta = 2/(m+2) - \delta$, $\delta = 3(p-1)/p$, $k_8 = k_8(\beta,\ k_3, H, D_0)$ is a non-negative constant. The above points $Z_0 = x_0$, $Z_1 = x_1 + iY_1$ can be replaced by $Z_0 = x_0 + iY_0$, $Z_1 = x_1 + iY_1 \in \overline{D_0}$, $0 < Y_0 < Y_1$ and $2(Y_1 - Y_0)/\sqrt{3} \le |Z_1 - Z_0| \le 2(Y_1 - Y_0)$.

Finally we consider any two points $Z_1 = x_1 + iY_1$, $Z_2 = x_2 + iY_1$ and $x_1 < x_2$, from the above estimates, the following estimate can be derived

$$\begin{aligned}
|\Psi(Z_1) - \Psi(Z_2)| &\le |\Psi(Z_1) - \Psi(Z_3)| + |\Psi(Z_3) - \Psi(Z_2)| \\
&\le k_8|Z_1 - Z_3|^{\beta} + k_8|Z_3 - Z_2|^{\beta} \le k_9|Z_1 - Z_2|^{\beta},
\end{aligned} \tag{2.17}$$

where $Z_3 = (x_1 + x_2)/2 + i[Y_1 + (x_2 - x_1)/(2\sqrt{3})]$. If $Z_1 = x_1 + iY_1$, $Z_2 = x_1 + iY_2$, $Y_1 < Y_2$, and we choose $Z_3 = x_1 + (Y_2 - Y_1)/2\sqrt{3} + i(Y_2 + Y_1)/2$, and can also get (2.17). If $Z_1 = x_1 + iY_1$, $Z_2 = x_2 + iY_2$, $x_1 < x_2, Y_1 < Y_2$, and we choose $Z_3 = x_2 + iY_1$, obviously

$$|\Psi(Z_1) - \Psi(Z_2)| \le |\Psi(Z_1) - \Psi(Z_3)| + |\Psi(Z_3) - \Psi(Z_2)|, \tag{2.18}$$

and $|\Psi(Z_1)-\Psi(Z_3)|$, $|\Psi(Z_3)-\Psi(Z_2)|$ can be estimated by the above way, hence we can obtain the estimate of $|\Psi(Z_1)-\Psi(Z_2)|$. For other case, the similar estimate can be also derived. Hence we have the estimate (2.13).

Remark 2.1 If the condition $H(y) = y^{m/2}$ in Lemma 2.1 is replaced by $H(y) = y^\eta$, herein η is a positive constant satisfying the inequality $\eta < (m+2)/2$, then by the same method we can prove that the integral $\Psi(Z) = T(g/H)$ satisfies the estimate

$$C_\beta[\Psi(Z), D_Z] \le M_1,$$

where $\beta = 1-2\eta/(m+2)-\delta$, δ is a sufficiently small positive constant, and $M_1 = M_1(\beta, k_3, H, D_Z)$ is a positive constant. In particular if $H(y) = y$, i.e. $\eta = 1$, then we can choose $\beta = m/(m+2)-\delta$, δ is a sufficiently small positive constant.

Now we give two representation theorems of solutions of Problem A for system (2.2) or equation (2.5).

Theorem 2.2 *Suppose that the equation* (2.5) *satisfies Condition* C. *Then any solution of Problem* A *for* (2.5) *can be expressed as*

$$w[z(Z)] = [\tilde{\Phi}(Z)+\tilde{\psi}(Z)]e^{\tilde{\phi}(Z)} \text{ in } D_Z, \tag{2.19}$$

where $\tilde{\psi}(Z)$, $\tilde{\phi}(Z)$ *possess the form*

$$\tilde{\phi}(Z)=T\tilde{h}=-\frac{1}{\pi}\iint_{D_0}\frac{\tilde{h}(t)}{t-Z}d\sigma_t \text{ in } D_Z,$$

$$\tilde{h}(Z)=\begin{cases}\dfrac{1}{H(y)}\{A_1[z(Z)]+A_2[z(Z)]\dfrac{\overline{w[z(Z)]}}{w[z(Z)]}\} \text{ if } w[z(Z)]\neq 0, Z\in D_Z,\\ 0 \text{ if } w[z(Z)]=0,\ Z\in D_Z, \text{ or } Z\in D_0\backslash D_Z,\end{cases}$$

$$\tilde{\psi}(Z)=T\tilde{f}=-\frac{1}{\pi}\iint_{D_0}\frac{\tilde{f}(t)}{t-Z}d\sigma_t,\ \tilde{f}(Z)=\frac{A_3[z(Z)]}{H(y)}e^{-\tilde{\phi}(Z)},$$

in which D_0 *is as stated before*, $\tilde{\phi}(Z)$, $\tilde{\psi}(Z)$ *satisfy the estimate similar to that in* (2.13), $Z = x+iY = x+iG(y)$, *and* $\tilde{\Phi}(Z)$ *is an analytic function in* D_Z *satisfying the estimate*

$$C_\delta[X(Z)\tilde{\Phi}(Z), \overline{D_Z}] \le M_2, \tag{2.20}$$

where $X(Z) = |Z-t_1|^{\eta_1}|Z-t_2|^{\eta_2}$, *here* $\eta_j = \max(-4\gamma_j, 0)+8\delta$, $j = 1,2$, $\gamma_j\,(j = 1,2)$ *are as stated in* (2.10), *and* $t_1 = -1, t_2 = 1$, δ *is a sufficiently small positive constant,* $k = (k_0, k_1, k_2)$*, and* $M_2 = M_2(\delta, k, H, D_Z)$ *is a non-negative constant..*

Proof On the basis of Lemma 2.1, we see that $\tilde{\phi}(Z), \tilde{\psi}(Z)$ in $\overline{D_Z}$ satisfy the similar estimate as in (2.13). Next it is easy to derive that

$$\tilde{\Phi}_{\overline{Z}} = [w_{\overline{Z}} - w(A_1 + A_2\overline{w}/w)/H - A_3/H]e^{-\tilde{\phi}(Z)} = 0 \text{ in } D_Z,$$

namely $\tilde{\Phi}(Z)$ is an analytic function in D_Z, which satisfies the boundary conditions

$$\begin{aligned} &\mathrm{Re}[\overline{\lambda(z(Z))}e^{\tilde{\phi}(Z)}\Phi(Z)] = r[z(Z)] - \mathrm{Re}[\overline{\lambda(z(Z))}e^{\tilde{\phi}(Z)}\tilde{\psi}(Z)] \text{ on } \partial D_Z^*, \\ &\mathrm{Im}[\overline{\lambda(z_0)}e^{\tilde{\phi}(Z_0)}\tilde{\Phi}(Z_0)] = b_0 - \mathrm{Im}[\overline{\lambda(z_0)}e^{\tilde{\phi}(Z_0)}\tilde{\psi}(Z_0)], \end{aligned} \tag{2.21}$$

in which $z(Z)$ is the inverse function of $Z(z)$, $Z_0 = Z(z_0)$, $\partial D_Z^* = \partial D_Z \backslash \{-1, 1\}$, and the index of $\lambda[z(Z)]\exp[\overline{\tilde{\phi}(Z)}]$ on ∂D_Z is $K = 0$. Hence according to the proof of Theorems 1.1 and 1.8, Chapter IV, [87]1), we can derive that $\tilde{\Phi}(Z)$ in $\overline{D_Z}$ satisfies the estimate (2.20). This completes the proof.

Theorem 2.3 *Suppose that the equation* (2.5) *satisfies Condition C. Then any solution of Problem A for* (2.5) *can be expressed as*

$$w[z(Z)] = \Phi(Z)e^{\phi(Z)} + \psi(Z) \text{ in } D_Z, \tag{2.22}$$

where $\psi(Z)$, $\phi(Z)$ *possess the form*

$$\begin{aligned} &\psi(Z) = Tf = -\frac{1}{\pi}\iint_{D_0} \frac{f(t)}{t - Z} d\sigma_t, L_\infty[f(Z)H(y), D_Z] < \infty \\ &\phi(Z) = Th = -\frac{1}{\pi}\iint_{D_0} \frac{h(t)}{t - Z} d\sigma_t \text{ in } D_Z, \end{aligned}$$

$$h(Z) = \begin{cases} \dfrac{1}{H(y)}\{A_1[z(Z)] + A_2[z(Z)]\dfrac{\overline{W(Z)}}{W(Z)}\} \text{ if } W(Z) \neq 0, Z \in D_Z, \\ 0 \text{ if } W(Z) = 0,\ Z \in D_Z \cup \{D_0 \backslash D_Z\}, \end{cases}$$

in which $\psi(Z)$, $\phi(Z)$ *satisfy the estimate* (2.13), $W(Z) = w[z(Z)] - \psi(Z)$, $Z = x + iY = x + iG(y)$, *and* $\Phi[z(Z)]$ *is an analytic function in* D_Z.

Proof Firstly by using the method of parameter extension as stated in the proof of Theorem 2.5 below, Lemma 3.4, Chapter IV, [86]9), or Theorem

3.3, Chapter II, [87]1), we can find a solution of equation (2.5) in the form

$$\psi(Z) = -\frac{1}{\pi}\iint_{D_0}\frac{f(t)}{t-Z}d\sigma_t,\ H(y)f(Z)\in L_\infty(D_Z).$$

On the basis of Theorem 2.2, the solution of (2.5) in D_Z can be expressed by $\psi(Z)=\tilde{\psi}(Z)e^{\tilde{\phi}(Z)}$, where

$$\tilde{\phi}(Z)=T\tilde{h}=-\frac{1}{\pi}\iint_{D_0}\frac{\tilde{h}(t)}{t-Z}d\sigma_t \text{ in } D_Z,$$

$$\tilde{h}(Z)=\begin{cases}\dfrac{1}{H(y)}\{A_1[z(Z)]+A_2[z(Z)]\dfrac{\overline{\psi(Z)}}{\psi(Z)}\} \text{ if } \psi(Z)\neq 0, Z\in D_0,\\[2mm] 0 \text{ if } \psi(Z)=0,\ Z\in D_0,\end{cases}$$

$$\tilde{\psi}(Z)=T\tilde{f}=-\frac{1}{\pi}\iint_{D_0}\frac{\tilde{f}(t)}{t-Z}d\sigma_t,\ \tilde{f}(Z)=A_3[z(Z)]e^{-\tilde{\phi}(Z)},$$

it is clear that the functions $\phi(Z),\psi(Z)$ satisfy the estimate similar to (2.13).

Next let $w(z)$ be a solution of Problem A for equation (2.5), it is clear that $W(Z)=\Phi(Z)e^{\phi(Z)}=w[z(Z)]-\psi(Z)$ is a solution of the complex equation

$$W_{\overline{Z}}=A_1W(Z)+A_2\overline{W(Z)} \text{ in } D_Z,$$

where $\psi(Z)$ is as stated in (2.22), and we can verify that the function $\Phi(Z)$ is an analytic function in D_Z. Finally applying Theorem 1.3, we can find an analytic function $\Phi(Z)$ in D_Z satisfying the boundary conditions

$$\begin{gathered}\text{Re}[\overline{\lambda(z(Z))}e^{i\text{Im}\phi(Z)}\Phi(Z)]=\{r[z(Z)]-\text{Re}[\overline{\lambda(z(Z))}\psi(Z)]\}e^{-\text{Re}\phi(Z)}\\ \text{on } \partial D^*,\ \text{Im}[\overline{\lambda(z_0)}e^{i\text{Im}\phi(Z_0)}\Phi(Z_0)]=\{b_0-\text{Im}[\overline{\lambda(z_0)}\psi(Z_0)]\}e^{-\text{Re}\phi(Z_0)},\end{gathered}\tag{2.23}$$

herein $Z_0=Z(z_0)$, hence the function $w[z(Z)]=\Phi(Z)e^{\phi(Z)}+\psi(Z)$ in (2.22) is just the solution of Problem A in D_Z for equation (2.5).

On the basis of Lemma 2.1 and the above discussion, we can obtain the estimates of solutions of Problem A for equation (2.5), namely

Theorem 2.4 *Any solution $w[z(Z)]$ of Problem A for equation* (2.5) *satisfies the estimates*

$$\hat{C}_\delta[w(z),\overline{D}]=C_\delta[X(Z)w(z(Z)),\overline{D_Z}]\le M_3,\ \hat{C}_\delta[w(z),\overline{D}]\le M_4(k_1+k_2),\tag{2.24}$$

in which $X(Z) = |Z-t_1|^{\eta_1}|Z-t_2|^{\eta_2}$, *here* $\eta_j = \max(-4\gamma_j, 0)+8\delta$, $j = 1, 2$, $\gamma_j (j = 1, 2)$ *are as stated in* (2.10), *and* $t_1 = -1, t_2 = 1$, δ *is a sufficiently small positive constant,* $k = (k_0, k_1, k_2)$, *and* $M_3 = M_3(\delta, k, H, D)$, $M_4 = M_4(\delta, k_0, H, D)$ *are non-negative constants.*

Proof Noticing the conditions (2.7), and using Lemma 2.1 and Theorem 2.3, we see that the functions $\psi(Z)$, $\phi(Z)$ in (2.22) satisfy the estimates

$$C_\beta[\psi(Z), \overline{D_Z}] \le M_5, \; C_\beta[\phi(Z), \overline{D_Z}] \le M_5, \tag{2.25}$$

where $\beta = 2/(m+2)-\varepsilon$, ε is a sufficiently small positive constant, and $M_5 = M_5(\beta, k, H, D)$ is a non-negative constant. Due to the analytic function $\Phi(Z)$ satisfies the boundary condition (2.23), and from (2.20) and Theorem 2.3, we can get the representation and estimate of the analytic function $\Phi(Z)$ in D_Z similar to those in (2.22) and (2.20), thus the first estimate of (2.24) is derived. Moreover we verify the second estimate in (2.24). If $k = k_1+k_2 > 0$, then the function $w^*(z) = u^*(z)+iv^*(z) = u(z)/k+iv(z)/k$ is a solution of Problem A for equation

$$w^*_{\overline{Z}} = g^*(Z), g^*(Z) = \frac{g(Z)}{kH(y)} = \frac{1}{H(y)}[A_1w^* + A_2\overline{w^*} + \frac{A_3}{k}] \text{ in } D_Z. \tag{2.26}$$

By the proof of the first estimate in (2.24), we can derive the estimate of the solution $w^*(z)$:

$$\hat{C}_\beta[w^*(z), \overline{D}] \le M_4 = M_4(\beta, k_0, H, D). \tag{2.27}$$

From the above estimate it follows that the second estimate of (2.24) holds with $k > 0$. If $k = 0$, we can choose any positive number ε to replace $k = 0$. By using the same proof as before, we have

$$\hat{C}_\beta[w(z), \overline{D}] \le M_4\varepsilon.$$

Let $\varepsilon \to 0$, it is obvious that the second estimate in (2.24) with $k = 0$ is derived.

2.3 Solvability of Riemann-Hilbert problem for degenerate elliptic complex equations

Theorem 2.5 *Suppose that equation* (2.2) *satisfies Condition* C. *Then Problem* A *for* (2.5) *has a unique solution in* D.

Proof We first verify the uniqueness of the solution of Problem A for system (2.2) or equation (2.5). Let $w_1(z), w_2(z)$ be any two solutions of

Problem A for (2.5). It is easy to see that $w(z) = w_1(z) - w_2(z)$ satisfy the homogeneous equation and boundary conditions

$$\begin{gathered} w_{\overline{Z}} = [A_1 w + A_2 \overline{w}]/H(y) \text{ in } D_Z, \\ \operatorname{Re}[\overline{\lambda((Z))} w(z(Z))] = 0 \text{ in } \partial D^*, \ \operatorname{Im}[\overline{\lambda(z_0)} \Phi(Z_0)] = 0. \end{gathered} \tag{2.28}$$

Due to the solution $w[z(Z)]$ possesses the expression (2.22), but $\psi(Z) = 0$ in D_Z, and the index $K = 0$ of $\lambda[z(Z)]$ on ∂D_Z, from Theorem 1.1, Chapter IV, [87]1), it is not difficult to derive that $\Phi(Z) = 0$ in D_Z, hence $w(z) = w_1(z) - w_2(z) = 0$ in D.

As for the existence of solutions of Problem A for equation (2.5), we can prove by using the method of parameter extension. In fact, the complex equation (2.5) can be rewritten as

$$\begin{gathered} w_{\overline{Z}} = F(Z, w), \\ F(Z, w) = \frac{1}{H(y)} \{A_1[z(Z)]w + A_2[z(Z)]\overline{w} + A_3[z(Z)]\} \text{ in } D_Z. \end{gathered} \tag{2.29}$$

In order to find a solution $w(z)$ of Problem A in D, we can express $w(z)$ in the form (2.22), and consider the equation with the parameter $t \in [0, 1]$:

$$w_{\overline{Z}} - tF(z, w) = S(z) \text{ in } \overline{D_Z}, \tag{2.30}$$

in which the function $S(z)$ satisfies the condition

$$H(y)X(Z)S(z) \in L_\infty(\overline{D_Z}), \tag{2.31}$$

where $X(Z)$ is as stated in (2.20). This problem is called Problem A_t.

When $t = 0$, the complex equation (2.30) becomes the equation

$$w_{\overline{Z}} = S(z) \text{ in } \overline{D_Z}. \tag{2.32}$$

It is clear that the unique solution of Problem A_0, i.e. Problem A for $w_{\overline{Z}} = S(z)$ can be found, namely $X(Z)w[z(Z)] = \Phi(Z) + TXS$. Suppose that when $t = t_0$ $(0 \le t_0 < 1)$, Problem A_{t_0} is solvable, i.e. Problem A_{t_0} for (2.30) has a solution $w_0(z)$ $(w_0(z) \in \hat{C}(\overline{D})$, i.e. $X[Z(z)]w_0(z) \in C(\overline{D}))$. We can find a neighborhood $T_\varepsilon = \{|t - t_0| \le \varepsilon, 0 \le t \le 1\}(0 < \varepsilon < 1)$ of t_0 such that for every $t \in T_\varepsilon$, Problem A_t is solvable. In fact, Problem A_t can be written in the form

$$w_{\overline{Z}} - t_0 F(z, w) = (t_0 - t)F(z, w) + S(z) \text{ in } \overline{D_Z}, \tag{2.33}$$

Replacing $w_0(z)$ into the right-hand side of (2.33) by a function $w_0(z) \in \hat{C}(\overline{D})$, especially, we select $w_0(z) = 0$ and substitute it into the right-hand side of (2.33), it is obvious that the boundary value problem for such equation in (2.33) then has a solution $w_1(z) \in \hat{C}(\overline{D})$. Using successive iteration, we obtain a sequence of solutions $w_n(z)\,(w_n(z) \in \hat{C}(\overline{D}),\, n = 1, 2, ...)$, which satisfy the equations

$$w_{n+1\overline{Z}}-t_0F(z,w_{n+1})=(t-t_0)F(z,w_n)+S(z) \text{ in } \overline{D},$$

$$\mathrm{Re}[\overline{\lambda(z)}w_{n+1}(z)]=r(z) \text{ on } \partial D^*, \mathrm{Im}[\overline{\lambda(z_0)}w_{n+1}(z_0)]=b_0.$$

From the above formulas, it follows that

$$[w_{n+1}-w_n]_{\overline{Z}}-t_0[F(z,w_{n+1})-F(z,w_n)]$$

$$=(t-t_0)[F(z,w_n)-F(z,w_{n-1})] \text{ in } D,$$

$$\mathrm{Re}[\overline{\lambda(z)}(w_{n+1}(z)-w_n(z))]=0 \text{ on } \partial D^*,$$

$$\mathrm{Im}[\overline{\lambda(z_0)}(w_{n+1}(z_0)-w_n(z_0))]=0.$$

Noting that

$$L_\infty[H(y)X(Z)(F(z,w_n)-F(z,w_{n-1})),\overline{D_Z}] \le 2k_0\hat{C}[w_n-w_{n-1},\overline{D_Z}],$$

and then by Theorem 2.4, we can derive

$$\hat{C}[w_{n+1}-w_n,\overline{D_Z}] \le 2|t-t_0|M_4\hat{C}[w_n-w_{n-1},\overline{D_Z}],$$

where the constant $M_4 = M_4(\beta, k_0, H, D)$ is as stated in (2.24). Choosing the constant ε so small such that $2\varepsilon M_4 \le 1/2$ and $|t-t_0| \le \varepsilon$, it follows that

$$\hat{C}[w_{n+1}-w_n,\overline{D_Z}]\le 2\varepsilon M_4\hat{C}[w_n-w_{n-1},\overline{D_Z}]\le\frac{1}{2}\hat{C}[w_n-w_{n-1},\overline{D_Z}],$$

and when $n, m \ge N_0+1$ (N_0 is a positive integer),

$$\hat{C}[w_{n+1}-w_n,\overline{D_Z}]\le 2^{-N_0}\sum_{j=0}^{\infty}2^{-j}\hat{C}[w_1-w_0,\overline{D_Z}]\le 2^{-N_0+1}\hat{C}[w_1-w_0,\overline{D_Z}].$$

Hence $\{w_n(z)\}$ is a Cauchy sequence. According to the completeness of the Banach space $\hat{C}(\overline{D_Z})$, there exists a function $w_*(z) \in \hat{C}(\overline{D_Z})$, so that $\hat{C}[w_n - w_*, \overline{D_Z}] \to 0$ as $n \to \infty$, we can see that $w_*(z)$ is a solution of Problem A_t for every $t \in T_\varepsilon = \{|t-t_0| \le \varepsilon\}$. Because the constant ε is independent of t_0 $(0 \le t_0 < 1)$, therefore from the solvability of Problem A_{t_0} when $t_0 = 0$, we can derive the solvability of Problem A_t when $t = \varepsilon, 2\varepsilon, ..., [1/\varepsilon]\varepsilon, 1$, where $[1/\varepsilon]$ means the integer part of $1/\varepsilon$. In particular, when $t = 1$ and $S(z) = 0$, Problem A_1, i.e. Problem A for (2.5) in D has a solution $w(z)$.

3 The Discontinuous Riemann-Hilbert Problem for Quasilinear Degenerate Elliptic Complex Equations of First Order

In this section we discuss the discontinuous Riemann-Hilbert Problem for quasilinear degenerate elliptic system of first order equations in a bounded simply connected domain. We first give the representation of solutions of the boundary value problem for the equations, and then prove the existence and uniqueness of solutions for the problem.

3.1 Formulation of discontinuous Riemann-Hilbert problem for degenerate elliptic complex equations

Let D be a simply connected bounded domain in the complex plane $\mathbf{C}$ with the boundary $\partial D = \Gamma \cup \gamma$, where $\Gamma(\subset \{y > 0\}) \in C^1_\alpha(0 < \alpha < 1)$ with the end points $z = -1, 1$ and $\gamma = (-1, 1)$ on the x-axis. As stated in Section 2, there is no harm in assuming that the boundary $\Gamma(\in C^1_\alpha)$ is a curve with the form $x = -1 + G(y)\,(-1 \le x \le 0)$ and $x = 1 - G(y)\,(0 \le x \le 1)$ near the points $z = -1, 1$. We consider the quasilinear degenerate elliptic system of first order equations

$$\begin{cases} H(y)u_x - v_y = a_1u + b_1v + c_1 \\ H(y)v_x + u_y = a_2u + b_2v + c_2 \end{cases} \text{in } D, \tag{3.1}$$

in which $H(y) = \sqrt{K(y)}, Y = G(y) = \int_0^y H(t)dt, G'(y) = H(y)$, $K(y)$ is the same as stated in (2.1), and $a_j, b_j, c_j (j = 1, 2)$ are functions of $(x, y)(\in D), u, v(\in \mathbf{R})$. The following degenerate elliptic system is a special case of system (3.1) with $H(y) = y^{m/2}$:

$$\begin{cases} y^{m/2}u_x - v_y = a_1u + b_1v + c_1 \\ y^{m/2}v_x + u_y = a_2u + b_2v + c_2 \end{cases} \text{in } D, \tag{3.2}$$

where m is a positive constant. According to Section 2, the system (3.1) can be written in the complex form

$$\begin{gathered} w_{\bar z} = F(z, w),\ F(z, w) \\ = A_1(z, w)w + A_2(z, w)\overline{w} + A_3(z, w) = g(Z) \text{ in } D, \text{ i.e.} \\ w_{\overline{Z}} = [A_1w + A_2\overline{w} + A_3]/H(y) = g(Z)/H(y) \text{ in } D_Z, \end{gathered} \tag{3.3}$$

where

$$A_1=\frac{1}{4}[a_1+ia_2-ib_1+b_2],\,A_2=\frac{1}{4}[a_1+ia_2+ib_1-b_2],\,A_3=\frac{1}{2}[c_1+ic_2],$$

in which $w=u+iv$, $Z=x+iG(y)$, D_Z is the image domain of D with respect to the mapping $Z=Z(z)$.

Suppose that equation (3.3) satisfies **Condition** C, namely

1) $A_j(z,w)\,(j=1,2,3)$ are measurable in D for all continuous functions $w(z)$ in $D^*=\bar{D}\backslash\{-1,1\}$, and satisfy

$$L_\infty[A_j,\overline{D}]\le k_0,\ j=1,2,\ L_\infty[A_3,\overline{D}]\le k_1\ \text{ in } D. \tag{3.4}$$

2) For any continuously differentiable functions $w_1(z),w_2(z)$ in D^*, the equality

$$F(z,w_1)-F(z,w_2)=\tilde{A}_1(w_1-w_2)+\tilde{A}_2(\overline{w_1}-\overline{w_2})\ \text{ in } D \tag{3.5}$$

holds, where $\tilde{A}_j=\tilde{A}_j(z,w_1,w_2)\,(j=1,2)$ satisfy the conditions

$$L_\infty[\tilde{A}_j,\overline{D}]\le k_0,\ j=1,2, \tag{3.6}$$

in (3.4), (3.6), k_0, k_1 are non-negative constants. In particular, when (3.3) is a linear equation, the condition (3.5) obviously holds.

Now we formulate the general discontinuous Riemann-Hilbert boundary value problem. Let $Z'=\{z_1=-1,...,z_n=1,...,z_m=z_0\}$ be m points on $\Gamma\cup\gamma$ arranged according to the positive direction successively. Denote by Γ_j the curve on Γ from z_{j-1} to z_j, and Γ_j does not include the end point $z_{j-1},z_j\,(j=1,2,...,m)$.

Problem B Find a continuous solution $w(z)$ of (3.3) in $D^*=\overline{D}\backslash Z'$, which satisfies the boundary conditions

$$\begin{aligned}&\mathrm{Re}[\overline{\lambda(z)}w(z)]=r(z)\ \text{ on } \partial D^*=\{\Gamma\cup\gamma\}\backslash Z',\\&\mathrm{Im}[\overline{\lambda(z_j')}w(z_j')]=b_j,\ j=1,...,2K+1=J,\end{aligned} \tag{3.7}$$

in which $\lambda(z)=\mathrm{Re}\lambda(z)+i\mathrm{Im}\lambda(z)\ne 0$ on $\Gamma\cup\gamma$, $z_j'\,(\notin Z',j=1,...,J)$ are distinct points on $\Gamma\cup\gamma$, $b_j\,(j=1,...,J)$ are real constants and $\lambda(z),r(z),b_j\,(j=1,...,J)$ satisfy the conditions

$$\begin{aligned}&C_\alpha[\lambda(z),\Gamma_j]\le k_0,\ C_\alpha[R_j(z)r(z),\Gamma_j]\le k_2,\ j=1,...,m,\\&C_\alpha[\lambda(z),\gamma]\le k_0,\ C_\alpha[r(z),\gamma]\le k_2,\ |b_j|\le k_2,\ j=1,...,J,\end{aligned} \tag{3.8}$$

where $R_j(z)=|z-z_{j-1}|^{\beta_{j-1}}|z-z_j|^{\beta_j}$, $\beta_j\,(j=1,...,m)$ are similar to those in (1.39) with $a_0=1/4$, $\alpha\,(0<\alpha<1), k_0, k_2$ are non-negative constants, and the number

$$K=\frac{1}{2}(K_1+\cdots+K_m) \tag{3.9}$$

is called the index of Problem B, where

$$\begin{aligned}K_j&=\left[\frac{\phi_j}{\pi}\right]+J_j,\ J_j=0\ \text{or}\ 1,\\ e^{i\phi_j}&=\frac{\lambda(z_j-0)}{\lambda(z_j+0)},\gamma_j=\frac{\phi_j}{\pi}-K_j, j=1,...,m.\end{aligned} \tag{3.10}$$

Here we choose the index $K\geq -1/2$. From Theorems 3.2 and 3.4 below, we shall see that Problem B for (3.3) is well-posed.

3.2 Representation and uniqueness of solutions of discontinuous Riemann-Hilbert problem for elliptic complex equations

Now we give the representation theorem of solutions for equation (3.3).

Theorem 3.1 *Suppose that equation* (3.3) *satisfies Condition C. Then any solution of Problem B for* (3.3) *can be expressed as*

$$w[z(Z)]=[\tilde{\Phi}(Z)+\tilde{\psi}(Z)]e^{\tilde{\phi}(Z)}=\Phi(Z)e^{\phi(Z)}+\psi(Z), \tag{3.11}$$

where

$$\tilde{\psi}(Z)=T\tilde{f}=-\frac{1}{\pi}\iint_{D_t}\frac{\tilde{f}(t)}{t-Z}d\sigma_t,\ \tilde{f}(Z)=\frac{A_3[z(Z)]}{H(y)}e^{-\tilde{\phi}(Z)},$$

$$\tilde{\phi}(Z)=T\tilde{h}=-\frac{1}{\pi}\iint_{D_t}\frac{\tilde{h}(t)}{t-Z}d\sigma_t\ \text{in}\ D_Z,$$

$$\tilde{h}(Z)=\begin{cases}\dfrac{1}{H(y)}\{A_1[z(Z)]+A_2[z(Z)]\dfrac{\overline{w(z(Z))}}{w(z((Z))}\}\ \text{if}\ w[z(Z)]\neq 0, Z\in D_Z,\\ 0\ \text{if}\ w[z(Z)]=0,\ Z\in D_Z,\end{cases}$$

$$\psi(Z)=Tf=-\frac{1}{\pi}\iint_{D_t}\frac{f(t)}{t-Z}d\sigma_t, L_\infty[f(Z)H(y), D_Z]\leq k_3,$$

$$\phi(Z)=Th=-\frac{1}{\pi}\iint_{D_t}\frac{h(t)}{t-Z}d\sigma_t \text{ in } D_Z,$$

$$h(Z)=\begin{cases}\dfrac{1}{H(y)}\{A_1[z(Z)]+A_2[z(Z)]\dfrac{\overline{W(Z)}}{W(Z)}\} \text{ if } W(Z)\neq 0, Z\in D_Z,\\[2mm] 0 \text{ if } W(Z)=0,\ Z\in D_Z,\end{cases}\tag{3.12}$$

in which $W(Z) = w[z(Z)] - \psi(Z)$, $k_3 = k_3(k_0,k_1,k_2,H,D)$ *is a non-negative constant,* $Z = x + iY = x + iG(y)$, *and* $\Phi(Z)$ *is an analytic function in* D_Z *satisfying the boundary conditions*

$$\begin{aligned}&\mathrm{Re}[\overline{\lambda(z(Z))}e^{\phi(Z)}\Phi(Z)]=r[z(Z)]-\mathrm{Re}[\overline{\lambda(z(Z))}\psi(Z)] \text{ on } \partial D^*,\\ &\mathrm{Im}[\overline{\lambda(z'_j)}e^{\phi(Z_j)}\Phi(Z_j)]=b_j-\mathrm{Im}[\overline{\lambda(z'_j)}\psi(Z_j)], j=1,...,2K+1,\end{aligned}\tag{3.13}$$

where $Z_j=Z(z'_j)$, $j=1,...,2K+1$, *hence the function* $w[z(Z)]=\Phi(Z)e^{\phi(Z)}+\psi(Z)$ *in* (3.11) *is just the solution of Problem* B *in* D_Z *for equation* (3.3).

Proof Let $w(z)$ be a solution of Problem B for equation (3.3), and be substituted in the positions of w in (3.3), thus the coefficients A_j $(j = 1,2,3)$ be determined. Moreover according to the method in the proof of Theorem 2.5, we can find the solution $\psi(Z)$ of the linear complex equation

$$w_{\overline{Z}} = [A_1w + A_2\overline{w} + A_3]/H(y) \text{ in } D_Z,\tag{3.14}$$

and the function $\psi(Z) = \tilde{\psi}(Z)e^{\tilde{\phi}(Z)}$, herein $\tilde{\phi}(Z)$, $\tilde{\psi}(Z)$ are two double integrals as stated in the proof of Theorem 2.3 and satisfy the similar estimate in (2.13). Moreover the function $\phi(Z)$ is determined as stated in (3.12), and $\Phi(Z)$ is an analytic function in D_Z satisfying the boundary condition (3.13). It is clear that $w[z(Z)]$ possesses the representation (3.11).

Theorem 3.2 *Suppose that equation* (3.3) *satisfies Condition* C. *Then Problem* B *for* (3.3) *has at most one solution in* D.

Proof Let $w_1(z), w_2(z)$ be any two solutions of Problem B for (3.3). It is easy to see that $w(z) = w_1(z) - w_2(z)$ satisfy the homogeneous equation

$$w_{\overline{Z}} = [\tilde{A}_1w + \tilde{A}_2\overline{w}]/H(y) \text{ in } D_Z,\tag{3.15}$$

and homogeneous boundary condition of (3.13), i.e.

$$\begin{aligned}&\mathrm{Re}[\overline{\lambda(z(Z))}e^{\phi(Z)}\Phi(Z)] = 0 \text{ in } \partial D_Z,\\ &\mathrm{Im}[\overline{\lambda(z'_j)}e^{\phi(Z'_j)}\Phi(Z'_j)]=0,\ j=1,...,2K+1,\end{aligned}\tag{3.16}$$

where $Z'_j = Z(z'_j)\,(j = 1, ..., m)$. According to the proof of Theorem 2.5, we can prove $\Phi(Z) = 0$ in D_Z, thus $w(z) = w_1(z) - w_2(z) = 0$ in D.

3.3 Estimates and existence of solutions of Riemann-Hilbert problem for degenerate elliptic complex equations

Now we shall give the estimates of the solutions of Problem B for (3.3) in $\overline{D}$. We rewrite equation (3.3) in the form

$$w_{\bar{z}} = F(z, w),\ F(z, w) = A_1 w + A_2 \overline{w} + A_3, \tag{3.17}$$

in which $A_j\,(j = 1, 2, 3)$ are as stated in (3.3).

Theorem 3.3 *Let equation* (3.3) *satisfy Condition* C. *Then any solution* $w(z)$ *of Problem* B *satisfies the estimates*

$$\begin{aligned} \hat{C}_\delta[w(z), \overline{D}] &= C_\delta[X(Z)w(z(Z)), \overline{D_Z}] \le M_1, \\ \hat{C}_\delta[w(z), \overline{D}] &\le M_2(k_1 + k_2), \end{aligned} \tag{3.18}$$

where

$$X(Z) = \prod_{j=0}^{m} |Z(z) - Z(t_j)|^{\eta_j},\ \eta_j = \begin{cases} \max(-4\gamma_j, \beta_j) + 8\delta, j = 1, n, \\ \max(-\gamma_j, \beta_j) + 2\delta, j = 2, ..., m, j \ne n \end{cases} \tag{3.19}$$

herein $\gamma_j\,(j = 1, ..., m)$ *are as stated in* (3.10), *and* $t_1 = z_1 = -1, ..., t_n = z_n = 1, ..., t_m = z_m$, $k = (k_0, k_1, k_2)$, *and* δ *is a sufficiently small positive constant, and* $M_1 = M_1(\delta, k, H, D)$, $M_2 = M_2(\delta, k_0, H, D)$ *are non-negative constants.*

Proof Taking into account $A_j[z, w(z)] \in L_\infty(D_Z)$, $j = 1, 2, 3$, and applying (2.25), we get

$$C_\beta[\psi(Z), \overline{D_Z}] \le M_3,\ C_\beta[\phi(Z), \overline{D_Z}] \le M_3, \tag{3.20}$$

where $\phi(z)$, $\psi(z)$ are the functions as in (3.11), β is as stated in (2.25), and $M_3 = M_3(\beta, k, H, D)$ is a non-negative constant. Moreover due to the analytic function $\Phi(z)$ satisfies the boundary condition (3.13), similarly to (2.20), we can obtain the estimate

$$\hat{C}_\delta[\Phi(z), \overline{D}] \le M_4 = M_4(\delta, k, H, D). \tag{3.21}$$

Combining (3.20), (3.21), the first estimate in (3.18) is derived.

As for the second estimate in (3.18), which can be verified according to the proof of Theorem 2.4.

Now we prove the existence of solutions of Problem B for equation (3.3) by the method of continuity.

Theorem 3.4 *Suppose that equation* (3.3) *satisfies Condition C. Then the discontinuous Riemann-Hilbert problem* (*Problem* B) *for* (3.3) *has a solution.*

Proof We discuss the complex equation (3.17), i.e.

$$w_{\overline{Z}}=F(z,w),\ F(z,w)=[A_1w+A_2\overline{w}+A_3]/H(y) \text{ in } D_Z. \tag{3.22}$$

In order to find a solution $w(z)$ of Problem B in D by the method of continuity, we consider Problem B for the complex equation with the parameter $t \in [0,1]$:

$$w_{\overline{Z}} - tF(z,w) = S(z) \text{ in } \overline{D_Z}, \tag{3.23}$$

in which the function $S(z)$ satisfies the condition

$$H(y)X(Z)S(z) \in L_\infty(\overline{D_Z}). \tag{3.24}$$

This problem is called Problem B_t.

Let T be a point set in the interval [0,1], such that for every $t \in T$, Problem B_t for equation (3.23) has a solution $w(Z) \in \hat{C}_\delta(\overline{D_Z})$ for every function $S(Z)$ satisfying the condition (3.24). It is clear that when $t = 0$, Problem B_0 for $w_{\overline{Z}} = S(Z)$ has a solution

$$X(Z)w(Z) = \Phi(Z) + TXS, \tag{3.25}$$

where $\Phi(Z)$ is an analytic function in D_Z. Hence T is non-empty. If we can prove that T is both open and closed in [0,1], then we can derive that $T = [0,1]$. In particular, when $t = 1$ and $S(Z) = 0$, Problem B_1 possesses a solution, i.e. Problem B for equation (3.22) is solvable.

In order to prove that T is a open set in [0,1], let $t_0 \in T$. We rewrite (3.23) in the form

$$w_{\overline{Z}} - t_0F(z,w) = (t-t_0)F(z,w) + S(z) \text{ in } \overline{D_Z}, \tag{3.26}$$

Replacing $w_0(z)(\in \hat{C}(\overline{D}))$ into the right-hand side of (3.26) by a function $w_0(z)$, especially, we select $w_0(z) = 0$ and substitute it into the right-hand side of (3.26), it is obvious that Problem B_{t_0} for such equation in (3.26) then has a solution $w_1(z)\,(w_1(z)\in \hat C(\overline{D}))$. Using successive iteration, we obtain a sequence of solutions $w_n(z)\,(w_n(z)\in \hat C(\overline{D}),\ n=1,2,...)$, which satisfy the equations and the boundary conditions

$$w_{n+1\overline{Z}}-t_0F(z,w_{n+1})=(t-t_0)F(z,w_n)+S(z) \text{ in } \overline{D}, \tag{3.27}$$

$$\text{Re}[\overline{\lambda(z)}w_{n+1}(z)]=r(z) \text{ on } \partial D^*, \text{Im}[\overline{\lambda(z_j')}w_{n+1}(z_j')]=b_j, j=1,...,2K+1. \tag{3.28}$$

From the above formulas, it follows that

$$\begin{aligned}
&[w_{n+1}-w_n]_{\overline{Z}}-t_0[F(z,w_{n+1})-F(z,w_n)]\\
&=(t-t_0)[F(z,w_n)-F(z,w_{n-1})] \text{ in } D,\\
&\text{Re}[\overline{\lambda(z)}(w_{n+1}(z)-w_n(z))]=0 \text{ on } \partial D^*,\\
&\text{Im}[\overline{\lambda(z_j')}(w_{n+1}(z_j')-w_n(z_j'))]=0, j=1,...,2K+1.
\end{aligned} \tag{3.29}$$

Noting that

$$L_\infty[H(y)X(Z)(F(z,w_n)-F(z,w_{n-1})),\overline{D_Z}]\le 2\hat C[w_n-w_{n-1},\overline{D_Z}], \tag{3.30}$$

and according to the proof of Theorem 2.5, we can derive

$$\hat C[w_{n+1}-w_n,\overline{D}]\le 2|t-t_0|M_2\hat C[w_n-w_{n-1},\overline{D}], \tag{3.31}$$

where the constant $M_2=M_2(\delta,k_0,H,D)$ is as stated in (3.18). Choosing the constant ε so small such that $2\varepsilon M_2\le 1/2$ and $|t-t_0|<\varepsilon$, it follows that

$$\hat C[w_{n+1}-w_n,\overline{D}]\le 2\varepsilon M_2\hat C[w_n-w_{n-1},\overline{D}]\le\frac{1}{2}\hat C[w_n-w_{n-1},\overline{D}], \tag{3.32}$$

and when $n,m\ge N_0+1\,(N_0$ is a positive integer),

$$\hat C[w_{n+1}-w_n,\overline{D}]\le 2^{-N_0}\sum_{j=0}^{\infty}2^{-j}\hat C[w_1-w_0,\overline{D}]\le 2^{-N_0+1}\hat C[w_1-w_0,\overline{D}].$$

Hence $\{w_n(z)\}$ is a Cauchy sequence. According to the completeness of the Banach space $\hat C(\overline{D})$, there exists a function $w_*(z)\in\hat C(\overline{D})$, so that $\hat C[w_n-w_*,\overline{D}]\to 0$ as $n\to\infty$. Obviously $w_*(z)$ is a solution of Problem B_t for every $t\in T_\varepsilon=\{|t-t_0|<\varepsilon\}$. Because the constant ε is independent of $t_0\,(0\le t_0<1)$, therefore from the solvability of Problem B_{t_0} when $t_0=0$, we can derive the solvability of Problem B_t for equation (3.23) when $t \in T_\varepsilon$. This shows that the set T in [0,1] is open.

Finally we verify that T is closed in [0,1]. Let $t_n \in T\,(n = 1, 2, ...)$, and $t_n \to t_0$ as $n \to \infty$. We shall prove that Problem B_{t_0} for equation (3.23) is solvable. Denote by $w_n(z)\,(n = 1, 2, ...)$ the solutions of Problems B_{t_n} $(t_n \in T,\ n = 1, 2, ...)$ for the corresponding equations (3.23), which can be expressed by

$$X(Z)w_n[z(Z)] = \Phi_n(Z)e^{\phi_n(Z)} + \psi_n(Z),\ n = 1, 2, ...$$

and satisfy the estimate (3.18). Hence from $\{w_n(z)\}$, we can choose a subsequence $\{w_{n_k}(z)\}$, such that $X(Z)w_{n_k}[z(Z)]$ uniformly converges a function $X(Z)w_0[z(Z)]$ in $\overline{D_Z}$, it is clear that the function $w_0(z)$ is just the solution of Problem B_{t_0} for equation (3.23) with $t = t_0$. This completes the proof.

4 The Riemann-Hilbert Problem for Degenerate Elliptic Complex Equations of First Order in Multiply Connected Domains

This section deals with the Riemann-Hilbert problem for degenerate elliptic complex equations of first order in multiply connected domains. We first give the representation of solutions of the boundary value problem for the equations, and then prove the uniqueness and existence of solutions for the problem.

4.1 Formulation of Riemann-Hilbert problem for degenerate elliptic complex equations in multiply connected domains

Let D be an $(N + 1)$-connected bounded domain in the upper half-plane with the boundary $\Gamma = \cup_{j=0}^{N}\Gamma_j \in C_\alpha(0 < \alpha < 1)$, where $\Gamma_j(j = 1, ..., N)$ are located in the domain D_0 bounded by $\Gamma_0 = \Gamma_{N+1}$, there is no harm in assuming that $\Gamma_0 = \Gamma' \cup \gamma$, herein $\gamma = \{-1 < x < 1, y = 0\}$ and $\Gamma'(\in \{y > 0\})$ is a curve with the end points $z = \pm 1$, and the inner angles of Γ' and γ at $z = \pm 1$ are equal to π, because otherwise through a conformal mapping the above requirement can be realized. We consider the quasilinear degenerate elliptic equation of first order: (3.1) with **Condition**

C, its complex form is as follows

$$\begin{gathered}w_{\overline{z}}=F(z,w), F=A_1(z,w)w+A_2(z,w)\overline{w}+A_3(z,w), \text{ i.e.}\\ H(y)w_{\overline{Z}}=g(Z) \text{ in } D,\ A_3=\frac{1}{2}[c_1+ic_2],\\ A_1=\frac{1}{4}[a_1+ia_2-ib_1+b_2], A_2=\frac{1}{4}[a_1+ia_2+ib_1-b_2],\end{gathered}\tag{4.1}$$

where the coefficients $A_j\,(j=1,2,3)$ in (3.1) satisfy

$$L_\infty[A_j,\overline{D}], L_\infty[\tilde{A}_j,\overline{D}]\le k_0, j=1,2,\ L_\infty[A_3,\overline{D}]\le k_1 \text{ in } D,\tag{4.2}$$

besides $\tilde{A}_j(j=1,2)$ are as stated in (3.6), and k_0, k_1 are non-negative constants. We mention that under Condition C, the above solution of equation (4.1) in D is a generalized solution, and if $A_j\in C_\alpha(D)$, then the solution of (4.1) is a classical solution.

The Riemann-Hilbert boundary value problem for equation (4.1) may be formulated as follows:

Problem A Find a continuous solution $w(z)$ of (4.1) in $\overline{D}$ satisfying the boundary condition

$$\mathrm{Re}[\overline{\lambda(z)}w(z)]=r(z),\ z\in\Gamma,\tag{4.3}$$

where $\lambda(z)\ne 0$, $z\in\Gamma$, and $\lambda(z)$, $r(z)$ satisfy the conditions

$$C_\alpha[\lambda(z),\Gamma]\le k_0,\ C_\alpha[r(z),\Gamma]\le k_2,\tag{4.4}$$

in which $\alpha\,(0<\alpha<1), k_2$ are non-negative constants.

The integer

$$K=\frac{1}{2\pi}\Delta_\Gamma\arg\lambda(z)$$

is the index of Problem A. When the index $K<0$, Problem A may not be solvable, when $K\ge 0$, the solution of Problem A is not necessarily unique. Hence we consider the well posedness of Problem A with the modified boundary conditions for the complex equation (4.1) as follows.

Problem B Find a continuous solution $w(z)$ of equation (4.1) satisfying the boundary condition

$$\mathrm{Re}[\overline{\lambda(z)}w(z)]=r(z)+h(z),\ z\in\Gamma,\tag{4.5}$$

where

$$h(z)=\begin{cases} 0,\ z\in\Gamma, & \text{if } K\geq N,\\ \left.\begin{array}{l} h_j,\ z\in\Gamma_j,\ j=1,...,N-K,\\ 0,\ z\in\Gamma_j, j=N-K+1,...,N+1 \end{array}\right\} & \text{if } 0\leq K<N,\\ \left.\begin{array}{l} h_j,\ z\in\Gamma_j,\ j=1,...,N,\\ h_0+\operatorname{Re}\displaystyle\sum_{m=1}^{-K-1}(h_m^+ + ih_m^-)[\zeta(z)]^m, z\in\Gamma_0 \end{array}\right\} & \text{if } K<0, \end{cases} \tag{4.6}$$

in which $h_j\,(j=0,1,...,N)$, $h_m^\pm\,(m=1,...,-K-1,K<0)$ are unknown real constants to be determined appropriately, $\zeta=\zeta(z)$ is a conformal mapping from the bounded domain with the boundary Γ_0 onto $|\zeta|<1$. In addition, for $K\geq 0$ the solution $w(z)$ is assumed to satisfy the point conditions

$$\operatorname{Im}[\overline{\lambda(a_j)}w(a_j)]=b_j, j\in J=\begin{cases} 1,...,2K-N+1, & \text{if } K\geq N,\\ N-K+1,...,N+1, & \text{if } 0\leq K<N, \end{cases} \tag{4.7}$$

where $a_j\in\Gamma_j\,(j=1,...,N)$, $a_j\in\Gamma_0\,(j=N+1,...,2K-N+1,K\geq N)$ are distinct points, and $b_j\,(j\in J)$ are all real constants satisfying the conditions

$$|b_j|\leq k_2,\ j\in J, \tag{4.8}$$

herein k_2 is a nonnegative constant.

4.2 Representation and uniqueness of solutions of Riemann-Hilbert problem for degenerate elliptic complex equations

It is easy to see that the complex equation

$$w_{\bar z}=0 \text{ in } \overline{D}, \text{ i.e. } w_{\overline{Z}}=0 \text{ in } D_Z \tag{4.9}$$

is a special case of equation (4.1). On the basis of the result in [86]9), we can find a unique solution of Problem B for equation (4.9) in $\overline{D_Z}$. Now we give the representation theorem of solutions for equation (4.1).

Theorem 4.1 *Suppose that the equation* (4.1) *satisfies Condition C. Then any solution of Problem B for* (4.1) *can be expressed as*

$$w[z(Z)] = W(Z) + \psi(Z) = \Phi(Z)e^{\phi(Z)} + \psi(Z), \tag{4.10}$$

where $\Phi(Z), \phi(Z), \psi(Z)$ *are as stated in* (2.22), $W(z)$ *is a solution of equation*

$$W_{\overline{Z}} = [A_1 W + A_2 \overline{W}]/H(y) \text{ in } D_Z, \tag{4.11}$$

and $\psi(Z)$ *is a solution of equation* (4.1) *in* D_Z *and possesses the form*

$$\psi(Z) = Tf = -\frac{1}{\pi}\iint_{D_t} \frac{f(t)}{t-Z} d\sigma_t \text{ in } D_Z, \tag{4.12}$$

$$f(Z) = [A_1\psi + A_2\overline{\psi} + A_3]/H(y) \text{ in } D_Z, \tag{4.13}$$

in which $Z = x + iY(y) = x + iG(y)$, *and* $W[Z(z)]$ *in* D *satisfies the boundary conditions*

$$\begin{aligned} &\mathrm{Re}[\overline{\lambda(z)}W(z(Z))] = r(z) + h(z) - \mathrm{Re}[\overline{\lambda(z)}\psi(Z(z))], \; z \in \Gamma, \\ &\mathrm{Im}[\overline{\lambda(a_j)}W(a_j)] = b_j - \mathrm{Im}[\overline{\lambda(a_j)}\psi(Z(a_j))], \; j \in J. \end{aligned} \tag{4.14}$$

Proof Let $w(z)$ be a solution of Problem B for equation (4.1), and be substituted in the coefficients of equation (4.1). By using the method in the proof of Theorem 3.4, we can find a solution $\psi(z)$ of such equation (4.1), and $\psi(z)$ possesses the form (4.12), (4.13). Moreover we can find the solution $W(z)$ in $\overline{D}$ of (4.11) with the boundary condition (4.14), thus

$$w[z(Z)] = W(Z) + \psi(Z) \text{ in } D \tag{4.15}$$

is the solution of Problem B in D_Z for equation (4.1), where $W(z) = \Phi(z)e^{\phi(z)}$ is as stated in (4.10).

Theorem 4.2 *Suppose that equation* (4.1) *satisfies Condition C. Then Problem B for* (4.1) *has at most one solution in D.*

Proof Let $w_1(z), w_2(z)$ be any two solutions of Problem B for (4.1). It is easy to see that $w(z) = w_1(z) - w_2(z)$ satisfies the homogeneous equation and boundary conditions

$$w_{\overline{z}} = \tilde{A}_1 w + \tilde{A}_2 \overline{w} \text{ in } D, \tag{4.16}$$

$$\mathrm{Re}[\overline{\lambda(z)}w(z)] = 0 \text{ on } \Gamma, \mathrm{Im}[\overline{\lambda(a_j)}w(a_j)] = 0, \; j \in J. \tag{4.17}$$

Noting the function $g(Z)$ in (4.1) with the condition $g(Z) \in L_\infty(D_Z)$, similarly to Theorem 2.5, by using the way as in Theorem 1.2, Chapter I and Theorem 4.1, Chapter II, [87]1), if the function $w(z) \not\equiv 0$ in $\overline{D}$, we can derive the absurd inequalities

$$\begin{gathered} 2K+1 \le 2N_D + N_\Gamma \le 2K, \text{ when } K \ge 0, \\ 2K-2N \le 2N_D+N_\Gamma \le 2N-2K-2, \text{ when } K<0, \end{gathered} \tag{4.18}$$

where N_D, N_Γ are denoted the totals of zero points of the solution $w(z)$ in D and Γ respectively. Hence $w(z) = w_1(z) - w_2(z) = 0$ in D. This proves the uniqueness of solutions of Problem B for (4.1).

4.3 Estimates of solutions of Riemann-Hilbert problem for degenerate elliptic equations

Now we shall give the estimates of the solutions of Problem B for (4.1) in $\overline{D}$, namely

Theorem 4.3 *If equation* (4.1) *satisfies Condition* C, *then any solution* $w(z)$ *of Problem* B *satisfies the estimates*

$$C_\delta[w(z(Z)), \overline{D_Z}] \le M_1, \; C_\delta[w(z(Z)), \overline{D_Z}] \le M_2(k_1+k_2), \tag{4.19}$$

here δ *is a sufficiently small positive constant, and* $M_1 = M_1(\delta, k, H, D)$, $M_2 = M_2(\delta, k_0, H, D)$ *are non-negative constants.*

Proof We first prove that if the solution $w(z)$ of Problem B satisfies the estimate of boundedness, i.e.

$$C[w(z(Z)), \overline{D_Z}] \le M_3, \tag{4.20}$$

where $M_3 = M_3(\delta, k, H, D)$ is a positive constant, then the first estimate of (4.19) will be derived, because from Lemma 2.1, it follows that $F(z, w) \in L_\infty(D_Z)$, hence $C_\beta[\psi(Z), \overline{D_Z})] \le M_4 = M_4(\beta, k, H, D_Z, M_3) < \infty$, β is as stated in (2.13). On basis of the representation (4.10), the function $W(Z) = w[z(Z)] - \psi(Z) = \Phi(Z)e^{\phi(Z)}$ in D_Z satisfies the boundary conditions

$$\begin{gathered} \mathrm{Re}[\overline{\lambda[z(Z)]}W(Z)] = r[z(Z)] - \mathrm{Re}[\overline{\lambda[z(Z)]}\psi(Z)] + h[z(Z)], Z \in \partial D_Z, \\ \mathrm{Im}[\overline{\lambda(a_j)}W[Z(a_j)] = b_j - \mathrm{Im}[\overline{\lambda(a_j)}\psi[Z(a_j)]], \; j \in J, \end{gathered} \tag{4.21}$$

where $\partial D_Z = Z(\Gamma)$, hence the analytic function $\Phi(Z) = W(Z)e^{-\phi(Z)}$ in D_Z satisfies the estimate $C_\delta[\Phi(Z), \overline{D_Z}] \le M_5 = M_5(\delta, k, H, D, M_3)$. Now we use the reduction to absurdity. Suppose that (4.20) is not true, then there exist sequences of coefficients $\{A_l^{(m)}\}$ $(l = 1,2,3)$, $\{\lambda^{(m)}(z)\}$, $\{r^{(m)}(z)\}$ and $\{b_j^{(m)}\}$, which satisfy the same conditions of coefficients as stated in (4.2), (4.4) and (4.8), such that $\{A_l^{(m)}\}$ weakly converge to $A_l^{(0)}$ $(l = 1,2,3)$ in D and $\{\lambda^{(m)}(z)\}$, $\{r^{(m)}(z)\}$, $\{b_j^{(m)}\}$ on Γ uniformly converge to $\lambda^{(0)}(z)$, $r^{(0)}(z)$, $b_j^{(0)}$ $(j \in J)$, and the solutions of the corresponding boundary value problems

$$w^{(m)}_{\bar{\bar{z}}} = F^{(m)}(z, w^{(m)}), F^{(m)}(z, w^{(m)}) = A_1^{(m)} w^{(m)} + A_2^{(m)} \overline{w^{(m)}} + A_3^{(m)} \text{ in } \overline{D}, \tag{4.22}$$

$$\text{Re}[\overline{\lambda^{(m)}(z)} w^{(m)}(z)] = r^{(m)}(z) \text{ on } \Gamma, \text{Im}[\overline{\lambda^{(m)}(a_j)} w^{(m)}(a_j)] = b_j^{(m)}, j \in J, \tag{4.23}$$

have the solutions $w^{(m)}(z)$, but $C[w^{(m)}(z), \overline{D}]$ $(m = 1, 2, ...)$ are unbounded, hence we can choose a subsequence of $\{w^{(m)}(z)\}$ denote by $\{w^{(m)}(z)\}$ again, such that $H_m = C[w^{(m)}(z), \overline{D}] \to \infty$ as $m \to \infty$, we can assume $H_m \ge \max[k_1, k_2, 1]$. It is obvious that $\hat{w}^{(m)}(z) = w^{(m)}(z)/H_m$ are solutions of the boundary value problems

$$\hat{w}^{(m)}_{\bar{\bar{z}}} = F^{(m)}(z, \hat{w}^{(m)}), F^{(m)}(z, \hat{w}^{(m)}) = A_1^{(m)} \hat{w}^{(m)} + A_2^{(m)} \overline{\hat{w}^{(m)}} + \frac{A_3^{(m)}}{H_m} \text{ in } \overline{D}, \tag{4.24}$$

$$\text{Re}[\overline{\lambda^{(m)}(z)} \hat{w}^{(m)}(z)] = \frac{r^{(m)}(z)}{H_m} \text{ on } \Gamma, \text{Im}[\overline{\lambda^{(m)}(a_j)} \hat{w}^{(m)}(a_j)] = \frac{b_j^{(m)}}{H_m}, j \in J. \tag{4.25}$$

It is easy to see that the functions in above boundary value problems satisfy the conditions

$$\begin{gathered} L_\infty[A_l, \overline{D}] \le k_0, l = 1, 2, L_\infty[A_3/H_m, \overline{D}] \le 1, C_\alpha[\lambda^{(m)}(z), \Gamma] \le k_0, \\ C_\alpha[r^{(m)}(z)/H_m, \Gamma] \le 1, \ |b_j^{(m)}/H_m| \le 1, \ j \in J. \end{gathered} \tag{4.26}$$

From the representation (4.10), the above solutions can be expressed as

$$\begin{gathered} \hat{w}^{(m)}(z) = \hat{W}^{(m)}(Z) + \hat{\psi}^{(m)}(Z), \\ \hat{\psi}^{(m)}(Z) = -\frac{1}{\pi} \iint_{D_t} \frac{\hat{f}^{(m)}(t)}{t - Z} d\sigma_t \text{ in } \overline{D_Z}, \end{gathered} \tag{4.27}$$

noting that $L_\infty[H(y)\hat{f}^{(m)}(Z), \overline{D_Z}] \le M_6 = M_6(k_0, H, D)$, we can derive that

$$C_\delta[\hat{\psi}^{(m)}(Z), D_Z] \le M_7 = M_7(\delta, k, H, D). \tag{4.28}$$

Due to the functions $\hat{W}^{(m)}(Z)$ are the solutions of the equation corresponding to (4.11) in $\overline{D_Z}$ and $\hat{w}^{(m)}(z) = \hat{W}^{(m)}(Z)+\hat{\psi}^{(m)}(Z)$ satisfy the boundary conditions as in (4.25), we can obtain the estimate

$$C_\delta[\hat{W}^{(m)}(Z), \overline{D_Z}] \le M_8 = M_8(\delta, k, H, D). \tag{4.29}$$

Thus from $\{\hat{w}^{(m)}(z)\}=\{\hat{W}^{(m)}(Z)+\hat{\psi}^{(m)}(Z)\}$, we can choose a subsequence denoted by $\{\hat{w}^{(m)}(z)\}$ again, and $\{\hat{w}^{(m)}(z)\} = \{\hat{W}^{(m)}(Z) + \hat{\psi}^{(m)}(Z)\}$ uniformly converge to $\hat{w}^{(0)}(z)$, it is clear that $\hat{w}^{(0)}(z)$ is a solution of the homogeneous problem of Problem B, on the basis of Theorem 4.2, the solution $\tilde{w}^{(0)}(z) = 0$ in $\overline{D}$, however, from $C[\hat{w}^{(m)}(z), \overline{D}] = 1$, we can derive that there exists a point $z^* \in \overline{D}$, such that $\hat{w}^{(0)}(z^*) \neq 0$, it is impossible. This shows that (4.20) is true. By using the method from (4.20) to (4.28), (4.29), we can obtain the first estimate in (4.19). Moreover we can verify the second estimate in (4.19).

4.4 Existence of solutions of Riemann-Hilbert problem for degenerate elliptic equations

In this section, we prove the existence of solutions of Problem B for equation (4.1).

Theorem 4.4 *Let equation* (4.1) *satisfy Condition C. Then the Riemann-Hilbert problem* (*Problem B*) *for* (4.1) *in the multiply connected domain D has a unique solution.*

Proof In order to find a solution $w(z)$ of Problem B for equation (4.1) in D by the Leray-Schauder theorem, we consider the equation (4.1) with the parameter $t \in [0, 1]$:

$$w_{\bar{z}}=tF(z,w),\ F(z,w)=G(Z)=A_1w+A_2\overline{w}+A_3 \text{ in } D_Z, \tag{4.30}$$

and introduce a bounded open set B_M of the Banach space $B = C_\delta(\overline{D_Z})$, whose elements are functions $w(z)$ satisfying the condition

$$w(z) \in C_\delta(\overline{D}),\ \ C_\delta[w(z(Z)), \overline{D_Z}] < M_9 = 1 + M_1, \tag{4.31}$$

where δ, M_1 are constants as stated in (4.19). We choose an arbitrary function $W(z) \in B_M$ and substitute it in the position of w in $F(z, w)$. By Theorem 4.1, a solution $w(z) = \Phi(Z) + \Psi(Z) = W(Z) + T(tF)$ of Problem B for the complex equation

$$w_{\bar{z}} = tF(z, W) \tag{4.32}$$

can be found. Noting that $tF[z(Z), W(z(Z))] \in L_\infty(\overline{D_Z})$, the above solution of Problem B for (4.32) is unique. Denote by $w(z) = T[W,t]$ $(0 \le t \le 1)$ the mapping from $W(z)$ to $w(z)$. From Theorem 4.3, we know that if $w(z)$ is a solution $w(z)$ of Problem B for the equation

$$w_{\bar{\bar{z}}} = tF(z,w) \text{ in } D_Z, \tag{4.33}$$

then the function $w(z)$ satisfies the estimate

$$C_\delta[w, \overline{D_Z})] < M_9. \tag{4.34}$$

Set $B_0 = B_M \times [0,1]$. Now we verify the three conditions of the Leray-Schauder theorem:

1. For every $t \in [0,1]$, $T[W,t]$ continuously maps the Banach space B into itself, and is completely continuous on B_M. In fact, arbitrarily select a sequence $W_n(z)$ in B_M, $n = 0,1,2,...$, such that $C_\delta[W_n - W_0, \overline{D_Z}] \to 0$ as $n \to \infty$. By Condition C, we see that $L_\infty[F(z,W_n) - F(z,W_0), \overline{D}] \to 0$ as $n \to \infty$. Moreover, from $w_n = T[W_n,t]$, $w_0 = T[W_0,t]$, it is easy to see that $w_n - w_0$ is a solution of Problem B for the following complex equation

$$(w_n - w_0)_{\bar{\bar{z}}} = t[F(z,W_n) - F(z,W_0)] \text{ in } D, \tag{4.35}$$

and then we can obtain the estimate

$$C_\delta[w_n - w_0, \overline{D}] \le 2k_0 C[W_n(z) - W_0(z), \overline{D}]. \tag{4.36}$$

Hence $C_\delta[w_n - w_0, \overline{D}] \to 0$ as $n \to \infty$. In addition for $W_n(z) \in B_M$, $n = 1,2,...$, we have $w_n = T[W_n,t]$, $w_m = T[W_m,t]$, $W_n, W_m \in B_M$, and then

$$(w_n - w_m)_{\bar{\bar{z}}} = t[F(z,W_n) - F(z,W_m)] \text{ in } D, \tag{4.37}$$

where $L_\infty[F(z,W_n) - F(z,W_m), D_Z] \le 2k_0 M_9$, hence from (4.19), we can obtain the estimate

$$C_\delta[w_n - w_m, \overline{D_Z}] \le 2M_2 k_0 M_9. \tag{4.38}$$

Thus there exists a function $w_0(z) \in B_M$, from $\{w_n(z)\}$ we can choose a subsequence $\{w_{n_k}(z)\}$ such that $C_\delta[w_{n_k} - w_0, \overline{D_Z}] \to 0$ as $k \to \infty$. This shows that $w = T[W,t]$ is completely continuous in B_M. Similarly we can also prove that for $W(z) \in B_M$, $T[W,t)$ is uniformly continuous with respect to $t \in [0,1]$.

2. For $t = 0$, it is evident that $w = T[W,0] = \Phi(Z) \in B_M$.

3. From the estimate (4.19), we see that $w = T[W, t]\,(0 \le t \le 1)$ does not have a solution $w(z)$ on the boundary $\partial B_M = \overline{B_M} \backslash B_M$.

Hence by the Leray-Schauder theorem, there exists a function $w(z) \in B_M$ such that $w(z) = T[w(z), t]$, and the function $w(z) \in C_\delta(\overline{D_Z})$ is just a solution of Problem B for the complex equation (4.1).

By a similar way as in the proof of Theorem 4.8, Chapter II, [87]1), from Theorem 4.4 the following result can be derived.

Theorem 4.5 *Under the same conditions as in Theorem 4.4, the following statements hold.*

(1) *If the index $K \ge N$, then Problem A for (4.1) is solvable.*

(2) *If $0 \le K < N$, then the total number of solvability conditions for Problem A does not exceed $N - K$.*

(3) *If $K < 0$, then Problem A has $N - 2K - 1$ solvability conditions.*

In latter chapters the notations $M_j = M_j(p_0, \delta, k, D)$, $M'_j = M'_j(p_0, \delta, k, D)$ (j is a positive integer) mean all non-negative constants dependent on p_0, δ, k, D.

CHAPTER II
ELLIPTIC COMPLEX EQUATIONS OF SECOND ORDER

This chapter mainly deals with the mixed boundary value problem and oblique derivative problem for several classes of linear and quasilinear elliptic equations of second order with parabolic degeneracy. We first reduce the boundary value problems for the degenerate elliptic equations of second order to the corresponding boundary value problems for elliptic complex equations of first order with singular coefficients, give the representation and a priori estimates of solutions for the above boundary value problems, and then the uniqueness and existence of solutions of the problems for second order equations can be proved.

1 The Discontinuous Oblique Derivative Problem for Uniformly Elliptic Complex Equations

In this section, we first introduce the discontinuous oblique derivative problem for linear and nonlinear uniformly elliptic equations of second order. Afterwards various properties of solutions for the equations are given, which include the extremum principle, representation theorem and a priori estimates of solutions for the above boundary value problem, and then the uniqueness and existence of solutions of the above problem are proved.

1.1 Formulation of discontinuous oblique derivative problem for elliptic equations

Let D be the upper half-unit disk with the boundary $\partial D = \Gamma \cup \gamma$, where $\Gamma = \{|z| = 1, \operatorname{Im} z > 0\}$ and $\gamma = \{-1 < x < 1, y = 0\}$. We consider the quasilinear uniformly elliptic equation of second order

$$au_{xx} + 2bu_{xy} + cu_{yy} + du_x + eu_y + fu = g \text{ in } D, \tag{1.1}$$

in which a, b, c, d, e, f, g are given functions of $(x, y) \in \overline{D}$ and $u, u_x, u_y \in \mathbf{R}$. Under certain conditions, equation (1.1) can be transformed into the complex form

$$\begin{array}{l} u_{z\bar z}=F(z,u,u_z,u_{zz}),\ F=\mathrm{Re}[Qu_{zz}+A_1u_z]+A_2u+A_3, \\ W_{\bar z}=F(z,u,W,W_z),\ F=\mathrm{Re}[QW_z+A_1W]+A_2u+A_3, \end{array}\ \text{in } D, \qquad (1.2)$$

where $Q = Q(z, u, u_z)$, $A_j = A_j(z, u, u_z)\,(j = 1, 2, 3)$, $W(z) = u_z$, and

$$z=x+iy,\ u_z=\frac{1}{2}[u_x-iu_y]=\overline{u_{\bar z}},\ u_{z\bar z}=\frac{1}{4}[u_{xx}+u_{yy}],$$

$$Q=\frac{-a+c-2bi}{a+c},\ A_1(z)=\frac{-d-ei}{a+c},$$

$$A_2=\frac{-f}{2(a+c)},\ A_3(z)=\frac{g}{2(a+c)},$$

and the uniformly elliptic condition

$$\Delta = ac - b^2 \ge \Delta_0 > 0,\ \ a > 0 \ \text{in}\ \bar D$$

is reduced to the inequality

$$|Q(z, u, u_z)| \le q_0 < 1 \ \text{in}\ \bar D,$$

in which Δ_0, q_0 are non-negative constants. For the nonlinear uniformly elliptic equation of second order

$$\Phi(x, y, u, u_x, u_y, u_{xx}, u_{xy}, u_{yy}) = 0 \ \text{in}\ D,$$

under certain conditions, we can transform it into the complex equation in the form (1.2), but in which $Q = Q(z, u, u_z, u_{zz})$, $A_j = A_j(z, u, u_z)\,(j = 1, 2, 3)$ (see [86]9), [87]1)). In order to the latter usage, we assume that the nonlinear complex equation (1.2) satisfies the conditions, namely **Condition** C

1) $Q(z, u, w, V), A_j(z, u, w)\,(j = 1, 2, 3)$ are continuous in $u \in \mathbf{R}$, $w \in \mathbf{C}$ for almost every point $z \in D$ and $V \in \mathbf{C}$, and $Q = 0$, $A_j = 0\,(j = 1, 2, 3)$ for $z \notin D$.

2) The above functions are measurable in $z \in D$ for all continuously differentiable function $u(z)$ in $D^* = \bar D\backslash Z$, and satisfy

$$\begin{array}{c} L_p[A_j(z, u, u_z), \bar D] \le k_0,\ j = 1, 2, \\ L_p[A_3(z, u, z_z), \bar D] \le k_1, A_2(z, u, u_z) \ge 0 \ \text{in}\ D, \end{array} \qquad (1.3)$$

in which $p\,(>2)$, k_0, k_1 are non-negative constants, and $Z=\{-1,1\}$.

3) Equation (1.2) satisfies the uniform ellipticity condition, namely for any number $u\in\mathbf{R}, w\in\mathbf{C}$, the inequality

$$|Q(z,u,u_z,V)|\le q_0,\ \text{ i.e. } |F(z,u,u_z,V_1)-F(z,u,u_z,V_2)|\le q_0, \tag{1.4}$$

for almost every point $z\in D$ and u,u_z,V_1,V_2 holds, where $q_0(<1)$ is a non-negative constant.

If equation (1.1) in $\overline{D}$ is linear uniformly elliptic, and the coefficients a,b,c,d,e,f,g satisfy some stronger condition in $\overline{D}$, then equation (1.1) can be transformed into the simpler equation. By the result in [81]1), [86]9), [87]1), we can find a homeomorphic solution $\zeta(z)$ of the Beltrami equation

$$\zeta_{\bar z}-q(z)\zeta_z=0,\ \ q(z)=\frac{Q(z)}{1+\sqrt{1-|Q(z)|^2}}\ \text{ in } \bar D,$$

which quasiconformally maps the domain D onto a domain G, if D is an $N+1$-connected domain with the boundary $\Gamma\in C^2_\mu(0<\mu<1)$, then the domain G may be an $N+1$-connected circular domain. If the coefficients $a,b,c\in W^1_p(D)$, $p>2$, then $Q(z)$, $q(z)\in W^1_p(D)$, and then $\zeta(z)\in W^2_p(D)$ and its inverse function $z(\zeta)\in W^2_p(G)$. Thus through the function $U(\zeta)=u[z(\zeta)]$, the equation (1.2) can be reduced to the canonical form

$$U_{\zeta\bar\zeta}-\text{Re}[B_1(\zeta)U_\zeta]-B_2(\zeta)U=B_3(\zeta)\ \text{ in } \bar G, \tag{1.5}$$

in which

$$B_1(\zeta)=\frac{1}{A}[-2\zeta_{z\bar z}+Q\zeta_{zz}+\overline{Q}\zeta_{\bar z\bar z}+A_1\zeta_z+\overline{A_1}\zeta_{\bar z}],$$

$$B_2(\zeta)=\frac{A_2}{A},\ B_3(\zeta)=\frac{A_3}{A},\ A=[1+|q|^2-2\text{Re}(Q\bar q)]|\zeta_z|^2.$$

It is clear that if $f\le 0$ in $\bar D$, then $A_2(z)\ge 0$ in $\bar D$ and $B_2(\zeta)\ge 0$ in $\bar G$. Under the condition, we can find the solutions $\psi(\zeta)$, $\Psi(\zeta)$ of equation (1.5) and homogeneous equation

$$U_{\zeta\bar\zeta}-\text{Re}[B_1(\zeta)U_\zeta]-B_2(\zeta)U=0\ \text{ in } \bar G$$

with the boundary conditions $\psi(\zeta)=0$, $\Psi(\zeta)=1$ on the boundary ∂G respectively (see Theorem 1.1 below), thus if the function $U(\zeta)$ is a solution of equation (1.5), then $V(\zeta)=[U(\zeta)-\psi(\zeta)]/\Psi(\zeta)$ is a solution of the equation

$$V_{\zeta\bar\zeta}-\text{Re}[C_1(\zeta)V_\zeta]=0\ \text{ in } \bar G, \tag{1.6}$$

where

$$C_1(\zeta) = -2(\ln \Psi)_{\bar\zeta} + B_1.$$

When $C_1(\zeta) = 0$ in $\bar G$, then the equation is just the complex form of harmonic equation, i.e. the Laplace equation

$$V_{\zeta\bar\zeta} = 0, \text{ i.e. } \frac{1}{4}(V_{\xi\xi} + V_{\eta\eta}) = 0 \text{ in } \bar G,$$

in which $\zeta = \xi + i\eta$ (see [86]9)).

The discontinuous oblique derivative boundary value problem for equation (1.2) may be formulated as follows:

Problem P Find a continuously differentiable solution $u(z)$ of (1.2) in $D^* = \bar D \backslash Z$, which is continuous in $\overline{D}$ and satisfies the boundary conditions

$$\frac{1}{2}\frac{\partial u}{\partial \nu} = \mathrm{Re}[\overline{\lambda(z)}u_z] = r(z), z \in \Gamma^* = \partial D \backslash Z, u(-1) = b_0, u(1) = b_1, \qquad (1.7)$$

where $Z = \{-1, 1\}$ is the set of discontinuous points of $\lambda(z)$ on Γ^*, ν is a given vector at every point on Γ^*, $\lambda(z) = a(x) + ib(x) = \cos(\nu, x) - i\cos(\nu, y)$, $\cos(\nu, n) \geq 0$ on Γ^*, here n is the outward normal vector at every point on Γ^*, b_0, b_1 are real constants, and $\lambda(z)$, $r(z)$, b_0, b_1 satisfy the conditions

$$C_\alpha[\lambda(z), \Gamma^*] \leq k_0, \ C_\alpha[r(z), \Gamma^*] \leq k_2, \ |b_0|, |b_1| \leq k_2. \qquad (1.8)$$

Herein $\alpha\,(1/2 < \alpha < 1), k_0, k_2$ are non-negative constants. Problem P with $A_3(z) = 0$ in D, $r(z) = 0$ on Γ, $b_0 = b_1 = 0$ is called Problem P_0. The index of Problem P is K, where K is defined as in (1.18), Chapter I, but $m = 2$, $z_1 = -1$, $z_2 = 1$, here we choose $K = 0$ and can require $-1/2 \leq \gamma_j < 1/2\,(j = 1, 2)$, because otherwise it is sufficient to multiply the solution $W(z)$ of the complex equation (1.2) and the boundary condition (1.7)($W(z) = u_z$) by the function $X_0(z) = z$ or $X_0(z) = (z - 1/2)(z + 1/2)$, similarly to (2.39) and (2.40) below, the index $\hat K$ of $X_0(z)\lambda(z)$ on ∂D should be increased by $1/2$ and 1 respectively, if $\hat K = 1/2$, except the point conditions $u(-1) = b_0, u(1) = b_1$, we can add another point condition on ∂D, for instance $u(0) = 0$ such that the above boundary value problem is well-posed. We mention that if $\cos(\nu, n) \equiv 0$ on Γ, then the condition $u(1) = b_1$ can be cancelled. In fact, if $\cos(\nu, n) \equiv 0$ on Γ, from the boundary condition (1.7), we can determine the value $u(1)$ by the value $u(-1)$, namely

$$u(1) = 2\mathrm{Re}\int_{-1}^{1} u_z dz + u(-1) = 2\int_0^S \mathrm{Re}[z'(s)u_z]ds + b_0 = 2\int_0^S r(z)ds + b_0,$$

in which $\overline{\lambda(z)} = z'(s)$ on Γ, $z(s)$ is a parameter expression of arc length s of Γ with the condition $z(-1) = 0$, and S is the length of Γ, and if $u_x = r(x)$ on γ, then $u(x) = \int_{-1}^{x} u_x dx + b_0$, $u(1) = \int_{-1}^{1} r(x)dx + b_0 = b_1$. If $A_2(z) = 0$ in D, the last point condition in (1.7) can be replaced by

$$\operatorname{Im}[\overline{\lambda(z)}u_z]|_{z=z_0} = b_2, \tag{1.9}$$

where z_0 is a fixed point on $\Gamma\backslash\{-1,1\}$, and we do not need the assumption $\cos(\nu, n) \geq 0$ on Γ, where b_2 is a real constant. Then the boundary value problem for (1.2) will be called Problem Q. Later on we shall only discuss the case of $K = 0$, and the other case can be similarly discussed.

1.2 The representation theorem of discontinuous oblique derivative problem for elliptic equations

We first introduce a theorem.

Theorem 1.1 *Suppose that equation* (1.2) *satisfies Condition C. Then there exist two solutions* $\psi(z), \Psi(z)$ *of the Dirichlet problem* (*Problem D*) *of the linear case of* (1.2) *and its related homogeneous equation*

$$u_{z\bar{z}} - \operatorname{Re}[Qu_{zz} + A_1u_z] - A_2u = 0 \ \text{ in } D, \tag{1.10}$$

satisfying the boundary conditions

$$\psi(z) = 0, \ \Psi(z) = 1 \ \text{on } \Gamma' = \partial D \tag{1.11}$$

respectively, and $\psi(z), \Psi(z)$ *satisfy the estimates*

$$\begin{gathered} C_\beta^1[\psi(z), \bar{D}] \leq M_1, \ C_\beta^1[\Psi(z), \bar{D}] \leq M_2, \\ L_{p_0}[\psi_{z\bar{z}}, \bar{D}] \leq M_3, \ L_{p_0}[\Psi_{z\bar{z}}, \bar{D}] \leq M_3, \Psi \geq M_4 > 0 \text{ in } D, \end{gathered} \tag{1.12}$$

where $\beta\,(0 < \beta \leq \min(\alpha, 1 - p_0/2)/2$, $p_0(2 < p_0 \leq p)$, $M_j = M_j(q_0, p_0, \beta, k_0, k_1, D)$ $(j = 1,2,3,4)$ *are non-negative constants.*

Proof We first assume that the coefficients $Q = A_j = 0\,(j = 1, 2, 3)$ of (1.2) in the ε-neighborhood of $z = -1, 1$, i.e. $D_\varepsilon = \{|z \pm 1| < \varepsilon, \operatorname{Im} z \geq 0\}$, where $\varepsilon = 1/m$, m is a positive integer. Introduce the transformation and its inversion

$$\zeta(z) = -i\frac{z^2 + 2iz + 1}{z^2 - 2iz + 1}, \ z(\zeta) = \frac{1}{\zeta + i}[1 + i\zeta - \sqrt{2(1 - \zeta^2)}]. \tag{1.13}$$

The function $\zeta(z)$ maps D onto $G=\{|\zeta|<1\}$, such that the boundary points $-1,0,1$ are mapped onto the points $-1,-i,1$ respectively. Through the transformation $\zeta=\zeta(z)$, equation (1.2) is reduced to the equation

$$\begin{gathered}u_{\zeta\bar{\zeta}}=|z'(\zeta)|^2\{\mathrm{Re}[Qu_{\zeta\zeta}/(z'(\zeta))^2\\+(A_1/z'(\zeta)-Qz''(\zeta)/(z'(\zeta))^3)u_\zeta]+A_2u+A_3\}\text{ in }G.\end{gathered}\tag{1.14}$$

It is clear that equation (1.14) in $\bar{G}$ satisfies conditions similar to Condition C. Hence equation (1.14) and its related homogeneous equation

$$\begin{gathered}u_{\zeta\bar{\zeta}}=|z'(\zeta)|^2\{\mathrm{Re}[Qu_{\zeta\zeta}/(z'(\zeta))^2\\+(A_1/z'(\zeta)-Qz''(\zeta)/(z'(\zeta))^3)u_\zeta]+A_2u\}\text{ in }G\end{gathered}\tag{1.15}$$

possess the solutions $\psi(\zeta)$, $\Psi(\zeta)$ satisfying the boundary conditions

$$\psi(\zeta)=0,\ \Psi(\zeta)=1\ \text{ on }\ L=\zeta(\Gamma'),$$

and $\psi[\zeta(z)],\Psi[\zeta(z)]$ in D are the solutions of Problem D of (1.2),(1.10) satisfying the boundary condition (1.11) respectively, and $\psi(z)$, $\Psi(z)$ satisfy the estimate (1.12), but the constants $M_j=M_j(q_0,p_0,\beta,k_0,\ k_1,D,\varepsilon)\,(j=2,3,4)$ (see [86]9),[87]1)). Now we consider

$$\tilde{\psi}(z)=\begin{cases}\psi(z)\ \text{ in }D,\\-\psi(\bar{z})\ \text{ in }\tilde{D}=\{|z|<1,\mathrm{Im}z<0\}.\end{cases}\tag{1.16}$$

It is not difficult to see that $\tilde{\psi}(z)$ in $\Delta=\{|z|<1\}$ is a solution of the elliptic equation

$$u_{z\bar{z}}-\mathrm{Re}[\tilde{Q}u_{zz}+\tilde{A}_1u_z]-\tilde{A}_2u=\tilde{A}_3\ \text{ in }\Delta,\tag{1.17}$$

where the coefficients are as follows

$$\tilde{Q}=\begin{cases}Q(z),\\\overline{Q(\bar{z})}\end{cases}\tilde{A}_1=\begin{cases}A_1(z),\\\overline{A_1(\bar{z})},\end{cases}\tilde{A}_2=\begin{cases}A_2(z),\\A_2(\bar{z}),\end{cases}\tilde{A}_3=\begin{cases}A_3(z)\\-A_3(\bar{z})\end{cases}\text{in}\begin{Bmatrix}D\\\tilde{D}\end{Bmatrix},$$

where $\tilde{D}$ is the symmetrical domain of D with respect to the real axis. It is clear that the above coefficients in Δ satisfy conditions similar to those from Condition C. Obviously the solution $\tilde{\psi}(z)$ satisfies the boundary condition $\tilde{\psi}(z)=0$ on $\partial\Delta=\{|z|=1\}$. Denote by $\tilde{\psi}_m(z)$ the solution of equation (1.2) with $Q=A_j=0\,(j=1,2,3)$ in the $\varepsilon=1/m$-neighborhood of $z=-1,1$, we can derive that the function $\tilde{\psi}_m(z)$ in $\overline{\Delta}$ satisfies estimates similar to $\psi(z)$ in (1.12), where the constants $M_j(j=2,3)$ are independent of $\varepsilon=1/m$. Thus we can choose a subsequence of $\{\tilde{\psi}_m(z)\}$, which uniformly converges to $\psi_*(z)$, and $\psi_*(z)$ is just a solution of Problem D for the original equation (1.2) in D. Noting that the solution $\Psi(z)=\psi(z)+1$ of Problem D for equation (1.10) is equivalent to the solution $\psi(z)$ of Problem D for the equation

$$u_{z\bar{z}}-\text{Re}[Qu_{zz}+A_1u_z]-A_2u=A_2 \text{ in } D \tag{1.18}$$

with the boundary condition $\psi(z)=0$ on Γ', by using the same way, we can prove that there exists a solution $\Psi(z)$ of Problem D for (1.10) with the boundary condition $\Psi(z)=1$ on Γ', and the solution satisfies the estimates in (1.12).

Theorem 1.2 *Suppose that equation* (1.2) *satisfies Condition C, and* $u(z)$ *is a solution of Problem P for* (1.2). *Then* $u(z)$ *can expressed as*

$$u(z)=U(z)\Psi(z)+\psi(z),U(z)=2\text{Re}\int_{z_1}^{z}w(z)dz+b_0,w(z)=\Phi[\zeta(z)]e^{\phi(z)}, \tag{1.19}$$

where $z_1=-1$, $\psi(z),\Psi(z)$ *are as stated in Theorem* 1.1 *satisfying the estimate* (1.12), $\zeta(z)$ *is a homeomorphism in* $\bar{D}$, *which quasiconformally maps* D *onto the unit disk* $G=\{|\zeta|<1\}$ *with the boundary* L, *where* $\zeta(1)=1,\zeta(i)=i,\zeta(-1)=-1$, $\Phi(\zeta)$ *is an analytic function in* G, $\phi(z),\zeta(z)$ *and its inverse function* $z(\zeta)$ *satisfy the estimates*

$$\begin{gathered}C_\beta[\phi(z),\bar{D}]\le k_3,\ C_\beta[\zeta(z),\bar{D}]\le k_3, C_\beta[z(\zeta),\bar{G}]\le k_3,\\ L_{p_0}[|\phi_{\bar{z}}|+|\phi_z|,\bar{D}]\le k_3,\ L_{p_0}[|\chi_{\bar{z}}|+|\chi_z|,\bar{D}]\le k_4,\end{gathered} \tag{1.20}$$

in which $\chi(z)$ *is as stated in* (1.27), *Chapter* I, $\beta=\min(\alpha,1-2/p_0)/2$, $p_0\,(2<p_0\le p), k_j=k_j(q_0,p_0,k_0,k_1,D)\,(j=3,4)$ *are non-negative constants.*

Proof We substitute the solution $u(z)$ of Problem P into the coefficients of equation (1.2). It is clear that (1.2) in this case can be seen as a linear equation. Firstly, on the basis of Theorem 1.1, there exist two solutions $\psi(z),\Psi(z)$ of above Problem D of (1.2) and its homogeneous equation (1.10) satisfying the estimate (1.12). Thus the function

$$U(z)=[u(z)-\psi(z)]/\Psi(z) \text{ in } D, \tag{1.21}$$

is a solution of the equation

$$U_{z\bar{z}}-\text{Re}[QU_{zz}+AU_z]=0,\ A=A_1-2(\ln\Psi)_{\bar{z}}+2Q(\ln\Psi)_z \text{ in } D, \tag{1.22}$$

and $w(z) = U_z$ is a solution of the first order equation

$$w_{\bar{z}} = \frac{1}{2}[Qw_z + \overline{Q}\bar{w}_{\bar{z}} + Aw + A\bar{w}] \text{ in } D \tag{1.23}$$

satisfying the boundary condition

$$\begin{aligned} &\frac{1}{2}[\frac{\partial U}{\partial \nu} + (\ln \Psi)_\nu U] = r(z) - \mathrm{Re}[\overline{\lambda(z)}\psi_z] \text{ on } \Gamma^*, \text{ i.e.} \\ &\mathrm{Re}[\overline{\lambda(z)}U_z + (\ln \Psi)_\nu U/2] = r(z) - \mathrm{Re}[\overline{\lambda(z)}\psi_z] \text{ on } \Gamma^*. \end{aligned} \tag{1.24}$$

By Lemma 1.3 below, we see that $(\ln \Psi)_\nu > 0$ on Γ^*, and similarly to Theorem 1.1 in Chapter I, the last formula in (1.19) can be derived, and $\phi(z), \zeta(z)$ and its inverse function $z(\zeta)$, $\chi(z)$ satisfy the estimates (1.20).

Now we consider the linear homogeneous equation

$$u_{z\bar{z}} - \mathrm{Re}[Qu_{zz} + A_1(z)u_z] - A_2(z)u = 0 \text{ in } D, \tag{1.25}$$

and give a lemma.

Lemma 1.3 *Let the equation* (1.25) *in D satisfy Condition C, and $u(z)$ be a continuously differentiable solution of* (1.25) *in $\overline{D}$. If $M = \max_{z\in\overline{D}} u(z) \geq 0$, then there exists a point $z_0 \in \partial D$, such that $u(z_0) = M$. If $z_0 \in \Gamma^*$, and $u(z) < u(z_0)$ in $\bar{D}\backslash\{z_0\}$, then*

$$\frac{\partial u}{\partial l} = \lim_{z(\in l)\to z_0} \frac{u(z_0) - u(z)}{|z - z_0|} > 0, \tag{1.26}$$

where $z\,(\in D)$ approaches z_0 along a direction l, such that $\cos(l, n) > 0$, *here n is the outward normal vector at z_0 of Γ^*.*

Proof From Theorem 2.2, Chapter III, [87]1), we see that the solution $u(z)$ in $\overline{D}$ attains its non-negative maximum M at a point $z_0 \in \partial D$. There is no harm in assuming that z_0 is a boundary point of $\Delta = \{|z| < R\}$, because we can choose a subdomain ($\subset \overline{D}$) with the smooth boundary and the boundary point z_0, and then make a conformal mapping, this requirement can be realized. By Theorem 1.1, we find a continuously differentiable solution $\Psi(z)$ of (1.25) in $\overline{\Delta}$ satisfying the boundary condition: $\Psi(z) = 1$, $z \in \partial\Delta = \{|z| = R\}$, and can derive that $0 < \Psi(z) \leq 1$, $z \in \overline{\Delta}$. Due to $V(z) = u(z)/\Psi(z)$ is a solution of the following equation

$$\begin{aligned} &LV = V_{z\bar{z}} - \mathrm{Re}[Q(z)V_{zz} + A(z)V_z] = 0, \\ &A(z) = -2(\ln \Psi)_{\bar{z}} + 2Q(\ln \Psi)_z + A_1(z) \text{ in } \Delta, \end{aligned} \tag{1.27}$$

it is clear that $V(z) < V(z_0)$, $z \in \Delta$, and $V(z)$ attains the maximum at the point z_0. Afterwards, we find a continuously differentiable solution $\tilde{V}(z)$ of (1.27) in $\tilde{\Delta} = \{R/2 \le |z| \le R\}$ satisfying the boundary condition

$$\tilde{V}(z) = 0,\ z \in \partial\Delta;\ \tilde{V}(z) = 1,\ |z| = R/2.$$

It is easy to see that $\partial\tilde{V}/\partial s = 2\text{Re}[iz\tilde{V}_z]/R$ if $|z| = R$, $\partial\tilde{V}/\partial s = -4\text{Re}[iz\tilde{V}_z]/R$ if $|z| = R/2$, and

$$\frac{\partial\tilde{V}}{\partial n} = 2\frac{\text{Re}[z\tilde{V}_z]}{R},\ z \in \partial\Delta,\ \frac{\partial\tilde{V}}{\partial n} = -4\frac{\text{Re}[z\tilde{V}_z]}{R},\ |z| = \frac{R}{2},$$

where s, n are the tangent vector and outward normal vector on the boundary $\partial\tilde{\Delta}$. Noting that $W(z) = \tilde{V}_z$ satisfies the equation

$$W_{\bar{z}} - \text{Re}[Q(z)W_z + A(z)W] = 0,\ z \in \tilde{\Delta},$$

and the boundary condition $\text{Re}[izW(z)] = 0$, $z \in \partial\tilde{\Delta}$, and the index of $\overline{iz}$ on the boundary $\partial\tilde{\Delta}$ equals to 0, hence $W(z)$ has no zero point on $\partial\Delta$, thus $\partial\tilde{V}/\partial n = 2\text{Re}[zW(z)/R] < 0$, $z \in \partial\Delta$. The auxiliary function

$$\hat{V}(z) = V(z) - V(z_0) + \varepsilon\tilde{V}(z),\ z \in \tilde{\Delta},$$

by selecting a sufficiently small positive number ε, such that $\hat{V}(z) < 0$ on $|z| = R/2$, obviously satisfies $\hat{V}(z) \le 0$, $z \in \partial\Delta$. Due to $L\hat{V} = 0$, $z \in \tilde{\Delta}$, on the basis of the maximum principle, we have

$$\hat{V}(z) \le 0,\ z \in \partial\Delta,\ \text{i.e. } V(z_0) - V(z) \ge -\varepsilon[\tilde{V}(z_0) - \tilde{V}(z)],\ z \in \tilde{\Delta}.$$

Thus at the point $z = z_0$ we have

$$\frac{\partial V}{\partial n} \ge -\varepsilon\frac{\partial\tilde{V}}{\partial n} > 0,\ \frac{\partial u}{\partial n} = \Psi\frac{\partial V}{\partial n} + V\frac{\partial\Psi}{\partial n} \ge -\varepsilon\frac{\partial\tilde{V}}{\partial n} + V\frac{\partial\Psi}{\partial n} > 0.$$

Moreover, from the conditions $\cos(l,n) > 0, \partial u/\partial n > 0, \partial u/\partial s = 0$ at the point z_0, where s is the tangent vector at z_0, it follows the inequality

$$\frac{\partial u}{\partial l} = \cos(l,n)\frac{\partial u}{\partial n} + \cos(l,s)\frac{\partial u}{\partial s} > 0. \tag{1.28}$$

Theorem 1.4 *If equation* (1.2) *satisfies Condition C and for any* $u_j(z) \in C^1(D^*)$, $j = 1, 2$, $u_{zz} \in \mathbf{C}$, *the following equality holds*:

$$F(z, u_1, u_{1z}, u_{1zz}) - F(z, u_2, u_{2z}, u_{2zz}) = -\text{Re}[\tilde{Q}u_{zz} + \tilde{A}_1u_z] - \tilde{A}_2u,$$

where $L_p[\tilde{A}_j, \bar{D}] < \infty$, $j = 1, 2$, *then the solution* $u(z)$ *of Problem* P *is unique.*

Proof Suppose that there exist two solutions $u_1(z), u_2(z)$ of Problem P for (1.2), we see that $u(z) = u_1(z) - u_2(z)$ satisfies the homogeneous equation and boundary conditions

$$\begin{aligned} &u_{z\bar{z}} = \operatorname{Re}[\tilde{Q}u_{zz} + \tilde{A}_1 u_z] + \tilde{A}_2 u \text{ in } D, \\ &\frac{1}{2}\frac{\partial u}{\partial \nu} = 0,\, z \in \Gamma^*,\ u(-1) = 0,\ u(1) = 0. \end{aligned} \tag{1.29}$$

If the maximum $M = \max_{\bar{D}} u(z) > 0$, it is clear that the maximum point $z^* \neq -1$ and 1. On the basis of Lemma 1.3, the maximum of $u(z)$ cannot attain on Γ^*, because if its maximum M attains at a point $z^* \in \Gamma^*$ and $\cos(\nu, n) > 0$ at z^*, from Lemma 1.3, we have $\partial u/\partial \nu > 0$ at z^*, this contradicts the boundary condition in (1.29); if $\cos(\nu, n) = 0$ at z^*, denote by $\tilde{\Gamma}$ the longest curve of Γ including the point z^*, so that $\cos(\nu, n) = 0$ and $u(z) = M$ on $\tilde{\Gamma}$, then there exists a point $z' \in \Gamma \backslash \tilde{\Gamma}$, such that at z', $\cos(\nu, n) > 0, \partial u/\partial n > 0, \cos(\nu, s) > 0\,(< 0)$, $\partial u/\partial s \geq 0\,(\leq 0)$, hence (1.28) at z' holds, it is impossible. This shows $z^* \notin \Gamma$. Hence $\max_{\overline{D}} u(z) = 0$. By the similar method, we can prove $\min_{\overline{D}} u(z) = 0$. Therefore $u(z) = 0$, $u_1(z) = u_2(z)$ in $\overline{D}$.

Theorem 1.5 *Suppose that equation* (1.2) *satisfies Condition* C. *Then the solution* $u(z)$ *of Problem* P *for* (1.2) *satisfies the estimates*

$$\begin{aligned} \hat{C}_\delta[u(z), \bar{D}] = C_\beta[u(z), \bar{D}] + C_\delta[X(z)u_z, \bar{D}] \leq M_5, \\ \hat{C}_\delta[u(z), \bar{D}] \leq M_6(k_1 + k_2), \end{aligned} \tag{1.30}$$

where $\beta = \min(\alpha, 1 - 2/p_0)/2$, $X(z) = |z+1|^{\eta_1}|z-1|^{\eta_2}$, $\eta_j = \max(-2\gamma_j, 0) + 4\delta$, $j = 1, 2$, $\gamma_j\,(j = 1, 2)$ *are real constants as stated in* (1.17), *Chapter* I *and* $\delta\,(0 < \delta < \min(\beta, \tau))$ *is a sufficiently small positive constant, and* $M_5 = M_5(q_0, p_0, \beta, \delta, k, D)$, $M_6 = M_6(q_0, p_0, \beta, \delta, k_0, D)$ *are two non-negative constants.*

Proof We first verify that any solution $u(z)$ of Problem P for (1.2) satisfies the estimate

$$\hat{C}[u, \overline{D}] = C[u(z), \bar{D}] + C[X(z)u_z, \bar{D}] \leq M_7 = M_7(q_0, p_0, \alpha, k, D). \tag{1.31}$$

Otherwise, if the above inequality is not true, there exist sequences of coefficients: $\{Q^m\}$, $\{A_j^m\}\,(j = 1, 2, 3), \{\lambda^m\}, \{r^m\}, \{b_j^m\}(j = 1, 2)$ satisfying the same conditions of $Q, A_j (j = 1, 2, 3)$, $\lambda, r, b_j (j = 1, 2)$, and

$\{Q^m\}, \{A_j^m\}(j = 1,2,3)$ weakly converge in D to $Q^0, A_j^0(j = 1,2,3)$, and $\{\lambda^m\}, \{r^m\}, \{b_j^m\}(j = 1,2)$ uniformly converge on Γ^* to $\lambda^0, r^0, b_j^0(j = 1,2)$ respectively. Let u^m is a solution of Problem P for (1.2) corresponding to $\{Q^m\}, \{A_j^m\}$ $(j = 1,2,3)$, $\{\lambda^m\}, \{r^m\}$, $\{b_j^m\}(j = 1,2)$, but $\hat{C}[u^m(z), \overline{D}] = H_m \to \infty$ as $m \to \infty$. There is no harm in assuming that $H_m \geq \max[1, k_1, k_2]$. Let $U^m = u^m/H_m$. It is clear that $U^m(z)$ is a solution of the boundary value problem

$$U_{z\bar{z}}^m - \mathrm{Re}[Q^m U_{zz}^m + A_1^m U_z^m] - A_2^m U^m = \frac{A_3^m}{H_m},$$

$$\frac{1}{2}\frac{\partial U^m}{\partial \nu_m} = \frac{r^m(z)}{H_m},\ z \in \Gamma^*,\ U^m(-1) = \frac{b_0^m}{H_m},\ U^m(1) = \frac{b_1^m}{H_m}.$$

From the conditions in the theorem, we have

$$L_p[A_3^m/H_m, \bar{D}] \leq 1,\ C_\alpha[\lambda^m, \Gamma_j] \leq k_0,$$

$$C_\alpha[r^m(z)/H_m, \Gamma^*] \leq 1,\ |b_j^m/H_m| \leq 1,\ j = 1,2.$$

According to the method in the proof of Theorem 1.3, Chapter I, and denoting

$$w_m(z) = U_z^m,\ U^m(z) = 2\mathrm{Re}\int_{-1}^{z} w_m(z)dz + \frac{b_0^m}{H_m},$$

we can obtain that $U_m(z)$ satisfies the estimate

$$C_\delta[U^m(z), \bar{D}] + C_\delta[X(z)U_z^m, \bar{D}] \leq M_8, \tag{1.32}$$

in which $M_8 = M_8(q_0, p_0, \delta, \alpha, k, D)$, $\delta\,(> 0)$ are non-negative constants. Hence from $\{U^m(z)\}$ and $\{X(z)U_z^m\}$, we can choose subsequences $\{U^{m_k}(z)\}$ and $\{X(z)U_z^{m_k}\}$, which uniformly converge to $U^0(z)$ and $X(z)U_z^0$ in $\bar{D}$ respectively, and $U^0(z)$ is a solution of the following boundary value problem

$$U_{z\bar{z}}^0 = \mathrm{Re}[Q^0 U_{zz}^0 + A_1^0 u_z] + A_2^0 U^0 \text{ in } D,$$

$$\frac{\partial U^0}{\partial \nu} = 0 \text{ on } \Gamma^*,\ U^0(-1) = 0,\ U^0(1) = 0.$$

By the result in Theorem 1.4, we see that the solution $U^0(z) = 0$. However, from $\hat{C}[U^m, \bar{D}] = 1$, the inequality $\hat{C}[U^0, \bar{D}] > 0$ can be derived. Hence the estimate (1.31) is true. Moreover, by using the method from $\hat{C}[U^m, \bar{D}] = 1$ to (1.32), we can prove the first estimate in (1.30). The second estimate in (1.30) can be derived from the first one.

1.3 Existence of solutions of discontinuous oblique derivative problem for elliptic equations in upper half-unit disk

Theorem 1.6 *If equation* (1.2) *satisfies Condition* C, *then Problem* P *for* (1.2) *is solvable.*

Proof Noting that the index $K = 0$, we introduce the boundary value problem (Problem P_t) for the linear elliptic equation with a parameter $t(0 \le t \le 1)$:

$$Lu = u_{z\bar{z}} - \mathrm{Re}[Qu_{zz} + A_1(z)u_z] = G(z,u) + A(z),\ G(z,u) = tA_2(z)u \quad (1.33)$$

for any $A(z) \in L_{p_0}(\bar{D})$ and the boundary condition (1.7). It is evident that when $t = 1$, $A(z) = A_3(z)$, Problem P_t is just Problem P. When $t = 0$, the equation in (1.33) is

$$Lu = u_{z\bar{z}} - \mathrm{Re}[Qu_{zz} + A_1 u_z] = A(z),\ \text{i.e. } w_{\bar{z}} - \mathrm{Re}[Qw_z + A_1 w] = A(z), \quad (1.34)$$

where $w = u_z$. By Theorem 1.7 below, we see that Problem P for the first equation in (1.34) has a unique solution $u_0(z)$, which is just a solution of Problem P for equation (1.33) with $t = 0$. Suppose that when $t = t_0\,(0 \le t_0 < 1)$, Problem P_{t_0} is solvable, i.e. Problem P_t for (1.33) has a unique solution $u(z)$ such that $X(z)u_z \in C_\delta(\bar{D})$. We can find a neighborhood $T_\varepsilon = \{|t - t_0| < \varepsilon, 0 \le t \le 1, \varepsilon > 0\}$ of t_0, such that for every $t \in T_\varepsilon$, Problem P_t is solvable. In fact, Problem P_t can be written in the form

$$Lu - t_0 G(z,u) = (t - t_0)G(z,u) + A(z),\ z \in D \quad (1.35)$$

and (1.7). Replacing $u(z)$ in the right-hand side of (1.35) by a function $u_0(z)$ with the condition $X(z)u_{0z} \in C_\delta(\bar{D})$, especially, by $u_0(z) = 0$, it is obvious that the boundary value problem (1.35), (1.7) then has a unique solution $u_1(z)$ satisfying the condition $X(z)u_{1z} \in C_\delta(\bar{D})$. Using successive iteration, we obtain a sequence of solutions: $\{u_n(z)\}$ satisfying the conditions $X(z)u_{nz} \in C_\delta(\bar{D})(n = 1, 2, ...)$ and

$$Lu_{n+1} - t_0 G(z, u_{n+1}) = (t - t_0)G(z, u_n) + A(z),\ z \in D,$$

$$\mathrm{Re}[\overline{\lambda(z)}u_{n+1z}] = r(z), z \in \Gamma, u_{n+1}(-1) = b_0, u_{n+1}(1) = b_1, n = 1, 2, ...$$

From the above formulas, it follows that

$$\begin{aligned}&L(u_{n+1}-u_n)_{\bar{z}}-t_0\,[G(z,u_{n+1})-G(z,u_n)]\\&=(t-t_0)[G(z,u_n)-G(z,u_{n-1})],\ z\in D,\\&\mathrm{Re}[\overline{\lambda(z)}(u_{n+1z}-u_{nz})]=0,\ z\in\Gamma,\\&u_{n+1}(-1)-u_n(-1)=0,u_{n+1}(1)-u_n(1)=0.\end{aligned}\tag{1.36}$$

Noting that

$$L_p[(t-t_0)(G(z,u_n)-G(z,u_{n-1})),\bar{D}]\le|t-t_0|k_0\hat{C}_\delta[u_n-u_{n-1},\bar{D}],\tag{1.37}$$

where $\hat{C}_\delta[u_n-u_{n-1},\bar{D}]=C_\beta[u_n-u_{n-1},\bar{D}]+C_\delta[X(z)(u_{nz}-u_{n-1z}),\bar{D}]$, and applying Theorem 1.5, we get

$$\hat{C}_\delta[u_{n+1}-u_n,\bar{D}]\le|t-t_0|k_0M_6\hat{C}_\delta[u_n-u_{n-1},\bar{D}].\tag{1.38}$$

Choosing the constant ε so small that $2\varepsilon k_0M_6<1$, it follows that

$$\hat{C}_\delta[u_{n+1}-u_n,\bar{D}]\le\hat{C}_\delta[u_n-u_{n-1},\bar{D}]/2,\tag{1.39}$$

and when $n,m\ge N_0+1\,(N_0$ is a positive integer),

$$\begin{aligned}\hat{C}_\delta[u_{n+1}-u_n,\bar{D}]&\le 2^{-N_0}\sum_{j=0}^{\infty}2^{-j}\hat{C}_\delta[u_1-u_0,\bar{D}]\\&\le 2^{-N_0+1}\hat{C}_\delta[u_1-u_0,\bar{D}].\end{aligned}\tag{1.40}$$

Hence $\{u_n(z)\}$ is a Cauchy sequence. According to the completeness of the Banach space $\hat{C}_\delta(\bar{D})$, there exists a function $u_*(z)\in\hat{C}_\delta(\bar{D})$, so that $\hat{C}_\delta[u_n-u_*,\bar{D}]\to0$ for $n\to\infty$. From (1.40), we can see that $u_*(z)$ is a solution of Problem P_t for every $t\in T_\varepsilon=\{|t-t_0|\le\varepsilon\}$. Because the constant ε is independent of $t_0\,(0\le t_0<1)$, therefore from the solvability of Problem P_t when $t=0$, we can derive the solvability of Problem P_t when $t=\varepsilon,2\varepsilon,...,[1/\varepsilon]\,\varepsilon,1$. In particular, when $t=1$ and $A(z)=A_3(z)$, Problem P_1 for the linear case of equation (1.2) is solvable.

Next, we discuss the nonlinear equation (1.2) satisfying Condition C, but we first assume that the coefficients $Q=0$, $A_j(j=1,2,3)=0$ in $D_m=\{z\in\bar{D},\mathrm{dist}(z,\Gamma')<1/m\}$, here $m\,(\ge2)$ is a positive integer, namely consider

$$u_{z\bar{z}}=\mathrm{Re}[Q^mu_{zz}+A_1^mu_z]+A_2^mu+A_3^m\ \text{in}\ D,\tag{1.41}$$

where

$$Q^m = \begin{cases} Q(z,u,u_z), \\ 0, \end{cases} \quad A_j^m = \begin{cases} A_j(z,u,u_z) \\ 0 \end{cases} \text{in} \begin{Bmatrix} \tilde{D}_m = D\backslash D_m \\ D_m \end{Bmatrix}, \; j=1,2,3.$$

Now, we introduce a bounded, closed and convex set B_M in the Banach space $B = \hat{C}_\delta(\bar{D})$, any element of which satisfies the inequality

$$\hat{C}_\delta[u(z), \bar{D}] \le M_5, \tag{1.42}$$

where M_5 is a non-negative constant as stated in (1.30). We are free to choose an arbitrary function $U(z) \in B_M$ and insert it into the coefficients of equation (1.41). It is clear that the equation can be seen as a linear equation, hence there exists a unique solution $u(z)$ of Problem P, and by Theorem 1.5, we see $u(z) \in B_M$. Denote by $u(z) = S[U(z)]$ the mapping from $U(z) \in B_M$ to $u(z)$, obviously $u(z) = S[U(z)]$ maps B_M onto a compact subset of itself. It remains to verify that $u(z) = S[U(z)]$ continuously maps the set B_M onto a compact subset. In fact, we arbitrarily select a sequence of functions: $\{U_n(z)\}$, such that $\hat{C}_\delta[U_n(z) - U_0(z), \bar{D}] \to 0$ as $n \to \infty$. Setting $u_n(z) = S[U_n(z)]$, and subtracting $u_0(z) = S[U_0(z)]$ from $u_n(z) = S[U_n(z)]$, we obtain the equation for $\tilde{u}_n = u_n(z) - u_0(z)$:

$$\begin{gathered} \tilde{u}_{n\bar{z}} - \text{Re}[Q^m(z, U_n, U_{nz})\tilde{u}_{nzz} + A_1^m(z, U_n, U_{nz})\tilde{u}_{nz}] - A_2^m(z, U_n, U_{nz})\tilde{u}_n \\ = C_n(z, U_n, U_0, u_0), C_n = \tilde{A}_3^m - \text{Re}[\tilde{Q}^m u_{0zz} + \tilde{A}_1^m u_{0z}] - \tilde{A}_2^m u_0, \end{gathered} \tag{1.43}$$

in which $\tilde{Q}^m = Q^m(z, U_n, U_{nz}) - Q^m(z, U_0, U_{0z})$, $\tilde{A}_j^m = A_j^m(z, U_n, U_{nz}) - A_j^m(z, U_0, U_{0z})$, $j = 1, 2, 3$, and the solution $\tilde{u}_n(z)$ satisfies the homogeneous boundary conditions

$$\text{Re}[\overline{\lambda(z)}u_z] = 0, \; z \in \Gamma^* = \Gamma\backslash Z, \; u(-1) = 0, \, u(1) = 0. \tag{1.44}$$

Noting that the function $C_n = 0$ in D_m, according to the method in the formula (2.43), Chapter II, [87]1), we can prove that

$$L_{p_0}[C_n, \bar{D}] \to 0 \text{ as } n \to \infty.$$

On the basis of the second estimate in (1.30), we obtain

$$\hat{C}_\delta[u_n(z) - u_0(z), \bar{D}] \le M_6 L_{p_0}[C_n, \bar{D}], \tag{1.45}$$

thus $\hat{C}_\delta[u_n(z) - u_0(z), \bar{D}] \to 0$ as $n \to \infty$. This shows that $u(z) = S[U(z)]$ in the set B_M is a continuous mapping. Hence by the Schauder fixed-point theorem, there exists a function $u(z) \in B_M$, such that $u(z) = S[u(z)]$, and the function $u(z)$ is just a solution of Problem P for the nonlinear equation (1.41).

Finally we cancel the conditions: the coefficients $Q = 0$, $A_j = 0\,(j = 1, 2, 3)$ in $D_m = \{z, \text{dist}(z, \Gamma') < 1/m\}$. Denote by $u^m(z)$ a solution of Problem P for equation (1.41). By Theorem 1.5, we see that the solution satisfies the estimate (1.30). Hence from the sequence of solutions: $u^m(z), m = 2, 3, ...$, we can choose a subsequence $\{u^{m_k}(z)\}$, for convenience denote $\{u^{m_k}(z)\}$ by $\{u^m(z)\}$ again, which uniformly converges to a function $u_0(z)$ in $\bar{D}$, and $u_0(z)$ satisfies the boundary condition (1.7) of Problem P. At last we need to verify that the function $u_0(z)$ is a solution of equation (1.2). Construct a twice continuously differentiable function $g_n(z)$ as follows

$$g_n(z) = \begin{cases} 1 & z \in \tilde{D}_n = \bar{D} \backslash D_n, \\ 0, & z \in D_{2n}, \end{cases} \quad 0 \le g_n(z) \le 1 \text{ in } D_n \backslash D_{2n}, \tag{1.46}$$

where $n(\ge 2)$ is a positive integer. It is not difficult to see that the function $u_n^m(z) = g_n(z) u^m(z)$ is a solution of the following Dirichlet boundary value problem

$$u^m_{nz\bar{z}} - \text{Re}[Q^m u^m_{nzz}] = C^m_n \text{ in } D, \tag{1.47}$$

$$u^m_n(z) = 0 \text{ on } \Gamma^*, \tag{1.48}$$

where

$$\begin{aligned} C^m_n &= g_n[\text{Re}(A^m_1 u^m_z) + A^m_2 u^m + A^m_3] \\ &+ u^m[g_{nz\bar{z}} - \text{Re}(Q^m g_{nzz})] + 2\text{Re}[g_{nz} u^m_{\bar{z}} - Q^m g_{nz} u^m_z]. \end{aligned} \tag{1.49}$$

By using the method in the proof of Theorem 1.5, we can obtain the estimates of $u^m_n(z) = u^m(z)$ in $\tilde{D}_n$, namely

$$C^1_\beta[u^m_n, \tilde{D}_n] \le M_9, ||u^m_n||_{W^2_{p_0}(\tilde{D}_n)} \le M_{10}, \tag{1.50}$$

where $\beta = \min(\alpha, 1 - 2/p_0)/2, 2 < p_0 \le p, M_j = M_j(q_0, p_0, \alpha, k_0, k_1, M'_n, g_n, D_n), j = 9, 10$, here $M'_n = \max_{1 \le m < \infty} C^{1,0}[u^m, \tilde{D}_{2n}]$. Hence from $\{u^m_n(z)\}$, we can choose a subsequence $\{u_{nm}(z)\}$, such that $\{u_{nm}(z)\}$, $\{u_{nmz}(z)\}$ uniformly converge to $u_0(z), u_{0z}(z)$ and $\{u_{nmzz}(z)\}$, $\{u_{nmz\bar{z}}(z)\}$ weakly converge to $u_{0zz}(z), u_{0z\bar{z}}(z)$ in $\tilde{D}_n$, respectively. For instance, we take $n = 2$, $u^m_2(z) = u^m(z)$ in $\tilde{D}_2$, $\{u^m_2(z)\}$ has a subsequence $\{u_{m2}(z)\}$ in $\tilde{D}_2$, the limit function of which is $u_0(z)$ in $\tilde{D}_2$. Next, we take $n = 3$, from $\{u^m_3(z)\}$ we can select a subsequence $\{u_{m3}(z)\}$ in $\tilde{D}_3$, the limit function is $u_0(z)$ in

$\tilde{D}_3$. Similarly, from $\{u_n^m(z)\}(n>3)$, we can choose a subsequence $\{u_{mn}(z)\}$ in $\tilde{D}_n$ and the limit of which is $u_0(z)$ in $\tilde{D}_n$. Finally from $\{u_n^m(z)\}$ in $\tilde{D}_n$, we choose the diagonal sequence $\{u_{mm}(z)\}$ $(m=2,3,4,...)$, such that $\{u_{mm}(z)\},\{u_{mmz}(z)\}$ uniformly converge to $u_0(z),u_{0z}(z)$ and $\{u_{mmzz}(z)\}$, $\{u_{mmz\bar{z}}(z)\}$ weakly converge to $u_{0zz}(z),u_{0z\bar{z}}(z)$ in any closed subset of D respectively, the limit function $u(z)=u_0(z)$ is just a solution of equation (1.2) in D.

Theorem 1.7 *If equation* (1.2) *with* $A_2(z)=0$ *satisfies Condition* C, *then Problems* Q *and* P *for* (1.2) *have a unique solution.*

Proof According to the proof of Theorem 1.3 in Chapter I, we choose $D'=D,n=m=2,z_1=-1,z_2=1$ and $K=0$, the second linear equation in (1.34) with $\Lambda(z)=\Lambda_3(z)$ has a unique solution $w_0(z)$, and the function

$$u_0(z)=2\,\mathrm{Re}\int_{-1}^{z}w_0(z)dz+b_0 \tag{1.51}$$

is a solution of Problem Q for the first linear equation in (1.34). If $u_0(1)=b_1'=b_1$, then the solution is just a solution of Problem P for the linear equation (1.2) with $A_2(z)=0$. Otherwise, $u_0(1)=b_1'\neq b_1$, we find a solution $u_1(z)$ of Problem Q with the boundary conditions

$$\mathrm{Re}[\overline{\lambda(z)}u_{1z}]=0\text{ on }\Gamma,\ \mathrm{Im}[\overline{\lambda(z)}u_{1z}]|_{z=0}=1,\ u_1(-1)=0.$$

On the basis of Theorem 1.4, it is clear that $u_1(1)\neq 0$, hence there exists a real constant $d\neq 0$, such that $b_1=b_1'+du_1(1)$, thus $u(z)=u_0(z)+du_1(z)$ is just a solution of Problem P for the linear equation (1.2) with $A_2(z)=0$. As for the nonlinear equation (1.2) with $A_2=0$, the existence of solutions of Problem Q and Problem P can be proved by the way as stated in the proof of last theorem.

1.4 The discontinuous oblique derivative problem for elliptic equations in general domains

In this subsection, let D' be a general simply connected domain, whose boundary $\Gamma'=\Gamma_1'\cup\Gamma_2'$, herein $\Gamma_1',\Gamma_2'\in C_\alpha^2(1/2<\alpha<1)$ have two intersection points z',z'' with the inner angles $\alpha_1\pi,\alpha_2\pi\,(0<\alpha_1,\alpha_2<1)$, respectively. We discuss the nonlinear uniformly elliptic equation

$$u_{z\bar{z}}=F(z,u,u_z,u_{zz}),\ F=\mathrm{Re}[Qu_{zz}+A_1u_z]+A_2u+A_3,\ z\in D', \tag{1.52}$$

in which $F(z,u,u_z,u_{zz})$ satisfy Condition C in D'. There are m points $Z=\{z_1=z'\ ...,z_n=z'',...,z_m\}$ on Γ' arranged according to the positive direction successively. Denote by Γ_j the curve on Γ' from z_{j-1} to z_j, $j=1,2,...,m$, $z_0=z_m$, and $\Gamma_j(j=1,2,...,m)$ does not include the end points.

Problem P' The discontinuous oblique derivative boundary value problem for (1.52) is to find a continuous solution $w(z)$ in $D^*=\overline{D'}\backslash Z$ satisfying the boundary condition:

$$\frac{1}{2}\frac{\partial u}{\partial \nu}=\mathrm{Re}[\overline{\lambda(z)}u_z]=r(z),\ z\in\Gamma^*=\Gamma'\backslash Z,\ u(z_j)=b_j,\ j=0,1,...,2K+1, \tag{1.53}$$

where $\cos(\nu,n)\ge 0$, $\lambda(z),r(z)$ are given functions satisfying

$$C_\alpha[\lambda(z),\Gamma_j]\le k_0,\ C_\alpha[R_j(z)r(z),\Gamma_j]\le k_2,\ |b_j|\le k_2,\ j=0,1,...,2K+1, \tag{1.54}$$

in which $\alpha\,(1/2<\alpha<1),k_0,k_2$ are non-negative constants, $R_j(z)=|z-z_{j-1}|^{\beta_{j-1}}|z-z_j|^{\beta_j}$, and assume that $(\beta_j+\gamma_j)/\beta<1,\beta=\alpha_0\min(\alpha,1-2/p_0)$, $\beta_j\,(<1,\ j=1,...,m)$ are non-negative constant, and $\alpha_0=\min(\alpha_1,\alpha_2)$. Problem P' with $A_3(z)=0$ in D, $r(z)=0$ on Γ' is called Problem P_0'. Similarly to Subsection 1.1, if $\cos(\nu,n)\not\equiv 0$ on each of $\Gamma_j(j=1,...,m)$, we can choose $-1/2\le\gamma_j<1/2\,(j=1,...,m)$ and the index $K\ge 0$ of Problem P', which are defined as that in Section 1, Chapter I. If $A_2=0$ in D, the point conditions in (1.53) can be replaced by

$$u(z_0)=b_0,\ \mathrm{Im}[\overline{\lambda(z)}u_z]|_{z=z_j'}=b_j,\ j=1,...,2K+1. \tag{1.55}$$

Here $z_j'(\notin Z,j=1,...,2K+1)\in\Gamma'$ are distinct points and the condition $\cos(\nu,n)\ge 0$ on Γ' can be cancelled. This boundary value problem is called Problem Q'.

Applying a similar method as before, we can prove the following theorem.

Theorem 1.8 *Let equation* (1.52) *in D' satisfy Condition C similar to before. Then Problem P' and Problem Q' for* (1.52) *are solvable, and the solution $u(z)$ can be expressed by* (1.19), *but where $\beta=\alpha_0\min(\alpha,1-2/p_0)$. Moreover, if $Q=0$ in D', then the solution $u(z)$ of equation* (1.52) *possesses the form in* (1.19), *where $w(z)=2\Psi(z)\mathrm{Re}\int_{z_0}^{z}\Phi(z)e^{\phi(z)}dz+\psi(z)$ and $u(z)$ satisfies the estimate*

$$\hat{C}_\delta[u,\overline{D'}]=C_\delta[u(z),\overline{D'}]+C_\delta[X(z)w(z),\overline{D'}]\le M_{11}, \tag{1.56}$$

in which $M_{11}=M_{11}(p_0,\delta,k,D')$ is a non-negative constant, and $X(z)$ is given as

$$X(z)=\prod_{j=2,j\ne n}^{m}|z-z_j|^{\eta_j}|z-z_1|^{\eta_1/\alpha_1}|z-z_n|^{\eta_n/\alpha_2}, \tag{1.57}$$

where $\eta_j = 2\delta$ *if* $\gamma_j \geq 0$ *and* $\eta_j = -\gamma_j + 2\delta$ *if* $\gamma_j < 0\,(1 \leq j \leq m)$. *Besides the solution of Problem* P' *and Problem* Q' *for* (1.52) *are unique, if the following condition holds: For any real functions* $u_j(z) \in C^1(D^*), V_j(z) \in L_{p_0}(D^*)(j = 1, 2)$, *the equality*

$$F(z, u_1, u_{1z}, V_1) - F(z, u_1, u_{1z}, V_2) = \mathrm{Re}[\tilde{Q}(V_1 - V_2) + \tilde{A}_1(u_1 - u_2)_z] + \tilde{A}_2(u_1 - u_2) \text{ in } D',$$

holds, where $|\tilde{Q}| \leq q_0$ in D', $\tilde{A}_1, \tilde{A}_2 \in L_{p_0}(D')$.

Proof We first consider the case of $A_2 = 0$ in (1.52). It is clear that Problem Q' for (1.52) with $A_2 = 0$ is equivalent to Problem A' for the complex equation

$$w_{\bar{z}} = \mathrm{Re}[Qw_z + A_1 w] + A_3 \tag{1.58}$$

with the boundary conditions

$$\mathrm{Re}[\overline{\lambda(z)}w(z)] = r(z),\, z \in \Gamma^*,\, \mathrm{Im}[\overline{\lambda(z_j')}w(z_j')] = b_j,\quad j = 1, ..., 2K + 1, \tag{1.59}$$

and the relation

$$u(z) = 2\mathrm{Re}\int_{z_0}^{z} w(z)dz + b_1 \text{ in } D'. \tag{1.60}$$

From Theorem 1.3, Chapter I, we see that Problem A' for (1.58) has a solution $w(z)$, and $u(z)$ in (1.60) is the solution of Problem Q' for (1.52) with $A_2 = 0$. Next let $u_0(z)$ be a solution of Problem Q' for the linear equation (1.52) with $A_2 = 0$, if $u_0(z)$ satisfies the point conditions in (1.53), then the solution is also a solution of Problem P' for the equation. Otherwise we can find $2K + 1$ solutions $u_k(z)\,(k = 1, ..., 2K + 1)$ of Problem Q' for the homogeneous equation of (1.52) satisfying the boundary conditions

$$\mathrm{Re}[\overline{\lambda(z)}u_{kz}] = 0,\, z \in \Gamma^*,\, \mathrm{Im}[\overline{\lambda(z)}u_{kz}]|_{z=z_j'} = \delta_{jk},$$
$$j = 1, ..., 2K + 1,\ u_n(z_0) = 0,$$

and $2K + 1$ real constants $d_k\,(k = 1, ..., 2K + 1)$, such that

$$u(z) = u_0(z) + \sum_{k=1}^{2K+1} d_k w_k(z)$$

is a solution of Problem P' for (1.52) with $A_2 = 0$. Moreover by using the method of parameter extension and the Schauder fixed-point theorem, which are similar to the proof of Theorem 1.6, we can find a solution of Problem P' for the general equation (1.52). Besides we can also prove the other part of the theorem (see [86]33),[92]6)).

2 The Mixed Boundary Value Problem for Degenerate Elliptic Equations of Second Order

Many authors posed and discussed some boundary value problems, mainly the Dirichlet problem and mixed boundary value problem for second order elliptic equations with parabolic degeneracy, for instance [22], [23]1), [65], [82] and so on. In this section we mainly discuss the mixed boundary value problem for the degenerate elliptic equations of second order. We first give a priori estimates of solutions of the boundary value problem for the equations, and then by using the above estimates of solutions and the compactness principle, the existence of solutions for the above mixed boundary value problem is proved, which will be used to solving Tricomi problem for second order equations of mixed type. By using the similar method, we can discuss some oblique derivative problem for the above elliptic equations (see Section 3 below).

2.1 Formulation of mixed boundary value problem for degenerate elliptic equations of second order

Let D be a simply connected bounded domain in the complex plane $\mathbf{C}$ with the boundary $\partial D=\Gamma\cup\gamma$, where $\Gamma\,(\in C^2_\mu,\ 0<\mu<1)$ is a curve in the upper-half plane with the end points $z=-1,1$, and $\gamma=\{-1<x<1, y=0\}$, later on we sometimes simply write $\gamma=(-1,1)$. We consider the linear elliptic equation of second order with parabolic degenerate line

$$Lu=K(y)u_{xx}+u_{yy}+a(x,y)u_x+b(x,y)u_y+c(x,y)u=-d(x,y) \text{ in } D. \tag{2.1}$$

Denote $H(y)=\sqrt{K(y)}$, $K(y)=y^m h(y)$, m is a positive number and $h(y)$ is a continuously differentiable positive function in $\overline{D}$. Suppose that the coefficients of (2.1) satisfy **Condition** C, namely

$$L_\infty[a,\overline{D}], L_\infty[b,\overline{D}], L_\infty[c,\overline{D}]\le k_0,\ L_\infty[d,\overline{D}]\le k_1,\ c\le 0 \text{ in } \overline{D}. \tag{2.2}$$

If the above conditions are replaced by

$$C_\alpha[a,\overline{D}], C_\alpha[b,\overline{D}], C_\alpha[c,\overline{D}]\le k_0,\ C_\alpha[d,\overline{D}]\le k_1,\ c\le 0 \text{ in } \overline{D}, \tag{2.3}$$

in which $\alpha\,(0<\alpha<1)$, k_0, k_1 are non-negative constants, then the conditions will be called **Condition** C'. It is clear that the solution of equation (2.1) with Condition C is the generalized solution, and the solution of equation (2.1) with Condition C' is the classical solution in D.

If the function $K(y)=y^m$, $H(y)=y^{m/2}$, here m is as stated before, then

$$G(y)=\int_0^y H(t)dy=\frac{2}{m+2}y^{(m+2)/2} \text{ in } \overline{D},$$

and the inverse function of $Y=G(y)$ is

$$y=G^{-1}(Y)=\left(\frac{m+2}{2}\right)^{2/(m+2)}Y^{2/(m+2)}=JY^{2/(m+2)} \text{ in } \overline{D}.$$

Denote

$$\begin{aligned}&W(z)=U+iV=[H(y)u_x-iu_y]/2=u_{\tilde{z}}=H(y)[u_x-iu_Y]/2=H(y)u_Z,\\&W_{\bar{\tilde{z}}}=[H(y)W_x+iW_y]/2=H(y)[W_x+iW_Y]/2=H(y)W_{\overline{Z}},\end{aligned} \tag{2.4}$$

where $Z=x+iY=x+iG(y)$, $G'(y)=H(y)$, we can get

$$\begin{aligned}&K(y)u_{xx}+u_{yy}=H[Hu_x-iu_y]_x+i[Hu_x-iu_y]_y-iH_yu_x\\&=2\{H[U+iV]_x+i[U+iV]_y\}-iH_yu_x=4W_{\bar{\tilde{z}}}-i[H_y/H]Hu_x\\&=4H(y)W_{\overline{Z}}-i[H_y/H]Hu_x=-[au_x+bu_y+cu+d], \text{ i.e.}\\&W_{\overline{Z}}=\{i\,[H_y/H(y)]H(y)u_x-[au_x+bu_y+cu+d]\}/4H(y)\\&=\{[iH_y/H-a/H](W+\overline{W})-ib(W-\overline{W})-cu-d\}/4H(y)\\&=[A_1(z)W+A_2(z)\overline{W}+A_3(z)u+A_4(z)]/H(y)=g(Z)/H(y),\end{aligned} \tag{2.5}$$

where

$$\begin{aligned}A_1(z)&=\frac{1}{4}[\frac{iH_y}{H}-\frac{a}{H}-ib],\ A_3(z)=-\frac{c}{4},\\A_2(z)&=\frac{1}{4}[\frac{iH_y}{H}-\frac{a}{H}+ib],\ A_4(z)=-\frac{d}{4}.\end{aligned}$$

Obviously the complex equation

$$W_{\bar{\tilde{z}}}=0 \text{ in } \overline{D}, \text{ i.e. } W_{\overline{Z}}=0 \quad In\ \overline{D_Z}, \tag{2.6}$$

is a special case of equation (2.5).

The mixed boundary value problem for equation (2.1) may be formulated as follows:

Problem M Find a real continuous solution $u(z)$ of (2.1) in $\overline{D}$, where u_x, u_y are continuous in $D^* = \overline{D}\backslash T$, and satisfy the boundary conditions

$$u(z) = \phi(x) \text{ on } \Gamma, \ u_y = r(x) \text{ on } \gamma, \tag{2.7}$$

where $T = \{a_1, a_2\}$ $(a_1 = t_1 = -1, a_2 = t_2 = 1)$, and the known functions $\phi(z)$, $r(x)$ satisfy the conditions

$$C^2_\alpha[\phi(z), \Gamma] \leq k_1, \ C^1_\alpha[r(x), \gamma] \leq k_1. \tag{2.8}$$

Problem M with the conditions $d(z) = 0$, $z \in \bar{D}$, $\phi(z) = 0$ on Γ and $r(z) = 0$ on γ will be called Problem M_0.

Now we give some explanations about the domain D_Z. As stated in Section 2, Chapter I, denote by $t_1 = -1$, $t_2 = 1$ the corner points of D, and if the inner angle $\alpha_j\pi$ of D at the point $z = t_j (1 \leq j \leq 2)$ is greater than $\pi/2$ and not greater than π, i.e. the slopes $\partial y/\partial x$ of the boundary Γ at $z = t_1, t_2$ are satisfied the conditions $-\infty < \partial y/\partial x \leq 0$, $0 \leq \partial y/\partial x < \infty$ respectively, then $\partial Y/\partial x = (\partial Y/\partial y)(\partial y/\partial x) = H(y)\partial y/\partial x = 0$ at $Z = t_j\ (j = 1, 2)$, i.e. the inner angles of D_Z are equal to π at $Z = t_j\ (j = 1, 2)$. If the slopes of the boundary Γ at $z = t_1, t_2$ are satisfied the conditions $0 \leq \partial y/\partial x < \infty$, $-\infty < \partial y/\partial x \leq 0$ respectively, then $\partial Y/\partial x = (\partial Y/\partial y)(\partial y/\partial x) = H(y)\partial y/\partial x = 0$ at $Z = t_j (j = 1, 2)$, i.e. the inner angles of D_Z are equal to 0 at $Z = t_j\ (j = 1, 2)$. Besides we can also discuss the case of other slopes.
1. We first consider the case: the inner angles of D at $z = t_1 = -1, t_2 = 1$ are satisfied the conditions $\alpha_j\pi\ (1/2 < \alpha_j \leq 1, j = 1, 2)$. In this case, from the mixed boundary condition (2.7), if $\phi = \phi(x)$ on Γ near $z = \pm 1$, we can find the derivative according to the parameter $s = \text{Re}z = x$ on Γ, and obtain

$$\begin{aligned} &u_s = u_x x_s + u_y y_s = \phi'(x) x_s, \text{ i.e. } H(y)u_x + H(y)u_y y_x \\ &= H(y)\phi'(x) \text{ on } \Gamma \text{ near } x = -1, \\ &u_s = u_x x_s + u_y y_s = \phi'(x) x_s, \text{ i.e. } H(y)u_x + H(y)u_y y_x \\ &= H(y)\phi'(x) \text{ on } \Gamma \text{ near } x = 1, \end{aligned} \tag{2.9}$$

then the complex form of (2.9) can be written as

$$\text{Re}[\overline{\lambda(z)}(U + iV)] = \text{Re}[\overline{\lambda(z)}(H(y)u_x - iu_y)]/2 = R(z) \text{ on } \Gamma \cup \gamma, \tag{2.10}$$

where $-y_x|_{x=-1} = -dy/dx|_{x=-1}$, $y_x|_{x=1} = dy/dx|_{x=1}$ are positive numbers,

$H(0)y_x|_{x=\pm1}=0$, $H(y)y_x\to 0$ as $z(\in\Gamma)\to\pm1$, and

$$\lambda(z)=\begin{cases}1-iH(y)y_x,\\ 1-iH(y)y_x,\\ i,\end{cases} R(z)=\begin{cases}R_1(z) \text{ on } \Gamma \text{ near } z=-1,\\ R_2(z) \text{ on } \Gamma \text{ near } z=1,\\ -r(x)/2=R_3(x) \text{ on } \gamma,\end{cases}$$

where $R_1(z)=H(y)\phi'(x)/2, R_2(z)=H(y)\phi'(x)/2$. Denote $t_1=-1, t_2=1$, we have

$$\begin{aligned}&e^{i\phi_1}=\frac{\lambda(t_1-0)}{\lambda(t_1+0)}=e^{-\pi i-\pi i/2}=e^{-3\pi i/2}, \gamma_1=\frac{-3\pi/2}{\pi}-K_1=-\frac{1}{2}, K_1=-1,\\ &e^{i\phi_2}=\frac{\lambda(t_2-0)}{\lambda(t_2+0)}=e^{\pi i/2-\pi i}=e^{-\pi i/2}, \gamma_2=-\frac{\pi/2}{\pi}-K_2=-\frac{1}{2}, K_2=0,\end{aligned} \tag{2.11}$$

hence the index of $\lambda(z)$ on $\Gamma\cup\gamma$ is $K=(K_1+K_2)/2=-1/2$.

2. Next the inner angle of D at $z=-1$ is $\alpha_1\pi\,(0\le\alpha_1<1/2)$, and the inner angle of D at $z=1$ is $\alpha_2\pi\,(0\le\alpha_2<1/2)$. In this case, the mixed boundary conditions can be written as in (2.7)-(2.10), and we can compute the numbers

$$\begin{aligned}&e^{i\phi_1}=\frac{\lambda(t_1-0)}{\lambda(t_1+0)}=e^{0\pi i-\pi i/2}=e^{-\pi i/2}, \gamma_1=\frac{-\pi/2}{\pi}-K_1=\frac{1}{2},\ K_1=-1,\\ &e^{i\phi_2}=\frac{\lambda(t_2-0)}{\lambda(t_2+0)}=e^{\pi i/2-0\pi i}=e^{\pi i/2}, \gamma_2=\frac{\pi/2}{\pi}-K_2=\frac{1}{2}, K_2=0,\end{aligned}$$

hence the index of $\lambda(z)$ on $\Gamma\cup\gamma$ is $K=(K_1+K_2)/2=-1/2$. If we consider $\phi=\phi(s)$ on Γ, noting $ds=\sqrt{(dx)^2+(dy)^2}=\sqrt{1+(dy/dx)^2}|dx|=\sqrt{1+(dx/dy)^2}|dy|$, it suffices to replace the above function $H(y)\phi'(x)$ by $H(y)\sqrt{1+(dy/dx)^2}\phi'(s)$.

3. The inner angles of D at $z=t_1=-1, t_2=1$ are equal to $\pi/2$ and whose slopes equal $\alpha_j/H(y)\,(0<|\alpha_j|<\infty, j=1,2)$. In this case, the mixed boundary conditions are as stated in (2.7), and we have (2.9), (2.10), where $H(y)\partial y/\partial x|_{z=-1}=\alpha_1, H(y)dy/dx|_{x=1}=\alpha_2$, and the functions $\lambda(z)$, $R(z)$ are as follows

$$\lambda(z)=\begin{cases}1-i\alpha_1,\\ 1-i\alpha_2,\\ i,\end{cases} R(z)=\begin{cases}H(y)\phi'(x)/2 \text{ on } \Gamma \text{ near } z=-1,\\ H(y)\phi'(x)/2 \text{ on } \Gamma \text{ near } z=1,\\ -r(x)/2=R_3(x) \text{ on } \gamma,\end{cases}$$

moreover we have

$$e^{i\phi_1}=\frac{\lambda(t_1-0)}{\lambda(t_1+0)}=e^{-i\tan^{-1}\alpha_1-\pi i/2},-\frac{1}{2}\leq\gamma_1=-\frac{\tan^{-1}\alpha_1+\pi/2}{\pi}$$

$$-K_1=\frac{1}{2}-\frac{\tan^{-1}\alpha_1}{\pi}\leq\frac{1}{2},\ K_1=-1,\ \alpha_1\geq 0,$$

$$e^{i\phi_2}=\frac{\lambda(t_2-0)}{\lambda(t_2+0)}=e^{\pi i/2-i\tan^{-1}(-\alpha_2)},-\frac{1}{2}\leq\gamma_2=\frac{\pi/2-\tan^{-1}(-\alpha_2)}{\pi}$$

$$-K_2=\frac{1}{2}-\frac{\tan^{-1}(-\alpha_2)}{\pi}\leq\frac{1}{2},\ K_2=0,\ \alpha_2\leq 0,$$

and if $\alpha_1\leq 0$, then the above number $-\tan^{-1}\alpha_1$ should be replaced by $-\pi+\tan^{-1}(-\alpha_1)$, and if $\alpha_2\geq 0$, then the above number $\tan^{-1}(-\alpha_2)$ should be replaced by $\pi-\tan^{-1}\alpha_2$, here $\tan^{-1}\alpha_l$ is the main value of $\mathrm{Tan}^{-1}\alpha_l$ satisfying the condition $0\leq\tan^{-1}\alpha_l\leq\pi/2\,(1\leq l\leq 2)$, hence the index of $\lambda(z)$ on $\Gamma\cup\gamma$ is $K=(K_1+K_2)/2=-1/2$.

4. Finally the inner angles of D at $z=t_1=-1,t_2=1$ are equal to $\pi/2$, and $\lim_{z(\in\Gamma)\to t_j}x_y/H(y)=0\,(j=1,2)$, which includes $x_y=0$ and $H^2(y)$ near $Z=t_j(j=1,2)$. In this case, the mixed boundary condition is as follows:

$$u(z)=\phi(z)\ \text{on}\ \Gamma,\ u_y=r(x)\ \text{on}\ \gamma, \tag{2.12}$$

where $\phi(z),r(x)$ satisfy the conditions

$$C^2_\alpha[\phi(z),\Gamma]\leq k_2,\ C^1_\alpha[r(x),\gamma]\leq k_2, \tag{2.13}$$

in which $\alpha\,(0<\alpha<1)$, k_2 are non-negative constants. If $\phi(z)=\phi(y)$ on Γ near $z=\pm 1$, we find the derivative for (2.12) according to the parameter $s=\mathrm{Im}z=y$ on Γ, and obtain

$$u_s=u_xx_y+u_y=H(y)u_xx_y/H(y)+u_y=\phi'(y),\ \text{i.e.}$$

$$H(y)u_xx_y/H(y)+u_y=\phi'(y)\ \text{ on }\Gamma\text{ near }x=-1,$$

$$u_s=u_xx_y+u_y=H(y)u_xx_y/H(y)+u_y=\phi'(y),\ \text{i.e.}$$

$$-H(y)u_xx_y/H(y)-u_y=-\phi'(y)\ \text{ on }\Gamma\text{ near }x=1,$$

then the complex form of the above formula can be written as

$$\mathrm{Re}[\overline{\lambda(z)}(U+iV)]=\mathrm{Re}[\overline{\lambda(z)}(H(y)u_x-iu_y)]/2=R(z)\ \text{ on }\Gamma\cup\gamma,$$

in which

$$\lambda(z)=\begin{cases} x_y/H(y)-i, \\ -x_y/H(y)+i, \\ i, \end{cases} R(z)=\begin{cases} \phi'(y)/2 \text{ on } \Gamma \text{ at } z=-1, \\ -\phi'(y)/2 \text{ on } \Gamma \text{ at } z=1, \\ -r(x)/2=R_3(x) \text{ on } \gamma. \end{cases}$$

Denote $z=t_1=-1, t_2=1$, we have

$$e^{i\phi_1}=\frac{\lambda(t_1-0)}{\lambda(t_1+0)}=e^{-\pi i/2-0\pi i}=e^{-\pi i/2}, \gamma_1=\frac{-\pi/2}{\pi}-K_1=-\frac{1}{2}, K_1=0,$$
$$e^{i\phi_2}=\frac{\lambda(t_2-0)}{\lambda(t_2+0)}=e^{0\pi i-\pi i/2}=e^{-\pi i/2}, \gamma_2=\frac{-\pi i/2}{\pi}-K_2=-\frac{1}{2}, K_2=0, \tag{2.14}$$

where we consider

$$\text{Re}[\overline{\lambda(x)}W(z)]=0, \ \lambda(x)=1 \ \text{ on } \gamma,$$

hence the index of $\lambda(z)$ on $\Gamma\cup\gamma$ is $K=(K_1+K_2)/2=0$. If we choose $\gamma_1=1/2$, $\gamma_2=-1/2$, $K_1=-1, K_2=0$, then $K=-1/2$. If we consider the boundary condition

$$\text{Re}[\overline{\lambda(x)}(U+iV)]=R(x), \ \lambda(x)=i=e^{i\pi/2} \ \text{ on } \gamma,$$

then the corresponding numbers $\gamma_1=\gamma_2=0$, and $K_1=-1$, $K_2=0$, $K=-1/2$. Later on the index $K=0$ or $-1/2$ will be chosen according to our requirement.

We can assume that the boundary the Γ of the domain D is a smooth curve with the form $x-G(y)=-1, x+G(y)=1$ near the points $z=-1,1$ respectively as stated in Chapter I. Actually we can discuss any other case, for instance the boundary Γ is a smooth curve with the form $x-\tilde{G}(y)=-1, x+\tilde{G}(y)=1$ are two curves vertical to the axis Im$z=0$ at $z=-1,1$ respectively, which can be seen as in the last part of Section 5 below. But more simply we can use the method of conformal mappings as follows.

Later on we can give the estimate of solutions of Problem M in the neighborhood of every corner point $Z=t_j\,(1\le j\le 2)$ of D_Z separately. Because for equation (2.5) we can give a conformal mapping $\zeta=\zeta(Z)$, which maps the domain D_Z onto D_ζ such that the line segment $\gamma=(-1,1)$ on the real axis and boundary points $-1,1$ are mapped onto themselves respectively, and the boundary $\partial D_\zeta\backslash\gamma\,(\in C^1_\alpha)$ is a curve with the form $\text{Re}\,\zeta=\mp1\pm\tilde{G}(\text{Im}\zeta)$ including the line segments $\text{Re}\zeta=\mp1$ near the points

$\zeta = \mp 1$ respectively. Denote by $Z = Z(\zeta)$ the inverse function of $\zeta = \zeta(Z)$, thus equation (2.5) is reduced to

$$W_{\bar{\zeta}} = g[Z(\zeta)]\overline{Z'(\zeta)}/H(y), \text{ i.e.}$$
$$W_{\bar{\zeta}} = [A_1(z)w + A_2(z)\overline{w} + A_3(z)]\overline{Z'(\zeta)}/H(y) \text{ in } \overline{D_\zeta}. \tag{2.15}$$

It is not difficult to see that if the inner angle of the boundary ∂D_ζ in $z-$plane at $z = -1$ or 1 is greater that $\pi/2$, then $Z'(\zeta)$ has a zero point of order $\alpha_1 \leq 1$ at the point; and if the inner angle of D is less than $\pi/2$, then in general $Z'(\zeta)$ has a pole of order $\alpha_2 \leq 1$ at the point. Through the above conformal mapping $\zeta = \zeta(Z)$, the index K of $\lambda[Z(\zeta)]$ on ∂D_ζ is unchanged. In this case we can choose an appropriate function $X(\zeta)$ as stated below, and multiply it on the both sides of equation (2.15), thus the equation (2.15) is transformed into the form

$$[X(\zeta)W]_{\bar{\zeta}} = X(\zeta)g[Z(\zeta)]\overline{Z'(\zeta)}/H(y), \text{ i.e.}$$
$$[X(\zeta)W]_{\bar{\zeta}} = X(\zeta)[A_1(z)w + A_2(z)\overline{w} + A_3(z)]\overline{Z'(\zeta)}/H(y) \text{ in } \overline{D_\zeta}, \tag{2.16}$$

the index of the boundary condition of the new function $X(Z)W(Z)$ maybe increase, but we can add some point conditions such that the solution of Problem M is unique. Hence later on we can only consider equation (2.5) (or (2.16)), and suppose that these equations satisfy Condition C in D_Z (or D_ζ) and the boundary Γ_Z (or Γ_ζ) of the domain D_Z (or D_ζ) including the line segments $\text{Re}Z = t_j$ (or $\text{Re}\zeta = t_j$) $(j = 1, 2)$ near the points $Z = t_j$ (or $\zeta = t_j$) $(j = 1, 2)$.

In addition there is no harm in assuming that $u(z) = 0$ on Γ in (2.10), because otherwise we can find a twice continuously differentiable function in $\overline{D}$, for instance a harmonic function $u_0(z)$ in D satisfying the boundary conditions $u_0(z) = \phi(z)$ on Γ, denote

$$v(z) = u(z) - u_0(z),$$

then the function $v(z)$ is a solution of the equation

$$Lv = K(y)v_{xx} + v_{yy} + a(x, y)v_x + b(x, y)v_y$$
$$+ c(x, y)v = F(x, y), \ F = -d - Lu_0 \text{ in } D \tag{2.17}$$

satisfying the boundary conditions

$$v(z) = 0 \text{ on } \Gamma, \ v_y = r(x) - u_{0y} \text{ on } \gamma, \text{ i.e.}$$
$$\text{Re}[\overline{\lambda(z)}v_{\bar{z}}] = R(z) \text{ on } \Gamma \cup \gamma, \ v(-1) = 0, \tag{2.18}$$

in which $R(z) = 0$ on Γ, and $\lambda(z)$ is as stated before. In next subsection, we only discuss the boundary value problem with the boundary condition $\mathrm{Re}[\overline{\lambda(z)}v_{\bar z}]=\mathrm{Re}[\overline{\lambda(z)}W(z)]=0$ on Γ.

2.2 Representation of solutions of mixed boundary value problem for degenerate elliptic equations

The boundary value problem for equation (2.5) with the boundary condition (2.10) ($W(z) = u_{\bar z}$) and the relation (2.20) below will be called Problem A. On the basis of the result in [86]9) and the way in the proof in Theorems 1.4 and 1.6, we can find a unique solution of Problem A for equation (2.6) in $\overline{D}$.

Now, we give the representation of solutions for the mixed boundary value problem (Problem M) for equation (2.1) in $\overline{D}$. For this, we first introduce the Riemann-Hilbert problem (Problem A) for equation (2.5) in $\overline{D}$ with the boundary conditions

$$\mathrm{Re}[\overline{\lambda(z)}(U+jV)] = R(z) \text{ on } \Gamma\cup\gamma,\ u(-1)=\phi(-1)=b_1, \tag{2.19}$$

in which $\lambda(z) = a(z) + jb(z)$ is as stated before, $R(z) = 0$ on Γ, $\lambda(z) = i$ on γ.

The representation of solutions of Problem M for equation (2.1) is as follows.

Theorem 2.1 *Under Condition C, any solution $u(z)$ of Problem M for equation (2.1) in $\overline{D}$ can be expressed as follows*

$$\begin{gathered}u(z)=u(x)-2\int_0^y V(z)dy=2\mathrm{Re}\int_{-1}^z\Big[\frac{\mathrm{Re}W}{H(y)}+i\mathrm{Im}W\Big]dz+b_0 \text{ in } \overline{D},\\ W(z)=\Phi(Z)+\Psi(Z)=\hat\Phi(Z)+\hat\Psi(Z) \text{ in } D_Z,\\ \Psi(Z)=2\mathrm{Re}Tf,\ \hat\Psi(Z)=2i\mathrm{Im}Tf,\ Tf=-\frac{1}{\pi}\iint_{D_t}\frac{f(t)}{t-Z}d\sigma_t,\end{gathered}\tag{2.20}$$

where $\Phi(Z)$, $\hat\Phi(Z)$ are analytic functions in D_Z, and

$$f(Z)=\frac{g(Z)}{H(y)}=\frac{1}{H(y)}[A_1(z)W+A_2(z)\overline{W}+A_3(z)u+A_4(z)] \text{ in } \overline{D_Z},$$

herein $Z = x + iG(y)$.

Proof It is clear that

$$[Tf]_{\overline{Z}} = f(Z),\ [\overline{Tf}]_{\overline{Z}} = 0,\ [\Phi(Z)]_{\overline{Z}} = 0,\ [\hat{\Phi}(Z)]_{\overline{Z}} = 0 \text{ in } D, \tag{2.21}$$

the above formulas show that $\overline{Tf}$, $\Phi(Z)$, $\hat{\Phi}(Z)$ are analytic functions in D, hence we have the representation (2.20) of solutions for equation (2.1).

Now we prove the uniqueness of solutions of Problem M for equation (2.1).

Theorem 2.2 *Suppose that equation* (2.1) *satisfies Condition C. Then Problem M for* (2.1) *has at most one solution in D.*

Proof Let $u_1(z), u_2(z)$ be any two solutions of Problem M for (2.1). It is easy to see that $u(z) = u_1(z) - u_2(z)$ and $W(z) = u_{\bar{z}}$ satisfy the homogeneous equation and boundary conditions

$$\begin{gathered} u_{z\bar{z}} = A_1 u_{\bar{z}} + A_2 \overline{u_{\bar{z}}} + A_3 u \text{ in } D, \\ u(z) = 0 \text{ on } \Gamma,\ u_y = 0 \text{ on } \gamma. \end{gathered} \tag{2.22}$$

We shall verify that the above solution $u(z) \equiv 0$ in D. If the maximum $M = \max_{\overline{D}} u(z) > 0$, it is clear that the maximum point $z^* \notin \overline{D}\backslash\gamma$, here $\gamma = (-1, 1)$. If the maximum M attains at a point $z^* = x^* \in \gamma$, from Lemma 2.3 below, we can derive that $u_x(x^*) = 0$, $u_y(x^*) < 0$, this contradicts the equality in (2.22) on γ. Hence $\max_{\overline{D}} u(z) = 0$. By the similar way, we can prove $\min_{\overline{D}} u(z) = 0$. Thus $u(z) = 0$, $u_1(z) = u_2(z)$ in $\overline{D}$. This completes the proof.

Lemma 2.3 *Suppose that equation* (2.1) *satisfies Condition C and $Lu \geq 0$ (or $Lu \leq 0$) in D, if the solution $u(z) \in C^2(D) \cap C(\overline{D})$ of* (2.1) *attains its positive maximum (or negative minimum) at a point $x_0 \in \gamma$, and $\max_\Gamma u(z) < u(x_0)$ (or $\min_\Gamma u(z) > u(x_0)$) on Γ, then*

$$\overline{\lim_{y\to 0}} \frac{\partial u(x_0, y)}{\partial y} < 0 \ (or\ \underline{\lim_{y\to 0}} \frac{\partial u(x_0, y)}{\partial y} > 0). \tag{2.23}$$

Proof Assume that the first inequality is not true, namely

$$\overline{\lim_{y\to 0}} \frac{\partial u(x_0, y)}{\partial y} = M' \geq 0.$$

Obviously $M' = 0$. Denote $M = u(x_0)$, $B = \max_{\bar{D}} |b(z)|$ and by d the diameter of D. Thus there exists a small positive constant $\varepsilon < M$ such that $\max_\Gamma u(z) \le M - \varepsilon$. Making a function

$$v(z) = \varepsilon u(z)/(Me^{Bd} - \varepsilon e^{By}),$$

we have

$$v(z) \le \frac{\varepsilon(M-\varepsilon)}{Me^{Bd}-\varepsilon e^{Bd}} < \frac{\varepsilon M}{Me^{Bd}-\varepsilon} \text{ on } \Gamma, v(x) \le v(x_0) = \frac{\varepsilon M}{Me^{Bd}-\varepsilon} \text{ on } \gamma.$$

Noting that $Lu \ge 0$, the function $v(x,y)$ satisfies the inequality

$$K(y)v_{xx} + v_{yy} + a(x,y)v_x + \tilde{b}(x,y)v_y + \tilde{c}(x,y)v \ge 0 \text{ in } D, \qquad (2.24)$$

where $\tilde{b} = b - 2\varepsilon Be^{By}/(Me^{Bd} - \varepsilon e^{By})$, $\tilde{c}(x,y) = c - \varepsilon(B+b)Be^{By}/(Me^{Bd} - \varepsilon e^{By}) \le 0$ in D. According to the above assumption, we get

$$\overline{\lim_{y\to 0}} \frac{\partial v(x_0,y)}{\partial y} = \frac{\varepsilon^2 BM}{(Me^{Bd}-\varepsilon)^2} > 0.$$

Hence $v(x,y)$ attains its maximum in D, but from (2.24), it is impossible. This proves the first inequality in (2.23). Similarly the second inequality in (2.23) can be proved.

2.3 Estimates and existence of solutions of mixed problem for degenerate elliptic equations

We first consider the uniformly elliptic equation of second order

$$K_n(y)u_{xx} + u_{yy} + a(x,y)u_x + b(x,y)u_y + c(x,y)u + d(x,y) = 0 \text{ in } D, \quad (2.25)$$

where $K_n(y) = K(y+1/n)$, $H_n(y) = \sqrt{K_n(y)}$, $K(y)$ is as stated in (2.1), n is a positive integer, and equation (2.25) can be satisfied the conditions similar to Condition C. Denote

$$W_n(z) = [H_n(y)u_x - iu_y]/2 = u_{\tilde{z}}, W_{n\bar{\tilde{z}}} = [H_n(y)W_{nx} + iW_{ny}]/2 \text{ in } \overline{D},$$

where $Z = x + iY$, $Y = G_n(y) = \int_0^y H_n(t)dt$, $G_n'(y) = H_n(y)$. Similarly to (2.4), (2.5), we can derive

$$\begin{aligned} &K_n(y)u_{xx}+u_{yy}=H_n[H_nu_x-iu_y]_x+i[H_nu_x-iu_y]_y\\ &-iH_{ny}u_x = 4W_{n\bar{z}} - i[H_{ny}/H_n]H_nu_x = 4H_n(y)W_{n\overline{Z}}\\ &-i[H_{ny}/H_n]H_nu_x = -[au_x + bu_y + cu + d], \text{ i.e.}\\ &W_{n\overline{Z}}=\{i[H_{ny}/H_n(y)]H_n(y)u_x-[au_x+bu_y+cu+d]\}/4H_n(y)\\ &=[A_1^n(z)W_n+A_2^n(z)\overline{W}_n+A_3^n(z)u+A_4^n(z)]/H_n=g_n(Z)/H_n \text{ in } D_Z, \end{aligned} \tag{2.26}$$

where D_Z is the image domain of D with respect to the mapping $Z = Z(z)$.

In the meantime, we consider the corresponding boundary value problem (Problems M_n) for equation (2.25) with the boundary conditions

$$\text{Re}[\overline{\lambda(z)}u_{n\bar{z}}] = R_n(z) \text{ on } \Gamma\cup\gamma,\ u_n(-1) = b_1, \tag{2.27}$$

in which $\lambda(z) = a(z) + ib(z)$ on Γ and $\lambda(x) = 1, R_n(x) = 0$ or $\lambda(x) = i, R_n(x) = -r(x)/2$ on γ. It is clear that Problem M_n is equivalent to the boundary value problem (Problem A_n), i.e. equation (2.26), the boundary conditions

$$\text{Re}[\overline{\lambda(z)}W_n(z)] = R_n(z) \text{ on } \Gamma\cup\gamma,\ u_n(-1) = b_1, \tag{2.28}$$

and the relation

$$u_n(z)=u_n(x)-2\int_0^y \text{Im}W_n(z)dy=2\text{Re}\int_{-1}^z [\frac{\text{Re}W_n}{H_n(y)}+i\text{Im}W_n]dz \text{ in } \overline{D}. \tag{2.29}$$

On the basis of Theorem 1.8, we see that Problem A_n for (2.26) has a solution $[W_n(z), u_n(z)]$ satisfying the estimate

$$\hat{C}_\delta[W_n(z),\overline{D}]=C_\delta[X_0(Z)(\text{Re}W_n/H_n+i\text{Im}W_n),\overline{D_Z}]+C_\delta[u_n(z),\overline{D}]\le M_0, \tag{2.30}$$

where $X_0(Z) = (Z-t_1)^{\eta_1}(Z-t_2)^{\eta_2}$, $\eta_j = 2\tau$ if $\gamma_j \ge 0$, and $\eta_j = -2\gamma_j + 2\tau$ if $\gamma_j < 0$, τ, $\delta(<\tau)$ are sufficiently small positive constants, here we choose a branch of multi-valued function $X_0(Z)$ such that $\arg X_0(x) = 0$ on γ, and $M_0 = M_0(\delta, k, H, D, n)$ is a non-negative constant dependent on δ, k, H, D, n, and $k = (k_0, k_1, k_2)$. Obviously $X_0(Z)W_n[z(Z)]$ satisfies the complex equation

$$\begin{aligned} &[X_0(Z)W_n]_{\overline{Z}} = X_0(Z)[A_1^n(z)W_n + A_2^n\overline{W}_n\\ &+A_3^n(z)u + A_4^n(z)]/H_n=X_0(Z)g_n(Z)/H_n \text{ in } D_Z, \end{aligned} \tag{2.31}$$

and the boundary conditions

$$\mathrm{Re}[\overline{\Lambda(Z)}X_0(Z)W_n(z)]=|X_0(Z)|R_n(z) \text{ on } \Gamma\cup\gamma, u_n(-1)=b_1. \tag{2.32}$$

Noting that

$$\begin{aligned} e^{i\tilde{\phi}_j} &= \frac{\Lambda(t_j-0)}{\Lambda(t_j+0)} = \frac{\lambda(t_j-0)}{\lambda(t_j+0)}\frac{e^{i\arg X_0(t_j-0)}}{e^{i\arg X_0(t_j+0)}} \\ &= e^{i(\phi_j+\eta_j\pi/2)},\ \tau_j=\frac{\tilde{\phi}_j}{\pi}-K_j,\ j=1,2, \end{aligned} \tag{2.33}$$

hence $\tau_j=\gamma_j+\tau$ if $\gamma_j\geq 0$, and $\tau_j=\tau$ if $\gamma_j<0$ about $\Lambda[z(Z)]=\lambda[z(Z)]e^{i\arg X_0(Z)}$, which are corresponding to $\gamma_j\ (j=1,2)$.

Next we consider the equation

$$K_1^n(y)u_{xx}+K_2(y)u_{yy}+\hat{a}u_x+\hat{b}u_y+\hat{c}u+\hat{d}=0 \text{ in } D, \tag{2.34}$$

in which $K_2(y)=y^{m_2}, K_1^n(y)=K_2(y)K(y+1/n), \hat{a}=K_2(y)a, \hat{b}=K_2(y)b$, $\hat{c}=K_2(y)c, \hat{d}=K_2(y)d$, herein $m_2(<1)$ is a small positive number. Now we shall give the estimates of solutions of Problem M_n for (2.34) in D_Z. Similarly to (2.5) and (5.11) below, we see that Problem M is equivalent to Problem A_n for the complex equation

$$\begin{aligned} &W_{\overline{Z}}=[A_1(z)W+A_2\overline{W}+A_3(z)u+A_4(z)]/H_1(y) \text{ in } D_Z, \\ &A_1=\frac{1}{4}[-\frac{\hat{a}}{H_1}+\frac{iH_2H_{1y}}{H_1}-i\frac{\hat{b}}{H_2}+iH_{2y}],\ A_3=-\frac{\hat{c}}{4}, \\ &A_2=\frac{1}{4}[-\frac{\hat{a}}{H_1}+\frac{iH_2H_{1y}}{H_1}+i\frac{\hat{b}}{H_2}-iH_{2y}],\ A_4=-\frac{\hat{d}}{4}, \end{aligned} \tag{2.35}$$

with the boundary condition

$$\mathrm{Re}[\overline{\lambda(z)}W(z)]=R(x) \text{ on } \Gamma\cup\gamma,\ u(-1)=b_1, \tag{2.36}$$

and the relation

$$u(z)=u(x)-2\int_0^y V(z)dy=2\mathrm{Re}\int_{-1}^z[\frac{\mathrm{Re}W}{H_1(y)}+i\frac{\mathrm{Im}W}{H_2(y)}]dz+b_1 \text{ in } \overline{D}, \tag{2.37}$$

where $H_1(y)=y^{m_2/2}\sqrt{K_1(y+1/n)},\ H_2(y)=y^{m_2/2}$.

Now we consider the boundary Γ including the line segments $\mathrm{Re}z=\pm1$ near $z=\pm1$ respectively and $\mathrm{Re}[\overline{\lambda(x)}W(z)]=0,\ \lambda(x)=1$ on γ, hence

$\gamma_1 = \gamma_2 = -1/2$, $K = 0$ as stated in (2.14). Introduce a function

$$X(Z) = \prod_{j=1}^{2}(Z - t_j)^{\eta_j}, \tag{2.38}$$

where $t_1 = a_1 = -1, t_2 = a_2 = 1$, and $\eta_j = 1\,(j = 1,2)$. Obviously $X(Z)W[z(Z)]$ satisfies the complex equation

$$\begin{aligned}&[X(Z)W]_{\overline{Z}} = X(Z)[A_1(z)W + A_2(z)\overline{W}\\ &+A_3(z)u + A_4(z)]/H(y) = X(Z)g(Z)/H_1(y) \text{ in } D_Z,\end{aligned} \tag{2.39}$$

and the boundary conditions

$$\mathrm{Re}[\overline{\hat{\lambda}(z)}X(Z)W(z)] = |X(Z)|R(z) \text{ on } \Gamma\cup\gamma, u(a_j) = b_j, j = 1,2, \tag{2.40}$$

where $R(z) = 0$ on Γ, $b_1 = \phi(-1), b_2 = \phi(1)$ are real constants, as stated before we can assume that $R(z) = 0$ on Γ and $b_1 = b_2 = 0$. It is easy to see that

$$\begin{aligned}e^{i\tilde{\phi}_j} &= \frac{\hat{\lambda}(t_j - 0)}{\hat{\lambda}(t_j + 0)} = \frac{\lambda(t_j - 0)}{\lambda(t_j + 0)}\frac{e^{i\arg X(t_j-0)}}{e^{i\arg X(t_j+0)}}\\ &= e^{i(\phi_j + \eta_j\pi/2)}, \tau_j = \frac{\tilde{\phi}_j}{\pi} - \tilde{K}_j, j = 1,2,\end{aligned} \tag{2.41}$$

here $\tau_j\,(j = 1,2)$ corresponding to $\hat{\lambda}[z(Z)] = \lambda[z(Z)]e^{i\arg X(Z)}$ are equal to 0, $\tilde{K}_1 = 0$, $\tilde{K}_2 = 0$, and the index $\tilde{K} = 0$. Hence we have two point conditions in (2.40), which is similar to the manner as stated in Section 1.

Theorem 2.4 *Let equation* (2.1) *satisfy Condition C. Then any solution of Problem A_n for* (2.35) – (2.40) *satisfies the estimate*

$$\begin{aligned}&\hat{C}_\delta[W(z), \overline{D}] = C_\delta[X(Z)(\mathrm{Re}W/H_1 + i\mathrm{Im}W/H_2), \overline{D_Z}]\\ &+ C_\delta[u(z), \overline{D}] \le M_1, \quad \hat{C}_\delta[W(z), \overline{D}] \le M_2(k_1 + k_2),\end{aligned} \tag{2.42}$$

in which $X(Z)$ is as stated in (2.38), *and δ is a sufficiently small positive constant, $M_1 = M_1(\delta,\ k, H, D)$, $M_2 = M_2(\delta, k_0, H, D)$ are non-negative constants, and $H = (H_1, H_2), k = (k_0, k_1, k_2)$.*

Proof We first assume that any solution $[W(z),\ u(z)]$ of Problem A_n satisfies the estimate

$$\hat{C}[W(z), \overline{D}] = C[X(Z)(\mathrm{Re}W/H_1 + i\mathrm{Im}W/H_2), \overline{D_Z}] + C[u(z), \overline{D}] \le M_3, \tag{2.43}$$

in which M_3 is a non-negative constant. Substituting the solution $[W(z), u(z)]$ into equation (2.35) and noting $\text{Re}X(Z)W(Z) = 0$ on γ, $b_1 = b_2 = 0$, we can extend the function $X(Z)W[z(Z)]$ onto the symmetrical domain $\tilde{D}_Z$ of D_Z with respect to the real axis $\text{Im}Z = 0$, namely set

$$\tilde{W}(Z) = \begin{cases} X(Z)W[z(Z)] \text{ in } D_Z, \\ -\overline{X(\overline{Z})W[z(\overline{Z})]} \text{ in } \tilde{D}_Z, \end{cases}$$

which satisfies the boundary conditions

$$\text{Re}[\overline{\tilde{\lambda}(Z)}\tilde{W}(Z)] = \tilde{R}(Z) \text{ on } \Gamma \cup \tilde{\Gamma},$$

$$\tilde{\lambda}(Z) = \begin{cases} \hat{\lambda}[z(Z)], \\ \overline{\hat{\lambda}[z(\overline{Z})]}, \end{cases} \quad \tilde{R}(Z) = \begin{cases} |X(Z)|R[z(Z)] \text{ on } \Gamma, \\ -|X(\overline{Z})|R[z(\overline{Z})] \text{ on } \tilde{\Gamma}, \end{cases}$$

where $\tilde{\Gamma}$ is the symmetrical curve of Γ about $\text{Im}Z = 0$. It is clear that the corresponding function $u(z)$ in (2.37) can be extended to the function $\tilde{u}(Z)$, where $\tilde{u}(Z) = u[z(Z)]$ in D_Z and $\tilde{u}(Z) = -u[z(\overline{Z})]$ in $\tilde{D}_Z$. Noting Condition C and the condition (2.43), the function $\tilde{f}(Z) = X(Z)g(Z)/H(y)$ in D_Z and $\tilde{f}(Z) = -\overline{X(\overline{Z})g(\overline{Z})/H(y)}$ in $\tilde{D}_Z$ satisfies the condition $L_\infty[y^\tau \tilde{f}(Z), D'_Z] \leq M_4$, in which $D'_Z = D_Z \cup \tilde{D}_Z \cup \gamma$, $\tau = \max(1 - m_1/2 - m_2/2, 1 - m_2, 0)$, $M_4 = M_4(\delta, k, H, D, M_3)$ is a positive constant. By using the result in Lemma 2.1, Chapter I, we can see that the function $\tilde{\Psi}(Z) = 2i\text{Im}T\tilde{f} = 2i\text{Im}(-1/\pi)\iint_{D_t}[\tilde{f}(t)/(t-Z)]d\sigma_t$ satisfies the estimates

$$C_\beta[\tilde{\Psi}(Z), D_Z \cup \tilde{D}_Z] \leq M_5,\ \tilde{\Psi}(Z) - \tilde{\Psi}(t_j) = O(|Z - t_j|^{\beta_j}),\ 1 \leq j \leq 2,$$

in which $\tilde{f}(Z) = X(Z)f(Z)/H(y)$, $\beta = (2-m_2)/(2+m_1-m_2) - 2\delta = \beta_j\ (1 \leq j \leq 2)$, δ is a constant as stated in (2.42), and $M_5 = M_5(\delta, k, H, D, M_3)$ is a positive constant. On the basis of Theorem 2.1, the solution $\tilde{W}(Z)$ can be expressed as $\tilde{W}(Z) = \tilde{\Phi}(Z) + \tilde{\Psi}(Z)$, $\tilde{\Psi}(Z) = 2i\text{Im}T\tilde{f}$, where $\tilde{\Phi}(Z)$ is an analytic function in D_Z satisfying the boundary conditions

$$\text{Re}[\overline{\tilde{\lambda}(Z)}\tilde{\Phi}(Z)] = \tilde{R}(Z) - \text{Re}[\overline{\tilde{\lambda}(Z)}\tilde{\Psi}(Z)] = \hat{R}(Z) \text{ on } \Gamma, u(a_j) = 0,\ j = 1, 2.$$

Obviously $\text{Re}\tilde{\Psi}(Z) = 0$, $\text{Re}\tilde{W}(Z) = \text{Re}\tilde{\Phi}(Z)$ in D_Z, $\text{Re}\tilde{W}(x) = \text{Re}\tilde{\Phi}(x) = X(x)H_1(y)u_x(x)/2$ on γ, there is no harm in assuming that $\tilde{\Psi}(t_j) = 0$, otherwise it suffices to replace $\tilde{\Psi}(Z)$ by $\tilde{\Psi}(Z) - \tilde{\Psi}(t_j)$, where the boundedness of $\tilde{\Psi}(t_j)\ (1 \leq j \leq 2)$ is as stated before. For giving the estimates of $\tilde{\Phi}(Z)$ in $\overline{D}_Z \cap \{|Z \pm 1| > \varepsilon(> 0)\}$, from the integral expression of solutions of the discontinuous Riemann-Hilbert problem for analytic functions, we can write the representation of the solution $\tilde{\Phi}(Z)$ of Problem A_n for analytic functions, namely

$$\tilde{\Phi}[Z(\zeta)]=\frac{X_1(\zeta)}{2\pi i}\left[\int_{\partial D_t}\frac{(t+\zeta)\tilde{\lambda}[Z(t)]\hat{R}[Z(t)]dt}{(t-\zeta)tX_1(t)}+ic_*\frac{\zeta_1+\zeta}{\zeta_1-\zeta}\right],$$

$$c_*=c_0\frac{\zeta_1-\zeta}{\zeta_1+\zeta}\text{ if }\tilde{K}=0,\text{ and }c_*=i\int_{\partial D_t}\frac{\tilde{\lambda}[Z(t)]\hat{R}[Z(t)]dt}{tX_1(t)},\text{ if }\tilde{K}=-\frac{1}{2},$$

(see [86]11),[87]), where $X_1(\zeta)=\Pi_{j=1}^2(\zeta-t_j)^{\tau_j}$, $\tau_j\,(j=1,2)$ are as stated before, $Z=Z(\zeta)$ is the conformal mapping from the unit disk $D_\zeta=\{|\zeta|<1\}$ onto the domain D_Z, such that the three points $\zeta=-1,i,1$ are mapped onto $Z=-1,Z_0(\in\Gamma\backslash\{-1,1\}),1$ respectively, and ζ_1 is a point on $|\zeta|=1$, if $\tilde{K}=0$, then the real constant c_0 is determined by last one point condition in the boundary condition. According to the result in [87]1), we see that the function $\tilde{\Phi}(Z)$ determined by the above integral in $\overline{D}_Z\cap\{|Z\pm1|>\varepsilon(>0)\}$ is Hölder continuous.

For giving the estimates of $X(Z)u_x$, $X(Z)u_y$ in $\tilde{D}_j=D_j\cap D_Z\,(D_j=\{|Z-t_j|<\varepsilon(>0)\})$, $j=1$ or 2 separately, we can locally handle the problem in $D_j\,(1\le j\le2)$. We first reduce the function $\tilde{\lambda}(Z)=1$ on $\Gamma'=\Gamma\cup\tilde{\Gamma}$ near $Z=t_j\,(1\le j\le2)$. It is sufficient to find an analytic function $S(Z)$ in $D_j\cap D'_Z$ satisfying the boundary conditions

$$\mathrm{Re}S(Z)=-\arg\tilde{\lambda}(Z)\ \text{ on }\Gamma'=\Gamma\cup\tilde{\Gamma}\ \text{ near }t_j,\ \mathrm{Im}S(t_j)=0,$$

and the estimate

$$C_\delta[S(Z),D_j\cap D'_Z]\le M_6=M_6(\delta,k,H,D,M_3)<\infty,$$

then the function $e^{jS(Z)}\tilde{W}(Z)$ is satisfied the boundary condition

$$\mathrm{Re}[e^{iS(Z)}\tilde{W}(Z)]=0\ \text{ on }\Gamma'=\Gamma\cup\tilde{\Gamma}\ \text{ near }Z=t_j\,(1\le j\le2).$$

Moreover we symmetrically extend the function $\tilde{W}(Z)$ in D'_Z onto the symmetrical domain D^*_Z with respect to $\mathrm{Re}Z=t_j\,(1\le j\le2)$, namely let

$$\hat{W}(Z)=\begin{cases}e^{iS(Z)}\tilde{W}(Z)\ \text{ in }D'_Z,\\ -\overline{e^{iS(Z^*)}\tilde{W}(Z^*)}\ \text{ in }D^*_Z,\end{cases}$$

here $Z^*=-\overline{(Z-t_j)}+t_j$. Actually by the above condition, we may choose

$S(Z) = \pi/2$ when $t_1 = -1$ and $S(Z) = -\pi/2$ when $t_2 = 1$, and can get

$$C_\delta[\tilde{\Phi}(Z), D_\varepsilon] \leq M_7, C_\delta[X(Z)u_x, D_\varepsilon] \leq M_7, C_\delta[X(Z)u_y, D_\varepsilon] \leq M_7,$$
$$C_\delta[u_x, D'_\varepsilon] \leq M_8,\ C_\delta[u_y, D'_\varepsilon] \leq M_8, \tag{2.44}$$

in which $D_\varepsilon = \overline{D_Z} \cap \{\text{dist}(Z,\gamma) \geq \varepsilon\}$, $D'_\varepsilon = \overline{D_Z} \cap \{\text{dist}(Z, \Gamma \cup \tilde{\Gamma}) \geq \varepsilon\}$, ε is arbitrary small positive constant, $M_7 = M_7(\delta, k, H, D_\varepsilon, M_3)$, $M_8 = M_8(\delta, k, H, D'_\varepsilon, M_3)$ are non-negative constants. In fact the first three estimates in (2.44) can be derived by the above integral representation of $\tilde{\Phi}(Z)$. In addition from (2.20) and (2.43), $W(Z) = [H_1(y)u_x - iH_2(y)u_y]/2 = \Phi^*(Z) + \Psi^*(Z) = \hat{\Phi}^*(Z) + \hat{\Psi}^*(Z)$ in D_Z, the integral $\hat{\Psi}^*(Z) = 2i\text{Im}Tg/H_1$ over D'_ε is bounded and Hölder continuous in D'_ε, $\hat{\Phi}^*(Z)$ in D_Z is an analytic function. Noting that $\text{Re}W(Z) = \text{Re}\hat{\Phi}^*(Z) = H_1(y)u_x/2 = 0$ on $D'_\varepsilon \cap \{Y = 0\}$, we can extend the function $\hat{\Phi}^*(Z)$ from D'_ε onto the symmetrical domain $\tilde{D}'_\varepsilon$ about the real axis $\text{Im}Z = 0$, the extended function is denoted by $\hat{\Phi}^*(Z)$ again, obviously $\text{Re}\hat{\Phi}^*(Z)$ is a harmonic function in D_Z, thus $\text{Re}\hat{\Phi}^*(Z) = \text{Re}W(Z) = H_1(y)u_x/2 - YF$, here F in D'_ε is Hölder continuous, and then $u_x = O(y^{(2-m_2)/2})$ in D'_ε. Similarly we can derive the estimate of u_y. Hence the last two estimates in (2.44) are true.

Moreover as stated in (2.20), the solution $\hat{W}(Z)$ can be also expressed as $\hat{W}(Z) = \Phi(Z) + \Psi(Z) = \hat{\Phi}(Z) + \hat{\Psi}(Z)$ in $\hat{D}'_Z = \{D'_Z \cup D^*_Z\} \cap \{Y > 0\}$ as stated in (2.20), and the corresponding functions $\Psi(Z)$, $\hat{\Psi}(Z)$ in $\hat{D}'_Z$ are also Hölder continuous, $\text{Im}\Psi(Z) = 0$, $\text{Re}\hat{\Psi}(Z) = 0$ in $\hat{D}'_Z$, and $\Phi^*(Z) = \Phi(Z)$ or $\Phi^*(Z) = \hat{\Phi}(Z)$ as an analytic function can be extended in $\hat{D}_j = D_j \cap \hat{D}'_Z$ satisfying the boundary conditions in the form

$$\text{Re}[\Phi^*(Z)] = \hat{R}(Z) \text{ on } \Gamma' = \Gamma \cup \tilde{\Gamma},\ u(a_j) = 0, j = 1, 2,$$

because in the above case $\tilde{\lambda}(Z) = 1$ on $\partial D'_Z$ near $Z = t_j\,(1 \leq j \leq 2)$. Noting that the analytic function $\Phi^*(Z)$ in $D_j = \{|Z - t_j| < \varepsilon(> 0)\}$ satisfies the condition $\Phi^*(Z) = O(|Z - t_j|)$ near $Z = t_j$ in $\hat{D}_j$, it is clear that $\text{Im}\Phi(Z) = \text{Im}\hat{W}(Z)$ and $\text{Re}\hat{\Phi}(Z) = \text{Re}\hat{W}(Z)$ extended are harmonic functions in $\hat{D}'_Z$, and $\text{Re}\hat{\Phi}(Z)$, $\text{Im}\Phi(Z)$ can be expressed as

$$2\text{Re}\hat{\Phi}(Z) = \sum_{j,k=0}^{\infty} c^{(1)}_{jk} X^j Y^k,\ 2\text{Im}\Phi(Z) = \sum_{j,k=0}^{\infty} c^{(2)}_{jk} X^j Y^k$$

in D_j, here $X = x - t_j$, and we use equation (2.34) and denote

$$H_2(y) = \sqrt{K_2(y)}, H_1(y) = \sqrt{K_1^n(y)},\ W(Z) = [H_1(y)u_x - iH_2(y)u_y]/2.$$

Denote $X(Z)e^{iS(z)} = \tilde{X} + i\tilde{Y}$, where $X(Z)$ is as stated in (2.38). Noting that

$$\begin{array}{l} 2\mathrm{Re}\hat{\Phi}(Z) = \tilde{X}H_1(y)u_x + \tilde{Y}H_2(y)u_y = 0 \\ 2\mathrm{Im}\Phi(Z) = \tilde{Y}H_1(y)u_x - \tilde{X}H_2(y)u_y = 0 \end{array} \quad \text{on } \mathrm{Im}Z = Y = 0,$$

we have

$$2\mathrm{Re}\hat{\Phi}(Z) = 2\mathrm{Re}X(Z)\hat{W}(Z) = \tilde{X}H_1u_x + \tilde{Y}H_2u_y = YF_1,$$

$$2\mathrm{Im}\Phi(Z) = 2\mathrm{Im}X(Z)\hat{W}(Z) = \tilde{Y}H_1u_x - \tilde{X}H_2u_y = YF_2$$

in $\tilde{D}_j = D_j \cap D_Z$, where F_1, F_2 are continuous in D_j $(1 \le j \le 2)$. From the system of algebraic equations, we can solve u_x, u_y as follows

$$u_x = YH_2(y)(\tilde{X}F_1 + \tilde{Y}F_2)/H_1(y)H_2(y)|X(Z)|^2,$$

$$u_y = YH_1(y)(\tilde{Y}F_1 - \tilde{X}F_2)/H_1(y)H_2(y)|X(Z)|^2, \text{ i.e.}$$

$$X(Z)H_1(y)u_x = Y(\tilde{X}F_1 + \tilde{Y}F_2)/\overline{X(Z)},$$

$$X(Z)H_2(y)u_y = Y(\tilde{Y}F_1 - \tilde{X}F_2)/\overline{X(Z)},$$

and then

$$X(Z)u_x = O(y^{(2-m_2)/2}),\ X(Z)u_y = O(y^{(2-m_2)/2})$$

can be derived. Thus

$$C_\delta[X(Z)u_x, \tilde{D}_j] \le M_9, C_\delta[X(Z)u_y, \tilde{D}_j] \le M_9, 1 \le j \le 2, \tag{2.45}$$

in which $X(Z)$, δ are as stated in (2.42), and $M_9 = M_9(\delta,\ k, H, D, M_3)$ is a non-negative constant. Hence the first estimate in (2.42) is derived, but the constant M_1 is dependent on M_3.

Finally we use the reduction to absurdity, suppose that (2.43) is not true, then there exist sequences of coefficients $\{A_j^{(m)}(z)\}$ $(j = 1,2,3,4)$, $\{\lambda^{(m)}(z)\}$, $\{R^{(m)}(z)\}$ and $\{b_j^{(m)}\}(j = 1,2)$, which satisfy the same conditions of coefficients of (2.35), (2.36), which are corresponding to those as in (2.2), (2.8), such that $\{A_j^{(m)}(z)\}$, $\{\lambda^{(m)}(z)\}$, $\{R^{(m)}(z)\}$, $\{b_j^{(m)}\}(j = 1,2)$ in $\overline{D}$, Γ, γ weakly converge or uniformly converge to $A_j^{(0)}(z)$ $(j = 1,2,3,4)$, $\lambda^{(0)}(z)$, $R^{(0)}(z)$ and $b_j^{(0)}$ $(j = 1,2)$, and the solutions of the corresponding boundary value problem

$$W_{\bar{\bar{z}}}^{(m)}=F^{(m)}(z,u^{(m)},W^{(m)}),F^{(m)}(z,u^{(m)},W^{(m)})$$
$$=A_1^{(m)}W^{(m)}+A_2^{(m)}W^{(m)}+A_3^{(m)}u^{(m)}+A_4^{(m)} \text{ in } \overline{D},$$
$$\text{Re}[\overline{\lambda^{(m)}(z)}W(z)]=R^{(m)}(z) \text{ on } \Gamma\cup\gamma, u^{(m)}(a_j)=b_j^{(m)},\ j=1,2,$$

and

$$u^{(m)}(z)=u^{(m)}(x)-2\int_0^y\frac{V^{(m)}}{H_2(y)}dy$$
$$=2\text{Re}\int_{-1}^z[\frac{\text{Re}W^{(m)}}{H_1(y)}+i\frac{\text{Im}W^{(m)}}{H_2(y)}]dz+b_1^{(m)} \text{ in } \overline{D}$$

have the solutions $[W^{(m)}(z),u^{(m)}(z)]$, but $\hat{C}[W^{(m)}(z),\overline{D}]$ $(m=1,2,...)$ are unbounded, hence we can choose a subsequence of $[W^{(m)}(z),u^{(m)}(z)]$ denoted by $[W^{(m)}(z),u^{(m)}(z)]$ again, such that $h_m=\hat{C}[W^{(m)}(z),\overline{D}]\to\infty$ as $m\to\infty$, we can assume $h_m\ge\max[k_1,k_2,1]$. It is obvious that $[\tilde{W}^{(m)}(z),\tilde{u}^{(m)}(z)_m]=[W^{(m)}(z)/h_m,\ u^{(m)}(z)/h_m]$ are solutions of the boundary value problems

$$\tilde{W}_{\bar{\bar{z}}}^{(m)}=\tilde{F}^{(m)}(z,\tilde{u}^{(m)},\tilde{W}^{(m)}),\tilde{F}^{(m)}(z,\tilde{u}^{(m)},\tilde{W}^{(m)})$$
$$=A_1^{(m)}\tilde{W}^{(m)}+A_2^{(m)}\tilde{W}^{(m)}+A_3^{(m)}\tilde{u}^{(m)}+A_4^{(m)}/h_m \text{ in } \overline{D_Z},$$
$$\text{Re}[\overline{\lambda^{(m)}(z)}\tilde{W}^{(m)}(z)]=\frac{R^{(m)}(z)}{h_m} \text{ on } \Gamma\cup\gamma,\tilde{u}^{(m)}(a_j)=\frac{b_j^{(m)}}{h_m},j=1,2,\tag{2.46}$$

and

$$\tilde{u}^{(m)}(z)=\frac{u^{(m)}(x)}{h_m}-2\int_0^y\frac{\tilde{V}^{(m)}}{H_2(y)}dy$$
$$=2\text{Re}\int_{-1}^z[\frac{\text{Re}\tilde{W}^{(m)}}{H_1(y)}+i\frac{\text{Im}\tilde{W}^{(m)}}{H_2(y)}]dz+\frac{b_1^{(m)}}{h_m} \text{ in } D.$$

We can see that the functions in above boundary value problems satisfy the conditions

$$L_\infty[H_1\text{Re}A_1/H_2,\overline{D}],L_\infty[y^{1-m_2/2}(\text{Im}(A_1\pm A_2),\overline{D}]\le k_3,$$
$$L_\infty[A_3,\overline{D}]\le k_3,\ L_\infty[A_4/h_m,\overline{D}]\le 1,$$
$$C_\alpha[\lambda^{(m)}(z),\Gamma]\le k_3,C_\alpha[r^{(m)}(x)/h_m,\gamma]\le 1,$$
$$|b_j^{(m)}/h_m|\le 1,\ j=1,2,$$

where $k_3 = k_3(\delta, k, H, D)$ is a positive constant. From the representation (2.20), the above solutions can be expressed as

$$\begin{aligned}
&\tilde{u}^{(m)}(z)=\frac{u^{(m)}(x)}{h_m}-2\int_0^y \tilde{V}^{(m)}dy\\
&=2\mathrm{Re}\int_{-1}^{z}[\frac{\mathrm{Re}\tilde{W}^{(m)}}{H_1(y)}+i\frac{\mathrm{Im}\tilde{W}^{(m)}}{H_2(y)}]dz+\frac{b_1^{(m)}}{h_m} \text{ in } \overline{D},\\
&\tilde{W}^{(m)}(Z)=\tilde{\Phi}^{(m)}(Z)+\tilde{\Psi}^{(m)}(Z),\\
&\tilde{\Psi}^{(m)}(Z)=-\mathrm{Re}\frac{2}{\pi}\iint_{D_Z}\left[\frac{\tilde{f}^{(m)}(t)}{t-Z}\right]d\sigma_t \text{ in } \overline{D_Z},
\end{aligned}$$

in which $\tilde{f}^{(m)}(Z) = f^{(m)}(Z)/H_1(y)$. Similarly to the proof of (2.43) to (2.45), and notice that $y^\tau f^{(m)}(Z) = y^\tau X(Z)g^{(m)}(Z) \in L_\infty(D_Z)$, we can verify that

$$C_\beta[2\mathrm{Re}T(\tilde{f}^{(m)}(Z)),\overline{D_Z}] \leq M_{10},$$
$$\mathrm{Re}[T(\tilde{f}^{(m)}(Z))-T(\tilde{f}^{(m)}(Z))|_{Z=t_j}]=O(|Z-t_j|^{\beta_j}), j=1,2,$$

in which β, $\beta_j\,(j = 1, 2)$ are as before, $M_{10} = M_{10}(\delta, k, H, D)$ is a non-negative constant, and we can obtain the estimate

$$\hat{C}_\delta[\tilde{W}^{(m)}(Z),\overline{D_Z}] \leq M_{11} = M_{11}(\delta, k, H, D).$$

Hence from the sequence $\{X(Z)[\mathrm{Re}\tilde{W}^{(m)}(z)/H_1(y)+i\mathrm{Im}\tilde{W}^{(m)}(z)/H_2(y)]\}$ and the corresponding sequence $\{\tilde{u}^{(m)}(z)\}$, we can choose the subsequences denoted by $\{X(Z)\,[\mathrm{Re}\tilde{W}^{(m)}(z)/H_1(y) + i\mathrm{Im}\tilde{W}^{(m)}(z)/H_2(y)]\}$, $\{\tilde{u}^{(m)}(z)\}$ again, which in the closed domain $\overline{D}$ uniformly converge to the functions $X(Z)\,[\mathrm{Re}\tilde{W}^{(0)}(z)/H_1(y)+ i\mathrm{Im}\tilde{W}^{(0)}(z)/H_2(y)]$, $\tilde{u}^{(0)}(z)$ respectively, it is obvious that $[\tilde{W}^{(0)}(z), \tilde{u}^{(0)}(z)]$ is a solution of the homogeneous problem of Problem A_n, namely which satisfies the homogeneous equation of (2.35) and the homogeneous boundary conditions of (2.40). On the basis of Theorem 2.2, the solution $\tilde{W}^{(0)}(z) = 0$, $\tilde{u}^{(0)}(z) = 0$ in $\overline{D}$, however, from $\hat{C}[\tilde{W}^{(m)}(z),\overline{D}] = 1$, we can derive that there exists a point $z^* \in \overline{D}$, such that $\hat{C}[\tilde{W}^{(0)}(z^*),\overline{D}] = 1$, it is impossible. This shows the first estimate in (2.42) is true. Moreover it is not difficult to verify the second estimate in (2.42).

Theorem 2.5 *Suppose that equation* (2.1) *satisfy Condition C. Then there exists a solution of Problem M for equation* (2.1).

Proof From the estimates of solutions $u_n(z)$ of Problem M_n for equation (2.34), we can choose a subsequence of $\{u_n(z)\}$, which uniformly converges to $u_*(z)$ in $\overline{D}$, and $u_*(z)$ is just a solution of Problem M for equation (2.1).

Next we consider second order quasilinear elliptic equation with parabolic degeneracy

$$K(y)u_{xx}+u_{yy}+au_x+bu_y+cu+d=0 \text{ in } D, \tag{2.47}$$

where $K(y)$ are as stated in (2.1), and a, b, c, d are real functions of $z(\in D), u, u_x, u_y\,(\in \mathbf{R})$, its complex form is the following complex equation of second order

$$u_{\bar{z}\bar{z}} - iH_y u_x/4 = F(z,u,u_z),\ F = \mathrm{Re}[B_1 u_{\bar{z}}] + B_2 u + B_3 \text{ in } D, \tag{2.48}$$

where $B_j = B_j(z,u,u_z)\,(j=1,2,3)$ and

$$u_{\bar{z}}=[H(y)u_x-iu_y]/2,\, u_{\bar{z}\bar{z}} = [H(y)(u_{\bar{z}})_x+i(u_{\bar{z}})_y]/2=iH_y u_x/4$$
$$+[K(y)u_{xx}+u_{yy}]/4,\, B_1=-a/2H-ib/2,\, B_2=-c/4,\, B_3=-d/4 \text{ in } D.$$

Suppose that equation (2.47) satisfies the following conditions, namely
Condition C

1) For any continuously differentiable function $u(z)$ in $\overline{D}$, $B_j(z,u,u_z)$ $(j=1,2,3)$ are measurable in D and satisfy

$$\begin{gathered} L_\infty[H\mathrm{Re}B_1,\overline{D}],\, L_\infty[\mathrm{Im}B_1,\overline{D}],\, L_\infty[B_2,\overline{D}]\le k_0, \\ L_\infty[B_3,\overline{D}] \le k_1,\ B_2 \ge 0 \text{ in } D. \end{gathered} \tag{2.49}$$

2) For any continuously differentiable functions $u_1(z), u_2(z)$ in $D^* = \overline{D}\backslash\{-1,1\}$, the equality

$$F(z,u_1,u_{1z})-F(z,u_2,u_{2z})=\mathrm{Re}[\tilde{B}_1(u_1-u_2)_z]+\tilde{B}_2(u_1-u_2) \text{ in } D$$

holds, where $\tilde{B}_j = \tilde{B}_j(z,u_1,u_2)\,(j=1,2)$ satisfy the conditions

$$L_\infty[H\mathrm{Re}\tilde{B}_1,\overline{D}] \le k_0,\, L_\infty[\mathrm{Im}\tilde{B}_1,\overline{D}] \le k_0,\, L_\infty[\tilde{B}_2,\overline{D}] \le k_0, \tag{2.50}$$

in (2.49), (2.50), k_0, k_1 are non-negative constants. In particular, when (2.47) is a linear equation, the condition (2.50) obviously holds. Moreover we formulate the discontinuous mixed boundary value problem for equation (2.47) as follows.

Problem M' Find a real continuous solution $u(z)$ of (2.1) in $\overline{D}$, where u_x, u_y are continuous in $D^* = \overline{D}\backslash T$, and satisfy the boundary conditions

$$u(z) = \phi(x) \text{ on } \Gamma,\ u_y = r(x) \text{ on } \gamma^* = \gamma\backslash T, \tag{2.51}$$

where $T = \{a_1, ..., a_n\}$, $a_j (j = 1, 2, ..., n,\ a_1 = -1 < a_2 < ... < a_n = 1)$ are real numbers, $\gamma_j = (a_{j-1}, a_j)\,(j = 2, ..., n)$ on the $x-$axis, and the known functions $\phi(z)$, $r(x)$ satisfy the conditions

$$C^2_\alpha[\phi(z), \Gamma] \le k_1,\ C^1_\alpha[r(x), \gamma_j] \le k_1, j = 2, ..., n. \tag{2.52}$$

Problem M' with the conditions $d(z) = 0$, $z \in \bar{D}$, $\phi(z) = 0$ on Γ and $r(z) = 0$ on γ^* will be called Problem M'_0.

By using the similar method, we have the numbers $\gamma_j\,(j{=}1, ..., n)$, which are defined by

$$\begin{aligned}
&e^{i\phi_1} = \frac{\lambda(t_1-0)}{\lambda(t_1+0)} = e^{-\pi i/2-0\pi i} = e^{-\pi i/2}, \gamma_1 = \frac{-\pi/2}{\pi} - K_1 = -\frac{1}{2}, K_1 = 0,\\
&e^{i\phi_j} = \frac{\lambda(t_j-0)}{\lambda(t_j+0)} = e^{0\pi i-0\pi i} = e^{0\pi i}, \gamma_j = \frac{0\pi}{\pi} - K_j = 0, K_j = 0, j = 2, ..., n-1,\\
&e^{i\phi_n} = \frac{\lambda(t_n-0)}{\lambda(t_n+0)} = e^{0\pi i-\pi i/2} = e^{-\pi i/2}, \gamma_n = -\frac{\pi/2}{\pi} - K_n = -\frac{1}{2}, K_n = 0,
\end{aligned} \tag{2.53}$$

hence the index of $\lambda(z)$ on $\Gamma \cup \gamma$ is $K = (K_1 + \cdots + K_n)/2 = 0$, where $\lambda(t_l + 0) = \lambda(t_{l+1} - 0) = e^{i0\pi}$, $l = 1, 2, ..., n-1$, because we consider $\text{Re}[\overline{\lambda(z)}W(z)] = 0$, $\lambda(x) = 1$ on γ^*. Now the function $X(Z)$ in (2.38) is replaced by $X(Z) = \Pi^n_{j=1}(Z - t_j)^{\eta_j}$, $\eta_j = 1\,(j = 1, ..., n)$, and then multiply the complex equation (2.35) by the function $X(Z)$ and obtain the complex equation about the function $\tilde{W}(Z) = X(Z)W(z)$ in the form (2.39) and the boundary condition possesses the form (2.40), it is not difficult to see that the index $\tilde{K} = (n-2)/2$ of $\tilde{\lambda}(z) = X(Z)\lambda(z)$ on ∂D, because we have the numbers $\tilde{\gamma}_j = 0$, $\tilde{K}_j = 0$, $j = 1, n$, and $\tilde{\gamma}_j = 0$, $\tilde{K}_j = 1$, $j = 2, ..., n-1$. In the proof of Theorem 2.4, the above conditions about $j = 1, 2$ are replaced by those about $j = 1, ..., n$, and the point conditions $u(a_j) = b_j\,(j = 1, 2)$ should be replaced by $u(a_j) = b_j\,(j = 1, ..., n)$ respectively, where we can choose $b_j = 0\,(j = 2, ..., n-1)$, then the following theorem can be similarly proved.

Theorem 2.6 *Let equation* (2.47) *satisfy Condition* C'. *Then the mixed problem* (*Problem* M') *for* (2.47) *with the boundary condition* (2.51) *has a unique solution.*

Finally we mention that the coefficients $K(y)$ in equation (2.1) can be replaced by functions $K(x,y)$ with some conditions, for instance $K(x,y) = |y|^m h(x,y)$, $h(x,y)$ is a continuously differentiable positive function. Besides in [23]1), G. C. Dong investigated the unique solvability of mixed boundary value problem for linear degenerate elliptic equations of second order with the boundary conditions $u(z) = \phi(z)$ on Γ, $u_y = \psi(x)$ on γ, but he assumes that equation (2.1) is homogeneous, the coefficients of equation (2.1) are smooth enough and $a(x,y)$ satisfies the condition $a(x,y) = o(y^{m/2-1+\varepsilon})$ as $y \to 0$, $m \geq 2$, ε is a sufficiently small positive constant. Under the above conditions, he gave the estimates of solutions of the above problem, namely $|y|^{m/2+\varepsilon}|u_x|$, $|u_y|$ in the domain D are bounded. In this section we consider the equation (2.1) is non-homogeneous and only assume that the coefficients of (2.1) are bounded for almost every point in D, and gave the estimate (2.42) of solutions of the mixed problem for equation (2.1), which shows that the derivatives u_x, u_y of the solution $u(z)$ in $\overline{D}\backslash\{|z \pm 1| < \varepsilon(>0)\}$ are bounded and Hölder continuous.

3 The Oblique Derivative Problem for Second Order Elliptic Equations with Nonsmooth Degenerate Line

In [72]1),2), the authors posed and discussed some boundary value problems of second order mixed equations with nonsmooth degenerate line, but the coefficients of equations have strong restrictions. The present section deals with the oblique derivative problem for elliptic equations with nonsmooth degenerate line, where the coefficients satisfy weaker conditions. We first state the formulation of the problem for the equations, give the representation and estimates of solutions for the boundary value problem, and then prove the existence of solutions for the above problem by the Leray-Schauder theorem.

3.1 Formulation of oblique derivative problem for degenerate elliptic equations

Let D be a simply connected bounded domain in the complex plane $\mathbf{C}$ with the boundary $\partial D = \Gamma \cup \gamma$, where $\Gamma(\subset \{y > 0\}) \in C^2_\mu\,(0 < \mu < 1)$ is a curve with the end points $z = h_1, h_2$, and $\gamma = \{h_1 < x < h_2, y = 0\}$, here $-h_1, h_2$ are positive constants. There is no harm in assuming that the boundary Γ of the domain D is a smooth curve, which possesses the form $\tilde{G}_2(x) - \tilde{G}_1(y) = -1$ and $\tilde{G}_2(x) + \tilde{G}_1(y) = 1$ including the line segments

$\mathrm{Re}z = h_j\,(j = 1,2)$ near the points $z = h_j\,(j = 1,2)$ and the line segment $\mathrm{Im}z = y_0$ near $z = iy_0$ respectively, here $\tilde{G}_1(y), \tilde{G}_2(x)$ are similar to $\tilde{G}(y)$ as stated in Section 2, iy_0 is the intersection of Γ and the imaginary axis: $\{\mathrm{Re}z = 0\}$.

We first consider the linear elliptic equation with nonsmooth degenerate line

$$\begin{gathered} Lu = K_1(y)u_{xx} + |K_2(x)|u_{yy} + a(x,y)u_x \\ +b(x,y)u_y + c(x,y)u = -d(x,y) \text{ in } D. \end{gathered} \tag{3.1}$$

Denote $H_1(y)=\sqrt{K_1(y)}$, $K_1(0)=0$, $H_2(x)=\sqrt{|K_2(x)|}$, $K_2(0) = 0$, $K_1(y) = y^{m_1}h_1(y)$, $K_2(x) = |x|^{m_2}h_2(x)$, m_1, m_2 are positive numbers, $h_1(y), h_2(x)$ are continuously differentiable positive functions in $\overline{D}$. Suppose that the coefficients of (3.1) satisfy the following conditions, namely **Condition** C:

$$\begin{gathered} L_\infty[\eta, \overline{D}] \le k_0, \eta = a,b,c,\ L_\infty[d,\overline{D}] \le k_1,\ c \le 0 \text{ in } \overline{D}, \\ \eta|x|^{-m_2/2} = O(1) \text{ as } z = x+iy \to 0,\quad \eta = a,b,c,d, \end{gathered} \tag{3.2}$$

where k_0, k_1 are non-negative constants. If $H_1(y) = y^{m_l/2}$, $H_2(x) = |x|^{m_2/2}$, m_1, m_2 are positive numbers, thus

$$\begin{gathered} Y = G_1(y) = \int_0^y H_1(t)dt = \frac{2}{m_1+2}y^{(m_1+2)/2} \text{ in } D, \\ X=G_2(x)=\int_0^x H_2(t)dt=\pm\frac{2}{m_2+2}|x|^{(m_2+2)/2} \text{ in } \overline{D^\pm}, \end{gathered} \tag{3.3}$$

here $D_1=D\cap\{x<0\}$,$D_2=D\cap\{x>0\}$, and their inverse functions are

$$\begin{gathered} y=G_1^{-1}(Y)=\left(\frac{m_1+2}{2}\right)^{2/(m_1+2)} Y^{2/(m_1+2)}=J_1Y^{2/(m_1+2)}, \\ x=\pm|G_2^{-1}(X)|=\pm\left(\frac{m_2+2}{2}\right)^{2/(m_2+2)}|X|^{2/(m_2+2)}=\pm J_2|X|^{2/(m_2+2)}. \end{gathered} \tag{3.4}$$

The oblique derivative boundary value problem for equation (3.1) may be formulated as follows:

Problem P Find a continuous solution $u(z)$ of (3.1) in $\bar{D}$, where u_x, u_y

are continuous in $D^* = \overline{D}\backslash\{h_1, 0, h_2\}$, and satisfy the boundary conditions

$$\begin{gathered}\frac{1}{2}\frac{\partial u}{\partial \nu} = \frac{1}{H(x,y)}\mathrm{Re}[\overline{\lambda(z)}u_{\tilde{z}}] = \mathrm{Re}[\overline{\Lambda(z)}u_z] = r(z) \text{ on } \Gamma\cup\gamma^*, u(h_1) = b_0,\\ u(h_2) = b_1, \text{ or } \frac{1}{H(x,y)}\mathrm{Im}[\overline{\lambda(z)}u_{\tilde{z}}]|_{z=z_1} = \mathrm{Im}[\overline{\Lambda(z)}u_{\tilde{z}}]|_{z=z_1} = b_1,\end{gathered} \tag{3.5}$$

in which z_1 is a fixed point on $\Gamma\backslash\{h_1, h_2\}$, ν is a given vector at every point $z \in \Gamma \cup \gamma$, $u_{\tilde{z}} = [H_1(y)u_x - iH_2(x)u_y]/2$, $\Lambda(z) = \cos(\nu, x) - i\cos(\nu, y)$ and $\lambda(z) = \mathrm{Re}\lambda(z) + i\mathrm{Im}\lambda(z), H(x,y) = H_1(y)$ or $H_2(x)$, for instance $\lambda(z) = i, H(x,y) = H_2(x)$ if $z \in \gamma^*$, $\gamma^* = \{h_1 < x < 0, y = 0\} \cup \{0 < x < h_2, y = 0\}$, b_0, b_1 are real constants, and $r(z), b_0, b_1$ satisfy the conditions

$$\begin{gathered}C^1_\alpha[\lambda(z), \Gamma] \le k_0,\ C^1_\alpha[r(z), \Gamma] \le k_2,\\ C^1_\alpha[r(z), \gamma^*] \le k_2,\ \cos(\nu, n) \ge 0 \text{ on } \Gamma,\ |b_0|, |b_1| \le k_2,\end{gathered} \tag{3.6}$$

in which n is the outward normal vector at every point on Γ, $\alpha\,(0 < \alpha < 1), k_0, k_2$ are non-negative constants. Problem P with the conditions $A_3(z) = 0,\ z \in \bar{D},\ r(z) = 0,\ z \in \Gamma \cup \gamma^*$ and $b_0 = b_1 = 0$ will be called Problem P_0. The number

$$K = \frac{1}{2}(K_1 + K_2 + K_3) \tag{3.7}$$

is called the index of Problem P and Problem P_0, where

$$K_j = \left[\frac{\phi_j}{\pi}\right] + J_j, J_j = 0 \text{ or } 1, e^{i\phi_j} = \frac{\lambda(t_j - 0)}{\lambda(t_j + 0)}, \gamma_j = \frac{\phi_j}{\pi} - K_j, j = 1, 2, 3, \tag{3.8}$$

in which $t_1 = h_1$, $t_2 = h_2, t_3 = 0$. For Problem P, in general the index $K = 0$ on the boundary ∂D of D can be chosen. It is clear that $\gamma_3 = 0$, we can require that $-1/2 \le \gamma_j < 1/2\,(j = 1, 2)$, especially for the Neumann boundary condition on Γ, we have $\gamma_1 = \gamma_2 = -1/2$, because $\lambda(t_1 - 0) = e^{-\pi i}$ and $\lambda(t_2 + 0) = e^{\pi i}$ and consider $\mathrm{Re}[\overline{\lambda(x)}W(x)] = R(x) = H_2(x)r(x)$, $\lambda(x) = i$ on γ^*. Besides similarly to Sections 1 and 2, it suffices to multiply the complex equation (3.13) below by the function $X_0(Z) = Z$ or $X_0(Z) = Z(Z-1)(Z+1)$, then the index of the function $\hat{\lambda}[z(Z)] = \lambda[z(Z)]X_0(Z)$ on ∂D_Z in the boundary condition is $\hat{K} = 0$ or $1/2$. More simply if we rewrite the boundary condition on γ in an appropriate form, the above requirement can be realized, for instance setting that $\theta(x) = \arg\lambda(h_1 - 0) + [\arg\lambda(h_2 + 0) - \arg\lambda(h_1 - 0)](x - h_1)/(h_2 - h_1)$, for the function $\lambda(x) = e^{i\theta(x)}$ and $\mathrm{Re}[\overline{\lambda(x)}W(x)] = -\sin\theta(x)H_2(x)r(x)/2$ on γ^*, we have $\gamma_1 = \gamma_2 = \gamma_3 = 0$ and $K = 0$. When $\cos(\nu, n) \equiv 0$ on Γ, then we can choose $K = 0$ or $-1/2$. If $K = -1/2$, then the condition $u(h_2) = b_1$ can be cancelled. In fact, if $\cos(\nu, n) \equiv 0$ on Γ, from the boundary condition (3.5), we can determine the value $u(h_2)$ by the value $u(h_1)$, namely

$$u(h_2)=2\mathrm{Re}\int_{h_1}^{h_2} u_z dz+b_0=2\int_0^S \mathrm{Re}[z'(s)u_z]ds+b_0=2\int_0^S r(z)ds+b_0, \qquad (3.9)$$

in which $\overline{\Lambda(z)} = z'(s)$ on Γ, $z(s)$ is a parameter expression of arc length s of Γ with the condition $z(0) = h_1$, and S is the length of the curve Γ. In this section, we can choose the case $K = 0$, the other cases can be similarly discussed. We mention that when the last condition in (3.5) is chosen, it needs to assume $c = 0$ in equation (3.1). For the similar cases in (3.50), (5.5) below, we need to give the similar assumption, which can be handled by using the method in the proofs of Theorems 1.7 and 1.8.

For the boundary condition of the mixed problem (Problem M):

$$u(z) = \phi(x) \text{ on } \Gamma,\ u_y(x) = r(x) \text{ on } \gamma^*, \qquad (3.10)$$

where $C^2_\alpha[\phi(x), \Gamma] \le k_2$, $C^1_\alpha[r(x), \gamma^*] \le k_2$. We find the derivative for (3.10) according to the parameter $s = \mathrm{Im}\, z = y$ on Γ, and obtain

$$\begin{aligned}
&u_s=u_x x_y+u_y=\tilde H_1(y)u_x+H_2(x)u_y/H_2(x)=\phi'(y), \text{ i.e.}\\
&\tilde H_1(y)H_2(x)u_x+H_2(x)u_y=H_2(x)\phi'(y) \text{ on } \Gamma \text{ near } x=h_1,\\
&u_s=u_x x_y+u_y=-\tilde H_1(y)u_x + H_2(x)u_y/H_2(x)=\phi'(y), \text{ i.e.} \qquad (3.11)\\
&\tilde H_1(y)H_2(x)u_x-H_2(x)u_y=-H_2(x)\phi'(y) \text{ on } \Gamma \text{ near } x=h_2,\\
&H_1(y)u_x - H_2(x)u_y(x) = -H_2(x)r(x) \text{ on } \gamma^*,
\end{aligned}$$

where $\tilde H(y) = \tilde G'(y)$, it is clear that the complex form of (3.11) is as follows

$$\begin{aligned}
\mathrm{Re}[\overline{\lambda(z)}(U+iV)] &= \mathrm{Re}[\overline{\lambda(z)}(H_1(y)u_x - iH_2(x)u_y)]/2\\
&= R(x) \text{ on } \Gamma\cup\gamma,\ u(h_1) = b_0,
\end{aligned}$$

in which $H_1(y) = G_1'(y)$, $H_2(x) = G_2'(x)$, $b_0 = \phi(h_1)$, and

$$\lambda(z)=\begin{cases} -i, \\ i, \\ i, \end{cases} \quad R(z)=\begin{cases} H_2(x)\phi'(y)/2 \text{ on } \Gamma \text{ at } z=h_1,\\ -H_2(x)\phi'(y)/2 \text{ on } \Gamma \text{ at } z=h_2,\\ -H_2(x)r(x)/2 \text{ on } \gamma^*. \end{cases}$$

We have

$$
\begin{aligned}
e^{i\phi_1} &= \frac{\lambda(t_1-0)}{\lambda(t_1+0)} = e^{-\pi i/2-\pi i/2} = e^{-\pi i},\ \gamma_1 = \frac{-\pi}{\pi} - K_1 = 0,\ K_1 = -1,\\
e^{i\phi_2} &= \frac{\lambda(t_2-0)}{\lambda(t_2+0)} = e^{\pi i/2-\pi i/2} = e^{0\pi i},\ \gamma_2 = \frac{0\pi}{\pi} - K_2 = 0,\ K_2 = 0,\\
e^{i\phi_3} &= \frac{\lambda(t_3-0)}{\lambda(t_3+0)} = e^{\pi i/2-\pi i/2} = e^{0\pi i},\ \gamma_3 = \frac{0\pi}{\pi} - K_3 = 0,\ K_3 = 0,
\end{aligned}
$$

hence the index of $\lambda(z)$ on $\Gamma \cup \gamma^*$ is $K = (K_1 + K_2 + K_3)/2 = -1/2$. Besides if the boundary condition $u_y = r(x)$ on γ^* is replaced by $\mathrm{Re}W(z) = \mathrm{Re}[\overline{\lambda(x)}W(x)] = H_1(y)u_x/2 = 0$, $\lambda(x) = 1 = e^{i0\pi}$ on γ^*, then $\gamma_1 = 1/2$, $\gamma_2 = -1/2, \gamma_3 = 0$, and $K_1 = -1, K_2 = K_3 = 0, K = -1/2$. If we choose $\gamma_1 = \gamma_2 = -1/2, \gamma_3 = 0, K_1 = K_2 = K_3 = 0$, then the index $K = 0$, in this case, we can add one point condition $u(0) = 0$ or $u(h_2) = \phi(h_2) = b_1$ in the boundary condition. Similarly to Section 2, we can reduce the above boundary condition to the homogeneous boundary condition on Γ, i.e. $R(z) = 0$ on Γ in (3.5), and $b_0 = b_1 = 0$.

3.2 Representation of solutions of oblique derivative problem for elliptic equations

In this section, we first write the complex form of equation (3.1).

$$
\begin{aligned}
W(z) &= U+iV = \frac{1}{2}[H_1(y)u_x - iH_2(x)u_y] = u_{\bar z},\ H_1(y)H_2(x)W_{\overline{Z}}\\
&= \frac{H_1(y)H_2(x)}{2}[W_X + iW_Y] = \frac{1}{2}[H_1(y)W_x + iH_2(x)W_y] = W_{\tilde{\bar z}},
\end{aligned}
\tag{3.12}
$$

we have

$$
\begin{aligned}
&K_1(y)u_{xx} + |K_2(x)|u_{yy} = H_1(y)[H_1(y)u_x - iH_2(x)u_y]_x\\
&+iH_2(x)[H_1(y)u_x - iH_2(x)u_y]_y - iH_2(x)H_{1y}u_x + iH_1(y)H_{2x}u_y\\
&= 4W_{\tilde{\bar z}} - iH_2(x)H_{1y}u_x + iH_1(y)H_{2x}u_y = 2\{H_1[U+iV]_x\\
&+iH_2[U+iV]_y\} - i[H_2H_{1y}/H_1]H_1u_x + i[H_1H_{2x}/H_2]H_2u_y
\end{aligned}
$$

$$
\begin{aligned}
&=4H_1(y)H_2(x)W_{\overline{Z}}-i[H_2H_{1y}/H_1]H_1u_x+i[H_1H_{2x}/H_2]H_2u_y\\
&=-[au_x+bu_y+cu+d],\text{ i.e. } H_1(y)H_2(x)W_{\overline{Z}}=H_1H_2[W_X+iW_Y]/2\\
&=\{[iH_2H_{1y}/H_1-a/H_1](W+\overline{W})-i[iH_1H_{2x}/H_2+b/H_2](W-\overline{W})\\
&-[cu+d]\}/4=\{[iH_2H_{1y}/H_1-a/H_1+H_1H_{2x}/H_2-ib/H_2]W\\
&+[iH_2H_{1y}/H_1-a/H_1-H_1H_{2x}/H_2+ib/H_2]\overline{W}-cu-d]\}/4\\
&=A_1(z)W+A_2(z)\overline{W}+A_3(z)u+A_4(z)=g(Z)\text{ in } D_Z,
\end{aligned}
\tag{3.13}
$$

in which $2U=W+\overline{W}=H_1(y)u_x$, $2V=-i(W-\overline{W})=-H_2(x)u_y$, D_Z is the image domain of D with respect to the mapping $Z=G_2(x)+iG_1(y)=Z(z)$. If $g(Z)$ satisfies the condition

$$|x|^{\tau_2}y^{\tau_1}X(Z)g(Z)\in L_\infty(D_Z), \tag{3.14}$$

in which $\tau_j=\max(1-m_j/2,0)$ $(j=1,2)$, from Lemma 2.1, Chapter I, we see that the integral satisfies

$$Tg=-\frac{1}{2\pi}\iint_{D_t}\frac{f(t)}{t-Z}d\sigma_t\in C_\beta(D_Z), \tag{3.15}$$

in which $Z=Z(z)$ is the mapping from D to D_Z, $f(Z)=X(Z)g(Z)/H_1H_2$, $\beta=\min_{\{l=1,2\}}[2/(m_l+2),m_l/(m_l+2)]-\varepsilon$, ε is a sufficiently small positive constant, and $X(Z)$ is as stated in (3.26) below, which is chosen such that the condition (3.14) is satisfied.

It is obvious that a special case of (3.13) is the complex equation

$$W_{\overline{Z}}=0\text{ in }\overline{D_Z}. \tag{3.16}$$

The boundary value problem for equation (3.13) with the boundary condition (3.5) ($W(z)=H_1(y)H_2(x)u_Z=u_{\tilde{z}}$) and the relation (3.18) below will be called Problem A.

Now, we prove that there exists a solution of the Riemann-Hilbert problem (Problem A) for equation (3.16) in $\overline{D}$ with the boundary conditions

$$
\begin{aligned}
&\mathrm{Re}[\overline{\lambda(z)}W(Z(z))]=H_1(y)r(z)=R(z)\text{ on }\Gamma,\ u(h_1)=b_0,\\
&\mathrm{Re}[\overline{\lambda(z)}W(Z(z))]=r(x)=R(x)\text{ on }\gamma,\ u(h_2)=b_1,
\end{aligned}
\tag{3.17}
$$

where $\lambda(z) = a(z) + jb(z)$ on $\Gamma \cup \gamma$, b_0, b_1 are two real constants. Taking the index of $\lambda(z)$ is $K = 0$ into account, hence the boundary value problem (3.16), (3.17) has a unique solution $W(Z)$ in D_Z. By the property of solutions of the above Problem A (see [86]9), [87]1)), we see that the function $X(Z)W(Z) \in C_\delta(D_Z)$, where $X(Z), \delta$ are as stated in (3.26), (3.27) below.

Now we state and verify the representation of solutions of Problem P for equation (3.1).

Theorem 3.1 *Under Condition C, any solution $u(z)$ of Problem P for equation* (3.1) *in $\overline{D}$ can be expressed as follows*

$$u(z)=u(x)-2\int_0^y \frac{V(z)}{H_2(x)}dy=2\mathrm{Re}\int_{h_1}^z\left[\frac{\mathrm{Re}w}{H_1(y)}+i\frac{\mathrm{Im}w}{H_2(x)}\right]dz+b_0 \text{ in } \overline{D}, \tag{3.18}$$

and

$$\begin{gathered} w[z(Z)] = \Phi(Z) + \Psi(Z) = \hat{\Phi}(Z) + \hat{\Psi}(Z) \text{ in } D_Z, \\ \Psi(Z)=Tf+\overline{Tf}=2\mathrm{Re}Tf,\ \hat{\Psi}(Z)=Tf-\overline{Tf}=2i\mathrm{Im}Tf, \\ Tf = -\frac{1}{\pi}\iint_{D_t}\frac{f(t)}{t-Z}d\sigma_t \text{ in } D_Z, \end{gathered} \tag{3.19}$$

where $\Phi(Z)$, $\hat{\Phi}(Z)$ are analytic functions in D_Z, herein $Z = G_1(y)+iG_2(x)$, $f(Z) = g(Z)/H_1H_2$.

Proof Noting that

$$[Tf]_{\overline{Z}} = f(Z), [\overline{Tf}]_{\overline{Z}} = 0, [\Phi(Z)]_{\overline{Z}} = 0, [\hat{\Phi}(Z)]_{\overline{Z}} = 0 \text{ in } D_Z. \tag{3.20}$$

From (3.13) we see that equation (3.1) in $\overline{D}$ can be reduced to the formula (3.19). This shows that the formulas in (3.18), (3.19) are true.

In the following we prove the uniqueness of solutions of Problem P for (3.1).

Theorem 3.2 *Suppose that equation* (3.1) *satisfies Condition C. Then Problem P for* (3.1) *has at most one solution in D.*

Proof Let $u_1(z), u_2(z)$ be any two solutions of Problem P for (3.1). It is easy to see that $u(z) = u_1(z) - u_2(z)$ and $w(z) = H_1H_2u_Z = u_{\tilde{z}}$ satisfy the homogeneous equation and boundary conditions

$$\begin{gathered} K_1(y)u_{xx} + |K_2(x)|u_{yy} + au_x + bu_y + cu = 0, \text{ i.e.} \\ w_{\overline{Z}} = [A_1w + A_2\overline{w} + A_3u]/H_1H_2 \text{ in } D_Z, \end{gathered} \tag{3.21}$$

$$\frac{\partial u}{\partial \nu}=\frac{1}{2H(x,y)}\mathrm{Re}[\overline{\lambda(z)}w(z)]=0, z\in\Gamma\cup\gamma, u(h_1)=0, u(h_2)=0, \qquad (3.22)$$

where $\lambda(x) = i$, $x \in \gamma^* = (h_1, 0) \cup (0, h_2)$ on $x-$axis.

Now we verify that the above solution $u(z) \equiv 0$ in D. If the maximum $M = \max_{\overline{D}} u(z) > 0$, it is clear that the maximum point $z^* \notin (D\cup\{h_1, h_2\})$. If the maximum M attains at a point $z^* \in \Gamma$ and $\cos(\nu, n) > 0$ at z^*, we get $\partial u/\partial\nu > 0$ at z^*, this contradicts the formula of (3.22) on Γ; if $\cos(\nu, n) = 0$ at z^*, denote by Γ' the longest curve of Γ including the point z^*, so that $\cos(\nu, n) = 0$ and $u(z) = M$ on Γ', then there exists a point $z' \in \Gamma\backslash\Gamma'$, such that at z', $\cos(\nu, n) > 0, \partial u/\partial n > 0, \cos(\nu, s) > 0\,(< 0)$, $\partial u/\partial s \geq 0\,(\leq 0)$, hence

$$\frac{\partial u}{\partial \nu} = \cos(\nu, n)\frac{\partial u}{\partial n} + \cos(\nu, s)\frac{\partial u}{\partial s} > 0 \text{ at } z'$$

holds, where s is the tangent vector of Γ at $z' \in \Gamma$, it is impossible. This shows $z^* \notin \Gamma$. If $u(z)$ attains its maximum at a point $z^* = x^* \in \gamma^*$, by Lemma 2.3, we can derive that $u_x(x_*) = 0$, $u_y(x^*) < 0$, this contradicts the equality in (3.22) on γ^*. If $u(z)$ attains its maximum at a point $z^* = iy^* \in L = D \cap \{\mathrm{Re}z = 0\}$, by using the method in the proof of Theorem 3.3 below, we can add a point condition $u(0) = 0$. Because in this case, the above solution $u(z)$ can be expressed as in (3.19), where

$$W(Z)=\Phi(Z)+\Psi(Z) \text{ in } U(iy_*),$$

$$\Psi(Z)=2\mathrm{Re}Tf, Tf=-\frac{1}{\pi}\iint_{U(iy_*)}\frac{g(t)}{H_1H_2(t-Z)}d\sigma_t,$$

where $U(iy_*) = D \cap \{|Z - iY_*| < \varepsilon, X > 0\}$, $iY_* = Z(iy_*)$, ε is a sufficiently small positive constant, $\Psi(Z) \in C_\beta(U(iy_*))$, $\beta = \min(2, m_2)/(2 + m_2) - \varepsilon$, and $\Phi(Z)$ is an analytic function in $U(iy_*)$. Noting that $\mathrm{Im}\Psi(Z) = 0$ in $U(iy_*)$, $\mathrm{Im}\Phi(Z) = \mathrm{Im}W(Z) = -H_2(x)u_y/2$ is a harmonic function in $U(iy_*)$, and $\mathrm{Im}\Phi(Z) = -H_2(x)u_y/2 = 0$ on L, we can extend $\Phi(Z)$ onto the symmetrical set of $U(iy_*)$ with respect to $\mathrm{Re}Z = 0$, hence $\mathrm{Im}\Phi(Z) = XF$ in $U(iy_*)$, here F is continuous in $U(iy_*)$, thus $u_y/2 = O(|X|^{2/(2+m_2)})$ in $U(iy_*)$, this shows that $u_y = 0$ in $U(iy_*)$, and then $u(iy) = M > 0$ on L. From this we can extend that $u(z) = M$ on L, however $u(0) = 0$, this contradiction proves that $u(z)$ cannot attain the positive maximum in L. Thus $\max_{\overline{D}} u(z) = 0$. By the similar method, we can prove $\min_{\overline{D}} u(z) = 0$. Therefore $u(z) = 0$, $u_1(z) = u_2(z)$ in $\overline{D}$. This completes the proof.

3.3 Estimates of solutions of oblique derivative problem for elliptic equations

In the following we shall give the estimates of the solutions of Problem P for (3.1) in $\overline{D_Z}$. It is not difficult to see that Problem P is equivalent to Problem A for the complex equation

$$W_{\overline{Z}}=\frac{1}{H_1H_2}[A_1(z)W+A_2(z)\overline{W}+A_3(z)u+A_4(z)] \text{ in } D_Z, \tag{3.23}$$

$$\begin{aligned} A_1&=\frac{iH_2H_{1y}}{4H_1}+\frac{H_1H_{2x}}{4H_2}-\frac{a}{4H_1}-\frac{ib}{4H_2},\ A_3=\frac{-c}{4},\\ A_2&=\frac{iH_2H_{1y}}{4H_1}-\frac{H_1H_{2x}}{4H_2}-\frac{a}{4H_1}+\frac{ib}{4H_2},\ A_4=\frac{-d}{4}, \end{aligned} \tag{3.24}$$

with the boundary condition and the relation

$$u(z)=u(x)-2\int_0^y\frac{V(z)}{H_2(x)}dy=2\text{Re}\int_{h_1}^z\left[\frac{\text{Re}W}{H_1(y)}+i\frac{\text{Im}W}{H_2(x)}\right]dz+b_0 \text{ in } \overline{D}. \tag{3.25}$$

Moreover we can find a twice continuously differentiable function $u_0(z)$ in $\overline{D}$, for instance we can find a harmonic function in D satisfying the above boundary condition (3.5) on Γ, $u_0(h_1)=b_0$ and $u_0(h_2)=b_1$, denote $v(z)=u(z)-u_0(z)$, then the function $v(z)$ is a solution of the equation

$$\begin{aligned} Lv=&K_1(y)v_{xx}+|K_2(x)|v_{yy}+a(x,y)v_x+b(x,y)v_y\\ &+c(x,y)v=F(x,y),\ F=-d-Lu_0 \text{ in } D \end{aligned}$$

and $W(Z)=v_{\tilde{z}}$ satisfies the complex equation in the form (3.23) and the boundary condition

$$\begin{gathered} \text{Re}[\overline{\lambda(z)}W(z)]=R(z)=0 \text{ on } \Gamma,\ \text{Re}[\overline{\lambda(z)}W(z)]=R(x) \text{ on } \gamma,\\ u(h_1)=0,\ u(h_2)=0 \text{ or } \text{Im}[\overline{\lambda(z)}W(z)]|_{z=z_1}=0. \end{gathered}$$

Hence we shall discuss the case later on.

Introduce a function

$$X(Z)=\prod_{j=1}^4(Z-t_j)^{\eta_j}, \tag{3.26}$$

where $t_1 = G_2(h_1) = -1, t_2 = G_2(h_2) = 1, t_3 = 0, t_4 = iY_0 = iG_1(y_0)$, $\eta_j = 1 - 2\gamma_j$ if $\gamma_j \geq 0$, and $\eta_j = -2\gamma_j$ if $\gamma_j < 0$, $\eta_3 = 1, \eta_4 = 0$, here $\gamma_j\,(j = 1, 2)$ are as stated in (3.8), and a branch of multi-valued function $X(Z)$ such that $\arg X(x) = 0$ on γ is chosen. Obviously that $X(Z)W[z(Z)]$ satisfies the complex equation

$$[X(Z)W]_{\overline{Z}} = X(Z)[A_1W + A_2\overline{W}$$
$$+A_3u + A_4]/H_1H_2 = X(Z)g(Z)/H_1H_2 \text{ in } D_Z,$$

and the boundary conditions

$$\mathrm{Re}[\overline{\hat{\lambda}(z)}X(Z)W(z)] = |X(Z)|R(z) \text{ on } \Gamma\cup\gamma^*,$$
$$u(h_j) = b_j,\ j = 1, 2,\ u(0) = 0,$$

where $|X(x)|R(x) = 0$ on γ^*. It is easy to see that

$$e^{i\hat{\phi}_j} = \frac{\hat{\lambda}(t_j - 0)}{\hat{\lambda}(t_j + 0)} = \frac{\lambda(t_j - 0)e^{i\arg X(t_j-0)}}{\lambda(t_j + 0)e^{i\arg X(t_j+0)}}$$
$$= e^{i(\phi_j+\eta_j\pi/2)},\ \tau_j = \frac{\hat{\phi}_j}{\pi} - \hat{K}_j,\ j = 1, 2,$$

hence the numbers $\tau_j\,(j = 1, 2)$ about $\hat{\lambda}[z(Z)] = \lambda[z(Z)]e^{i\arg X(Z)}$ are equal to $1/2$ if $\gamma_j \geq 0$, or equal to 0 if $\gamma_j < 0\,(1 \leq j \leq 2)$, $\tau_3 = \tau_4 = 0$, which are corresponding to the numbers $\gamma_j\,(j = 1, 2, 3)$ in (3.8). and $\hat{K}_1 = \hat{K}_2 = 0, \hat{K}_3 = 1, \hat{K}_4 = 0, \hat{K} = 1/2$, hence we add one point condition $u(0) = 0$.

Theorem 3.3 Let equation (3.1) satisfy Condition C. Then any solution of Problem A for (3.23) satisfies the estimates

$$\hat{C}_\delta[W(z),\overline{D}] = C_\delta[X(Z)(\mathrm{Re}W/H_1 + i\mathrm{Im}W/H_2),\overline{D}] + C_\delta[u(z),\overline{D}] \leq M_1,$$
$$\hat{C}_\delta[W(z),\overline{D}] \leq M_2(k_1 + k_2), \tag{3.27}$$

where $X(Z)$ is as stated in (3.26), δ is a sufficiently small positive constant, $M_1 = M_1(\delta, k, H, D)$, $M_2 = M_2(\delta, k_0, H, D)$ are non-negative constants, $H = (H_1, H_2)$, and $k = (k_0, k_1, k_2)$.

Proof We first assume that any solution $[W(z),\ u(z)]$ of Problem A satisfies the estimate of boundedness

$$\hat{C}[W(z),\overline{D}] = C[X(Z)(\mathrm{Re}W/H_1 + i\mathrm{Im}W/H_2),\overline{D_Z}] + C[u(z),\overline{D}] \leq M_3, \tag{3.28}$$

in which M_3 is a non-negative constant. Now, substituting the solution $[W(z), u(z)]$ into equation (3.23) and noting $\mathrm{Re}W(Z) = 0$ on γ, $b_0 = 0$, we can extend the function $X(Z)W[z(Z)]$ onto the symmetrical domain $\tilde{D}_Z$ of D_Z with respect to the real axis $\text{Im}Z = 0$, namely set

$$\tilde{W}(Z) = \begin{cases} X(Z)W[z(Z)] \text{ in } D_Z, \\ -\overline{X(\overline{Z})W[z(\overline{Z})]} \text{ in } \tilde{D}_Z, \end{cases}$$

which satisfies the boundary conditions

$$\text{Re}[\overline{\tilde{\lambda}(Z)}\tilde{W}(Z)] = 0 \text{ on } \Gamma \cup \tilde{\Gamma} \cup \gamma,$$

$$\tilde{\lambda}(Z) = \begin{cases} \hat{\lambda}(z(Z)), \\ \overline{\hat{\lambda}(z(\overline{Z}))}, \\ 1, \end{cases} \quad R^*(Z) = \begin{cases} |X(Z)|R[z(Z)] \text{ on } \Gamma, \\ -|X(\overline{Z})|R[z(\overline{Z})] \text{ on } \tilde{\Gamma}, \\ 0 \text{ on } \gamma = (h_1, h_2), \end{cases}$$

where $\tilde{\Gamma}$ is the symmetrical curve of Γ about $\text{Im}Z = 0$. It is clear that the corresponding function $u(z)$ in (3.18) can be extended to the function $\tilde{u}(Z)$, where $\tilde{u}(Z) = u[z(Z)]$ in D_Z and $\tilde{u}(Z) = -u[z(\overline{Z})]$ in $\tilde{D}_Z$. Noting Condition C and the condition (3.28), we see that the function $\tilde{f}(Z) = X(Z)g(Z)/H_1H_2$ in D_Z and $\tilde{f}(Z) = -\overline{X(\overline{Z})g(\overline{Z})}/H_1H_2$ in $\tilde{D}_Z$ satisfies the condition

$$L_\infty[|x|^{\tau_2}|y|^{\tau_1}\tilde{f}(Z), D'_Z] \le M_4,$$

where $D'_Z = D_Z \cup \tilde{D}_Z \cup \gamma$, $\tau_j = \max(1 - m_j/2, 0)$, $j = 1, 2$, $M_4 = M_4(\delta, k, H, D, M_3)$ is a positive constant. On the basis of Lemma 2.1, Chapter I, we can verify that the function $\tilde{\Psi}(Z) = 2i\text{Im}T\tilde{f} = -2i\text{Im}(1/\pi)\iint_{D_t}[\tilde{f}(t)/(t-Z)]d\sigma_t$ satisfies the estimates

$$C_\beta[\tilde{\Psi}(Z), \overline{D_Z}] \le M_5,\ \tilde{\Psi}(Z) - \tilde{\Psi}(t_j) = O(|Z - t_j|^{\beta_j}),\ 1 \le j \le 2,$$

in which $\tilde{f}(Z) = X(Z)f(Z)/H_1H_2$, $\beta = \min(2, m_1)/(m_1+2) - 2\delta = \beta_j$ $(1 \le j \le 2)$, δ is a constant as stated in (3.27), and $M_5 = M_5(\delta, k, H, D, M_3)$ is a positive constant. On the basis of Theorem 3.1, the solution $\tilde{W}(Z)$ can be expressed as $\tilde{W}(Z) = \tilde{\Phi}(Z) + \tilde{\Psi}(Z)$, where $\tilde{\Phi}(Z)$ is an analytic function in D_Z satisfying the boundary conditions

$$\text{Re}[\overline{\tilde{\lambda}(Z)}\tilde{\Phi}(Z)] = R^*(Z) - \text{Re}[\overline{\tilde{\lambda}(Z)}\tilde{\Psi}(Z)]$$

$$= \hat{R}(Z) \text{ on } \Gamma \cup \gamma,\ u(h_1) = 0,\ u(h_2) = 0,\ u(0) = 0.$$

Obviously $\text{Re}\tilde{\Psi}(Z) = 0$, $\text{Re}\tilde{W}(z) = \text{Re}\tilde{\Phi}(Z)$ in D_Z and $\text{Re}\tilde{\Phi}(x) = \text{Re}\tilde{W}(x) = X(x)H_1(0)u_x(x)/2 = 0$ on γ, there is no harm in assuming that $\tilde{\Psi}(t_j) = 0$, otherwise it suffices to replace $\tilde{\Psi}(Z)$ by $\tilde{\Psi}(Z) - \tilde{\Psi}(t_j)$, because $\tilde{W}(t_j), \tilde{\Psi}(t_j)(1 \le j \le 2)$ are bounded. For giving the estimates of

$\tilde{\Phi}(Z)$ in the neighborhood $D_j=\{|Z-t_j|<\varepsilon(>0)\}$ of $t_j\,(1\le j\le 2)$ in D_Z, from the integral expression of solutions of the discontinuous Riemann-Hilbert problem for analytic functions, we can write the representation of the solution $\tilde{\Phi}(Z)$ of Problem A for analytic functions, namely

$$\tilde{\Phi}[Z(\zeta)]=\frac{X_0(\zeta)}{2\pi i}\left[\int_{\partial D_t}\frac{(t+\zeta)\tilde{\lambda}[Z(t)]\hat{R}[Z(t)]dt}{(t-\zeta)tX_0(t)}+Q(z)\right],$$

$$Q(z)=i\sum_{k=0}^{[\tilde{K}]}(c_k\zeta^k+\overline{c_k}\zeta^{-k})+\begin{cases}0, \text{ when } 2\tilde{K} \text{ is even},\\ ic_*\dfrac{\zeta_1+\zeta}{\zeta_1-\zeta},\, c_*=i\displaystyle\int_{\partial D_t}\frac{\tilde{\lambda}_n[Z(t)]\hat{R}[Z(t)]dt}{X_0(t)t},\\ \text{when } 2\tilde{K} \text{ is odd},\end{cases}$$

(see Chapter III, [86]33) and Chapter IV, [87]1)), where $X_0(\zeta)=\Pi_{j=1}^4(\zeta-t_l)^{\tau_l}$, $\tau_l\,(l=1,2,3,4)$ are as stated before, $Z=Z(\zeta)$ is the conformal mapping from the unit disk $D_\zeta=\{|\zeta|<1\}$ onto the domain D_Z such that the three points $\zeta=-1,i,1$ are mapped onto $Z=-1,Z_0(\in\Gamma\backslash\{-1,1\}),1$ respectively, and ζ_1 is a point on $|\zeta|=1$, if $\tilde{K}=1/2$, then the constants c_0,c_1 are determined by last two point conditions in the boundary condition. According to the result in [86]33),[87]1), we see that the function $\tilde{\Phi}(Z)$ determined by the above integral in $\overline{D}_Z\cap\{|Z\pm1|>\varepsilon(>0)\}$ is Hölder continuous.

1. For giving the estimates of $X(Z)u_x$, $X(Z)u_x$ in the neighborhood $\tilde{D}_j=D_j\cap D_Z$ of $t_j\,(j=1,2)$ separately, similarly to the proof of Theorem 2.4, we can locally handle the problem in $D_j\,(1\le j\le 2)$, and replace $X(Z)$ in (3.26) by the function $X_j(Z)=(Z-t_j)^{\eta_j}=\tilde{X}+i\tilde{Y}\,(1\le j\le 1)$. We first reduce the boundary condition such that $\tilde{\lambda}(z)=1$ on $\Gamma\cup\tilde{\Gamma}$ near $Z=t_j\,(1\le j\le 2)$. In fact we find an analytic function $S(Z)$ in $D_j\cup D'_Z$ satisfying the boundary conditions

$$\text{Re}S(Z)=-\arg\tilde{\lambda}(Z) \text{ on } \Gamma'=\Gamma\cup\tilde{\Gamma} \text{ near } t_j,\ \text{Im}S(t_j)=0,$$

and the estimate

$$C_\delta[S(Z),D_j\cap D'_Z]\le M_6=M_6(\delta,k,H,D,M_3)<\infty,$$

then the function $e^{jS(Z)}\tilde{W}(Z)$ satisfies the boundary condition

$$\text{Re}[e^{iS(Z)}\tilde{W}(Z)]=0 \text{ on } \Gamma\cup\tilde{\Gamma} \text{ near } Z=t_j\,(1\le j\le 2).$$

Next we symmetrically extend the function $\tilde{W}(Z)$ in D'_Z onto the symmetrical domain D^*_Z with respect to $\mathrm{Re}Z = t_j$ $(1 \le j \le 2)$, namely let

$$\hat{W}(Z) = \begin{cases} e^{iS(Z)}\tilde{W}(Z) \text{ in } D'_Z, \\ -\overline{e^{iS(Z^*)}\tilde{W}(Z^*)} \text{ in } D^*_Z, \end{cases}$$

where $Z^* = -\overline{(Z-t_j)}+t_j$, later on we shall omit the secondary part $e^{iS(Z)}$, and can get

$$C_\delta[\tilde{\Phi}(Z), D_\varepsilon] \le M_7, C_\delta[X_j(Z)u_x, D_\varepsilon] \le M_7, C_\delta[X_j(Z)u_y, D_\varepsilon] \le M_7,$$
$$C_\delta[u_x, D'_\varepsilon] \le M_8,\ C_\delta[u_y, D'_\varepsilon] \le M_8, \tag{3.29}$$

in which $D_\varepsilon = \overline{D_Z} \cap \{\mathrm{dist}(Z, \gamma) \ge \varepsilon\}$, $D'_\varepsilon = \overline{D_Z} \cap \{\mathrm{dist}(Z, \Gamma \cup \tilde{\Gamma} \cup \{0\}) \ge \varepsilon\}$, ε is arbitrary small positive constant, $M_7 = M_7(\delta, k, H, D_\varepsilon, M_3)$, $M_8 = M_8(\delta, k, H, D'_\varepsilon, M_3)$ are non-negative constants. In fact the first three estimates in (3.29) can be derived by the above integral representation of $\tilde{\Phi}(Z)$. Moreover there is no harm in assuming that $u_y = 0$ on γ in (3.5), because we can find a harmonic function $u_0(z)$ in D satisfying the boundary conditions $u_{0y} = r(x)$ on γ, which is twice continuously differentiable in $\overline{D}$, thus the function $v(z) = u(z) - u_0(z)$ satisfies the boundary condition $v_y = 0$ on γ, from (3.19) and (3.28), $W(Z) = [H_1(y)u_x - iH_2(x)u_y]/2 = \Phi^*(Z) + \Psi^*(Z)$ in D_Z, the integral $\Psi^*(Z) = 2i\mathrm{Im}Tg/H_1$ over D'_ε is bounded and Hölder continuous in D'_ε, $\Phi^*(Z)$ in D_Z is an analytic function. Noting that $\mathrm{Re}W(Z) = \mathrm{Re}\hat{\Phi}^*(Z) = H_1(y)u_x/2 = 0$ in $D'_\varepsilon \cap \{Y = 0\}$, we can extend the function $\Phi^*(Z)$ from D'_ε onto the symmetrical domain $\tilde{D}'_\varepsilon$ about the real axis $\mathrm{Im}Z = 0$, the extended function is denoted by $\Phi^*(Z)$ again, obviously $\mathrm{Re}\Phi^*(Z)$ is a harmonic function in D_Z, thus $\mathrm{Re}\Phi^*(Z) = \mathrm{Re}W(Z) = H_1(y)u_x/2 = YF_1$ herein F_1 is a Hölder continuous in D'_ε, and then $u_x = O(Y^{2/(2+m_1)}F_1)$ in D'_ε. Moreover the estimate of u_y can be derived as stated in Section 2. This shows that the last two estimates in (3.29) are true.

Now we can use the way as stated in Section 2, the solution $X_j(Z)W(Z)$ can be also expressed as $X_j(Z)W(Z) = \Phi(Z) + \Psi(Z) = \hat{\Phi}(Z) + \hat{\Psi}(Z)$ in $\hat{D}'_Z = \hat{D}_Z \cap \{Y > 0\}$, where $\hat{D}_Z = D'_Z \cup D^*_Z$, $X_j(Z) = (Z - t_j)^{\eta_j} = \tilde{X} + i\tilde{Y}$ $(1 \le j \le 2)$, $\Psi(Z), \hat{\Psi}(Z)$ in $\hat{D}'_Z$ are Hölder continuous, $\mathrm{Im}\Psi(Z) = 0$, $\mathrm{Re}\hat{\Psi}(Z) = 0$ in $\hat{D}'_Z$, $\Phi^*(Z) = \Phi(Z)$ and $\Phi^*(Z) = \hat{\Phi}(Z)$ are analytic functions in $\hat{D}'_Z$ satisfying the boundary conditions in the form

$$\mathrm{Re}[\overline{\lambda(Z)}\Phi^*(Z)] = 0 \text{ on } \Gamma \cup \tilde{\Gamma}, u(h_1) = 0, u(h_2) = 0, u(0) = 0.$$

From the boundary condition and $H_1(y)u_x/2 = 0$ on $\overline{\hat{D}_Z} \cap \{Y = 0\}$, we see that he function $\Phi^*(Z)$ in $D_j = \{|Z - t_j| < \varepsilon(> 0)\}$ is analytic, and $\Phi^*(t_j) = 0$, hence $\Phi^*(Z) = O(|Z - t_j|)$ near $Z = t_j$, it is clear that $\text{Im}\Phi(Z) = \text{Im}X_j(Z)W(Z)$ and $\text{Re}\hat{\Phi}(Z) = \text{Re}X_j(Z)W(Z)$ extended are harmonic functions in $\hat{D}'_Z$, and $\text{Re}\hat{\Phi}(Z)$, $\text{Im}\Phi(Z)$ can be expressed as

$$2\text{Re}\hat{\Phi}(Z) = \sum_{j,k=0}^{\infty} c_{jk}^{(1)} X^j Y^k, \ 2\text{Im}\Phi(Z) = \sum_{j,k=0}^{\infty} c_{jk}^{(2)} X^j Y^k$$

in D_j, herein $X = x - t_j$. Noting that

$$2\text{Re}\hat{\Phi}(Z) = \tilde{X}H_1(y)u_x + \tilde{Y}H_2(x)u_y = 0 \ \text{ on } \text{Im}Y = 0,$$
$$2\text{Im}\Phi(Z) = \tilde{Y}H_1(y)u_x - \tilde{X}H_2(x)u_y = 0 \ \text{ at } Z = t_j,$$

we have

$$2\text{Re}\hat{\Phi}(Z) = \tilde{X}H_1u_x + \tilde{Y}H_2u_y = YF_1,$$
$$2\text{Im}\Phi(Z) = \tilde{Y}H_1u_x - \tilde{X}H_2u_y = |Z - t_j|F_2$$

in $\tilde{D}_j = D_j \cap D_Z$, where F_1, F_2 are continuous functions in D_j. From the system of algebraic equations, we can solve u_x, u_y as follows

$$u_x = H_2(Y\tilde{X}F_1 + |Z - t_j|\tilde{Y}F_2)/H_1H_2|X(Z)|^2,$$
$$u_y = H_1(Y\tilde{Y}F_1 - |Z - t_j|\tilde{X}F_2)/H_1H_2|X(Z)|^2, \text{ i.e.}$$
$$X(Z)H_1u_x = (Y\tilde{X}F_1 + |Z - t_j|\tilde{Y}F_2)/\overline{X(Z)} = O(|Y|),$$
$$X(Z)H_2u_y = (Y\tilde{Y}F_1 - |Z - t_j|\tilde{X}F_2)/\overline{X(Z)} = O(|Z - t_j|), \text{ i.e.}$$
$$X(Z)u_x = O(|Y|^{2/(2+m_1)}), \ X(Z)u_y = O(|Z - t_j|), 1 \le j \le 2,$$

thus we have

$$\begin{gathered} C_\delta[X_j(Z)u_x, \tilde{D}_j] \le M_9, \ C_\delta[X_j(Z)u_y, \tilde{D}_j] \le M_9, \text{ i.e.} \\ C_\delta[X(Z)u_x, \tilde{D}_j] \le M_{10}, C_\delta[X(Z)u_y, \tilde{D}_j] \le M_{10}, 1 \le j \le 2, \end{gathered} \tag{3.30}$$

where $M_j = M_j(\delta, k, H, D, M_3)$ $(j = 9, 10)$ are non-negative constants.

2. For giving the estimate of $X(Z)u_x$, $X(Z)u_y$ in the neighborhood D_3 of $Z = t_3 = 0$, we choose $X_3(Z) = Z = X + iY$ to replace the function $X(Z)$ in (3.26). In this case, as stated in (3.19), the solution $X_3(Z)W(Z)$ can be also expressed as $X_3(Z)W(Z) = \Phi(Z) + \Psi(Z) = \hat{\Phi}(Z) + \hat{\Psi}(Z)$, where $\Psi(Z), \hat{\Psi}(Z)$ in D_Z are Hölder continuous, and $\Phi^*(Z) = \Phi(Z)$ and $\Phi^*(Z) =$

$\hat{\Phi}(Z)$ are analytic functions in D_Z satisfying the boundary conditions in the form

$$\mathrm{Re}[\overline{\tilde{\lambda}(z)}\Phi^*(Z)] = \hat{R}(z) \text{ on } \Gamma\cup\gamma,\ u(h_1)=0,$$
$$u(h_2)=0, u(0)=0 \text{ or } \mathrm{Im}[\overline{\hat{\lambda}(Z)}\Phi^*[Z(z)]|_{z=z_j}=b_j'', j=1,2,$$

where $\hat{R}(Z)=0$ on $\Gamma\cup\gamma$, $b_j''=|X[Z(z_j)]|H_1(\mathrm{Im}z_j)b_j$, $z_j(\in\Gamma\backslash\{h_1,h_2\}, j=1,2)$ are two points. For $t_3=0$, we have $\Phi^*(Z)=O(|Z|)$, $\Phi'^*(Z)=O(1)$ near $Z=t_3=0$, it is clear that $\mathrm{Re}\Phi^*(Z)$, $\mathrm{Im}\Phi^*(Z)$ extended are harmonic functions in $D_Z'=D_Z\cup\tilde{D}_Z\cup\gamma$, and $\mathrm{Re}\Phi^*(Z), \mathrm{Im}\Phi^*(Z)$ can be expressed as

$$\mathrm{Re}\Phi^*(Z)=\sum_{j,k=0}^{\infty}c_{jk}^{(1)}X^jY^k,\ \mathrm{Im}\Phi^*(Z)=\sum_{j,k=0}^{\infty}c_{jk}^{(2)}X^jY^k \text{ in } D_3.$$

Noting that

$$\mathrm{Re}\hat{\Phi}(Z)=XH_1(y)u_x+YH_2(x)u_y=0$$
$$\text{on } \mathrm{Im}Z=0 \text{ and } \mathrm{Re}Z=X=0,$$
$$\mathrm{Im}\Phi(Z)=YH_1(y)u_x-XH_2(x)u_y=0 \text{ at } Z=t_j,$$

we have

$$\mathrm{Re}\hat{\Phi}(Z)=XH_1u_x+YH_2u_y=XY\sum_{j,k=1}^{\infty}c_{jk}^{(1)}X^jY^{k-1}=XYF_1,$$
$$\mathrm{Im}\Phi(Z)=YH_1u_x-XH_2u_y=|Z-t_j|F_2 \text{ in } D_3.$$

From the system of algebraic equations, we can solve u_x, u_y as follows

$$u_x=YH_2(|Z-t_j|F_2+X^2F_1)/H_1H_2|Z|^2,$$
$$u_y=XH_1(Y^2F_1-|Z-t_j|F_2)/H_1H_2|Z|^2, \text{ i.e.}$$
$$Zu_x=Y(|Z-t_j|F_2+X^2F_1)/\overline{Z}H_1=O(|Y|^{2/(2+m_1)}),$$
$$Zu_y=X(Y^2F_1-|Z-t_j|F_2)/\overline{Z}H_2=O(|X|^{2/(2+m_2)}).$$

Now we explain that from the above estimates and the last condition in (3.2), we can obtain $\hat{\Psi}(z)$, $\Psi(Z)\in C_\beta(\overline{D_Z})$, here $\beta=\min_{l=1,2}(2,m_l)/(2+m_l)-2\delta$. Moreover we can derive that u_x, u_y satisfy the estimates

$$C_\delta[X_3(Z)u_x,\tilde{D}_3]\le M_{11},\ C_\delta[X_3(Z)u_y,\tilde{D}_3]\le M_{11},$$

and then

$$C_\delta[X(Z)u_x,\tilde{D}_3]\le M_{12},\ C_\delta[X(Z)u_y,\tilde{D}_3]\le M_{12}, \tag{3.31}$$

in which $X(Z)$, δ are as stated in (3.26), (3.27), and $M_j = M_j(\delta,k,H,D,M_3)$ $(j=11,12)$ are non-negative constants.

3. Now we shall give the estimate in the neighborhood D_4 of $t_4=iY_0=iY(y_0)$. As stated before there is no harm in assuming that the intersectional angle between Γ and $\text{Re}z=0$ is equal to $\pi/2$ and Γ includes the line segment $\text{Im}z=0$ near $z=iy_0$. We can extend the domain D_Z onto the symmetric domain $\tilde{D}_Z$. In this case, the problem in D_4 is discussed as the inner part of the domain extended. Similarly to the case 1, the corresponding analytic function $\Phi(Z)$ satisfies the Hölder continuous condition, and then the function u_x,u_y in D_4 is also Hölder continuous. Thus we can derive the estimate in D_4 similar to (3.31).

Next we use the reduction to absurdity, suppose that (3.28) is not true, then there exist sequences of coefficients $\{A_j^{(m)}\}\,(j=1,2,3,4)$, $\{\lambda^{(m)}\}$, $\{r^{(m)}\}$ and $\{b_j^{(m)}\}\,(j=0,1)$, which satisfy the same conditions of coefficients as stated in (3.2), (3.6), such that $\{A_j^{(m)}\}\,(j=1,2,3,4)$, $\{\lambda^{(m)}\}$, $\{r^{(m)}\}$, $\{b_j^{(m)}\}\,(j=0,1)$ in $\overline{D}$, Γ, γ^* weakly converge or uniformly converge to $A_j^{(0)}\,(j=1,2,3,4)$, $\lambda^{(0)}$, $r^{(0)}$ respectively, and $b_j^{(0)}\,(j=0,1)$, and the solutions of the corresponding boundary value problems

$$W_{\overline{Z}}^{(m)}=F^{(m)}(z,u^{(m)},W^{(m)}),\,F^{(m)}(z,u^{(m)},W^{(m)})$$

$$=A_1^{(m)}W^{(m)}+A_2^{(m)}\overline{W^{(m)}}+A_3^{(m)}u^{(m)}+A_4^{(m)}\text{ in }\overline{D},$$

$$\text{Re}[\overline{\lambda^{(m)}(z)}W(z)]=R^{(m)}(z)\text{ on }\Gamma\cup\gamma^*,$$

$$u^{(m)}(h_1)=b_0^{(m)},\,u^{(m)}(h_2)=b_1^{(m)},\text{ or}$$

$$\text{Im}[\overline{\lambda^{(m)}(z)}W(z)]|_{z=z_1}=b_1'^{(m)},$$

and

$$u^{(m)}(z)=u^{(m)}(x)-2\int_0^y\frac{V^{(m)}}{H_2(x)}dy$$

$$=2\text{Re}\int_{h_1}^z\left[\frac{\text{Re}W^{(m)}}{H_1(y)}+i\frac{\text{Im}W^{(m)}}{H_2(x)}\right]dz+b_0^{(m)}\text{ in }\overline{D}$$

have the solutions $[W^{(m)}(z),u^{(m)}(z)]$, but $\hat{C}[W^{(m)}(z),\overline{D}]$ $(m=1,2,...)$ are unbounded, hence we can choose a subsequence of $[W^{(m)}(z),u^{(m)}(z)]$ denoted by $[W^{(m)}(z), u^{(m)}(z)]$ again, such that $h_m = \hat{C}[W^{(m)}(z), \overline{D}] \to \infty$ as $m \to \infty$, we can assume $h_m \geq \max[k_1, k_2, 1]$. It is obvious that $[\tilde{W}^{(m)}(z), \tilde{u}^{(m)}(z)_m] = [W^{(m)}(z)/h_m,\ u^{(m)}(z)_m/h_m]$ are solutions of the boundary value problems

$$\begin{aligned}&\tilde{W}^{(m)}_{\overline{Z}} = \tilde{F}^{(m)}(z, \tilde{u}^{(m)}, \tilde{W}^{(m)}),\ \tilde{F}^{(m)}(z, \tilde{u}^{(m)}, \tilde{W}^{(m)})\\ &= A_1^{(m)}\tilde{W}^{(m)} + A_2^{(m)}\overline{\tilde{W}^{(m)}} + A_3^{(m)}\tilde{u}^{(m)} + A_4^{(m)}/h_m \text{ in } \overline{D_Z},\end{aligned} \tag{3.32}$$

$$\begin{aligned}&\text{Re}[\overline{\lambda^{(m)}(z)}\tilde{W}^{(m)}(z)] = R^{(m)}(z)/h_m \text{ on } \Gamma \cup \gamma^*,\\ &\tilde{u}^{(m)}(h_1) = b_0^{(m)}/h_m,\ \tilde{u}^{(m)}(h_2) = b_1^{(m)}/h_m, \text{ or}\\ &\text{Im}[\overline{\lambda^{(m)}(z)}\tilde{W}^{(m)}(z)]|_{z=z_1} = b_1'^{(m)}/h_m,\end{aligned} \tag{3.33}$$

and

$$\begin{aligned}\tilde{u}^{(m)}(z) &= \frac{u^{(m)}(x)}{h_m} - 2\int_0^y \frac{\tilde{V}^{(m)}}{H_2(x)}dy\\ &= 2\text{Re}\int_{h_1}^z \left[\frac{\text{Re}\tilde{W}^{(m)}}{H_1(y)} + i\frac{\text{Im}\tilde{W}^{(m)}}{H_2(x)}\right]dz + \frac{b_0^{(m)}}{h_m} \text{ in } \overline{D}.\end{aligned} \tag{3.34}$$

We can see that the functions in above boundary value problems satisfy the conditions

$$\begin{aligned}&L_\infty[H_1\text{Re}(A_1+A_2),\overline{D}],\ L_\infty[x\text{Re}(A_1-A_2),\overline{D}] \leq k_3,\\ &L_\infty[y\text{Im}(A_1+A_2),\overline{D}],\ L_\infty[H_2\text{Im}(A_1-A_2),\overline{D}] \leq k_3,\\ &L_\infty[A_3(z),\overline{D}] \leq k_3,\ \ L_\infty[A_4(z)/h_m,\overline{D}] \leq 1,\\ &C_\alpha[\lambda^{(m)}(z),\Gamma] \leq k_3,\ \ C_\alpha[r^{(m)}(z)/h_m,\Gamma] \leq 1,\\ &C_\alpha[r^{(m)}(x)/h_m,\gamma] \leq 1,\ |b_j^{(m)}/h_m| \leq 1,\ j=0,1,\end{aligned}$$

where $k_3 = k_3(\delta, k, H, D)$ is a positive constant. From the representation (3.18), the above solutions can be expressed as

$$\begin{aligned}\tilde{u}^{(m)}(z) &= \frac{u^{(m)}(x)}{h_m} - 2\int_0^y \frac{\tilde{V}^{(m)}}{H_2(x)}dy\\ &= 2\text{Re}\int_{h_1}^z \left[\frac{\text{Re}\tilde{W}^{(m)}}{H_1(y)} + i\frac{\text{Im}\tilde{W}^{(m)}}{H_2(y)}\right]dz + \frac{b_0^{(m)}}{h_m} \text{ in } \overline{D},\end{aligned}$$

$$
\begin{aligned}
&\tilde{W}^{(m)}(z) = \tilde{\Phi}^{(m)}(t) + \tilde{\Psi}^{(m)}(t),\\
&\tilde{\Psi}^{(m)}(Z) = -\mathrm{Re}\frac{2}{\pi}\iint_{D_Z}\frac{f^{(m)}(t)}{t-Z}d\sigma_t,\\
&\tilde{W}^{(m)}(z) = \hat{\Phi}^{(m)}(t) + \hat{\Psi}^{(m)}(t),\\
&\hat{\Psi}^{(m)}(Z) = -\mathrm{Im}\frac{2i}{\pi}\iint_{D_Z}\frac{f^{(m)}(t)}{t-Z}d\sigma_t \text{ in } \overline{D_Z},
\end{aligned} \tag{3.35}
$$

Similarly to the proof of (3.28) and (3.31), and notice that $|x|^{\tau_2}y^{\tau_1}H_1H_2$ $\tilde{f}^{(m)}(Z) = |x|^{\tau_2}y^{\tau_1}X(Z)g^{(m)}(Z) \in L_\infty(D_Z)$, $\tau_j = \max(0, 1-m_j/2), j = 1, 2$, we can verify that

$$
\begin{aligned}
&C_\beta[2\mathrm{Re}T(\tilde{f}^{(m)}(Z)), \overline{D_Z}] \le M_{13},\\
&\mathrm{Re}[T(\tilde{f}^{(m)}(Z)) - T(\tilde{f}^{(m)}(Z))|_{Z=t_j}] = O(|Z-t_j|^{\beta_j}), j=1,2,
\end{aligned} \tag{3.36}
$$

where β is as stated in (3.15), $\beta_j = \min[2/(m_1+2), m_1/(m_1+2)] - 2\delta$ $(j = 1, 2)$, $M_{13} = M_{13}(\delta, k, H, D)$ is a non-negative constant, and we can obtain the estimate

$$
\hat{C}_\delta[\tilde{W}^{(m)}(Z), \overline{D_Z}] \le M_{14} = M_{14}(\delta, k, H, D) < \infty. \tag{3.37}
$$

Hence from $\{X(Z)[\mathrm{Re}\tilde{W}^{(m)}(z)/H_1 + i\mathrm{Im}\tilde{W}^{(m)}(z)/H_2]\}$ and the sequence of corresponding functions $\{\tilde{u}^{(m)}(z)\}$, we can choose the subsequences denoted by $\{X(Z)[\mathrm{Re}\tilde{W}^{(m)}(z)/H_1 + i\mathrm{Im}\tilde{W}^{(m)}(z)/H_2]\}$, $\{\tilde{u}^{(m)}(z)\}$ again, which uniformly converge to $X(Z)[\mathrm{Re}\tilde{W}^{(0)}(z)/H_1 + i\mathrm{Im}\tilde{W}^{(0)}(z)/H_2]$, $\tilde{u}^{(0)}(z)$ respectively, it is clear that $[\tilde{W}^{(0)}(z), \tilde{u}^{(0)}(z)]$ is a solution of the homogeneous problem of Problem A. On the basis of Theorem 3.2, the solution $\tilde{W}^{(0)}(z) = 0$, $\tilde{u}^{(0)}(z) = 0$ in $\overline{D}$, however, from $\hat{C}[\tilde{W}^{(m)}(z), \overline{D}] = 1$, we can derive that there exists a point $z^* \in \overline{D}$, such that $\hat{C}[\tilde{W}^{(0)}(z^*), \overline{D}] = 1$, it is impossible. This shows that (3.28) and the first estimate in (3.27) are true. Moreover we can verify the second estimate in (3.27) by using the similar method in the proof of Theorem 2.4, Chapter I.

3.4 Existence of solutions of oblique derivative problem for elliptic equations

In this section, we prove the existence of solutions of Problem P for equation (3.1). Firstly we discuss the complex equation

$$
W_{\overline{Z}} = \frac{1}{H_1H_2}[A_1(z)W + A_2(z)\overline{W} + A_3(z)u + A_4(z)] = \frac{g(Z)}{H_1H_2} \text{ in } D, \tag{3.38}
$$

with the relation

$$u(z) = 2\mathrm{Re}\int_{h_1}^{z}\left[\frac{\mathrm{Re}W(z)}{H_1(y)} + \frac{i\mathrm{Im}W(z)}{H_2(y)}\right]dz + b_0 \text{ in } \overline{D}, \tag{3.39}$$

where $H_1(y)$, $H_2(x)$ are as stated in (3.1), and the coefficients in (3.38) satisfy the corresponding conditions as those in (3.2), i.e.

$$\begin{gathered}L_\infty[H_1\mathrm{Re}(A_1+A_2),\overline{D}], L_\infty[x\mathrm{Re}(A_1-A_2),\overline{D}]\le k_3,\\ L_\infty[y\mathrm{Im}(A_1+A_2),\overline{D}], L_\infty[H_2\mathrm{Im}(A_1-A_2),\overline{D}], L_\infty[A_j,\overline{D}] \le k_3, j=3,4,\end{gathered} \tag{3.40}$$

in which $k_3 = k_3(\delta, k, H, D)$ is a non-negative constant, and the boundary value problem (3.38) with the boundary conditions

$$\begin{gathered}\mathrm{Re}[\overline{\lambda(z)}W(z)] = R(z) \text{ on } \Gamma\cup\gamma,\ u(h_1) = b_0,\\ u(h_2) = b_1 \text{ or } \mathrm{Im}[\overline{\lambda(z_1)}W(z_1)] = b_1',\end{gathered} \tag{3.41}$$

is called Problem A, where $\lambda(z), r(z), b_0, b_1$ are as stated in (3.5), (3.6), and $R(z) = 0$ on Γ, $b_0 = b_1 = 0$.

Theorem 3.4 *Let equation* (3.1) *satisfy Condition C. Then the oblique derivative problem* (*Problem P*) *for* (3.1) *has a unique solution.*

Proof We first write the complex equation (3.38) in the form

$$W_{\overline{Z}}=F(z,u,W), F=[A_1(z)W+A_2(z)\overline{W}+A_3(z)u+A_4(z)]/H_1H_2 \text{ in } \overline{D}, \tag{3.42}$$

where the coefficients satisfy (3.40). In order to find a solution $W(z)$ of Problem A for (3.42) in D, we can express $[W(z), u(z)]$ in the form (3.18), (3.19). In the following, we use the Leray-Schauder theorem to prove that Problem A for equation (3.42) has a unique solution. We consider the equation and boundary conditions with the parameter $t \in [0, 1]$:

$$W_{\overline{Z}} - tF(z, u, W) = 0 \text{ in } \overline{D_Z}, \tag{3.43}$$

and introduce a bounded open set B_M of the Banach space $B = \hat{C}_\delta(\overline{D_Z})$, whose elements are functions $w(z)$ satisfying the condition

$$w(Z) \in \hat{C}_\delta(\overline{D}),\quad \hat{C}_\delta[w(Z), \overline{D_Z}] < M_{15} = 1 + M_1, \tag{3.44}$$

where δ, M_1 are constants as stated in (3.27). Choose an arbitrary function $w(Z) \in B_M$ and substitute it into the position of W in $F(Z, u, W)$, by

Theorem 3.1, we can find a solution $w(z)=\Phi(Z)+\Psi(Z)=w_0(Z)+T(tF)$ of Problem A for the complex equation

$$W_{\overline{Z}}=tF(z,u,W). \tag{3.45}$$

Noting that $|x|^{\tau_2}y^{\tau_1}X(Z)H_1H_2F[z(Z),u(z(Z)),w(z(Z))]\in L_\infty(\overline{D_Z})$, where $\tau_1=\max(0,1-m_1/2)$, $\tau_2=\max(0,1-m_2/2)$, the above solution of Problem A for (3.45) is unique. Denote by $W(z)=T[w,t]$ $(0\le t\le 1)$ the mapping from $w(z)$ to $W(z)$. From Theorem 3.3, we know that if $W(z)$ is a solution of Problem A for the equation

$$W_{\overline{Z}}=tF(Z,u,W) \text{ in } D_Z, \tag{3.46}$$

then the function $W(Z)$ satisfies the estimate

$$\hat{C}_\delta[W(Z),\overline{D_Z}]<M_{15}. \tag{3.47}$$

Set $B_0=B_M\times[0,1]$. Now we verify the three conditions of the Leray-Schauder theorem:

1. For every $t\in[0,1]$, $T[w,t]$ continuously maps the Banach space B into itself, and is completely continuous on B_M. In fact, we can arbitrarily select a sequence $w_n(z)$ in B_M, $n=0,1,2,\ldots$, such that $\hat{C}_\delta[w_n-w_0,\overline{D_Z}]\to 0$ as $n\to\infty$. By Condition C, it is easy to see that $L_\infty[|x|^{\tau_2}y^{\tau_1}X(Z)H_1(y)H_2(x)(F(z,u_n,w_n)-F(z,u_0,w_0)),\overline{D}]\to 0$ as $n\to\infty$. Moreover, from $W=T[w_n,t]$, $W_0=T[w_0,t]$, it is easy to see that W_n-W_0 is a solution of Problem A for the complex equation

$$(W_n-W_0)_{\overline{Z}}=t[F(z,u_n,w_n)-F(z,u_0,w_0)] \text{ in } D_Z, \tag{3.48}$$

and then we can obtain the estimate

$$\hat{C}_\delta[W_n-W_0,\overline{D_Z})]\le 2k_0\hat{C}[w_n-w_0,\overline{D_Z}]. \tag{3.49}$$

Hence $\hat{C}_\delta[W-W_0,\overline{D_Z}]\to 0$ as $n\to\infty$. Moreover for $w_n(z)\in B_M$, $n=1,2,\ldots$, we can choose a subsequence $\{w_{n_k}(z)\}$ of $\{w_n(z)\}$, such that $\hat{C}[w_{n_k}-w_0,\overline{D_Z}]\to 0$ as $k\to\infty$, where $w_0(z)\in B_M$. Let $W_{n_k}=T[w_{n_k},t]$, $W_0=T[w_0,t]$, similarly we have (3.49) with $n=n_k, k=1,2,\ldots$, hence $C_\delta[W_{n_k}-W_0,\overline{D_Z}]\to 0$ as $k\to\infty$. This shows that $W=T[w,t]$ is completely continuous in B_M. Applying the similar method, we can also prove that for $w(Z)\in B_M$, $T[w,t)$ is uniformly continuous with respect to $t\in[0,1]$.

2. For $t=0$, it is evident that $W=T[w,0]=\Phi(Z)\in B_M$.

3. From the estimate (3.27), we see that $W = T[w,t]\,(0 \le t \le 1)$ does not have a solution $w(z)$ on the boundary $\partial B_M = \overline{B_M}\backslash B_M$.

Hence by the Leray-Schauder theorem, we know that there exists a function $W(z) \in B_M$, such that $W(z) = T[W(z),1]$, and the function $W(z) \in \hat{C}_\delta(\overline{D_Z})$ is just a solution of Problem A for the complex equation (3.42), and then the solution of Problem P for equation (3.1) is found.

Next we can consider second order quasilinear elliptic equation with parabolic degeneracy

$$K_1(y)u_{xx}+|K_2(x)|u_{yy}+au_x+bu_y+cu+d=0 \text{ in } D, \tag{3.50}$$

where $K_1(y), K_2(x)$ are as stated in (3.1), and a,b,c,d are functions of $z(\in D), u, u_x, u_y\,(\in \mathbf{R})$, its complex form is the following complex equation of second order

$$\begin{gathered}u_{z\bar{z}}-iH_{1y}H_2(x)u_x/4+iH_1(y)H_{2x}u_y/4=F(z,u,u_z),\\ F(z,u,u_z)=\text{Re}[B_1u_{\bar{z}}]+B_2u+B_3 \text{ in } D,\end{gathered} \tag{3.51}$$

where $B_j = B_j(z,u,u_z)\,(j=1,2,3)$ and

$$\begin{aligned}u_{\bar{z}}&=[H_1(y)u_x-iH_2(x)u_y]/2,\, u_{z\bar{z}}=[H_1(y)(u_{\bar{z}})_x+iH_2(x)(u_{\bar{z}})_y]/2\\ &=[K_1(y)u_{xx}+|K_2(x)|u_{yy}]/4+iH_2H_{1y}u_x/4-iH_1H_{2x}u_y/4,\\ B_1&=-\frac{1}{2}[\frac{a}{H_1}+i\frac{b}{H_2}],\, B_2=-\frac{c}{4},\, B_3=-\frac{d}{4} \text{ in } D.\end{aligned} \tag{3.52}$$

This boundary value problem (3.51), (3.5) will be called Problem P.

Suppose that equation (3.50) satisfies the following conditions, namely **Condition C**

1) For any continuously differentiable function $u(z)$ in $D^* = \overline{D}\backslash\{-1,0,1\}$, $B_j(z,u,u_z)$ $(j=1,2,3)$ are measurable in D and satisfy

$$\begin{gathered}L_\infty[H_1\text{Re}B_1,\overline{D}], L_\infty[H_2\text{Im}B_1,\overline{D}], L_\infty[B_2,\overline{D}]\le k_0,\\ L_\infty[B_3,\overline{D}]\le k_1,\ B_2\ge 0 \text{ in } D.\end{gathered} \tag{3.53}$$

2) For any continuously differentiable functions $u_1(z), u_2(z)$ in D^*, the equality

$$F(z,u_1,u_{1z})-F(z,u_2,u_{2z})=\text{Re}[\tilde{B}_1(u_1-u_2)_z]+\tilde{B}_2(u_1-u_2) \text{ in } D \tag{3.54}$$

holds, where $\tilde{B}_j = \tilde{B}_j(z, u_1, u_2)$ $(j = 1, 2)$ satisfy the conditions

$$L_\infty[H_1 \mathrm{Re}\tilde{B}_1, \overline{D}], L_\infty[H_2 \mathrm{Im}\tilde{B}_1, \overline{D}], L_\infty[\tilde{B}_2, \overline{D}] \le k_0, \tilde{B}_2 \ge 0 \text{ in } D, \tag{3.55}$$

in (3.53),(3.55), k_0, k_1 are non-negative constants. In particular, when (3.50) is a linear equation, the condition (3.54) obviously holds.

By using the similar method, we can prove the following theorem.

Theorem 3.5 *Suppose that equation* (3.50) *satisfies Condition C. Then the oblique derivative problem (Problem P) for* (3.50) *with the boundary condition* (3.5) *has a solution.*

Proof We can use the method of parameter extension to prove the unique solvability of Problem P for equation (3.51). Firstly we consider the linear equation and boundary conditions with the parameter $t \in [0, 1]$:

$$\begin{aligned} &u_{\bar{z}\bar{z}} - \frac{i}{4} H_{1y} H_2(x) u_x + \frac{i}{4} H_1(y) H_{2x} u_y \\ &= \mathrm{Re}[B_1 u_z] + t B_2 u + B_3 \text{ in } D, \end{aligned} \tag{3.56}$$

and the boundary condition (3.5), the above boundary value problem is called Problem P_t. Next we can find a solution $u(z)$ of Problem P_0. Moreover from the solvability of Problem P_0 and the estimate of solution of Problem P_t $(0 \le t \le 1)$, we can obtain the solvability of Problem P_1 by the method of parameter extension. As for Problem P for the quasilinear equation (3.50), we can prove its solvability by the Schauder fixed point theorem.

4 Boundary Value Problems for Homogeneous Elliptic Equations of Second Order with Degenerate Rank 0

4.1 Boundary value problems for homogeneous elliptic equations of second order with degenerate rank 0

In this section, we first discuss the second order linear elliptic equation with degenerate rank 0:

$$Lu = y^{m_1} u_{xx} + y^{m_2} u_{yy} + a u_x + b u_y + c u + d = 0 \text{ in } D, \tag{4.1}$$

where the domain D is a bounded domain with the boundary $\partial D = \Gamma \cup \gamma$, herein $-1, 1$ are the end points of $\Gamma (\in C^2_\mu, 0 < \mu < 1)$ in $\{y > 0\}$, $\gamma =$

$\{-1 < x < 1, y = 0\}$, m_1 and m_2 are positive constants, $C_\alpha[\eta, \bar{D}] \leq k_0 < \infty$, $\eta = a, b, c, d$, $0 < \alpha < 1$, $c \leq 0$ in D. The complex form of (4.1) is

$$\begin{cases} u_{z\bar{z}} = \mathrm{Re}[Qu_{zz} + A_1 u_z] + A_2 u + A_3, \\ Q = -\dfrac{y^{m_1} - y^{m_2}}{y^{m_1} + y^{m_2}}, \ A_1 = -\dfrac{a + ib}{y^{m_1} + y^{m_2}}, \\ A_2 = -\dfrac{c}{2(y^{m_1}+y^{m_2})}, A_3 = -\dfrac{d}{2(y^{m_1}+y^{m_2})}. \end{cases}$$

The so-called mixed boundary value problem (Problem M) is to find a continuously differentiable solution $u(z)$ of (4.1) in $D^* = \overline{D}\backslash\{-1, 1\}$ satisfying the boundary conditions

$$\begin{aligned} lu = \frac{1}{2}\frac{\partial u}{\partial \nu} + \sigma(z)u(z) &= \frac{1}{H_1(y)}\mathrm{Re}[\overline{\lambda(z)}u_{\bar{z}}] + \sigma(z)u(z) \\ = \mathrm{Re}[\overline{\Lambda(z)}u_z] + \sigma(z)u(z) &= \tau(z), \ z \in \Gamma, \ u(x) = r(x), \ x \in \gamma, \end{aligned} \tag{4.2}$$

where $H_1(y) = y^{m_1/2}$, ν is a direction at every point on Γ, $\lambda = \cos(\nu, x) - i\cos(\nu, y)$. We assume that the known functions $\lambda(z), \sigma(z), \tau(z)$, $r(x)$ and the real constants $b_0 = r(-1), b_1 = r(1)$ satisfy the conditions

$$\begin{gathered} C_\alpha[\eta, \Gamma] \leq k_0, \ \eta = \lambda, \sigma, \ C_\alpha[\tau, \Gamma] \leq k_1, \\ C_\alpha[r(x), \gamma] \leq k_1, \cos(\nu, n) \geq 0 \text{ on } \Gamma, \ |b_0|, |b_1| \leq k_1, \end{gathered} \tag{4.3}$$

in which $\alpha\,(0 < \alpha < 1)$, k_0, k_1 are non-negative constants, and n is the outward normal direction on Γ. Problem M with the conditions: $\cos(\nu, n) = 0, \sigma(z) = 0$ on Γ is called the Dirichlet problem (Problem M_1), because if $\cos(\nu, n) = 0$ and $\sigma(z) = 0$ on Γ, then from the first formula in (4.2), we can derive

$$u(z) = 2\,\mathrm{Re}\int_{-1}^{z} u_z dz + b_0 = r(z) \ \text{ on } \Gamma,$$

$$u(1) = 2\,\mathrm{Re}\int_{-1}^{1} u_z dz + b_0 = b_1.$$

Besides we consider the case $\cos(\nu, n) > 0$ on Γ and assume $\sigma(z) \geq \sigma_0 > 0$ on Γ, where σ_0 is a positive constant, the boundary value problem is called Problem M_2. If $\tau(z) = 0$ on Γ, we can only require $\sigma(z) \geq 0$ on Γ, the boundary value problem is called Problem M_3. For Problems M_2 and M_3, we assume that $(\nu, y) < \pi/2$ at the corner points $z = -1, 1$ of $\partial D = \Gamma \cup \gamma$.

If the last condition in (4.2) is replaced by the boundedness of the solution $u(z)$ in $\overline{D}$, then the boundary value problem is called Problem E. In the following we shall discuss the case of $d=0$.

Lemma 4.1 *If there exists a real valued function $V(z)$ in $\overline{D}$, which is twice continuously differentiable in a neighborhood of any point $z_0=x_0\in\gamma$ and satisfies the conditions*

(A) $V(x_0)=0$;

(B) $V(z)>0$ *for* $z\neq x_0$;

(C) $LV\le -c_0<0$ in $D_\delta=\overline{D}\cap\{0\le y\le\delta\}$;

(D) $lV\ge c_0>0$ in $U=\overline{D}\cap\{|x-x_0|\le\delta, 0\le y\le\delta\}$, $x_0=\pm 1$ *for Problems M_2 and M_3; where δ is a sufficiently small positive constant, then Problem M for equation* (4.1) *has a unique classical solution $u(z)$.*

Proof We first continuously extend $r(z)$, $\sigma(z)$ in the closure $\overline{D}$ of the domain D such that $C_\alpha[r,\overline{D}]\le k_1+1$, $C_\alpha[\sigma,\overline{D}]\le k_0+1$, $\sigma(z)\ge\sigma_0/2>0$ in $\overline{D}$ for Problem M_2 and $\sigma(z)\ge 0$ in $\overline{D}$ for Problem M_3, and choose a sequence of equations

$$Lu=(y+1/n)^{m_1}u_{xx}+(y+1/n)^{m_2}u_{yy}+au_x+bu_y+cu=0 \text{ in } D, \tag{4.4}$$

similar to (2.25), where $n\,(>1)$ is an integer. Obviously the above equation is uniformly elliptic in $\overline{D}$. According to the proof of Theorem 1.6, we can prove that there exists a solution $u_n(z)$ of equation (4.4) in D, which satisfies the boundary conditions:

$$\operatorname{Re}[\overline{\Lambda(z)}u_{nz}]+\sigma(z)u_n(z)=\tau(z),\, z\in\Gamma,\ u_n(x)=r(x),\ \ x\in\gamma.$$

By using the extremum principle of solutions of equation (4.4), the positive maximum or negative minimum of $u_n(z)$ in $\overline{D}$ cannot attain at a point in D. For Problem M_1, it is clear that $C[u_n(z),\overline{D}]\le\max_{\partial D}|r(z)|$. For Problem M_3, the positive maximum or negative minimum of $u_n(z)$ in $\overline{D}$ cannot attain at a point in Γ, hence $C[u_n(z),\overline{D}]\le\max_\gamma|r(x)|$. For Problem M_2, if the positive maximum or negative minimum of $u_n(z)$ in $\overline{D}$ attains at a point in Γ, then $\partial u_n^2/4\partial\nu=\tau(z)u_n-\sigma(z)u_n^2\ge 0$ at the extremum point, hence $C[u_n(z),\overline{D}]\le\max\{\max_\gamma|r(x)|,\max_\Gamma|\tau(z)|/\sigma_0\}$. The situations show that the solution $u_n(z)$ satisfies the estimate

$$C[u_n(z),\overline{D}]\le M_1=M_1(\alpha,k_0,k_1,\sigma_0,D). \tag{4.5}$$

Denote $D_{\delta_0}=\overline{D}\cap\{y\ge\delta_0\}$, δ_0 is a small positive constant, as stated in [23]1), or Chapter VI, [86]9), let $P(z,t)$ be the Poisson kernel of the

Dirichlet problem for equation (4.4) in D_{δ_0} with the boundary condition $u_n(z) = u_0(z)$ on ∂D_{δ_0}, where $u_0(z)$ is a continuously differentiable function in $\overline{D}$. Then the solution $u_n(z)$ of Problem D for equation (4.4) can be expressed as

$$u_n(z) = \frac{1}{2\pi}\int_{\partial D_{\delta_0}} P(z,t)\psi(t)ds.$$

From (4.2), (4.5) and the above formula, we can obtain that the solution $u_n(z)$ satisfies the estimate

$$C^1_\delta[u_n, D_{2\delta_0}] \le M_2 = M_2(\alpha, k_0, k_1, \sigma_0, D_{2\delta_0}),$$

where δ is a sufficiently small positive constant, M_2 is a non-negative constant dependent of δ_0. Hence we can choose a subsequence of $\{u_n(z)\}$, which uniformly converges to a continuous function $u_0(z)$ in $D_{2\delta_0}$. Noting the arbitrariness of δ_0, there exists a subsequence of $\{u_n(z)\}$, which converges to the continuous function $u_0(z)$ in $\overline{D}\backslash\gamma$. We can prove that $u_0(z)$ is a solution of equation (4.1) in D and satisfies the boundary condition (4.2) on Γ. It remains to prove that $u_0(z)$ satisfies the boundary condition (4.2) on γ. We choose an arbitrary point $x_0 \in [-1,1]$ such that $|r(z)-r(x_0)| < \beta$ in $U = \overline{D}\cap\{|x-x_0| \le \delta, 0 \le y \le \delta\}$, where β is an arbitrary small positive number, and then introduce the function

$$W^{\pm}(z) = hV(z) + \beta \pm r(x_0) \mp u_n(z),$$

in which h and β are undetermined positive constants. It is evident that

$$LW^{\pm} = hLV + c[\beta \pm r(x_0)] < 0 \ \text{ in } U,$$

when the constant h is large enough. This shows that $W^{\pm}(z)$ cannot attain a negative minimum in U. Moreover, it is not difficult to see that

$$W^{\pm}(z) \ge hV(z) + \beta - |r(z) - r(x_0)| > 0 \ \text{ in } \partial D\cap\overline{U},$$

and

$$W^{\pm}(z) \ge hV(z) + \beta \pm r(x_0) - M_1 > 0 \ \text{ in } \partial U\backslash\partial D,$$

for the sufficiently large constant h. Hence

$$W^{\pm}(z) = hV(z) + \beta \pm r(x_0) \mp u_n(z) \ge 0 \ \text{ in } U.$$

It follows that

$$|u_n(z) - r(x_0)| \le hV(z) + \beta \ \text{ in } U.$$

Letting n tend to ∞ and z tend to x_0, we obtain

$$\overline{\lim_{n\to\infty}}\ \overline{\lim_{z\to x_0}} |u_n(z) - r(x_0)| \le \beta.$$

Noting the arbitrariness of β, the above formula implies $u_0(x_0) = r(x_0)$ (see [47]). As for Problems M_2 and M_3, we have

$$\begin{aligned} lW^{\pm}(z) &= \frac{1}{2}\frac{\partial W^{\pm}(z)}{\partial\nu} + \sigma(z)W^{\pm}(z) \\ &= hlV \mp \frac{1}{2}\frac{u_n(z)}{\partial\nu} + \sigma[\beta \pm r(x_0) \mp u_n(z)] \\ &= hlV \mp \tau(z) + \sigma[\beta \pm r(x_0)] > 0 \text{ in } U, \end{aligned}$$

where ν is continuously extended in U appropriately, provided that h is chosen large enough, we can get

$$W^{\pm}(z) = hV(z) + \beta \pm r(x_0) \mp u_n(z) \geq 0 \text{ in } U,$$

similarly to before, the required result can be derived.

The uniqueness of solutions of Problem M can be derived by the extremum principle.

Lemma 4.2 *If there exists a function $V(z) \in C(\overline{D}) \cap C^2(D)$, such that*

(A') $V(z) \geq 0$ in $\overline{D}$;

(B') $\lim_{y\to+0} V(z) = +\infty$ *uniformly for* $z \in \gamma$;

(C') $LV < 0$ in D;

then Problem E for (4.1) *has a unique solution.*

Proof Similarly to the proof of Lemma 4.1, we can find a solution of Problem M for equation (4.4), and then the solvability of Problem E for (4.1) can be derived by the method as stated in the proof of Lemma 4.1. To prove the uniqueness of solutions of Problem E, it is sufficient to prove that Problem E for (4.1) with homogeneous boundary condition $lu = 0$ on Γ has only the trivial solution. Since $L(hV - u) < 0$ in D, where h is any positive number, we can see that $hV(z) - u(z)$ cannot attain a negative minimum in D. Thus we have

$$\underline{\lim}_{z\to t}[hV(z) - u(z)] \geq 0, \text{ for } z \in \partial D,$$

and then

$$hV(z) - u(z) \geq 0, \text{ i.e. } u(z) \leq hV(z) \text{ in } D.$$

Letting h tend to 0 we have $u(z) \leq 0$ in D. Similarly, we consider the function $-hV(z) - u(z)$. From $L(-hV - u) > 0$,

$$-hV(z) - u(z) \leq 0, \text{ i.e. } -hV(z) \leq u(z) \text{ in } D$$

can be obtained. Thus $u(z) \geq 0$ in D. So $u(z) = 0$ in D (see [47]).

Theorem 4.3 *Suppose that one of the following conditions holds:*

(*a*) $m_2 < 1$;

(*b*) $m_2 = 1$, $b(x,0) < 1$, $x \in (-1,1)$;

(*c*) $1 < m_2 < 2$, $b(x,y) \in C^1(\overline{D^*})$, $D^* = \{-1-\delta_0 < x < 1+\delta_0, 0 < y < \delta_0\}$, *where* δ_0 *is a positive constant,* $b(x,0) \leq 0$, $x \in (-1-\delta_0, 1+\delta_0)$, *in which* $b(x,0)$ *is extended to* $(-1-\delta_0, 1+\delta_0)$;

(*d*) $m_2 \geq 2$, $b(x,y) < 0$, $x \in (-1,1)$.

Then Problem M for (4.1) *has a unique solution.*

Proof We introduce a function

$$V(z) = (x-x_0)^2 + y^\beta, \ -1 \leq x_0 \leq 1,$$

where $\beta\,(0 < \beta < 1)$ is an unknown constant to be determined appropriately. It is clear that $V(z)$ satisfies the conditions (A) and (B) in Lemma 4.1, and

$$LV \leq 2y^{m_1} + \beta(\beta-1)y^{m_2+\beta-2} + 2a(x-x_0) + \beta b y^{\beta-1}. \tag{4.6}$$

Moreover noting the condition: $0 \leq (\nu, y) < \pi/2$ at the corner points $x_0 = -1, 1$ of D, there exists a sufficiently small positive constant δ_0 such that $\cos(\nu, y) \geq \delta_0 > 0$, when $(x,y) \in U \cap \Gamma$, we have

$$lV = \frac{1}{2}\frac{\partial V}{\partial \nu} + \sigma V = (x-x_0)\cos(\nu, x)$$
$$+\beta y^{\beta-1}\cos(\nu,y)/2 + \sigma[(x-x_0)^2 + y^\beta] > 0 \text{ in } U$$

for Problems M_2 and M_3, provided that δ is chosen small enough, where $U = \overline{D} \cap \{|x-x_0| \leq \delta, 0 \leq y \leq \delta\}$. Thus the condition (D) in Lemma 4.1 is satisfied.

(*a*) $m_2 < 1$. We choose $\beta = 1/2$ and denote $X = \max_{z\in\bar{D}}|x|$, $Y = \max_{z\in\bar{D}} y$. From (4.6), it follows that

$$\begin{aligned} LV &\leq 2y^{m_1} - \frac{1}{4}y^{m_2-3/2} + 2|a(x-x_0)| + \frac{1}{2}|b|y^{-1/2} \\ &\leq 2Y^{m_1} + 2k_0(X+1) - \frac{1}{4}y^{m_2-3/2}(1-2k_0y^{1-m_2}) \to -\infty \text{ as } y \to +0. \end{aligned} \tag{4.7}$$

Hence there exists a positive constant δ, such that $LV \le -1$, if $0 < y < \delta$. This shows that $V(z)$ satisfies condition (C) in Lemma 4.1.

(*b*) $m_2 = 1$, $b(x,0) < 1$, $x \in \gamma = (-1,1)$. Obviously, we can find a small positive constant $\eta\,(0 < \eta < 1/2)$, so that

$$b(x,0) \le 1 - 2\eta < 1,\ \ x \in (-1,1).$$

Then there exists a positive number δ_0, such that $b(x,y) \le 1-\eta$, if $(x,y) \in \{-1 \le x \le 1, 0 < y < \delta_0\}$. Choosing $\beta = \eta/2$, we see that

$$\begin{aligned} LV &\le 2y^{m_1} + 2k_0|x-x_0| + \beta(\beta-1)y^{\beta-1} + \beta(1-\eta)y^{\beta-1} \\ &\le 2Y^{m_1} + 2k_0(X+1) - \eta^2 y^{\eta/2-1}/4 \to -\infty \text{ as } y \to +0. \end{aligned} \tag{4.8}$$

Consequently there is a positive number $\delta < \delta_0$, so that $LV \le -1$, if $0 < y < \delta$. Hence, condition (C) is satisfied.

(*c*) $1 < m_2 < 2$, $b(x,y) \in C^1(\overline{D^*})$, $b(x,0) \le 0$ for $-1-\delta_0 < x < 1+\delta_0$. When $y < \delta_0$, we have

$$b(x,y) = b(x,0) + b'(x,\theta y)y \le C^1[b,\bar{D}]y,\ \ 0 < \theta < 1.$$

Assuming that $C^1[b,\bar{D}] = k_2 < \infty$ and choosing that $\beta = 1 - m_2/2\,(0 < \beta < 1)$, we find

$$\begin{aligned} LV &\le 2y^{m_1} + 2k_0|x-x_0| + \beta(\beta-1)y^{m_2+\beta-2} + \beta k_2 y^{\beta} \\ &\le 2Y^{m_1} + 2k_0(X+1) - \frac{m_2}{2}(1-\frac{m_2}{2})y^{m_2/2-1} \\ &+ (1-\frac{m_2}{2})k_2 y^{1-m_2/2} \to -\infty, \text{ as } y \to +0. \end{aligned} \tag{4.9}$$

Hence the condition (C) is satisfied.

(*d*) $m_2 \ge 2$, $b(x,0) < 0$, $x \in (-1,1)$. Similarly to (*b*), there exist two positive constants η and δ, such that $b(x,y) \le -\eta < 0$, if $(x,y) \in \{-1 \le x \le 1, 0 < y < \delta\}$. From (4.6), we can obtain

$$LV \le 2Y^{m_1} + 2k_0(X+1) - \beta\eta y^{\beta-1} \to -\infty, \text{ as } y \to +0. \tag{4.10}$$

Therefore we have condition (C).

On the basis of Lemma 4.1, Theorem 4.3 is proved.

Theorem 4.4 *Suppose that one of the following conditions holds:*

(a') $m_2 = 1$, $b(x,y) \in C^1(\overline{D^*})$, $b(x,0) \geq 1$, $x \in (-1-\delta_0, 1+\delta_0)$, *where* D^* *and* δ_0 *are as stated in Theorem* 4.3;

(b') $m_2 > 1$, $b(x,0) > 0$, $x \in \gamma = (-1,1)$.

Then Problem E for (4.1) *or* (4.2) *has a unique solution.*

Proof (a') Denote $Y = \max_{z\in\overline{D}} y$ and $k_2 = C^1[b,\overline{D}] + 1$, we introduce a function

$$V(y) = e^{nY} - e^{ny} + V_1(y),\ \ V_1(y) = \int_y^Y \frac{e^{k_2 t}}{t} dt, \tag{4.11}$$

where n is an undetermined positive constant. It is easy to see that $V(y)$ satisfies the condition (A') in Lemma 4.2. Due to

$$V_1(y) \geq \int_y^Y \frac{dt}{t} = \ln Y - \ln y \to +\infty,\ \text{ as } y \to +0,$$

the condition (B') in Lemma 4.2 is satisfied. In order to verify condition (C'), we note that because of the continuity of $b(x,y)$, from $b(x,0) \geq 1$, $x \in (-1-\delta_0, 1+\delta_0)$, there exists a positive constant $\delta < 1$, such that $b(x,y) \geq 1/2$ if $(x,y) \in \{-1 \leq x \leq 1, 0 < y < \delta\}$. Moreover,

$$\begin{aligned} LV &= L(e^{nY} - e^{ny}) + LV_1 \leq -n^2 y e^{ny} - b(x,y) n e^{ny} \\ &\quad + yV_1'' + bV_1' \leq yV_1'' + [b(x,0) + b'(x,\theta y)y]V_1' \\ &< y(-k_2\frac{e^{k_2 y}}{y} + \frac{e^{k_2 y}}{y^2}) - \frac{e^{k_2 y}}{y} + k_2 e^{k_2 y} = 0, \end{aligned} \tag{4.12}$$

if $0 < y < \delta$. If $\delta \leq y < Y$, there exists a positive constant M_3 independent of n, such that $LV_1 \leq M_3 < \infty$. Let n be large enough, so that

$$n > \max\{(k_0+1)/\delta, M_3\}.$$

We can immediately obtain

$$LV \leq -\delta n^2 e^{ny} + n k_0 e^{ny} + M_3 < 0, \tag{4.13}$$

where $C[b,\overline{D}] \leq k_0$. So condition ($C'$) in Lemma 4.2 holds. According to Lemma 4.2, under condition (a'), the result in Theorem 4.4 is true.

(b') $m_2 > 1$ and $b(x,0) > 0$, $x \in (-1,1)$. Similarly to (d) in the proof of Theorem 4.3, there exist two positive constants η and δ, such that $b(x,y) \geq \eta$, if $0 \leq y < \delta$. We introduce an auxiliary function

$$V(y) = e^{nY} - e^{ny} + V_2(y),\ \ V_2(y) = \int_y^Y e^{\frac{\eta}{m_2-1} t^{1-m_2}} dt. \tag{4.14}$$

Obviously $V(y)$ satisfies condition (A') in Lemma 4.2. Let $j = 1/(m_2 - 1) + 1$. It is clear that if $0 < t < 1$, then

$$e^{\frac{\eta}{m_2-1}t^{1-m_2}} \geq \frac{1}{j!}\left(\frac{\eta}{m_2-1}t^{1-m_2}\right)^j$$

$$= \frac{1}{j!}\left(\frac{\eta}{m_2-1}\right)^j \left(\frac{1}{t}\right)^{j(m_2-1)} \geq \frac{1}{j!}\left(\frac{\eta}{m_2-1}\right)^j \frac{1}{t}.$$

Consequently

$$V(y) \geq \int_y^{\min(1,Y)} \frac{1}{j!}\left(\frac{\eta}{m_2-1}\right)^j \frac{dt}{t} \to +\infty \text{ as } y \to +0.$$

This shows that the condition (B') is satisfied. If $0 < y < \delta$, then

$$\begin{aligned} LV &\leq -n^2 y^{m_2} e^{ny} - nbe^{ny} + y^{m_2} V_2'' + bV_2' \\ &< y^{m_2}\left(\frac{\eta}{y^{m_2}} e^{\frac{\eta}{m_2-1}y^{1-m_2}}\right) - \eta e^{\frac{\eta}{m_2-1}y^{1-m_2}} = 0. \end{aligned} \tag{4.15}$$

When $\delta \leq y \leq Y$, we choose n large enough such that

$$n > \max\{(k_0+1)/\delta^{m_2}, M_4\},$$

where $C[b, \overline{D}] \leq k_0$, $M_4 \,(\geq LV_2)$ is a positive constant independent of n. Thus

$$\begin{aligned} LV &\leq -n^2 y^{m_2} e^{ny} + nk_0 e^{ny} + LV_2 \\ &\leq -ne^{ny}(n\delta^{m_2} - k_0) + M_4 < 0. \end{aligned} \tag{4.16}$$

Therefore condition (C') holds. By Lemma 4.2, the result in Theorem 4.4 is justified.

4.2 Boundary value problems for axisymmetric filtration

Now we introduce a boundary value problem in axisymmetric filtration, i.e. a steady axisymmetric filtration with homogeneous medium. Denote by $k = k(z, r)$ the filtration coefficient, and $\rho = \rho(z, r)$ is the density of fluid, we can assume $\rho = 1$, then the potential function $\phi = \phi(z, r)$ and the stream function $\psi(z, r)$ in a filtration domain D satisfy the following degenerate elliptic system of first order equations

$$\begin{gathered} \psi_z = -r\phi_r, \ \psi_r = r\phi_z \ \text{ in } D, \\ D = \{0<z<a, 0<r<b\} \backslash \{a_0 \leq z < a, 0 < r \leq b_0\}, \end{gathered} \tag{4.17}$$

where a, a_0, b, b_0 are positive constants. If the coordinate (z, r) is replaced by (x, y), then the above system can be written as

$$\begin{gathered} y\phi_x = \psi_y, \ -y\phi_y = \psi_x \ \text{ in } D, \\ D = \{0<x<a, 0<y<b\} \backslash \{a_0 \le x < a, 0 < y \le b_0\}. \end{gathered} \tag{4.18}$$

It is obvious that the system (4.18) can be reduced to

$$y\psi_{xx} + y\psi_{yy} - \psi_y = 0 \ \text{ in } D. \tag{4.19}$$

The Dirichlet boundary condition of the equation (4.19) is as follows

$$\psi(x, y) = r(x, y) \ \text{ on } \partial D. \tag{4.20}$$

The boundary value problem (Problem D) for equation (4.19) has a unique solution, if the condition $b(x, 0) = -1 < 1$ for $x \in [0, a_0]$ holds (see Theorem 4.3)). For the steady axisymmetric filtration with homogeneous medium, due to $BC = \{0 \le x \le a, y = b\}, EFA = \{x = a_0, 0 \le y \le b_0\} \cup \{a_0 \le x \le a, y = b_0\}$ are equipotential lines, and $AB = \{x = a, b_0 \le y \le b\}, CDE = \{x = 0, 0 \le y \le b\} \cup \{0 \le x \le a_0, y = 0\}$ are stream lines, hence we can write the boundary conditions as follows

$$\phi(x, y) = \begin{cases} \phi_0 = 0 \ \text{ on } EFA, \\ \phi_1 \ \text{ on } BC, \end{cases} \qquad \psi(x, y) = \begin{cases} \psi_0 = 0 \ \text{ on } CDE, \\ \psi_1 \ \text{ on } AB, \end{cases} \tag{4.21}$$

where $\phi_0, \phi_1, \psi_0, \psi_1$ are real constants, we can assume that $\phi_0 = 0, \psi_0 = 0$, because it does not influence the speed. Moreover the boundary condition (4.21) can be rewritten as

$$\phi(x, y) = \begin{cases} \phi_0 = 0 \text{ on } EFA, \text{ i.e.} \\ y\phi_x = \psi_y = 0 \text{ on } FA, -y\phi_y = \psi_x = 0 \text{ on } EF, \\ \phi_1 \ \text{ on } BC, \ \text{i.e. } y\phi_x = \psi_y = 0 \ \text{ on } BC, \end{cases}$$

$$\psi(x, y) = \begin{cases} \psi_0 = 0, \ \text{i.e. } y\phi_x = \psi_y = 0 \text{ on } CD, -y\phi_y = \psi_x = 0 \text{ on } DE, \\ \psi_1 \ \text{ on } AB, \ \text{i.e. } y\phi_x = \psi_y = 0 \text{ on } AB, \end{cases} \quad \text{or}$$

$$\begin{cases} \psi_x = 0, \ z \in DEF, \\ \psi_y = 0, \ z \in FABCD, \\ \psi(0) = 0, \psi(a_0 + ib_0) = \psi_1 \text{ or } [\psi_x - i\psi_y]|_{z=a_0+ib_0} = 0. \end{cases} \tag{4.22}$$

This is a mixed boundary value problem, which will be called Problem F.

We can prove the following result.

Theorem 4.5 *Any solution $u(z)$ of Problem F satisfies the estimate*

$$|\psi(z)| \le M_5 = M_5(\phi_1, \psi_1, D), \tag{4.23}$$

where M_5 is a non-negative constant.

Proof For Problem F, we know that the solution $\psi(z)$ cannot attains its positive maximum M and its negative minimum m in D, hence if $\psi(z^*) = M$ or $\psi(z^*) = m$ at a point $z^* \in \Gamma_1 = AB \cup CDE$, then

$$|\psi(z)| \le |\psi(z^*)| \le \max(0, \max_{AB} \psi_1) \text{ in } \overline{D}. \tag{4.24}$$

If $\psi(z^*) = M$ or m at a point $z^* \in \Gamma_2 = BC \cup FA$, on the basis of Lemma 2.3, we get $\psi_y \neq 0$ at z^*, it is impossible. If $\psi(z^*) = M$ or m at $z^* \in \Gamma_3 = EF$, then $\psi_x \neq 0$ at z^*, this contradicts $\psi_x(z) = 0$ on Γ_3. Hence we have the estimate (4.23).

We consider the system of first order equations

$$(y + 1/n)\phi_x = \psi_y, \; -(y + 1/n)\phi_y = \psi_x, \tag{4.25}$$

and the corresponding equations

$$L\psi = (y + 1/n)\psi_{xx} + (y + 1/n)\psi_{yy} - \psi_y = 0 \text{ in } D, \tag{4.26}$$

where n is a sufficiently large positive integer. The boundary value problem (4.26), (4.22) will be called Problem F_n. Because the equation (4.26) in $\overline{D}$ is uniformly elliptic, on the basis of the results in Section 1, Problem F has a bounded solution in D. The complex form of Problem F is as follows

$$\psi_{nz\bar{z}} = \text{Re}[A(z)\psi_{nz}], \; A(z) = \frac{i}{2(y + 1/n)} \text{ in } D,$$

and

$$\text{Re}[\overline{\Lambda(z)}\psi_{nz}] = 0 \text{ on } \partial D,$$

where $\psi_{nz} = [\psi_{nx} - i\psi_{ny}]/2$, $\Lambda(z) = 1$ on DEF, $\Lambda(z) = -i$ on $FABCD$, according to the formulas (1.17)-(1.18), Chapter I, we can choose $m = 2$ and obtain $\gamma_1 = 1/2$ at $t_1 = 0$, $\gamma_2 = -1/2$ at $t_2 = a_0 + ib_0$, $K_1 = -1$, $K_2 = 1$, thus the index $K = (K_1 + K_2)/2 = 0$. Problem F is a well-posed mixed boundary value problem. Similarly to (1.30), we can get the estimate of solutions $\{\psi_n(z)\}$ of Problem F_n as follows

$$C_\delta[X(z)\psi_{nz}, D_{\delta_0}] \le M_6 = M_6(\delta, \phi_1, \psi_1, D_{\delta_0}),$$

in which $D_{\delta_0} = \overline{D} \cap \{y > \delta_0\}$, $X(z) = |z|^{2\delta}|z - a_0 - ib_0|^{1/3+2\delta}$, δ_0, δ are sufficiently small positive constants, and M_6 is a non-negative constant.

Theorem 4.6 *Under the above conditions, the boundary value problem (Problem F) for above axisymmetric filtration with homogeneous medium has a unique solution.*

Proof We continuously extend $\psi(z)$ in the closure $\overline{D}$ of the domain D such that $C^1_\alpha[\psi, \overline{D}] \le M_5 + 1$ holds, where $\alpha(0 < \alpha < 1)$ is a positive constant. Let x_0 is any point on $DE = \{0 < x < a_0, y = 0\}$, we introduce a function

$$V(z) = (x - x_0)^2 + y^\beta,\ 0 \le x \le a_0, y > 0, \tag{4.27}$$

where $\beta\,(0 < \beta < 1)$. It is clear that $V(z)$ satisfies the conditions $V(x_0) = 0$ and $V(z) > 0$ for $z(\in \overline{D}) \ne x_0$, and

$$LV \le 2y + \beta(\beta - 1)y^{\beta-1} - \beta y^{\beta-1}. \tag{4.28}$$

For arbitrary positive number $\eta(< 1/2)$, we can find a positive number δ, such that $|\psi_n(z) - \psi(x_0)| < \eta$, if $z \in U = D \cap \{|x - x_0| \le \delta, 0 < y < \delta\}$. Afterwards we introduce the function

$$W^\pm(z) = cV(z) + \eta \pm \psi(x_0) \mp \psi_n(z), \tag{4.29}$$

in which c is an undetermined positive constant, and similarly to the proof (b) of Theorem 4.3, we choose the constants $\beta = \eta/2$, then

$$LW^\pm = cLV < 0 \text{ in } U, \tag{4.30}$$

where $\delta\,(< 1)$ is a sufficiently small positive constant. This shows that $W^\pm(z)$ cannot attain a negative minimum in D. Moreover, it is not difficult to see that

$$W^\pm(z) \ge cV(z) + \eta - |\psi_n(z) - \psi(x_0)| > 0 \text{ in } \partial D \cap \overline{U},$$

and

$$W^\pm(z) \ge cV(z) + \eta \pm \psi(x_0) - M_1 > 0 \text{ in } \partial D \backslash \overline{U},$$

for a sufficiently large constant c, where $M_1 = \max_{\overline{D}} |\psi_n(z)|$. Therefore

$$W^\pm(z) = cV(z) + \eta \pm \psi(x_0) \mp \psi_n(z) \ge 0 \text{ in } \overline{D}.$$

It follows that

$$|\psi_n(z) - \psi(x_0)| \le cV(z) + \eta \text{ in } \overline{D}.$$

Letting n tend to ∞ and z tend to x_0, we obtain

$$\overline{\lim_{n\to\infty}}\ \overline{\lim_{z\to x_0}} |\psi_n(z) - \psi(x_0)| \le \eta.$$

Noting the arbitrariness of η, the above inequality implies

$$\overline{\lim_{n\to\infty}}\ \overline{\lim_{z\to x_0}} \psi_n(z) = \psi(x_0).$$

As for the corner points $z = 0, a_0$ of D, from the boundary condition $\psi_{nx} = 0$ on EF, we can extend the function $\psi_n(z)$ onto the symmetrical domain $\tilde{D}$ of D with respect to EF, namely define

$$\tilde{\psi}_n(z)=\begin{cases} \psi_n(z) \text{ in } D, \\ \psi_n[-(\bar{z}-a_0)+a_0] \text{ in } \tilde{D}, \end{cases}$$

then the function $\tilde{\psi}_n(z)$ satisfies the equation

$$L\tilde{\psi}_n=(y+1/n)\tilde{\psi}_{xnx}+(y+1/n)\tilde{\psi}_{nyy}-\tilde{\psi}_{ny}=0 \text{ in } \{D\cup\tilde{D}\}\cap\{0\le y<b_0\}.$$

Now $z = a_0$ is an inner point of the line segment $[0, 2a_0]$ on x-axis, δ is a small positive number, and the function satisfies the boundary condition $\tilde{\psi}_n(x) = 0$ on $DE = [0, a_0]$, therefore similarly to before, we can prove

$$\tilde{\psi}_n(z) = \psi_n(z) \to \psi(a_0) = 0 \text{ as } n \to \infty \text{ and } z\,(\in \overline{D}) \to a_0.$$

Applying the same method, we can prove

$$\psi_n(z) \to \psi(0) = 0 \text{ as } n \to \infty \text{ and } z\,(\in \overline{D}) \to 0.$$

This completes the proof of Theorem 4.6.

Besides the mathematical model of the explosion of cumulative energy (see Subsection 58, Section 4, Chapter III, [52]) is some boundary value problem for elliptic equations with parabolic degenerate line, which can be handled by using the above complex method.

5 The Oblique Derivative Problem for Nonhomogeneous Elliptic Equations of Second Order with Degenerate Rank 0

This section deals with the oblique derivative boundary value problem for second order quasilinear elliptic equations with degenerate rank 0, we first give estimates of solutions for the boundary value problem, and then prove the solvability of the problem.

5.1 Formulation of oblique derivative problems for degenerate elliptic equations

Let D be a simply connected bounded domain in the complex plane $\mathbf{C}$ with the boundary $\partial D = \Gamma \cup \gamma$ as stated in Section 2, and we can assume that the boundary Γ of the domain D possesses the form $x - \tilde{G}(y) = -1$ and $x + \tilde{G}(y) = 1$ including the line segments $\mathrm{Re}z = \pm 1$ near the points $z = \pm 1$. Denote $H_j(y) = \sqrt{K_j(y)}$, $j = 1,2$, in which $K_j(y) = |y|^{m_j} h_j(y)$ $(j = 1,2)$ in $\overline{D}$, here m_j $(j = 1,2, m_2 < 1)$ are positive numbers, $h_j(y)(j = 1,2)$ are continuously differentiable positive functions, and $H(y) = H_1(y)/H_2(y)$, $G(y) = \int_0^y H(t)dt$. We consider the second order quasilinear equation of mixed type with degenerate rank 0:

$$Lu = K_1(y)u_{xx} + K_2(y)u_{yy} + au_x + bu_y + cu = -d, \tag{5.1}$$

where a, b, c, d are real functions of $z(\in D)$, $u, u_x, u_y(\in \mathbf{R})$. Suppose that the coefficients of (5.1) satisfy **Condition** C, namely

1) a, b, c, d are measurable in D and continuous in $\overline{D}$ for any continuously differentiable function $u(z)$ in $D^* = \overline{D}\backslash\{-1, 1\}$, and satisfy

$$L_\infty[\eta, D] \le k_0, \eta = a, b, c, L_\infty[d, D] \le k_1, c \le 0 \text{ in } D. \tag{5.2}$$

2) For any two continuously differentiable functions $u_1(z)$, $u_2(z)$ in D^*, $F(z, u, u_z) = au_x + bu_y + cu + d$ satisfies the condition

$$F(z, u_1, u_{1z}) - F(z, u_2, u_{2z}) = \tilde{a}(u_1 - u_2)_x + \tilde{b}(u_1 - u_2)_y + \tilde{c}(u_1 - u_2) \text{ in } \overline{D},$$

in which $\tilde{a}, \tilde{b}, \tilde{c}$ satisfy the conditions

$$L_\infty[\tilde{a}, D], L_\infty[\tilde{b}, D], L_\infty[\tilde{c}, D] \le k_0, \tilde{c} \le 0 \text{ in } D, \tag{5.3}$$

where $H(y) = \sqrt{K(y)}, K(y) = K_1(y)/K_2(y)$, k_0, k_1 are non-negative constants.

If $H_j(y) = y^{m_j/2}, j = 1,2$, $H(y) = H_1(y)/H_2(y)$, here $m_1(\ge 0), m_2$

(> 0) are real numbers satisfying $m_1 - m_2 > -1$, then

$$G_j(y) = \int_0^y H_j(t)dt = \frac{2}{m_j+2} y^{(m_j+2)/2}, \ j = 1, 2,$$

$$Y = G(y) = \int_0^y H(t)dt = \frac{2}{m+2} y^{(m+2)/2}, m = m_1 - m_2 > -1 \text{ in } \overline{D},$$

and its inverse function of $Y = G(y)$ is

$$\begin{aligned} y = G_j^{-1}(Y) &= \left(\frac{m_j+2}{2}\right)^{2/(m_j+2)} Y^{2/(m_j+2)} = J_j Y^{2/(m_j+2)}, j = 1, 2, \\ y = G^{-1}(Y) &= \left(\frac{m+2}{2}\right)^{2/(m+2)} Y^{2/(m+2)} = J Y^{2/(m+2)} \text{ in } \overline{D}. \end{aligned} \tag{5.4}$$

The oblique derivative problem (Problem P or Q) for equation (5.1) is to find a continuous solution $u(z)$ of (5.1) in $\overline{D}$, and u_x, u_y are continuously differentiable $D^* = \overline{D} \backslash \{-1, 1\}$ satisfying the boundary conditions

$$\begin{aligned} &\frac{1}{2}\frac{\partial u}{\partial \nu} = \frac{1}{H_1(y)} \text{Re}[\overline{\lambda(z)} u_{\tilde{z}}] = \text{Re}[\overline{\Lambda(z)} u_z] = r(z) \text{ on } \Gamma \cup \gamma, \\ &u(-1) = b_0, u(1) = b_1 \text{ or } \frac{1}{H_1(y)} \text{Im}[\overline{\lambda(z)} u_{\tilde{z}}]|_{z=z_1} = b_1, \end{aligned} \tag{5.5}$$

where z_1 is a point on $\Gamma \backslash \{-1, 1\}$, ν is a given vector at every point $z \in \Gamma \cup \gamma$, $u_{\tilde{z}} = [H_1(y)u_x - iH_2(y)u_y]/2$, $\Lambda(z) = \cos(\nu, x) - i\cos(\nu, y)$, $\lambda(z) = \text{Re}\lambda(z) + i\text{Im}\lambda(z)$ if $z \in \Gamma$, $\lambda(z) = i$ if $z \in \gamma$, b_0, b_1 are real constants, and $r(z), b_0, b_1$ satisfy the conditions

$$\begin{aligned} &C_\alpha^1[\lambda(z), \Gamma] \le k_0, \ C_\alpha^1[r(z), \Gamma] \le k_2, \\ &C_\alpha^1[r(z), \gamma] \le k_2, \cos(\nu, n) \ge 0 \text{ on } \Gamma, |b_0|, |b_1| \le k_2, \end{aligned} \tag{5.6}$$

in which n is the outward normal vector at every point on $\Gamma \cup \gamma$, $\alpha \, (0 < \alpha < 1), k_0, k_2$ are non-negative constants. For the last point condition in (5.5), we need to assume $c = 0$ in equation (5.1). The number $K = (K_1 + K_2)/2$ is called the index of Problem P, where

$$K_j = \left[\frac{\phi_j}{\pi}\right] + J_j, J_j = 0 \text{ or } 1, e^{i\phi_j} = \frac{\lambda(t_j - 0)}{\lambda(t_j + 0)}, \gamma_j = \frac{\phi_j}{\pi} - K_j, j = 1, 2, \tag{5.7}$$

in which $t_1 = -1, t_2 = 1, \lambda(t) = \exp(i\pi/2)$ on $\gamma = \{-1 < x < 1, y = 0\}$ and $\lambda(t_1 + 0) = \lambda(t_2 - 0) = \exp(i\pi/2)$. Similarly to Section 1, we can choose $K=0$ and assume that $-1/2\le\gamma_j<1/2\,(j=1,2)$. For the mixed boundary value problem (Problem M) with the boundary condition (2.7), then $\lambda(t_1-0)=e^{-i\pi/2}$, $\lambda(t_2+0)=e^{i\pi/2}$, we can get

$$\begin{aligned}\gamma_1&=\frac{1}{\pi i}\ln\left[\frac{\lambda(t_1-0)}{\lambda(t_1+0)}\right]=\frac{-\pi/2-0\pi}{\pi}-K_1=-\frac{1}{2},\ K_1=0,\\ \gamma_2&=\frac{1}{\pi i}\ln\left[\frac{\lambda(t_2-0)}{\lambda(t_2+0)}\right]=\frac{0\pi-\pi/2}{\pi}-K_2=-\frac{1}{2},\ K_2=0,\end{aligned}\tag{5.8}$$

where we consider $\mathrm{Re}[\overline{\lambda(z)}(U+iV)]=0$, $\lambda(x)=1=e^{0\pi i}$ on γ, hence the index $K=(K_1+K_2)/2=0$, and we have the point conditions $u(-1)=\phi(-1)=b_0$, $u(1)=\phi(1)=b_1$ in the boundary condition.

If the last condition: $u_y(x)=r(x)$ on γ in (5.5) is replaced by

$$u(x)=r(x),\ \text{i.e.}\ u_x=r'(x),u(-1)=r(-1),u(1)=r(1),\tag{5.9}$$

where $r(x)$ satisfies the conditions

$$C^2_\alpha[r(x),\gamma]\le k_2,\tag{5.10}$$

in which k_2 is a non-negative constant, then the boundary value problem is called Problem M'. Similar to before, we can choose its index $K=-1/2$ or 0.

5.2 Representation of solutions of oblique derivative problem of elliptic equations

Now we denote

$$\begin{aligned}W(z)&=U+iV=[H_1(y)u_x-iH_2(y)u_y]/2\\&=H_1(y)[u_x-iu_Y]/2=H_1(y)u_Z=u_{\tilde z},\\ W_{\overline{\tilde z}}&=[H_1(y)W_x+iH_2(y)W_y]/2\\&=H_1(y)[W_x+iW_Y]/2=H_1(y)W_{\overline{Z}}\ \text{in}\ \overline{D},\end{aligned}$$

in which $Z=x+iY=x+iG(y)$, and then

$$W_{\tilde{\bar{z}}}=H_1(y)W_{\overline{Z}}=\frac{1}{4}\{H_1[H_1u_x-iH_2u_y]_x+iH_2[H_1u_x-iH_2u_y]_y\}$$
$$=\frac{1}{4}\{H_1^2u_{xx}+H_2^2u_{yy}-iH_1H_2[u_{yx}-u_{xy}]+iH_2[H_{1y}u_x$$
$$-iH_{2y}u_y]\}=\frac{1}{4}\{H_1^2u_{xx}+H_2^2u_{yy}+iH_2[H_{1y}u_x-iH_{2y}u_y]\}$$
$$=\frac{1}{4}\{H_1^2u_{xx}+H_2^2u_{yy}+iH_2\frac{H_{1y}}{H_1}(H_1u_x)+H_{2y}(H_2u_y)\}$$
$$=\frac{1}{4}\{[\frac{iH_2H_{1y}}{H_1}-\frac{a}{H_1}](H_1u_x)+[H_{2y}-\frac{b}{H_2}](H_2u_y)-cu-d\} \tag{5.11}$$
$$=\frac{1}{4}\{[\frac{iH_2H_{1y}}{H_1}-\frac{a}{H_1}](W+\overline{W})+i[H_{2y}-\frac{b}{H_2}](W-\overline{W})-cu-d\}$$
$$=\frac{1}{4}\{[\frac{iH_2H_{1y}}{H_1}-\frac{a}{H_1}-i\frac{b}{H_2}+iH_{2y}]W$$
$$+[\frac{iH_2H_{1y}}{H_1}-\frac{a}{H_1}+i\frac{b}{H_2}-iH_{2y}]\overline{W}-cu-d\}$$
$$=A_1(z)W+A_2(z)\overline{W}+A_3(z)u+A_4(z)=g(Z)\ \text{in}\ \overline{D_Z},$$

where D_Z is the image domain of D with respect to the mapping $Z=Z(z)$, and

$$A_1=\frac{1}{4}[-\frac{a}{H_1}+\frac{iH_2H_{1y}}{H_1}-i\frac{b}{H_2}+iH_{2y}],\ A_3=-\frac{c}{4},$$
$$A_2=\frac{1}{4}[-\frac{a}{H_1}+\frac{iH_2H_{1y}}{H_1}+i\frac{b}{H_2}-iH_{2y}],\ A_4=-\frac{d}{4},$$

if $H_j(y)=\sqrt{y^{m_j}h_j(y)}$, $l=1,2$, where m_j, $h_j(y)\,(j=1,2)$ are as stated before, and $m_1-m_2>-1$, then

$$A_1=\frac{1}{4}\left[-\frac{a}{H_1}-\frac{ib}{H_2}+iH_2\sum_{j=1}^{2}\left(\frac{h_{jy}}{2h_j}+\frac{m_j}{2y}\right)\right],\ A_3=-\frac{c}{4},$$
$$A_2=\frac{1}{4}\left[-\frac{a}{H_1}+\frac{ib}{H_2}+iH_2\sum_{j=1}^{2}(-1)^{j-1}\left(\frac{h_{jy}}{2h_j}+\frac{m_j}{2y}\right)\right],A_4=-\frac{d}{4},$$

hence the function

$$u(z)=2\text{Re}\int_0^z\left[\frac{U(z)}{H_1(y)}+\frac{iV(z)}{H_2(y)}\right]dz+b_0\ \text{in}\ \overline{D} \tag{5.12}$$

is a solution of equation (5.1), where b_0 is a real constant. The boundary value problems for the corresponding complex equation (5.11) and boundary conditions (5.5) ($W(z) = u_{\bar{z}}$) will be called Problems A and B respectively.

According to the method in the proof of Theorems 2.1 and 3.1, we can prove the following theorems.

Theorem 5.1 *Under Condition C, any solution $[W(z), u(z)]$ of Problem P or Q for equation* (5.1) *can be expressed as in* (5.12), *where $W(z)$ is a solution of Problem A or B for equation* (5.11) *as follows*

$$\begin{aligned} &W(z) = U(z)+iV(z)=\Phi(Z)+\Psi(Z)=\hat{\Phi}(Z)+\hat{\Psi}(Z) \text{ in } \overline{D_Z}, \\ &\Psi(Z) = 2\mathrm{Re}Tf,\ \hat{\Psi}(Z) = 2i\mathrm{Im}Tf,\ Tf =-\frac{1}{\pi}\iint_{D_t}\frac{f(t)}{t-Z}d\sigma_t, \end{aligned} \tag{5.13}$$

where $f(Z) = g(Z)/H_1(y)$, $\Phi(z)$, $\hat{\Phi}(Z)$ are analytic functions in D_Z, which are the solutions of the complex equation

$$W_{\bar{z}} = 0 \text{ in } D, \text{ i.e. } W_{\overline{Z}} = 0 \text{ in } D_Z. \tag{5.14}$$

5.3 Estimates and existence of solutions of oblique derivative derivative problems

Now we state and prove some results about Problem P or Q for equation (5.1). Similarly to Sections 2 and 3, we can only discuss the homogeneous boundary condition on Γ in (5.5), i.e. the conditions $R(z) = 0$ on Γ and $b_0 = b_1 = 0$, and first prove the uniqueness of solutions for Problems P and Q.

Theorem 5.2 *Suppose that equation* (5.11) *satisfies Condition C. Then the above boundary value problem* (*Problem P or Q*) *for* (5.11) *with the boundary condition* (5.5) *is unique.*

Proof In order to prove the uniqueness of the solution of Problem P or Q for equation (5.11), it suffices to verify that the corresponding homogeneous problem (Problem P_0 or Q_0) only has the trivial solution. The homogeneous equation of (5.11) can be written as

$$\begin{aligned} &K_1(y)u_{xx}+K_2(y)u_{yy}+\tilde{a}u_x+\tilde{b}u_y+\tilde{c}u = 0, \text{ i.e.} \\ &H_1(y)W_{\overline{Z}} = W_{\bar{z}} = A_1W + A_2\overline{W} + A_3u \text{ in } D, \end{aligned} \tag{5.15}$$

where $u_{\bar{z}} = W(z)$. Similarly to the proof of Theorem 3.2, we can prove that the solution $u(z)$ cannot attain the positive maximum in $\overline{D}\backslash\gamma$, and attain its positive maximum M of the solution at a point $x_0 \in \gamma$. Denote by $U(x_0) = \{|Z - x_0| < \varepsilon(> 0)\}$ the neighborhood of x_0, it is obvious that $u(z)$, u_x, u_y in $U(x_0) \cap \overline{D_Z}$ are bounded and then $\text{Re}W[z(Z)] = H_1(y)u_x/2 = 0$, $\text{Im}W[z(Z)] = H_2(y)u_y/2 = 0$ on $U(x_0) \cap \{y = 0\}$, thus we can extend the function $W(z)$ from $\hat{U}(x_0) = U(x_0) \cap \{y > 0\}$ onto the symmetrical domain $U(x_0) \cap \{y < 0\}$ about the real axis $\text{Im}z = y = 0$. From Theorem 5.1, the solution $u(z)$ can be expressed as in (5.12), (5.13), where

$$\begin{gathered}W[z(Z)] = \Phi(Z) + \Psi(Z) = \hat{\Phi}(Z) + \hat{\Psi}(Z) \text{ in } U(x_0),\\ \Psi(Z) = 2\text{Re}Tf,\ \hat{\Psi}(Z) = 2i\text{Im}Tf,\ Tf = -\frac{1}{\pi}\iint_{\hat{U}(x_0)} \frac{f(t)}{t - Z} d\sigma_t.\end{gathered} \tag{5.16}$$

From Lemma 2,1, Chapter I, we see that $\Psi(Z)$, $\hat{\Psi}(Z) \in C_\beta(U(x_0))$, $\beta = \min[2 - m_2, m_1/2]/(2 + m_1 - m_2) - 2\delta$, and $\Phi(Z)$, $\hat{\Phi}(Z)$ are analytic functions in $U(x_0)$. Noting that $\text{Re}\hat{\Psi}(Z) = 0$, $\text{Im}\Psi(Z) = 0$ in $U(x_0)$, similarly to the proof of Theorem 2.4, we know that $\text{Re}\hat{\Phi}(Z) = \text{Re}W[z(Z)]$, $\text{Im}\Phi(Z) = \text{Im}W[z(Z)]$ are harmonic functions in $U(x_0)$ satisfy the conditions $\text{Re}\hat{\Phi}(Z) = H_1(y)u_x/2 = 0$, $\text{Im}\Phi(Z) = -H_2(y)u_y/2 = 0$ on $U(x_0) \cap \{y = 0\}$, hence $\text{Re}\hat{\Phi}(Z) = H_1(y)u_x/2 = YF_1$, $\text{Im}\Phi(Z) = -H_2(y)u_y/2 = YF_2$ in $\tilde{U}(x_0) = U(x_0) \cap \overline{D_Z}$, here F_1, F_2 are continuous in $U(x_0)$, thus $u_x = O(Y^{(2-m_2)/(2+m_1-m_2)}F_1)$, $u_y = O(Y^{(2+m_1-2m_2)/(2+m_1-m_2)}F_2)$ in $\tilde{U}(x_0)$, this shows that $u_x = 0$ in $\tilde{U}(x_0)$, and then $u(x) = M > 0$ on $U(x_0) \cap \{y = 0\}$. From this we can extend that $u(z) = M$ on γ, however $u(-1) = 0$, this contradiction proves that $u(z)$ cannot attain the positive maximum in γ. Similarly we can prove that $u(z)$ cannot attain the negative minimum in $\overline{D_Z}$. Hence $u(z) = 0$ in D.

In order to prove the existence of solutions of Problems P and Q for equation (5.11), we first give estimates of solutions of Problems P and Q.

Theorem 5.3 *If equation* (5.1) *satisfies Condition C, then of Problem A or B for* (5.11) *has a solution $[W(z), u(z)]$ satisfying the estimates*

$$\begin{gathered}\hat{C}_\delta[W(z), \overline{D_Z}] = C_\delta[X(Z)(\text{Re}W/H_1 + i\text{Im}W/H_2), \overline{D_Z}]\\ + C_\delta[u(z), \overline{D}] \le M_1,\ \hat{C}_\delta[W(z), \overline{D_Z}] \le M_2(k_1 + k_2),\end{gathered} \tag{5.17}$$

where $X(Z) = (Z+1)^{\eta_1}(Z-1)^{\eta_2}$, $\eta_j = 1 - 2\gamma_j$ if $\gamma_j \ge 0$, $\eta_j = -2\gamma_j$ if $\gamma_j < 0$, and $\gamma_j (j = 1, 2)$ are similar to those as in (3.8), *δ is a sufficiently small* *positive constant, here we choose a branch of multi-valued function* $X(Z)$ *such that* $\arg X(x) = 0$ *on* γ, $M_1 = M_1(\delta, k, H, D)$, $M_2 = M_2(\delta, k_0, H, D)$ *are non-negative constants, and* $H = (H_1, H_2)$, $k = (k_0, k_1, k_2)$.

Proof As stated before Problem P or Q for (5.1) is equivalent to Problem A or B for the complex equation (5.11) with the boundary conditions

$$\begin{gathered}\operatorname{Re}[\overline{\lambda(z)}W(z)] = R(x) \text{ on } \Gamma\cup\gamma,\ u(-1) = b_0,\\ u(1)=b_1 \text{ or } \operatorname{Im}[\overline{\lambda(z)}W(z)]|_{z=z_1}=H(\operatorname{Im}z_1)b_1=b_1',\end{gathered}\tag{5.18}$$

and the relation

$$u(z)=u(x)-2\int_0^y \frac{V(z)}{H_2(y)}dy=2\operatorname{Re}\int_{-1}^z\left[\frac{\operatorname{Re}W}{H_1(y)}+i\frac{\operatorname{Im}W}{H_2(y)}\right]dz+b_0 \text{ in } \overline{D}. \tag{5.19}$$

Similarly to Theorem 3.3, it is easy to see that the function $\tilde{W}(Z) = X(Z)W[z(Z)]$ satisfies the complex equation

$$\begin{gathered}[X(Z)W]_{\overline{Z}}=X(Z)[A_1(z)W + A_2(z)\overline{W}\\ +A_3(z)u+A_4(z)]/H_1=X(Z)g(Z)/H_1 \text{ in } D_Z,\end{gathered}\tag{5.20}$$

and the boundary conditions

$$\begin{gathered}\operatorname{Re}[\overline{\tilde{\lambda}(z)}X(Z)W(z)] = |X(Z)|R(z) = \tilde{R}(z) \text{ on } \Gamma\cup\gamma,\ u(-1) = b_0,\\ u(1)=b_1 \text{ or } \operatorname{Im}[\overline{\tilde{\lambda}(z)}X(Z)W(z)]|_{z=z_1}=|X[Z(z_1)]|b_1'=b_1'',\end{gathered}\tag{5.21}$$

where $\tilde{R}(z) = 0$ on $\Gamma\cup\gamma$ and $b_0 = b_1 = 0$, and the numbers $\tau_j = 1/2$ if $\gamma_j \ge 0$ and 0 if $\gamma_j < 0\,(1 \le j \le 2)$ about $\tilde{\lambda}[z(Z)] = \lambda[z(Z)]e^{i\arg X(Z)}$ corresponding to the number $\gamma_j(j = 1, 2)$ in (3.8). By the reduction to absurdity, we can first verify that any solution $[W(z), u(z)]$ of Problem A satisfies the estimate

$$\hat{C}[W(z),\overline{D}]=C[X(Z)(\operatorname{Re}W/H_1+i\operatorname{Im}W/H_2),\overline{D_Z}]+C[u(z),\overline{D}]\le M_3, \tag{5.22}$$

in which $M_3 = M_3(\delta, k, H, D)$ is a non-negative constant. Moreover from the estimate (5.22), we can derive the estimates in (5.17). Next by using the Leray-Schauder theorem we can prove the unique solvability of Problem P for equation (5.1). But for obtaining the estimate (5.17) of solutions of Problem A or B for equation (5.11), we need to give the following discussion.

From Theorem 5.1, the solution $\tilde{W}(Z) = X(Z)W(z)$ can be expressed as $\tilde{W}(Z) = \Phi(Z) + \Psi(Z) = \hat{\Phi}(Z) + \hat{\Psi}(Z)$. According to the method in

Lemma 2.1, Chapter I, we can verify that the functions $\Psi^*(Z)=\Psi(Z)$ and $\Psi^*(Z)=\hat{\Psi}(Z)$ satisfy the estimates

$$C_\beta[\Psi^*(Z),\overline{D_Z}]\le M_4, \Psi^*(Z)-\Psi^*(t_j)=O(|Z-t_j|^{\beta_j}), 1\le j\le 2, \tag{5.23}$$

in which $\Psi(Z)=2\text{Re}Tf$, $\hat{\Psi}(Z)=2i\text{Im}Tf$, $f(Z)=X(Z)g(Z)/H_1(y)$, $\beta=\min[2-m_2,m_1/2]/(2+m_1-m_2)-2\delta=\beta_j\,(j=1,2)$, $t_1=-1,t_2=1$, and $M_4=M_4(\beta,k,H,D,M_3)$ is a positive constant, and $\Phi^*(Z)=\Phi(Z)$ or $\hat{\Phi}(Z)$ is an analytic function in D_Z satisfying the boundary conditions

$$\text{Re}[\overline{\tilde{\lambda}(z)}\Phi^*(Z)]=|X(Z)|R(z)-\text{Re}[\overline{\tilde{\lambda}(z)}\Psi^*(Z)]=\tilde{R}(z) \text{ on } \Gamma\cup\gamma, \tag{5.24}$$

because in the above case the index of $\tilde{\lambda}(z)$ on D_Z is $\tilde{K}=0$, we have two point conditions at $z=t_j\,(1\le j\le 2)$. Firstly we extend the function $W(Z)$ onto the symmetrical domain $\tilde{D}_Z$ of D_Z with respect to the real axis $\text{Im}Z=0$, and transform the function $\lambda(z)$ such that new function $\tilde{\lambda}(z)=1$ on Γ near the point $t_j(1\le j\le 2)$. Afterwards we symmetrically extend the function $\tilde{W}(z)$ in $D'_Z=D_Z\cup\tilde{D}_Z\cup\gamma\,(\gamma=(-1,1))$ onto the symmetrical domain D^*_Z with respect to $\text{Re}z=t_j$. Due to the solution $\tilde{W}(Z)$ extended can be also expressed as $X(Z)W(z)=\Phi(Z)+\Psi(Z)=\hat{\Phi}(Z)+\hat{\Psi}(Z)$ on $\hat{D}'_Z=\hat{D}_Z\cap\{Y>0\}$, $\hat{D}_Z=D'_Z\cup D^*_Z$, where $\Psi(Z),\hat{\Psi}(Z)$ in D_Z are Hölder continuous, $\text{Im}\Psi(Z)=0$, $\text{Re}\hat{\Psi}(Z)=0$ in $\hat{D}'_Z$, and $\Phi^*(Z)=\Phi(Z)$ are $\Phi^*(Z)=\hat{\Phi}(Z)$ are analytic functions in $\hat{D}'_Z$. It is clear that $\text{Im}\Phi(Z)=\text{Im}X(Z)W(z)$ and $\text{Re}\hat{\Phi}(Z)=\text{Re}X(Z)W(z)$ extended are harmonic functions in $\hat{D}'_Z$, and $\text{Re}\hat{\Phi}(Z)$, $\text{Im}\Phi(Z)$ can be expressed as

$$2\text{Re}\hat{\Phi}(Z)=\sum_{j,k=0}^{\infty}c^{(1)}_{jk}X^jY^k,\ 2\text{Im}\Phi(Z)=\sum_{j,k=0}^{\infty}c^{(2)}_{jk}X^jY^k \tag{5.25}$$

in the neighborhood D_j of t_j, here $X=x-t_j\,(1\le j\le 2)$. Noting that $2\text{Re}\hat{\Phi}(Z)=\tilde{X}H_1(y)u_x+\tilde{Y}H_2(y)u_y=0$ and $2\text{Im}\Phi(Z)=\tilde{Y}H_1(y)u_x-\tilde{X}H_2(y)u_y=0$ on $\text{Im}Z=Y=0$ in $\overline{\hat{D}_Z}$, we have

$$\begin{aligned}&2\text{Re}\hat{\Phi}(Z)=\tilde{X}H_1u_x+\tilde{Y}H_2u_y=YF_1,\\&2\text{Im}\Phi(Z)=\tilde{Y}H_1u_x-\tilde{X}H_2u_y=YF_2\end{aligned} \tag{5.26}$$

in $\tilde{D}_j=D_j\cap D_Z$, where $\tilde{X}=\text{Re}X(Z)$, $\tilde{Y}=\text{Im}X(Z)$, F_1,F_2 are continuous in $D_j\,(1\le j\le 2)$. From the system of algebraic equations, we can solve

u_x, u_y as follows

$$\begin{aligned} &u_x = YH_2(\tilde{X}F_1+\tilde{Y}F_2)/H_1H_2|X(Z)|^2,\\ &u_y = YH_1(\tilde{Y}F_1-\tilde{X}F_2)/H_1H_2|X(Z)|^2, \text{ i.e.}\\ &X(Z)H_1u_x = Y(\tilde{X}F_1+\tilde{Y}F_2)/\overline{X(Z)},\\ &X(Z)H_2u_y = Y(\tilde{Y}F_1-\tilde{X}F_2)/\overline{X(Z)}, \text{ i.e.}\\ &X(Z)u_x = O(|Y|^{(2-m_2)/(2+m_1-m_2)}),\\ &X(Z)u_y = O(|Y|^{(2+m_1-2m_2/(2+m_1-m_2}). \end{aligned} \tag{5.27}$$

Thus we have

$$C_\delta[X(Z)u_x,\tilde{D}_1]\le M_5,\ C_\delta[X(Z)u_y,\tilde{D}_1]\le M_5,\ 1\le j\le 2, \tag{5.28}$$

in which $X(Z)$, δ are as stated in (5.17), and $M_5=M_5(\delta,k,H,D,M_3)$ are non negative constants.

Theorem 5.4 *Under the conditions as in Theorem* 5.3, *Problem* P *for* (5.11) *with the boundary conditions* (5.5) *is solvable.*

Proof We first prove the solvability of Problem Q for the linear case of equation (5.11) with $A_3=0$. For this, we first consider the linear case of equation (5.11) with $A_3=0$, i.e.

$$W_{\overline{Z}}=[A_1W+A_2\overline{W}+A_4]/H_1. \tag{5.29}$$

Obviously Problem Q for the above equation is equivalent to Problem B for the linear complex equation (5.29) with the boundary conditions

$$\begin{aligned} &\mathrm{Re}[\overline{\lambda(z)}W(z)]=H_1(y)r(z)=R(z)=0 \text{ on } \Gamma,\ \mathrm{Im}W(x)=0 \text{ on } \gamma,\\ &u(-1)=0,\ \mathrm{Im}[\overline{\lambda(z)}W(z)]|_{z=z_1}=H_1(\mathrm{Im}z_1)b_1=b_1', \end{aligned} \tag{5.30}$$

and the relation

$$u(z)=u(x)-\int_0^y\frac{V(z)}{H_2(y)}dy=2\mathrm{Re}\int_{-1}^z\left[\frac{U(z)}{H_1(y)}+i\frac{V(z)}{H_2(y)}\right]dz+b_0 \text{ in } D. \tag{5.31}$$

From Theorem 2.5, Chapter I, we see that Problem B for (5.29) has a unique solution $W(z)$, and $u(z)$ in (5.31) is the solution of Problem Q for (5.11) with $A_3=0$. Now let $u_0(z)$ be a solution of Problem Q for the linear equation (5.11), if $u_0(z)$ satisfies the point condition $u_0(1)=b_1$, then the solution is also a solution of Problem P for the equation. Otherwise we can find a solutions u_1 of Problem Q for the homogeneous equation of (5.11) in D satisfying the boundary conditions

$$\mathrm{Re}[\overline{\lambda(z)}u_{1z}]=0 \text{ on } \Gamma, u_1(-1)=0, \mathrm{Im}[\overline{\lambda(z)}u_{1z}]|_{z=z_1}=1,$$

it is clear that there exists a real constant $d_1 \neq 0$, such that

$$u(z) = u_0(z) + d_1 u_1(z)$$

is a solution of Problem P for equation (5.11) satisfying the boundary conditions of Problem P. Moreover by using the method of parameter extension, namely we consider the equation with the parameter $t \in [0,1]$:

$$K_1(y)u_{xx}+K_2(y)u_{yy}+au_x+bu_y+tcu=-d. \tag{5.32}$$

From the solvability of Problem P for (5.32) with $t = 0$, we can find a solution of Problem P for the general linear equation (5.32) with $t = 1$, i.e. the linear case of equation (5.11).

Next we can use the Leray-Schauder theorem to prove the existence of solutions of Problem P for the quasilinear equation (5.11) as stated in the proof of Theorem 3.4. Introduce a bounded and open set B_M in the Banach space $B = \hat{C}_\delta(D_Z)$, the elements of which are all functions satisfying the following condition

$$\hat{C}_\delta[W(Z), \overline{D_Z}] < M_1 + 1,$$

where M_1 is a constant as stated in (5.17). Let us arbitrarily select a system of functions $[w(Z), u(z)] \in B$ and be substituted into the coefficients of (5.11), obviously such equation (5.11) can be seen a linear equation, i.e.

$$\begin{aligned} W_{\overline{Z}} = t[&A_1(z,u,w)W + A_2(z,u,w)\overline{W} + A_3(z,u,w)u \\ &+A_4(z,u,w)]/H_1 = \hat{G}(z,u,w,W)/H_1, \ 0\leq t\leq 1. \end{aligned} \tag{5.33}$$

Form the above discussion, we see that Problem P have a unique solution $W(z)$. Denote by $W = T(w,t)$ the mapping from $w(z)$ to $W(z)$, we can prove that $W = T(w,t)$ is continuous in B and completely continuous in B_M. In fact, arbitrarily select a sequence $w_n(z)$ in B, $n = 0,1,2,...$, such that $\hat{C}_\delta(w_n - w_0, \overline{D_Z}) \to 0$ as $n \to \infty$. By Condition C, we see that $L_\infty[y^\tau X(Z)(\hat{G}(z,u_n,w_n,W_0) - \hat{G}(z,u_0,w_0,W_0)), \overline{D_Z}] \to 0$ as $n \to \infty$, $\tau = \max(1 - m_1/2 - m_2/2, 1 - m_2)$. Moreover, from $W_n = T[w_n,t]$, $W_0 = T[w_0,t]$, it is easy to see that $W_n - W_0$ is a solution of Problem A for the following complex equation

$$H_1(y)(W_n-W_0)_{\overline{Z}}=\hat{G}(z,u_n,w_n,W_n)-\hat{G}(z,u_n,w_n,W_0)+\tilde{G}_n,$$

$$\tilde{G}_n(z,u_n,w_n,u_0,w_0,W_0)=\hat{G}(z,u_n,w_n,W_0)-\hat{G}(z,u_0,w_0,W_0) \text{ in } D_Z, \tag{5.34}$$

and then we can obtain the estimate

$$\hat{C}_\delta[W_n-W_0,\overline{D_Z})]\le 2k_0L_\infty[y^\tau X(Z)\tilde{G}_n,\overline{D_Z}],$$

in which $X(Z)=(Z+1)^{\eta_1}(Z-1)^{\eta_2}$, $\eta_j\ (j=1,2)$ are as stated in (5.17), and δ is a sufficiently small positive constant. Due to Condition C, $L_\infty[y^\tau X(Z)\tilde{G}_n,\overline{D_Z}]\to 0$ as $n\to\infty$, we get $\hat{C}_\delta[W_n-W_0,\overline{D_Z}]\to 0$ as $n\to\infty$. Moreover for $w_n(z)\in B_M$, $n=1,2,...$, from $\{w_n(z)\}$ we can select a subsequence denoted by $\{w_n(z)\}$ again and there exists a function $w_0(z)$, such that $\hat{C}[w_n-w_0,\overline{D_Z}]\to 0$ as $n\to\infty$. Similarly we have

$$H_1(y)(W_n-W_0)_{\overline{Z}}=\hat{G}(z,u_n,w_n,W_n)-\hat{G}(z,u_n,w_n,W_0),$$

$$\tilde{G}_n=\hat{G}(z,u_n,w_n,W_0)-\hat{G}(z,u_0,w_0,W_0) \text{ in } D_Z,$$

where $L_\infty[y^{\tau_1}X(Z)(\hat{G}(z,u_n,w_n,W_0)-\hat{G}(z,u_0,w_0,W_0)),\overline{D_Z}]\le 2k_0M_6$, herein $M_6=M_6(\delta,k,H,D)$ is a non-negative constant, hence from (5.17), and then obtain the estimate

$$\hat{C}_\delta[W_n-W_0,\overline{D_Z})]\le 2M_2k_0M_6.$$

Hence there exists a function $W_0(z)\in B_M$, such that $\hat{C}_\delta[W_n-W_0,\overline{D_Z}]\to 0$ as $n\to\infty$. This shows that $W=T(w,t)$ is completely continuous in B_M, and three conditions of Leray-Schauder theorem can be verified. Hence there exists a system of functions $[W,u]\in B_M$, such that $W=T(W,1)$. Thus the corresponding function $u(z)$ is just a solution of Problem P for (5.11). This completes the proof.

In particular we can obtain the unique solvability of the mixed boundary value problem (Problem M) for equation (5.11) with the boundary condition (2.7)($W(z)=u_{\bar{z}}$).

Remark 5.1 If $K_j(y)=y^{m_j}$, $m_j\ (j=1,2)$ are any positive constants, and the coefficients a,b,c,d of equation (5.1) satisfy the conditions

$$L_\infty[\eta/|y|^{[m_2]},\overline{D}]\le k_0, \eta=a,b,c, L_\infty[d/|y|^{[m_2]},\overline{D}]\le k_1, m_1-[m_2]>0,$$

where k_0,k_1 are non-negative constants, and $[m_2]$ means the integer part of m_2, then using the similar method, the results in Theorems 5.1-5.4 are proved.

Remark 5.2 For Problem P, if the boundary condition (5.5) are replaced by

$$\frac{1}{2}\frac{\partial u}{\partial \nu}+\sigma(z)u(z)=r(z) \text{ on } \Gamma,\ u_y=r(x) \text{ on } \gamma,\ u(-1)=b_0,\ u(1)=b_1,$$

where $\sigma(z) \geq 0$ on Γ is a non-negative function satisfying the condition $C_\alpha[\sigma(z),\Gamma] \leq k_0$, then the above corresponding theorems hold still.

Now we mention when $m_2 = 0$, then equation (5.1) becomes equation (2.1) in Section 2. Moreover the coefficients $K_j(y)\,(j=1,2)$ in equations (5.1) and (5.11) can be replaced by functions $K_j(x,y)\,(j=1,2)$ with some conditions, for instance $K_j(x,y)=y^{m_j}h_j(x,y)$, $m_j\ (j=1,2)$ are as stated before, and $h_j(x,y)\,(j=1,2)$ are continuously differentiable positive functions in $\overline{D}$.

Finally we introduce a mapping $Z = Z(z)$ of the domain D with the boundary $\Gamma\cup\gamma$, such that the boundary Γ maps the curve vertical to the axis $\text{Im}z = 0$ near $z = -1, 1$ respectively, and the function $Z = Z(z)$ and its inverse function $Z = Z(z)$ are Hölder continuous in the corresponding closed domains. We can choose the function $\tilde{G}(y) = py^q$, $p, q(> m)$ are two positive constants, m is as stated in (2.1), it is clear that $x = \tilde{G}(y) - 1$ and $x = 1 - \tilde{G}(y)$ are two curves vertical to the axis $\text{Im}z = 0$ near $z = -1, 1$ respectively. We can only consider the problem in the domain $D\cap\{y<\delta\}$. Let the partial boundary $\tilde{\Gamma} = Z(\Gamma\cap\{y<\delta\})$ of the domain $\tilde{D} = Z(D\cap\{y < \delta\})$ near $z = -1, 1$ is as follows

$$\tilde{\Gamma}_1=\{y=\gamma_1(s),0\leq s\leq s_1'\},\tilde{\Gamma}_2=\{y=\gamma_2(s),0\leq s\leq s_2'\}, \tag{5.35}$$

where s is the parameter of arc length of $\tilde{\Gamma}_1$ or $\tilde{\Gamma}_2$, $\gamma_k(0) = 0$, $\gamma_k(s) > 0$ on $\{0 < s \leq s_k'\}\,(k = 1,2)$ and $\gamma_k(s)$ on $\{0 \leq s \leq s_k'\}\,(k = 1,2)$ are continuously differentiable, the slopes of the curves $\tilde{\Gamma}_k\,(k = 1,2)$ at the points $z_k = x_k + iy_k\,(k = 1,2)$ are not equal to the slopes $dy/dx = (-1)^k/H(y_k)\,(k = 1,2)$ of the characteristic curves of families $s_k\,(k = 1,2)$ at those points respectively, where $z_k\,(k = 1,2)$ are the intersection points of $\tilde{\Gamma}_k\,(k = 1,2)$ and the characteristic curves, hence $\gamma_1(s)(0 \leq s \leq s_1')$, $\gamma_2(s)(0 \leq s \leq s_2')$ can be expressed by $\gamma_1[s(\tilde{\mu})]\,(-1 \leq \tilde{\mu} \leq 1)$, $\gamma_2[s(\tilde{\nu})]\,(-1 \leq \tilde{\nu} \leq 1)$ respectively.

Now we rewrite the another form of equation (5.11). It is clear that

$$\begin{aligned} W_{\tilde{\bar{z}}} &= [H_1(U+iV)_x+iH_2(U+iV)_y]/2=[H_1U_xH_2V_y+i(H_1V_x+H_2U_y)]/2 \\ &= H_1[U_x - H_2V_y/H_1 + i(V_x + H_2U_y/H_1)]/2 \end{aligned}$$

$$= H_1\{(U+V)_x/2-H_2(U+V)_y/2H_1+(U-V)_x/2+H_2(U-V)_y/2H_1$$
$$+i[(U+V)_x/2+H_2(U+V)_y/2H_1-(U-V)_x/2+H_2(U-V)_y/2H_1]\}/2$$
$$= H_1\{(U-V)_{\tilde\mu}+(U+V)_{\tilde\nu}+i[(U+V)_{\tilde\mu}-(U-V)_{\tilde\nu}]\}/2 = g(Z)$$
$$= iH_1[(U+V)-i(U-V)]_{\tilde\mu+i\tilde\nu}=iH_1\overline{[(U+V)+i(U-V)]_{\tilde\mu-i\tilde\nu}} \text{ in } \overline{D_Z}, \tag{5.36}$$

where $Z = x + iG(y)$, $\Omega(\tilde\tau) = U + V) + i(U - V)$, $\tilde\tau = \tilde\mu + i\tilde\nu$ and

$$\tilde\mu = x+G(y) = x + \int_0^y H(y)dy,\ \tilde\nu = x-G(y),$$
$$\frac{\partial x}{\partial\tilde\mu} = \frac{\partial y}{\partial\tilde\mu} = \frac{1}{2},\ \frac{\partial y}{\partial\tilde\mu} = -\frac{\partial y}{\partial\tilde\mu} = \frac{\tilde H_2(y)}{2\tilde H_1(y)} \text{ in } \overline{D}.$$

This shows that equation (5.33) or (5.36) can be rewritten as

$$\Omega_{\overline{\tilde\tau}} = i\overline{g[Z(\tilde\tau)]}/H_1 \text{ in } D_\tau = \tau(D_Z), \tag{5.37}$$

where D_τ is the image domain of D through the mapping $\tilde\tau = \tilde\mu + i\tilde\nu = x + G(y) + i[x - G(y)]$.

By the above conditions, the inverse function $x = (\tilde\mu + \tilde\nu)/2 = \sigma_1(\tilde\mu)$ of $\tilde\mu = x + G(y)(= x + \gamma_1(s))$, and then

$$\tilde\Gamma_1 : \tilde\nu = 2\sigma_1(\tilde\mu) - \tilde\mu = \sigma_1(x + \gamma_1(s)) - x - \gamma_1(s),\ 0 \le s \le s_1'.$$

Moreover we consider the inverse function $x = (\tilde\mu + \tilde\nu)/2 = \sigma_2(\tilde\nu)$ of $\tilde\nu = x - G(y)(= x - \gamma_2(s))$ can be found, which can be written as

$$\tilde\Gamma_2 : \tilde\mu = 2\sigma_2(\tilde\nu) - \tilde\nu = 2\sigma_2(x - \gamma_2(s)) - x + \gamma_2(s),\ 0 \le s \le s_2'.$$

We make a transformation

$$\mu = \frac{2[\tilde\mu - 2\sigma_2(\tilde\nu) + \tilde\nu]}{1 + 2\sigma_2(\tilde\nu) - \tilde\nu} + 1,\ \nu = \frac{2[\tilde\nu - 2\sigma_1(\tilde\mu) + \tilde\mu]}{1 - 2\sigma_1(\tilde\mu) + \tilde\mu} - 1,$$
$$\text{in } D_{\tilde\tau}=\{-1\le\tilde\mu\le 2\sigma_2(\tilde\nu)-\tilde\nu,\, 2\sigma_1(\tilde\mu)-\tilde\mu\le\tilde\nu\le 1\}, \tag{5.38}$$

where $\tilde\mu$, $\tilde\nu$ are real variables. From equation (5.36), denote $\xi = U + V$, $\eta = U - V$, we have

$$\xi_\mu = \frac{1 + 2\sigma_2(\tilde\nu) - \tilde\nu}{2}\xi_{\tilde\mu},\ \eta_\mu = \frac{1 + 2\sigma_2(\tilde\nu) - \tilde\nu}{2}\eta_{\tilde\mu},$$
$$\xi_\nu = \frac{1 - 2\sigma_1(\tilde\mu) + \tilde\mu}{2}\xi_{\tilde\nu},\ \eta_\nu = \frac{1 - 2\sigma_1(\tilde\mu) + \tilde\mu}{2}\eta_{\tilde\nu}, \tag{5.39}$$

and

$$\begin{aligned}\xi_{\tilde\mu}-\eta_{\tilde\nu}&=\frac{2}{1+2\sigma_2(\tilde\nu)-\tilde\nu}\xi_\mu-\frac{2}{1-2\sigma_1(\tilde\mu)+\tilde\mu}\eta_\nu\\&=\frac{2}{1-2\sigma_1(\tilde\mu)+\tilde\mu}[\frac{1-2\sigma_1(\tilde\mu)+\tilde\mu}{1+2\sigma_2(\tilde\nu)-\tilde\nu}\xi_\mu-\eta_\nu],\\\eta_{\tilde\mu}+\xi_{\tilde\nu}&=\frac{2}{1+2\sigma_2(\tilde\nu)-\tilde\nu}\eta_\mu+\frac{2}{1-2\sigma_1(\tilde\mu)+\tilde\mu}\xi_\nu\\&=\frac{2}{1+2\sigma_2(\tilde\nu)-\tilde\nu}[\frac{1+2\sigma_2(\tilde\nu)-\tilde\nu}{1-2\sigma_1(\tilde\mu)+\tilde\mu}\xi_\nu+\eta_\mu],\end{aligned}\tag{5.40}$$

thus according to (1.3)-(1.6), Chapter I, through the transformation (5.39), we can verify that the function $\Omega=\xi+i\eta$ satisfies the equation

$$\begin{aligned}\Omega_{\overline\tau}-Q(\tau)\Omega_\tau&=\frac{1-2\sigma_1(\tilde\mu)+\tilde\mu}{2}\mathrm{Re}\Omega_{\overline{\tilde\tau}}+i\frac{1+2\sigma_2(\tilde\nu)-\tilde\nu}{2}\mathrm{Im}\Omega_{\overline{\tilde\tau}}\\&=\frac{1-2\sigma_1(\tilde\mu)+\tilde\mu}{2}\mathrm{Re}\frac{\overline{ig(Z)}}{H(y)}+i\frac{1+2\sigma_2(\tilde\nu)-\tilde\nu}{2}\mathrm{Im}\frac{\overline{ig(Z)}}{H(y)}=F(\tau)\ \text{in}\ D_\tau,\end{aligned}\tag{5.41}$$

in which $\tau=\mu+i\nu$, $\tilde\tau=\tilde\mu+i\tilde\nu$, and

$$Q(\tau)=\frac{1+2\sigma_2(\tilde\nu)-\tilde\nu-1+2\sigma_1(\tilde\mu)-\tilde\mu}{1+2\sigma_2(\tilde\nu)-\tilde\nu+1-2\sigma_1(\tilde\mu)+\tilde\mu},\ |Q(\tau)|\le q_0<1\ \text{in}\ D_\tau,\tag{5.42}$$

the last inequality is the ellipticity condition, where q_0 is a non-negative constant. Moreover we can reduce equation (5.41) to the equation in the form

$$\Omega_{\overline\omega}=F\,\overline\tau_{\overline\omega}\ \text{ in }\ \omega(D_\tau)=D_\omega,$$

where $\omega(\tau)$ is a homeomorphism of the Beltrami equation

$$\omega_{\overline\tau}-Q\,\omega_\tau=0\ \text{ in }\ D_\tau,$$

such that $D_\tau,-1,0,1$ are mapped onto $D_\omega,-1,0,1$ respectively. Thus we can first estimate the solutions of the corresponding problem, and then get the estimates of solutions of the original problem.

By using the similar method, we can give a mapping $Z=Z(z)$ of the domain D with the boundary $\tilde\Gamma\cup\gamma$, which maps the curve $\tilde\Gamma$ to a smooth curve in the form $x=-1+G(y)$, $x=1-G(z)$ near the points $z=-1$ and 1 respectively, this is the simplest case of $Q(\tau)=0$ in D_τ in (5.42) and the inner angles of D_Z at $z=-1,1$ are equal to $\pi/4$. We can require that the function $Z=Z(z)$ and its inverse function $Z=Z(z)$ are Hölder continuous in the corresponding closed domains, hence it suffices to consider the boundary value problem of the corresponding domain, under the above mapping, the index of the boundary value problem for the corresponding functions remains to be unchanged.

CHAPTER III
HYPERBOLIC COMPLEX EQUATIONS OF FIRST AND SECOND ORDERS

In this chapter, we mainly discuss the Riemann-Hilbert boundary value problem for degenerate hyperbolic complex equations of first order, and oblique derivative boundary value problems for some classes of degenerate hyperbolic equations of second order.

1 The Riemann-Hilbert Problem for Uniformly Hyperbolic Complex Equations of First Order

In this section, we first reduce the hyperbolic systems of first order equations under some conditions to the complex forms, and then pose the Riemann-Hilbert boundary value problem for uniformly hyperbolic complex equations of first order in a simply connected domain. Moreover we give a representation theorem of solutions for the above boundary value problem, and prove the uniqueness and existence of solutions for the above problem by using the successive approximation.

1.1 Complex forms of linear and quasilinear hyperbolic system of first order equations

First of all, we introduce the hyperbolic number and hyperbolic complex function. The so-called hyperbolic number is $z = x + jy$, where x, y are two real numbers and j is called the hyperbolic unit such that $j^2 = 1$. Denote

$$e_1 = (1 + j)/2,\ e_2 = (1 - j)/2,$$

it is easy to see that

$$e_1 + e_2 = 1,\ e_k e_l = \begin{cases} e_k, & \text{if } k = l, \\ 0, & \text{if } k \neq l, \end{cases} \quad k, l = 1, 2,$$

and (e_1, e_2) will be called the hyperbolic element. Moreover, $w = f(z) = u(x, y) + jv(x, y)$ is called a hyperbolic complex function, where $u(x, y)$, $v(x, y)$ are two real functions of two real variables x, y, which are called the real part and imaginary part of $w = f(z)$ and denote $\mathrm{Re}\, w = u(z) = u(x, y)$, $\mathrm{Im}\, w = v(z) = v(x, y)$. Obviously,

$$z = x + jy = \mu e_1 + \nu e_2, \ w = f(z) = u + jv = \xi e_1 + \eta e_2, \tag{1.1}$$

in which

$$\mu = x + y, \ \nu = x - y, \ x = (\mu + \nu)/2, \ y = (\mu - \nu)/2,$$

$$\xi = u + v, \ \eta = u - v, \ u = (\xi + \eta)/2, \ v = (\xi - \eta)/2.$$

$\bar{z} = x - jy$ will be called the conjugate number of z. The absolute value of z is defined by $|z| = \sqrt{|x^2 - y^2|}$, and the hyperbolic model of z is defined by $\|z\| = \sqrt{x^2 + y^2}$. The operations of addition, subtraction and multiplication are the same as the real numbers, but $j^2 = 1$. It is clear that $|z_1 z_2| = |z_1||z_2|$, but the triangle inequality is not true. As for the hyperbolic model of z, we have the triangle inequality: $\| z_1 + z_2 \| \leq \| z_1 \| + \| z_2 \|$, and $\| z_1 z_2 \| \leq \sqrt{2} \| z_1 \| \| z_2 \|$. Later on the limit of the hyperbolic number will be defined by the hyperbolic model. The partial derivatives of a hyperbolic complex function $w = f(z)$ with respect to z and $\bar{z}$ are defined by

$$w_z = (w_x - jw_y)/2, \ w_{\bar{z}} = (w_x + jw_y)/2, \tag{1.2}$$

respectively, and then we have

$$\begin{aligned}
w_z &= (w_x - jw_y)/2 = [(u_x - v_y) + j(v_x - u_y)]/2 \\
&= [(w_x - w_y)e_1 + (w_x + w_y)e_2]/2 = w_\nu e_1 + w_\mu e_2 \\
&= [\xi_\nu e_1 + \eta_\nu e_2]e_1 + [\xi_\mu e_1 + \eta_\mu e_2]e_2 = \xi_\nu e_1 + \eta_\mu e_2, \\
w_{\bar{z}} &= [(u_x + v_y) + j(v_x + u_y)]/2 = w_\mu e_1 + w_\nu e_2 \\
&= (\xi e_1 + \eta e_2)_\mu e_1 + (\xi e_1 + \eta e_2)_\nu e_2 = \xi_\mu e_1 + \eta_\nu e_2.
\end{aligned}$$

Let D be a bounded domain in the (x, y)-plane. If $u(x, y), v(x, y)$ are continuously differentiable in D, then we say that the function $w = f(z)$ is continuously differentiable in D, and the following result can be derived.

Theorem 1.1 *Suppose that the hyperbolic complex function $w = f(z)$ is continuously differentiable. Then the following three conditions are equivalent.*

(1) $w_{\bar{z}} = 0;$ (1.3)

(2) $\xi_\mu = 0, \ \eta_\nu = 0;$ (1.4)

$$(3)\ u_x + v_y = 0,\ v_x + u_y = 0. \tag{1.5}$$

The system of equations (1.5) is the simplest hyperbolic system of first order equations, which corresponds to the Cauchy-Riemann system in the theory of elliptic equations. The continuously differentiable solution $w = f(z)$ of the complex equation (1.3) in D is called a hyperbolic regular function in D. In addition, we can obtain some properties about integrals of hyperbolic complex functions (see [86]33)).

Next, we transform the hyperbolic systems of first order equations into the complex forms. We first consider the linear hyperbolic system of first order partial differential equations

$$\begin{cases} a_{11}u_x + a_{12}u_y + b_{11}v_x + b_{12}v_y = a_1 u + b_1 v + c_1, \\ a_{21}u_x + a_{22}u_y + b_{21}v_x + b_{22}v_y = a_2 u + b_2 v + c_2, \end{cases} \tag{1.6}$$

where the coefficients $a_{kl}, b_{kl}, a_k, b_k, c_k\ (k,l=1,2)$ are known real functions in D, in which D is a bounded domain. System (1.6) is called hyperbolic at a point in D, if at the point, the inequality

$$I = (K_2 + K_3)^2 - 4K_1K_4 = (K_2 - K_3)^2 - 4K_5K_6 > 0 \tag{1.7}$$

holds, in which

$$K_1 = \begin{vmatrix} a_{11} & b_{11} \\ a_{21} & b_{21} \end{vmatrix},\ K_2 = \begin{vmatrix} a_{11} & b_{12} \\ a_{21} & b_{22} \end{vmatrix},\ K_3 = \begin{vmatrix} a_{12} & b_{11} \\ a_{22} & b_{21} \end{vmatrix},$$

$$K_4 = \begin{vmatrix} a_{12} & b_{12} \\ a_{22} & b_{22} \end{vmatrix},\ K_5 = \begin{vmatrix} a_{11} & a_{12} \\ a_{21} & a_{22} \end{vmatrix},\ K_6 = \begin{vmatrix} b_{11} & b_{12} \\ b_{21} & b_{22} \end{vmatrix}.$$

If the inequality (1.7) at every point (x, y) in D holds, then (1.6) is called a hyperbolic system in D. If the coefficients $a_{kl}, b_{kl}\ (k,l=1,2)$ in D are bounded, and the condition

$$I = (K_2 + K_3)^2 - 4K_1K_4 = (K_2 - K_3)^2 - 4K_5K_6 \geq I_0 > 0 \tag{1.8}$$

holds, in which I_0 is a positive constant, then (1.6) is called uniformly hyperbolic in D. Under certain conditions, by using the relations

$$\begin{cases} u_x = (w_{\bar z} + \overline{w}_z + w_z + \overline{w}_{\bar z})/2,\ u_y = j(w_{\bar z} - \overline{w}_z - w_z + \overline{w}_{\bar z})/2, \\ v_x = j(w_{\bar z} - \overline{w}_z + w_z - \overline{w}_{\bar z})/2,\ v_y = (w_{\bar z} + \overline{w}_z - w_z - \overline{w}_{\bar z})/2, \end{cases}$$

the system (1.6) can be reduced to the complex form:

$$w_{\bar{z}} - Q_1 w_z - Q_2 \overline{w}_{\bar{z}} = A_1 w + A_2 \bar{w} + A_3, \tag{1.9}$$

in which $Q_k = Q_k(z)\,(k = 1,2)$ are functions of $a_{kl}, b_{kl}\,(k,l = 1,2)$, and $A_k = A_k(z)\,(k = 1,2,3)$ are functions of $a_{kl}, b_{kl}, a_k, b_k, c_k\,(k,l = 1,2)$.

In addition, we discuss the quasilinear hyperbolic system of first order partial differential equations

$$\begin{cases} a_{11}u_x + a_{12}u_y + b_{11}v_x + b_{12}v_y = a_1 u + b_1 v + c_1, \\ a_{21}u_x + a_{22}u_y + b_{21}v_x + b_{22}v_y = a_2 u + b_2 v + c_2, \end{cases} \tag{1.10}$$

where the coefficients $a_{kl}, b_{kl}\,(k,l - 1,2)$ are known functions in $(x,y) \in D$ and $a_k, b_k, c_k\,(k = 1,2)$ are known functions of $(x,y) \in D$ and $u, v \in \mathbf{R}$. The conditions of hyperbolicity and uniform hyperbolicity of (1.10) are the same as for system (1.6), i.e. for any point $(x,y) \in D$, the inequalities (1.7) and (1.8) hold respectively. Hence we can obtain its complex form (1.9) (see [86]33)).

1.2 Formulation of the Riemann-Hilbert problem and uniqueness of its solutions for simplest hyperbolic complex equation

Let D be a simply connected bounded domain in the $x+jy$-plane with the boundary $\Gamma = L_1 \cup L_2 \cup L_3 \cup L_4$, where $L_1 = \{x = -y, 0 \le x \le R_1\}$, $L_2 = \{x = y + 2R_1, R_1 \le x \le R_2\}$, $L_3 = \{x = -y + 2R, R = R_2 - R_1 \le x \le R_2\}$, $L_4 = \{x = y, 0 \le x \le R = R_2 - R_1\}$, and denote $z_0 = 0$, $z_1 = (1-j)R_1, z_2 = R_2 + j(R_2 - 2R_1)$, $z_3 = (1+j)(R_2 - R_1) = (1+j)R$, and $L = L_1 \cup L_4$, where j is the hyperbolic unit. For convenience we only discuss the case $R_2 \ge 2R_1$, the other case can be discussed by a similar way. We consider the simplest hyperbolic complex equation of first order:

$$w_{\bar{z}} = 0 \text{ in } D. \tag{1.11}$$

The Riemann-Hilbert boundary value problem for the complex equation (1.11) may be formulated as follows:

Problem A Find a continuous solution $w(z)$ of (1.11) in $\overline{D}$ satisfying the boundary conditions

$$\operatorname{Re}[\overline{\lambda(z)}w(z)] = r(z), \;\; z \in L, \; \operatorname{Im}[\overline{\lambda(z_0)}w(z_0)] = b_1, \tag{1.12}$$

where $\lambda(z) = a(z) + jb(z) \neq 0$, $z \in L$, and $\lambda(z)$, $r(z)$, b_1 satisfy the conditions

$$C_\alpha[\lambda(z), L] = C_\alpha[\mathrm{Re}\lambda, L] + C_\alpha[\mathrm{Im}\lambda, L] \leq k_0,\ C_\alpha[r(z), L] \leq k_2, \tag{1.13}$$

$$|b_1| \leq k_2,\ \max_{z\in L_1} \frac{1}{|a(z) - b(z)|},\ \max_{z\in L_4} \frac{1}{|a(z) + b(z)|} \leq k_0, \tag{1.14}$$

in which b_1 is a real constant and $\alpha\,(0 < \alpha < 1), k_0, k_2$ are positive constants. In particular, when $a(z) = 1$, $b(z) = 0$, i.e. $\lambda(z) = 1$, $z \in L$, Problem A is the Dirichlet problem (Problem D), whose boundary condition is

$$\mathrm{Re}[w(z)] = r(z),\ z \in L,\ \mathrm{Im}[w(z_0)] = b_1. \tag{1.15}$$

Problem A with the conditions $r(z) = 0, z \in L$, $b_1 = 0$ is called Problem A_0.

On the basis of Theorem 1.1, it is clear that the complex equation (1.11) can be reduced to the form

$$\xi_\mu = 0,\ \eta_\nu = 0,\ \ (\mu, \nu) \in Q = \{0 \leq \mu \leq 2R, 0 \leq \nu \leq 2R_1\}, \tag{1.16}$$

where $\mu = x+y$, $\nu = x-y$, $\xi = u+v$, $\eta = u-v$. Hence the general solution of system (1.16) can be expressed as

$$\begin{aligned} &\xi = u+v = f(\nu) = f(x-y),\ \eta = u-v = g(\mu) = g(x+y),\ \text{ i.e.} \\ &u = [f(x-y) + g(x+y)]/2,\ v = [f(x-y) - g(x+y)]/2, \end{aligned} \tag{1.17}$$

in which $f(t)$, $g(t)$ are two arbitrary real continuous functions on $[0, 2R_1]$, $[0, 2R]$ respectively. From the boundary condition (1.12), we have

$$\begin{aligned} &a(z)u(z) - b(z)v(z) = r(z) \text{ on } L,\ \overline{\lambda(z_0)}w(z_0) = r(z_0) + jb_1,\ \text{ i.e.} \\ &[a((1-j)x) - b((1-j)x)]f(2x) + [a((1-j)x) + b((1-j)x)] \\ &\times g(0) = 2r((1-j)x), x \in [0, R_1], \\ &[a((1+j)x) - b((1+j)x)]f(0) + [a((1+j)x) + b((1+j)x)] \\ &\times g(2x) = 2r((1+j)x),\ x \in [0, R], \\ &(a(0) - b(0))f(0) = (a(0) - b(0))(u(0) + v(0)) = r(0) + b_1 \text{ or } 0, \\ &(a(0) + b(0))g(0) = (a(0) + b(0))(u(0) - v(0)) = r(0) - b_1 \text{ or } 0. \end{aligned} \tag{1.18}$$

The above formulas can be rewritten as

$$[a((1-j)t/2) - b((1-j)t/2)]f(t) + [a((1-j)t/2) + b((1-j)t/2)]$$

$$\times g(0) = 2r((1-j)t/2),\ t\in[0,2R_1],$$

$$[a((1+j)t/2)-b((1+j)t/2)]f(0)+[a((1+j)t/2)+b((1+j)t/2)]$$

$$\times g(t) = 2r((1+j)t/2),\ \ t\in[0,2R],\ \text{ i.e.}$$

$$\begin{aligned} f(x-y) &= \frac{2r((1-j)(x-y)/2)}{a((1-j)(x-y)/2)-b((1-j)(x-y)/2)}\\ &-\frac{[a((1-j)(x-y)/2)+b((1-j)(x-y)/2)]g(0)}{a((1-j)(x-y)/2)-b((1-j)(x-y)/2)},\ 0\le x-y\le 2R_1,\\ g(x+y) &= \frac{2r((1+j)(x+y)/2)}{a((1+j)(x+y)/2)+b((1+j)(x+y)/2)}\\ &-\frac{[a((1+j)(x+y)/2)-b((1+j)(x+y)/2)]f(0)}{a((1+j)(x+y)/2)+b((1+j)(x+y)/2)},\ 0\le x+y\le 2R. \end{aligned} \tag{1.19}$$

Thus the solution $w(z)$ of (1.11) can be expressed as

$$\begin{aligned} w(z) &= f(x-y)e_1+g(x+y)e_2\\ &=\frac{1}{2}\{f(x-y)+g(x+y)+j[f(x-y)-g(x+y)]\}, \end{aligned} \tag{1.20}$$

where $f(x-y)$, $g(x+y)$ are as stated in (1.19) and $f(0), g(0)$ are as stated in (1.18). It is not difficult to see that $w(z)$ satisfies the estimate

$$C_\alpha[w(z),\bar{D}]\le M_1,\ \ C_\alpha[w(z),\bar{D}]\le M_2k_2, \tag{1.21}$$

where $M_1=M_1(\alpha,k_0,k_2,D)$, $M_2=M_2(\alpha,k_0,D)$ are two positive constants only dependent on α,k_0,k_2,D and α,k_0,D respectively. The above results can be written as a theorem.

Theorem 1.2 *Any solution $w(z)$ of Problem A for the complex equation* (1.11) *possesses the representation* (1.20), *which satisfies the estimate* (1.21).

1.3 Uniqueness of solutions of the Riemann-Hilbert problem for linear hyperbolic complex equations

Now we discuss the following special case of the complex equation (1.9):

$$w_{\bar z} = A_1(z)w+A_2(z)\overline{w}+A_3(z), \tag{1.22}$$

and suppose that equation (1.22) satisfies the following conditions, i.e.
Condition C: $A_l(z)\,(l=1,2,3)$ are continuous in $\bar{D}$ and satisfy

$$\hat{C}[A_l,\bar{D}]=C[A_l,\bar{D}]+C[A_{lx},\bar{D}]\le k_0, l=1,2, \hat{C}[A_3,\bar{D}]\le k_1, \tag{1.23}$$

where $C[A_l,\bar{D}]=C[\mathrm{Re}A_l,\bar{D}]+C[\mathrm{Im}A_l,\bar{D}]$, and k_0, k_1 are positive constants.

Due to $w=u+jv=\xi e_1+\eta e_2$, $w_z=\xi_\nu e_1+\eta_\mu e_2$, $w_{\bar{z}}=\xi_\mu e_1+\eta_\nu e_2$, equation (1.22) can be rewritten in the form

$$\begin{gathered}\xi_\mu e_1+\eta_\nu e_2=[A(z)\xi+B(z)\eta+E(z)]e_1\\ +[C(z)\xi+D(z)\eta+F(z)]e_2,\ z\in D, \text{ i.e.}\\ \begin{cases}\xi_\mu=A(z)\xi+B(z)\eta+E(z),\\ \eta_\nu=C(z)\xi+D(z)\eta+F(z),\end{cases} z\in D,\end{gathered}\tag{1.24}$$

in which

$$\begin{gathered}A=\mathrm{Re}A_1+\mathrm{Im}A_1,\ B=\mathrm{Re}A_2+\mathrm{Im}A_2,\ C=\mathrm{Re}A_2-\mathrm{Im}A_2,\\ D=\mathrm{Re}A_1-\mathrm{Im}A_1,\ E=\mathrm{Re}A_3+\mathrm{Im}A_3, F=\mathrm{Re}A_3-\mathrm{Im}A_3.\end{gathered}\tag{1.25}$$

The boundary condition (1.12) can be reduced to

$$\mathrm{Re}[\overline{\lambda}(\xi e_1+\eta e_2)]=r(z),\ \mathrm{Im}[\overline{\lambda}(\xi e_1+\eta e_2)]|_{z=z_0}=b_1, \tag{1.26}$$

where $\lambda=(a+b)e_1+(a-b)e_2$. Moreover, the domain D is transformed into the rectangle $Q=\{0\le\mu\le 2R_1, 0\le\nu\le 2R, R=R_2-R_1\}$, and A,B,C,D,E,F are known functions of $(\mu,\nu)\in Q$. For convenience, sometimes we write $z\in D$ or $z\in Q$, and denote $L_1=\{\mu=0, 0\le\nu\le 2R\}$, $L_4=\{0\le\mu\le 2R_1, \nu=0\}$.

Now we give a representation of solutions of Problem A for equation (1.22).

Theorem 1.3 *If equation* (1.22) *satisfies Condition* C, *then any solution* $w(z)$ *of Problem* A *for* (1.22) *can be expressed as*

$$\begin{aligned}&w(z)=w_0(z)+\Phi(z)+\Psi(z) \text{ in } D,\\ &w_0(z)=f(x-y)e_1+g(x+y)e_2,\ \Phi(z)=\tilde{f}(x-y)e_1+\tilde{g}(x+y)e_2,\\ &\Psi(z)=\int_0^{x+y}[A\xi+B\eta+E]e_1d(x+y)+\int_0^{x-y}[C\xi+D\eta+F]e_2d(x-y),\end{aligned}\tag{1.27}$$

where $f(x-y)$, $g(x+y)$ *are as stated in* (1.19) *and* $\tilde{f}(x-y)$, $\tilde{g}(x+y)$ *are similar to* $f(x-y)$, $g(x+y)$ *in* (1.19), *but* $\Phi(z)$ *satisfies the boundary conditions*

$$\text{Re}[\overline{\lambda(z)}\Phi(z)]=-\text{Re}[\overline{\lambda(z)}\Psi(z)] \text{ on } L, \text{Im}[\overline{\lambda(z_0)}\Phi(z_0)]=-\text{Im}[\overline{\lambda(z_0)}\Psi(z_0)]. \tag{1.28}$$

Proof It is not difficult to see that the functions $w_0(z)$, $\Phi(z)$ are solutions of the complex equation (1.11) in $\bar{D}$, which satisfy the boundary conditions (1.12) and (1.28) respectively, $\Psi(z)$ satisfies the complex equation

$$[\Psi]_{\bar{z}} = [A\xi + B\eta + E]e_1 + [C\xi + D\eta + F]e_2, \tag{1.29}$$

and $\Phi(z) + \Psi(z)$ satisfies the boundary condition of Problem A_0. Hence $w(z) = w_0(z) + \Phi(z) + \Psi(z)$ satisfies the boundary condition (1.12) and is a solution of Problem A for (1.22).

Theorem 1.4 *Suppose that Condition C holds. Then Problem A for the complex equation* (1.22) *has at most one solution.*

Proof Let $w_1(z), w_2(z)$ be any two solutions of Problem A for (1.22) and substitute them into equation (1.22) and the boundary condition (1.12). It is clear that $w(z) = w_1(z) - w_2(z)$ satisfies the homogeneous complex equation and boundary conditions

$$w_{\bar{z}} = A_1 w + A_2 \overline{w} \text{ in } D, \tag{1.30}$$

$$\text{Re}[\overline{\lambda(z)}w(z)] = 0, \text{ if } (x,y) \in L, \ w(z_0) = 0. \tag{1.31}$$

On the basis of Theorem 1.3, the function $w(z)$ can be expressed in the form

$$\begin{gathered} w(z) = \Phi(z) + \Psi(z), \\ \Psi(z)=\int_0^{x+y}[A\xi+B\eta]e_1 d(x+y)+\int_0^{x-y}[C\xi+D\eta]e_2 d(x-y). \end{gathered} \tag{1.32}$$

Suppose $w(z) \not\equiv 0$ in the neighborhood($\subset \bar{D}$) of the point $z_0=0$. We may choose a sufficiently small positive number R_0, such that $8M_3MR_0 < 1$, where $M_3=\max\{C[A,Q_0], C[B,Q_0], C[C,Q_0], C[D,Q_0]\}$, $M=1+4k_0^2(1+2k_0^2)$ is a positive constant, and $m=C[w(z),\overline{Q_0}]>0$, herein $Q_0=\{0\le\mu\le R_0\}\cap\{0\le\nu\le R_0\}$. From (1.19),(1.20), (1.27),(1.28),(1.32) and Condition C, we have

$$||\Psi(z)|| \le 8M_3 m R_0, \ ||\Phi(z)|| \le 32M_3 k_0^2(1+2k_0^2)mR_0,$$

thus an absurd inequality $m \le 8M_3MmR_0 < m$ is derived. It shows $w(z) = 0$, $(x, y) \in Q_0$. Moreover, we extend along the positive directions of $\mu = x + y$ and $\nu = x - y$ successively, and finally obtain $w(z) = 0$ for $(x, y) \in D$, i.e. $w_1(z) - w_2(z) = 0$ in D. This proves the uniqueness of solutions of Problem A for (1.22).

1.4 Solvability of Riemann-Hilbert problem for linear hyperbolic complex equations

Theorem 1.5 *If the complex equation* (1.22) *satisfies Condition* C, *then Problem* A *for* (1.22) *has a solution.*

Proof In order to find a solution $w(z)$ of Problem A in D, we can express $w(z)$ in the form (1.27). In the following, by using successive approximation we can find a solution of Problem A for the complex equation (1.22). First of all, substituting the solution $w_0(z) = \xi_0 e_1 + \eta_0 e_2$ of Problem A for (1.11) into the position of $w = \xi e_1 + \eta e_2$ in the right-hand side of (1.22), the function

$$\begin{gathered} w_1(z) = w_0(z) + \Phi_1(z) + \Psi_1(z), \\ \Psi_1(z) = \int_0^\mu [A\xi_0 + B\eta_0 + E]e_1 d\mu + \int_0^\nu [C\xi_0 + D\eta_0 + F]e_2 d\nu, \end{gathered} \tag{1.33}$$

is determined, where $\mu = x + y$, $\nu = x - y$, $\Phi_1(z)$ is a solution of (1.11) in $\bar{D}$ satisfying the boundary conditions

$$\begin{gathered} \mathrm{Re}[\overline{\lambda(z)}\Phi_1(z)] = -\mathrm{Re}[\overline{\lambda(z)}\Psi_1(z)], z \in L, \\ \mathrm{Im}[\overline{\lambda(z_0)}\Phi_1(z_0)] = -\mathrm{Im}[\overline{\lambda(z_0)}\Psi_1(z_0)]. \end{gathered} \tag{1.34}$$

Thus from (1.33), (1.34), we have

$$||w_1(z) - w_0(z)|| = C[w_1(z) - w_0(z), \overline{D}] \le 2M_4M(4m + 1)R', \tag{1.35}$$

in which $M_4 = \max_{z\in\bar{D}}(|A|, |B|, |C|, |D|, |E|, |F|)$, $m = \| w_0 \|_{C(\bar{D})}$, $R' = \max(2R_1, 2R)$, $M = 1 + 4k_0^2(1 + 2k_0^2)$ is a positive constant as in the proof of Theorem 1.4. Moreover, we substitute $w_1(z) = w_0(z) + \Phi_1(z) + \Psi_1(z)$ and the corresponding functions $\xi_1(z) = \mathrm{Re}w_1(z) + \mathrm{Im}w_1(z)$, $\eta_1(z) = \mathrm{Re}w_1(z) - \mathrm{Im}w_1(z)$ into the positions of $w(z)$, $\xi(z)$, $\eta(z)$ in (1.27), and similarly to (1.33)-(1.35), we can find the corresponding functions $\Psi_2(z), \Phi_2(z)$ in $\overline{D}$ and the function

$$w_2(z) = w_0(z) + \Phi_2(z) + \Psi_2(z) \text{ in } \overline{D}.$$

Obviously the function $w_2(z)-w_1(z)$ satisfies the equality

$$w_2(z)-w_1(z)=\Phi_2(z)-\Phi_1(z)+\Psi_2(z)-\Psi_1(z)=\Phi_2(z)-\Phi_1(z)$$

$$\int_0^{\mu}[A(\xi_1-\xi_0)+B(\eta_1-\eta_0)]e_1d\mu+\int_0^{\nu}[C(\xi_1-\xi_0)+D(\eta_1-\eta_0)]e_2d\nu,$$

and then

$$\|w_2-w_1\|\le[2M_4M(4m+1)]^2\int_0^{R'}R'dR'\le\frac{[2M_4M(4m+1)R']^2}{2!},$$

where M_4 is a constant as stated in (1.35). Thus we can find a sequence of functions $\{w_n(z)\}$ satisfying

$$\begin{gathered}w_n(z)=w_0(z)+\Phi_n(z)+\Psi_n(z),\\ \Psi_n(z)=\int_0^{\mu}[A\xi_n+B\eta_n+E]e_1d\mu+\int_0^{\nu}[C\xi_n+D\eta_n+F]e_2d\nu,\end{gathered}\tag{1.36}$$

and $w_n(z)-w_{n-1}(z)$ satisfies

$$w_n(z)-w_{n-1}(z)=\Phi_n(z)-\Phi_{n-1}(z)+\Psi_n(z)-\Psi_{n-1}(z),$$

$$\begin{aligned}\Phi_n(z)-\Phi_{n-1}(z)=&\int_0^{\mu}[A(\xi_{n-1}-\xi_{n-2})+B(\eta_{n-1}-\eta_{n-2})]e_1d\mu\\ &+\int_0^{\nu}[C(\xi_{n-1}-\xi_{n-2})+D(\eta_{n-1}-\eta_{n-2})]e_2d\nu,\end{aligned}\tag{1.37}$$

and then

$$\|w_n-w_{n-1}\|\le[2M_4M(4m+1)]^n\int_0^{R'}\frac{R'^{n-1}}{(n-1)!}dR'\le\frac{[2M_4M(4m+1)R']^n}{n!}.\tag{1.38}$$

From the above inequality, we see that the sequence of functions $\{w_n(z)\}$, i.e.

$$w_n(z)=w_0(z)+[w_1(z)-w_0(z)]+\cdots+[w_n(z)-w_{n-1}(z)]\,(n=1,2,...)\tag{1.39}$$

uniformly converges to a function $w_*(z)$, and $w_*(z)$ satisfies the equality

$$\begin{gathered}w_*(z)=w_0(z)+\Phi_*(z)+\Psi_*(z),\\ \Psi_*(z)=\int_0^{\mu}[A\xi_*+B\eta_*+E]e_1d\mu+\int_0^{\nu}[C\xi_*+D\eta_*+F]e_2d\nu.\end{gathered}\tag{1.40}$$

It is easy to see that $w_*(z)$ satisfies equation (1.22) and the boundary condition (1.12), hence it is just a solution of Problem A for the complex equation (1.22) in the closed domain $\bar{D}$.

1.5 Another boundary value problem for linear hyperbolic complex equations of first order

Now we introduce another boundary value problem for equation (1.22) in D with the boundary conditions

$$\mathrm{Re}[\overline{\lambda(z)}w(z)] = r(z) \text{ on } L_1 \cup L_5, \ \mathrm{Im}[\overline{\lambda(z_1)}w(z_1)] = b_1, \tag{1.41}$$

in which $L_1 = \{y = -x, 0 \le x \le R\}$, $L_5 = \{y = (R+R_1)[x/(R-R_1) - 2R_1R/(R^2 - R_1^2)], R_1 \le x \le R = R_2 - R_1\}, R_2 > 2R_1$, $\lambda(z) = a(z) + jb(z)$, $z \in L_1$, $\lambda(z) = a(z) + jb(z) = 1 + j$, $z \in L_5$ and $\lambda(z)$, $r(z)$, b_1 satisfy the conditions

$$\begin{gathered} C_\alpha[\lambda(z), L_1] \le k_0, C_\alpha[r(z), L_1 \cup L_5] \le k_2, \\ |b_1| \le k_2, \max_{z\in L_1} \frac{1}{|a(z) - b(z)|} \le k_0, \end{gathered} \tag{1.42}$$

where b_1 is a real constant and $\alpha\,(0 < \alpha < 1), k_0, k_2$ are positive constants. The boundary value problem is called **Problem** B.

On the basis of Theorems 1.1 and 1.2, it is clear that the complex equation (1.11) can be reduced to the form (1.16) in $\bar{D}$. The general solution of system (1.16) can be expressed as

$$\begin{aligned} &w(z) = u(z) + jv(z) = [u(z) + v(z)]e_1 + [u(z) - v(z)]e_2 \\ &= f(x-y)e_1 + g(x+y)e_2 = \frac{1}{2}\{f(x-y) + g(x+y) + j[f(x-y) - g(x+y)]\}, \end{aligned} \tag{1.43}$$

where $f(t)\,(0 \le t \le 2R)$, $g(t)\,(0 \le t \le 2R_1)$ are two arbitrary real continuous functions. Noting that the boundary condition (1.41), namely

$$\begin{aligned} &a(z)u(z) - b(z)v(z) = r(z) \text{ on } L_1 \cup L_5, \\ &\overline{\lambda(z_1)}w(z_1) = r(z_1) + jb_1, \text{ i.e.} \\ &[a((1-j)x) - b((1-j)x)]f(2x) \\ &+[a((1-j)x) + b((1-j)x)]g(0) \\ &= 2r((1-j)x) \text{ on } [0, R_1], \\ &f(z_1) = u(z_1) + v(z_1) = \frac{r(z_1) + b_1}{a(z_1) - b(z_1)}, \end{aligned}$$

$$
\begin{aligned}
&\mathrm{Re}[\overline{\lambda(z)}w(z)] = u(z) - v(z)\\
&= g(x+y) = g\left(\frac{2Rx}{R-R_1} - \frac{2RR_1}{R-R_1}\right)\\
&= r\left[\left(1+\frac{R+R_1}{R-R_1}j\right)x - j\frac{2RR_1}{R-R_1}\right] \text{ on } [R_1, R].
\end{aligned} \tag{1.44}
$$

It is easy to see that the above formulas can be rewritten as

$$
\begin{aligned}
&[a((1-j)t/2)-b((1-j)t/2)]f(t)+[a((1-j)t/2)+b((1-j)t/2)]\\
&\times g(0) = 2r((1-j)t/2),\ t\in[0,2R_1],\\
&g(t) = r[((1+j)R-(1-j)R_1)\frac{t}{2R}+(1-j)R_1],\ t\in[0,2R],
\end{aligned}
$$

and then

$$
\begin{aligned}
&f(x-y) = \frac{2r((1-j)(x-y)/2)}{a((1-j)(x-y)/2)-b((1-j)(x-y)/2)}\\
&-\frac{[a((1-j)(x-y)/2)+b((1-j)(x-y)/2)]g(0)}{a((1-j)(x-y)/2)-b((1-j)(x-y)/2)}, 0\le x-y\le 2R,\\
&g(x+y)=r[((1+j)R-(1-j)R_1)\frac{x+y}{2R}+(1-j)R_1], 0\le x+y\le 2R_1.
\end{aligned} \tag{1.45}
$$

Substitute the above function $f(x-y), g(x+y)$ into (1.43), the solution $w(z)$ of (1.16) is obtained. We are not difficult to see that $w(z)$ satisfies the estimate

$$
C_\alpha[w(z),\bar{D}]\le M_5,\ C_\alpha[w(z),\bar{D}]\le M_6k_2, \tag{1.46}
$$

where $M_5 = M_5(\alpha, k_0, k_2, D)$, $M_6 = M_6(\alpha, k_0, D)$ are two positive constants. If $R_2 = 2R_1$, then $L_5 = \{x = R_1, -R_1\le y\le R_1\}$, we can similarly discussed.

Next we consider Problem B for equation (1.22). Similarly to before, we can derive the representation of solutions $w(z)$ of Problem B for equation (1.22) as stated in (1.27), where $f(x-y), g(x+y)$ possess the form (1.45), and $L = L_1\cup L_4$, z_0 in the formula (1.28) should be replaced by $L_1\cup L_5$, z_1. Moreover applying the successive approximation, the uniqueness and existence of solutions of Problem B for equation (1.22) can be proved, but L, z_0 in the formulas (1.31) and (1.34) are replaced by $L_1\cup L_5$, z_1. We write the results as a theorem.

Theorem 1.6 *Suppose that equation* (1.22) *satisfies Condition* C. *Then Problem B for* (1.22) *has a unique solution* $w(z)$, *which can be expressed in the form* (1.27), *where* $f(x-y), g(x+y)$ *possess the form* (1.45).

2 Boundary Value Problems for Degenerate Hyperbolic Complex Equations of First Order

In this section we discuss the Riemann-Hilbert boundary value problem for degenerate hyperbolic complex equations of first order in a simply connected domain. We first give a representation of solutions of the boundary value problem for the equations, and then prove the existence and uniqueness of solutions for the problem. Finally we discuss the Cauchy problem for the degenerate hyperbolic complex equations.

2.1 Formulation of Riemann-Hilbert problem for degenerate hyperbolic complex equations of first order

Let D be a simply connected bounded domain in the hyperbolic complex plane $\mathbf{C}$ with the boundary $\partial D = L_0 \cup L_1 \cup L_2$, where $L_0 = (0, 2R_1)$ on the real axis, $L = L_1 \cup L_2$, $L_1 = \{x + G(y) = 0, x \in (0, R_1)\}$, $L_2 = \{x - G(y) = 2R_1, x \in (R_1, 2R_1)\}$, $G(y) = \int_0^y \sqrt{|K(t)|}dt$, and $z_1 = x_1 + jy_1 = R_1 + jy_1$ is the intersection point of L_1 and L_2, where the hyperbolic number is used. We consider the linear degenerate hyperbolic system of first order equations

$$\begin{cases} H(y)u_x + v_y = a_1 u + b_1 v + c_1 \\ H(y)v_x + u_y = a_2 u + b_2 v + c_2 \end{cases} \text{in } D, \tag{2.1}$$

where $H(y) = G'(y) = \sqrt{|K(t)|}, K(y) = -|y|^m h(y)$ is continuous on $\overline{D}$, herein m is a positive number, $h(y)$ is a continuously differentiable positive function in $\overline{D}$. From (1.7), (2.1), we can obtain $I = (K_2 + K_3)^2 - 4K_1K_4 = 4H^2 > 0$ in D and $I = 0$ on L_0, this shows that system (2.1) is a hyperbolic system of first order equations with degenerate line L_0. The following degenerate complex equation is a special case of (2.1):

$$\begin{cases} |y|^{m/2}u_x + v_y = a_1 u + b_1 v + c_1 \\ |y|^{m/2}v_x + u_y = a_2 u + b_2 v + c_2 \end{cases} \text{in } D, \tag{2.2}$$

where m is a positive constant, $-K(y) = |y|^m$, $a_l, b_l, c_l\,(l = 1, 2)$ are functions of $z = x + jy(\in D)$. Denote

$$w(z)=u+jv, w_{\tilde{z}}=[H(y)w_x-jw_y]/2, w_{\bar{\tilde{z}}}=[H(y)w_x+jw_y]/2 \text{ in } D, \quad (2.3)$$

then system (2.1) in D can be reduced to the form

$$\begin{gathered} w_{\bar{\tilde{z}}} = A_1(z)w + A_2(z)\overline{w} + A_3(z) \text{ in } D, \\ A_1=\frac{a_1+b_2+ja_2+jb_1}{4}, A_2=\frac{a_1-b_2+ja_2-jb_1}{4}, A_3=\frac{c_1+jc_2}{2}. \end{gathered} \quad (2.4)$$

Suppose that equation (2.4) satisfies the following conditions: **Condition** C

The coefficients $A_l(z)\,(l = 1, 2, 3)$ in (2.4) are continuous in $\overline{D}$ and satisfy

$$\hat{C}[A_l,\overline{D}]=C[A_l,\overline{D}]+C[A_{lx},\overline{D}]\le k_0, l=1,2, \hat{C}[A_3,\overline{D}]\le k_1. \quad (2.5)$$

If the above conditions are replaced by

$$\hat{C}_\alpha[A_l,\overline{D}]=C_\alpha[A_l,\overline{D}]+C_\alpha[A_{lx},\overline{D}]\le k_0, l=1,2, \hat{C}_\alpha[A_3,\overline{D}] \le k_1, \quad (2.6)$$

where $\alpha\,(0 < \alpha < 1)$, k_0, $k_1(\ge \max(1, 6k_0))$ are positive constants, then the conditions will be called **Condition** C'.

Now we formulate the Riemann-Hilbert boundary value problem as follows:

Problem A Find a continuous solution $w(z)$ of (2.4) in $\overline{D}\backslash L_0$, which satisfies the boundary conditions

$$\text{Re}[\overline{\lambda(z)}w(z)] = r(z),\ z \in L = L_1 \cup L_2,\ \text{Im}[\overline{\lambda(z)}w(z)]|_{z=z_1} = b_1, \quad (2.7)$$

where $\lambda(z) = a(x) + jb(x)$, $z \in L$, b_1 is a real constant, and $\lambda(z)$, $r(z)$, b_1 satisfy the conditions

$$\begin{gathered} C_\alpha[\lambda(z),L_l]\le k_0, C_\alpha[r(z),L_l]\le k_2, l=1,2, |b_1|\le k_2, \\ \max_{z\in L_1}\frac{1}{|a(z)-b(z)|}\le k_0,\ \max_{z\in L_2}\frac{1}{|a(z)+b(z)|}\le k_0, \end{gathered} \quad (2.8)$$

in which $\alpha\,(0 < \alpha < 1), k_0, k_2$ are positive constants. For convenience, sometimes we assume that $w(z_1) = 0$. If $\lambda(z) = 1$ on $L_l, l = 1, 2$, then the problem is called Problem D.

2.2 Representation of solutions of Riemann-Hilbert problem for degenerate hyperbolic complex equations

Now we give the representation theorem of solutions for equation (2.4).

Theorem 2.1 *Suppose that the complex equation* (2.4) *satisfies Condition* C. *Then any solution of Problem A for* (2.4) *can be expressed as*

$$w(z) = u(z) + jv(z) = w_0(z) + W(z) \text{ in } D, \tag{2.9}$$

where $w_0(z)$ *is a solution of Problem A for the complex equation*

$$w_{\bar{\bar{z}}} = 0 \text{ in } D \tag{2.10}$$

with the boundary condition (2.7), *and* $W(z)$ *in* D *possesses the form*

$$W(z)=\Phi(z)+\Psi(z),\ \Psi(z)=\int_0^\mu g_1(z)d\mu e_1+\int_{2R_1}^\nu g_2(z)d\nu e_2 \text{ in } D, \tag{2.11}$$

in which $e_1 = (1+j)/2$, $e_2 = (1-j)/2$, $\mu = x + G(y)$, $\nu = x - G(y)$,

$$\begin{aligned}&g_1(z)=[\tilde{A}_1\xi+\tilde{B}_1\eta+\tilde{C}_1]/2H, \xi=\text{Re}w+\text{Im}w,\ \tilde{C}_1=c_1+c_2,\\&g_2(z)=[\tilde{A}_2\xi+\tilde{B}_2\eta+\tilde{C}_2]/2H, \eta=\text{Re}w-\text{Im}w,\ \tilde{C}_2=c_1-c_2,\\&\tilde{A}_1 = [a_1+a_2+b_1+b_2]/2,\ \ \tilde{B}_1 = [a_1+a_2-b_1-b_2]/2,\\&\tilde{A}_2=[a_1-a_2+b_1-b_2]/2,\ \tilde{B}_2 = [a_1-a_2-b_1+b_2]/2 \text{ in } D,\end{aligned} \tag{2.12}$$

$\Phi(z)$ *is the solution of equation* (2.10) *satisfying the boundary conditions*

$$\text{Re}[\overline{\lambda(z)}(\Phi(z)+\Psi(z))]=0, z\in L, \text{Im}[\overline{\lambda(z_1)}(\Phi(z_1)+\Psi(z_1))]=0. \tag{2.13}$$

Moreover $w_0(z)$ *is a solution of Problem A for equation* (2.10) *satisfying the estimate*

$$C_\alpha[w_0(z),\bar{D}]\le M_1=M_1(\alpha,k_0,k_2,D). \tag{2.14}$$

Proof First of all, from (2.1) or (2.4), we have

$$w_{\bar{\bar{z}}} = [H(y)w_x + jw_y]/2 = [H(y)(u+jv)_x + j(u+jv)_y]/2$$

$$
\begin{aligned}
&=\frac{e_1}{2}[H(y)u_x+v_y+H(y)v_x+u_y]+\frac{e_2}{2}[H(y)u_x+v_y-H(y)v_x-u_y]\\
&=H(y)[e_1\Big(\frac{u_x}{2}+\frac{v_y}{2H(y)}+\frac{v_x}{2}+\frac{u_y}{2H(y)}\Big)+e_2\Big(\frac{u_x}{2}+\frac{v_y}{2H(y)}-\frac{v_x}{2}-\frac{u_y}{2H(y)}\Big)]\\
&=H(y)e_1(u+v)_\mu+H(y)e_2(u-v)_\nu=A_1w+A_2\overline{w}+A_3\\
&=\frac{e_1}{2}[\tilde{A}_1\xi+\tilde{B}_1\eta+\tilde{C}_1]+\frac{e_2}{2}[\tilde{A}_2\xi+\tilde{B}_2\eta+\tilde{C}_2],
\end{aligned}
\tag{2.15}
$$

where $H(y)=\sqrt{-K(y)}$, and

$$
\begin{aligned}
&\mu=x+G(y)=x+\int_0^y\sqrt{-K(t)}dt, \nu=x-G(y)=x-\int_0^y\sqrt{-K(t)}dt,\\
&\mu+\nu=2x, \mu-\nu=2G(y), \frac{\partial x}{\partial\mu}=\frac{\partial x}{\partial\nu}=\frac{1}{2}, \frac{\partial y}{\partial\mu}=-\frac{\partial y}{\partial\nu}=\frac{1}{2H(y)}.
\end{aligned}
\tag{2.16}
$$

It is clear that equation (2.10) is equivalent to the first order system of $\xi=u+v, \eta=u-v$:

$$\xi_\mu=(u+v)_\mu=0,\ \eta_\nu=(u-v)_\nu=0 \text{ in } D_\tau, \tag{2.17}$$

where D_τ is the image domain of D under the mapping $\tau=\mu+j\nu=x+G(y)+j(x-G(y))$. Next, let $w(z)$ be a solution of Problem A for equation (2.4), and $w(z)$ be substituted in the positions of w in (2.11), (2.12), thus the functions $g_1(z)$, $g_2(z)$ and $\Psi(z)$ in $\overline{D}$ are determined. Moreover we can find the solution $\Phi(z)$ in $\overline{D}$ of (2.10) with the boundary condition (2.13), thus

$$w(z)=w_0(z)+W(z)=w_0(z)+\Phi(z)+\Psi(z) \text{ in } D \tag{2.18}$$

is the solution of Problem A in D for equation (2.4), and $W(z)$ can be expressed as in (2.11).

2.3 Existence and uniqueness of solutions of Riemann-Hilbert problem

In the following, we first prove the existence of solutions of Problem A for equation (2.10) in $\overline{D}$ and give its representation. For this, it is sufficient to find the solution of the system of first order equations

$$(u+v)_\mu=0,\ (u-v)_\nu=0 \text{ in } D_\tau \tag{2.19}$$

with the boundary condition (2.7), i.e.

$$\mathrm{Re}[\overline{\lambda(z)}(u+jv)]=r(z),\ z\in L,\ \mathrm{Im}[\overline{\lambda(z)}(u+jv)]|_{z=z_1}=b_1. \tag{2.20}$$

The general solution of the system (2.19) can be expressed as

$$\xi=u+v=f(\nu)=f(x-G(y)),\ \eta=u-v=g(\mu)=g(x+G(y)),\ \text{i.e.}$$
$$u=[f(x-G(y))+g(x+G(y))]/2, v=[f(x-G(y))-g(x+G(y))]/2, \tag{2.21}$$

in which $f(t), g(t)$ are two arbitrary real continuous functions on $[0,2R_1]$. Obviously the formula in (2.20) can be rewritten as

$$a(z)u(z)-b(z)v(z)=r(z)\ \text{on}\ L,\ \overline{\lambda(z_1)}w(z_1)=r(z_1)+jb_1,\ \text{i.e.}$$
$$[a(h(x))-b(h(x))]f(2x)+[a(h(x))+b(h(x))]g(0)=2r(h(x))\ \text{on}\ [0,R_1],$$
$$[a(k(x))-b(k(x))]f(2R_1)+[a(k(x))+b(k(x))]g(2x-2R_1)$$
$$=2r(k(x))\ \text{on}\ [R_1,2R_1],$$
$$(a(z_1)-b(z_1))f(2R_1)=(a(z_1)-b(z_1))(u(z_1)+v(z_1))=r(z_1)+b_1\ \text{or}\ 0,$$
$$(a(z_1)+b(z_1))g(0)=(a(z_1)+b(z_1))(u(z_1)-v(z_1))=r(z_1)-b_1\ \text{or}\ 0, \tag{2.22}$$

in which $h(x)=x+j(-G)^{-1}(x)$ and $k(x)=x+jG^{-1}(x-2R_1)$, herein $(-G)^{-1}(x)$ and $G^{-1}(x-2R_1)$ are the inverse functions of $x=-G(y)=\int_0^y H(t)dt$ and $x-2R_1=G(y)=\int_0^y H(t)dt$ respectively. The above formulas can be rewritten as

$$[a(h(t/2))-b(h(t/2))]f(t)+[a(h(t/2))+b(h(t/2))]g(0)=2r(h(t/2)),$$
$$[a(k(t/2+R_1))-b(k(t/2+R_1))]f(2R_1)+[a(k(t/2+R_1))$$
$$+b(k(t/2+R_1))]g(t)=2r(k(t/2+R_1)),\ t\in[0,2R_1],\ \text{i.e.}$$
$$f(x-G(y))=\frac{2r(h((x-G(y))/2))}{a(h((x-G(y))/2))-b(h((x-G(y))/2))}$$
$$-\frac{[a(h((x-G(y))/2))+b(h((x-G(y))/2))]g(0)}{a(h((x-G(y))/2))-b(h((x-G(y))/2))},\ 0\le x-G(y)\le 2R_1,$$
$$g(x+G(y))=\frac{2r(k((x+G(y))/2+R_1))}{a(k((x+G(y))/2+R_1))+b(k((x+G(y))/2+R_1))}$$
$$-\frac{[a(k((x+G(y))/2+R_1))-b(k((x+G(y))/2+R_1))]f(2R_1)}{a(k((x+G(y))/2+R_1))+b(k((x+G(y))/2+R_1))},$$
$$0\le x+G(y)\le 2R_1. \tag{2.23}$$

Thus the solution $w(z)$ of (2.10) can be expressed as

$$\begin{aligned} w(z) &= f(x-G(y))e_1 + g(x+G(y))e_2 \\ &= \frac{1}{2}\{f(x-G(y))+g(x+G(y))+j[f(x-G(y))-g(x+G(y))]\}, \end{aligned} \tag{2.24}$$

where $f(x-G(y))$, $g(x+G(y))$ are as stated in (2.23) and $f(2R_1), g(0)$ are as stated in (2.22). It is not difficult to see that $w(z)$ satisfies the estimate

$$C_\delta[w(z), \bar{D}] \le M_1, \; C_\delta[w(z), \bar{D}] \le M_2 k_2, \tag{2.25}$$

where $M_1 = M_1(\alpha, k_0, k_2, D)$, $M_2 = M_2(\alpha, k_0, D)$ are two positive constants only dependent on α, k_0, k_2, D and α, k_0, D respectively, $\delta = \delta(\alpha, k_0, k_2, D)$ is a sufficiently small positive constant. The above results can be written as a theorem.

Theorem 2.2 *If the domain D is as stated before, then Problem A for* (2.10) *in $\overline{D}$ has a unique solution as stated in* (2.24).

By using the method in Section 1, we first consider the case: $y_1 < y < \delta$, where δ is any negative number, and noting the arbitrariness of δ, the following result can be proved.

Theorem 2.3 *If equation* (2.4) *in D satisfies the above conditions, then Problem A for* (2.4) *in D has a unique solution.*

2.4 Unique solvability of Cauchy problem for degenerate hyperbolic complex equations

In this subsection, we prove the existence and uniqueness of solutions of the Cauchy problem for the degenerate hyperbolic equation (2.4) with the boundary conditions

$$u(x) = \phi(x), \; v(x) = \psi(x) \text{ on } L_0 = (0, 2R_1), \tag{2.26}$$

where $\phi(x), \psi(x)$ satisfy the condition $C^2_\alpha[\phi(x), L_0], C^2_\alpha[\psi(x), L_0] \le k_2$, the above boundary value problem is also called **Problem** C for (2.4). Making a transformation of function

$$W(z) = w(z) - \phi(x) - j\psi(x) \text{ in } D, \tag{2.27}$$

equation (2.4) and boundary condition (2.26) are reduced to the form

$$\begin{aligned}&W_{\bar{\bar{z}}}=B_1(z)W(z)+B_2(z)\overline{W(z)}+B_3(z),\\&B_1(z)=A_1(z),\ B_2(z)=A_2(z),\ B_3(z)=A_3-H(y)\\&\times(\phi_x+j\psi_x)/2+A_1(\phi+j\psi)+A_2(\phi-j\psi)\ \text{in}\ D,\end{aligned}\tag{2.28}$$

$$W(z)=U(x)+jV(x)=0\ \text{on}\ L_0.\tag{2.29}$$

Hence we may only discuss Cauchy problem (2.28),(2.29) and denote it by Problem C_0. It is clear that Problem C_0 for (2.28) is equivalent to the boundary value problem for the hyperbolic system of first order equations

$$\begin{aligned}&\xi_\mu=\frac{1}{2}[\tilde{A}_1\xi+\tilde{B}_1\eta+\tilde{C}_1],\ \eta_\nu=\frac{1}{2}[\tilde{A}_2\xi+\tilde{B}_2\eta+\tilde{C}_2],\\&\xi=U+V,\ \eta=U-V,\ U(x)=0,\ V(x)=0\ \text{on}\ L_0,\end{aligned}\tag{2.30}$$

which can be derived from (2.15). Denote s_1 by the member of the family $dx/dy=\sqrt{-K(y)}$ and by s_2 the member of the family $dx/dy=-\sqrt{-K(y)}$ passing through $z=x+jy\in\overline{D}$, namely

$$s_1:dx/dy=\sqrt{-K(y)},\ s_2:dx/dy=-\sqrt{-K(y)},\tag{2.31}$$

we have

$$ds_1=\sqrt{(dx)^2+(dy)^2}=-\sqrt{1+(dx/dy)^2}dy=-\sqrt{1-K}dy=-\frac{\sqrt{1-K}}{\sqrt{-K}}dx,$$

$$ds_2=\sqrt{(dx)^2+(dy)^2}=-\sqrt{1+(dx/dy)^2}dy=-\sqrt{1-K}dy=\frac{\sqrt{1-K}}{\sqrt{-K}}dx,$$

and

$$d\mu=d(x+G(y))=2dx=2H(y)dy\ \text{on}\ s_1,$$

$$d\nu=d(x-G(y))=dx=-2H(y)dy\ \text{on}\ s_2.$$

Integrating the hyperbolic system in (2.31) along the characteristic curves s_1,s_2, we obtain the system of integral equations as follows

$$\begin{aligned}&\xi(z)=\int_0^y[\tilde{A}_1\xi+\tilde{B}_1\eta+\tilde{C}_1]dy,\ z\in s_1,\\&\eta(z)=-\int_0^y[\tilde{A}_2\xi+\tilde{B}_2\eta+\tilde{C}_2]dy,\ z\in s_2,\end{aligned}\tag{2.32}$$

where the coefficients $\tilde{A}_l,\ \tilde{B}_l,\ \tilde{C}_l\,(l=1,2)$ are as stated in (2.12). In the following we may only discuss the case of $K(y)=-|y|^m h(y)$, herein $m,\ h(y)$

are as stated in (2.1). It is easy to see that for two characteristics s_1, s_2 passing through a point $z = x + jy \in D$ and x_1, x_2 are the intersection points with the axis $y = 0$ respectively, for any two points $\tilde{z}_1 = \tilde{x}_1 + j\tilde{y} \in s_1, \tilde{z}_2 = \tilde{x}_2 + j\tilde{y} \in s_2$, we have

$$\begin{aligned}|\tilde{x}_1 - \tilde{x}_2| &\le |x_1 - x_2| = 2|\int_0^y \sqrt{-K(t)}dt| \\ &\le \frac{2k_0}{m+2}|y|^{1+m/2} \le \frac{k_1}{6}|y|^{m/2+1} \le M|y|^{m/2+1}.\end{aligned} \tag{2.33}$$

for any points $z_1 = x_1 + jy \in s_1,\ z_2 = x_2 + jy \in s_2$, where $\max[2\sqrt{h(y)}, 1] \le k_0 (\le k_1/6)$, $M(> \max[1, k_1, M_1])$ is a positive constant. From (2.5), we can derive that the coefficients of (2.32) possess continuously differentiable with respect to $x \in L_0$ and satisfy the condition

$$|\tilde{A}_l|, |\tilde{A}_{lx}|, |\tilde{B}_l|, |\tilde{B}_{lx}|, |\tilde{C}_l|, |\tilde{C}_{lx}|, \left|\frac{1}{\sqrt{h}}\right|, \left|\frac{h_y}{h}\right| \le M, z \in \bar{D}, l = 1, 2. \tag{2.34}$$

According to the proof of Theorem 2.3, it is sufficient to find a solution of Problem C_0 for arbitrary segment $-\delta \le y \le 0$, where δ is a sufficiently small positive number, and choose a positive constant $\gamma\,(< 1)$ close to 1, such that the following inequalities hold:

$$4M|y|^\beta \le \gamma \ \text{ for } \ -\delta \le y \le 0, \tag{2.35}$$

where $\beta\,(< 1)$ is a positive constant. Next we shall find a solution of Problem C_0 for (2.32) on $-\delta < y < 0$. Firstly, setting that $z = x + jy \in \overline{D}_0$, here $D_0 = \{a_0 = \delta_0 \le x \le a_1 = 2R_1 - \delta_0, -\delta < y \le 0\}$, δ, δ_0 are sufficiently small positive constants, we choose $\xi_0 = 0, \eta_0 = 0$ and substitute them into the corresponding positions of ξ, η in the right-hand sides of (2.32), and obtain

$$\begin{aligned}\xi_1(z) &= \int_0^y [\tilde{A}_1\xi_0 + \tilde{B}_1\eta_0 + \tilde{C}_1]dy = \int_0^y \tilde{C}_1 dy,\ z \in s_1, \\ \eta_1(z) &= -\int_0^y [\tilde{A}_2\xi_0 + \tilde{B}_2\eta_0 + \tilde{C}_2]dy = -\int_0^y \tilde{C}_2 dy, z \in s_2.\end{aligned} \tag{2.36}$$

By the successive approximation, we find the sequences of functions $\{\xi_k\}, \{\eta_k\}$, which satisfy the relations

$$\begin{aligned}\xi_{k+1}(z) &= \int_0^y [\tilde{A}_1\xi_k + \tilde{B}_1\eta_k + \tilde{C}_1]dy,\ z \in s_1, \\ \eta_{k+1}(z) &= -\int_0^y [\tilde{A}_2\xi_k + \tilde{B}_2\eta_k + \tilde{C}_2]dy,\ z \in s_2, \\ & k = 0, 1, 2, \ldots.\end{aligned} \tag{2.37}$$

We can prove that $\{\xi_k\},\{\eta_k\}$ in D_0 satisfy the estimates

$$\begin{aligned}&|\xi_k(z)|,|\eta_k(z)|\le M\sum_{j=0}^{k}\gamma^j|y|^{\beta},\\&|\xi_k(z_1)-\xi_k(z_2)|\le M\sum_{j=0}^{k}\gamma^j|x_1-x_2|^{\beta_1}|y|^{\beta'},\\&|\eta_k(z_1)-\eta_k(z_2)|\le M\sum_{j=0}^{k}\gamma^j|x_1-x_2|^{\beta_1}|y|^{\beta'},\\&|\xi_{k+1}(z)-\xi_k(z)|,|\eta_{k+1}(z)-\eta_k(z)|\le M\gamma^k|y|^{\beta},\\&|\xi_{k+1}(z_1)-\xi_{k+1}(z_2)-\xi_k(z_1)+\xi_k(z_2)|\le M\gamma^k|x_1-x_2|^{\beta_1}|y|^{\beta'},\\&|\eta_{k+1}(z_1)-\eta_{k+1}(z_2)-\eta_k(z_1)+\eta_k(z_2)|\le M\gamma^k|x_1-x_2|^{\beta_1}|y|^{\beta'},\end{aligned}\tag{2.38}$$

in which $z=x+jy$ is the intersection point of s_1,s_2 in (2.31) passing through z_1,z_2, $\beta=1-\beta_1,\beta'=1+m/2-(2+m)\beta_1$, β_1 is a sufficiently small positive constant. In fact, from (2.36), it follows that the first formula with $k=1$ holds, namely

$$|\xi_1(z)|=|\int_0^y\tilde{C}_1dy|\le M|y|,|\eta_1(z)|\le M|y|=M\gamma^0|y|\le M\sum_{j=0}^{1}\gamma^j|y|.$$

Moreover we have

$$\begin{aligned}&|\xi_1(z_1)-\xi_1(z_2)|\le|\int_0^y[\tilde{C}_1(x_1+jt)-\tilde{C}_1(x_2+jt)]dt|\\&\le|\int_0^y|\tilde{C}_{1x}||x_1-x_2|dy|\le M|\int_0^y|x_1-x_2|dy|\\&\le M|x_1-x_2||y|\le M\sum_{j=0}^{1}\gamma^j|x_1-x_2|^{\beta_1}|y|^{\beta'},\\&|\eta_1(z_1)-\eta_1(z_2)|\le M\sum_{j=0}^{1}\gamma^j|x_1-x_2|^{\beta_1}|y|^{\beta'},\\&|\xi_1(z)-\xi_0(z)|=|\xi_1(z)|\le M\gamma^0|y|^{\beta},|\eta_1(z)-\eta_0(z)|=|\eta_1(z)|\le M\gamma^0|y|^{\beta},\\&|\xi_1(z_1)-\xi_1(z_2)-\xi_0(z_1)+\xi_0(z_2)|=|\xi_1(z_1)-\xi_1(z_2)|,|\eta_1(z_1)-\eta_1(z_2)\\&-\eta_0(z_1)+\eta_0(z_2)|=|\eta_1(z_1)-\eta_1(z_2)|\le M\gamma^0|x_1-x_2|^{\beta_1}|y|^{\beta'}.\end{aligned}$$

In addition, we use the inductive method, namely suppose the estimates in (2.38) for $k=n$ are valid, then they are also valid for $k=n+1$. In the following, we only give the estimates of $|\xi_{n+1}(z)|, |\xi_{n+1}(z_1)+\xi_{n+1}(z_2)|$, the other estimates can be similarly given. From (2.37), we have

$$|\xi_{n+1}(z)|\le\Big|\int_0^y[(|\tilde{A}_1|+|\tilde{B}_1|)M\sum_{j=0}^n\gamma^j|y|^\beta+|\tilde{C}_1|]dy\Big|$$

$$\le M\Big|\int_0^y[2M\sum_{j=0}^n\gamma^j|y|^\beta+1]dy\Big|\le M\{[2M\sum_{j=0}^n\gamma^j|y|^{1+\beta}+|y|\}\le M\sum_{j=0}^{n+1}\gamma^j|y|^\beta.$$

Moreover we consider

$$|\xi_{n+1}(z_1)-\xi_{n+1}(z_2)|\le\Big|\int_0^y[\tilde{A}_1(z_1)\xi_n(z_1)-\tilde{A}_1(z_2)\xi_n(z_2)$$

$$+\tilde{B}_1(z_1)\eta_n(z_1)-\tilde{B}_1(z_2)\eta_n(z_2)+\tilde{C}_1(z_1)-\tilde{C}_1(z_2)]dy\Big|,$$

noting that

$$I_1=|\tilde{A}_1(z_1)\xi_n(z_1)-\tilde{A}_1(z_2)\xi_n(z_2)+\tilde{B}_1(z_1)\eta_n(z_1)-\tilde{B}_1(z_2)\eta_n(z_2)|$$

$$\le|[\ddot{A}_1(z_1)-\ddot{A}_1(z_2)]\xi_n(z_1)+\ddot{A}_1(z_2)[\xi_n(z_1)-\xi_n(z_2)]+[\tilde{B}_1(z_1)-\tilde{B}_1(z_2)]$$

$$\times\eta_n(z_1)+\tilde{B}_1(z_2)[\eta_n(z_1)-\eta_n(z_2)]|\le 2M^2[|x_1-x_2|+|x_1-x_2|^{\beta_1}|y|^{\beta'}]$$

$$\times\sum_{j=0}^n\gamma^j,\ I_2=|\tilde{C}_1(z_1)-\tilde{C}_1(z_2)|\le M|x_1-x_2|\le M|x_1-x_2|^{\beta_1}|y|^{\beta'},$$

and then

$$|\xi_{n+1}(z_1)-\xi_{n+1}(z_2)|\le\Big|\int_0^y M[4M\sum_{j=0}^n\gamma^j+1]|x_1-x_2|^{\beta_1}|y|^{\beta'}dy\Big|$$

$$\le M[4M|y|\sum_{j=0}^n\gamma^j+1\}|x_1-x_2|^{\beta_1}|y|^{\beta'}\le M\sum_{j=0}^{n+1}\gamma^j|x_1-x_2|^{\beta_1}|y|^{\beta'},$$

and

$$|\eta_{n+1}(z_1)-\eta_{n+1}(z_2)|\le M\sum_{j=0}^{n+1}\gamma^j|x_1-x_2|^{\beta_1}|y|^{\beta'}.$$

On the basis of the estimates (2.38), we can derive that the sequences $\{\xi_k\},\{\eta_k\}$ in D_0 uniformly converge to ξ_*,η_* satisfying the system of integral equations

$$\xi_*(z)=\int_0^y[\tilde{A}_1\xi_*+\tilde{B}_1\eta_*+\tilde{C}_1]dy,\ z\in s_1,$$

$$\eta_*(z)=-\int_0^y[\tilde{A}_2\xi_*+\tilde{B}_2\eta_*+\tilde{C}_2]dy,\ z\in s_2,$$

and the function $W(z)=U(z)+jV(z)=[\xi_*+\eta_*]/2+[\xi_*-\eta_*]/2$ satisfies equation (2.28) and boundary condition (2.29), hence $w(z)=W(z)+\phi(x)+j\psi(x)$ is a solution of Problem C for (2.4). From the above discussion, we can see that the solution of Problem C for (2.4) in D is unique.

Theorem 2.4 *If equation* (2.4) *in D satisfies Condition C and the above other conditions, then Problem C for* (2.4) *in D has a unique solution.*

3 The Oblique Derivative Problem for Uniformly Hyperbolic Equations of Second Order

In this section, we mainly discuss the oblique derivative boundary value problem for uniformly hyperbolic equations of second order. Firstly, we transform some linear and nonlinear uniformly hyperbolic equations of second order with certain conditions into the complex forms, give the uniqueness theorem of solutions for the above boundary value problem. Moreover by using the successive approximation, the existence of solutions for the oblique derivative problem is proved. In the letter sections of this chapter, we shall introduce some boundary value problems for degenerate hyperbolic equations of second order with certain conditions.

3.1 Complex forms of hyperbolic equations of second order

Now, we first transform some linear uniformly hyperbolic equations of second order with certain conditions into the complex form, and then we state the conditions of some hyperbolic complex equations of second order.

Let D be a bounded domain, we consider the linear hyperbolic partial differential equation of second order

$$au_{xx} + 2bu_{xy} + cu_{yy} + du_x + eu_y + fu = g, \tag{3.1}$$

where the coefficients a, b, c, d, e, f, g are known real continuous functions of $(x, y) \in D$. The condition of hyperbolic type for (3.1) is that for any point (x, y) in D, the inequality

$$I = ac - b^2 < 0,\ a > 0 \tag{3.2}$$

holds. If a, b, c are bounded in D and

$$I = ac - b^2 \le I_0 < 0,\ a > 0 \text{ in } D, \tag{3.3}$$

where I_0 is a negative constant, then equation (3.1) is called uniformly hyperbolic in D. Introduce the notations as follows

$$\begin{aligned}&(\ )_z=\frac{1}{2}[(\ )_x-j(\ )_y],\ (\ )_{\bar z}=\frac{1}{2}[(\ )_x+j(\ )_y],\\&(\ )_{z\bar z}=\frac{1}{4}[(\ )_{xx}-(\ )_{yy}],\ (\ )_{zz}=\frac{1}{4}[(\ )_{xx}+(\ )_{yy}-2j(\ )_{xy}],\\&(\ )_{\bar z\bar z}=\frac{1}{4}[(\ )_{xx}+(\ )_{yy}+2j(\ )_{xy}],\ (\ )_{xy}=j[(\ )_{\bar z\bar z}-(\ )_{zz}],\\&(\ )_{xx}=(\ )_{zz}+(\ )_{\bar z\bar z}+2(\ )_{z\bar z},\ (\ )_{yy}=(\ )_{zz}+(\ )_{\bar z\bar z}-2(\ )_{z\bar z},\end{aligned}\tag{3.4}$$

equation (3.1) can be written in the form

$$\begin{aligned}&2(a-c)u_{z\bar z}+(a+c-2bj)u_{zz}+(a+c+2bj)u_{\bar z\bar z}\\&\quad+(d-ej)u_z+(d+ej)u_{\bar z}+fu=g\ \text{ in }\ D.\end{aligned}\tag{3.5}$$

If $a\neq c$ in D, then equation (3.5) can be reduced to the complex form

$$u_{z\bar z}-\operatorname{Re}[Q(z)u_{zz}+A_1(z)u_z]-A_2(z)u=A_3(z)\ \text{ in }\ D,\tag{3.6}$$

in which

$$Q=-\frac{a+c-2bj}{a-c},\ A_1=-\frac{d-ej}{a-c},\ A_2=-\frac{f}{2(a-c)},\ A_3=\frac{g}{2(a-c)}.$$

If $(a+c)^2\geq 4b^2$, then the conditions of hyperbolic type and uniformly hyperbolic type are transformed into

$$|Q(z)|<1\ \text{ in }\ D,\tag{3.7}$$

and

$$|Q(z)|\leq q_0<1\ \text{ in }\ D,\tag{3.8}$$

respectively.

For the quasilinear hyperbolic partial differential equation of second order

$$au_{xx}+2bu_{xy}+cu_{yy}+du_x+eu_y+fu=g,\tag{3.9}$$

where the coefficients a,b,c,d,e,f,g are known real continuous functions of $(x,y)\in D$ and $u,u_x,u_y\in\mathbf{R}$, if equation (3.9) satisfies the condition of hyperbolic type in D and other condition as stated before, then (3.9) can be reduced to the complex form

$$u_{z\bar z}-\operatorname{Re}[Qu_{zz}+A_1u_z]-A_2u=A_3\ \text{ in }\ D,\tag{3.10}$$

in which

$$Q=Q(z,u,u_z)=-\frac{a+c-2bj}{a-c}, A_1=A_1(z,u,u_z)=-\frac{d-ej}{a-c},$$

$$A_2=A_2(z,u,u_z)=-\frac{f}{2(a-c)}, A_3=A_3(z,u,u_z)=\frac{g}{2(a-c)}.$$

As stated in [12] 1),3), for the linear hyperbolic equation (3.1) or its complex form (3.6), if the coefficients a, b, c are sufficiently smooth, through a non-singular transformation of z, equation (3.1) can be reduced to the standard form

$$u_{xx}-u_{yy}+du_x+eu_y+fu=g \tag{3.11}$$

or its complex form

$$u_{z\bar{z}}-\mathrm{Re}[A_1(z)u_z]-A_2(z)u=A_3(z). \tag{3.12}$$

Let D be a simply connected bounded domain with the boundary $\Gamma = L_1\cup L_2\cup L_3\cup L_4$ as stated in Section 1, where $L_1=\{x=-y, 0\le x\le R_1\}$, $L_2=\{x=y+2R_1, R_1\le x\le R_2\}$, $L_3=\{x=-y+2R, R=R_2-R_1\le x\le R_2\}$, $L_4=\{x=y, 0\le x\le R\}$, and denote $z_0=0$, $z_1=(1-j)R_1, z_2=R_2+j(R_2-2R_1)$, $z_3=(1+j)R$, $L=L_3\cup L_4$, here there is no harm in assuming that $R_2\ge 2R_1$. Consider second order quasilinear hyperbolic equation in the form

$$u_{z\bar{z}}-\mathrm{Re}[A_1(z,u,u_z)u_z]-A_2(z,u,u_z)u=A_3(z,u,u_z), \tag{3.13}$$

whose coefficients satisfy the following conditions: **Condition** C

1) $A_l(z,u,u_z)(l=1,2,3)$ are continuous in $z\in\bar{D}$ for all continuously differentiable functions $u(z)$ in $\bar{D}$ and satisfy

$$\hat{C}[A_l,\bar{D}]=C[A_l,\bar{D}]+C[A_{lx},\bar{D}]\le k_0, l=1,2, \hat{C}[A_3,\bar{D}]\le k_1. \tag{3.14}$$

2) For any continuously differentiable functions $u_1(z), u_2(z)$ in $\bar{D}$, the equality

$$F(z,u_1,u_{1z})-F(z,u_2,u_{2z})=\mathrm{Re}[\tilde{A}_1(u_1-u_2)_z]+\tilde{A}_2(u_1-u_2) \text{ in } \bar{D} \tag{3.15}$$

holds, where

$$\hat{C}[\tilde{A}_l(z,u_1,u_2),\bar{D}]\le k_0,\ l=1,2, \tag{3.16}$$

in (3.14),(3.16), k_0, k_1 are non-negative constants. In particular, when (3.13) is a linear equation, from (3.14) it follows that the conditions (3.15), (3.16) hold.

It is clear that (3.13) is the complex form of the following real equation of second order

$$u_{xx} - u_{yy} = au_x + bu_y + cu + d \text{ in } D, \tag{3.17}$$

in which D is a bounded domain with the boundary $\Gamma = L_1 \cup L_2 \cup L_3 \cup L_4$ as stated before, a, b, c, d are functions of $(x, y)(\in D)$, $u, u_x, u_y(\in \mathbf{R})$ and

$$A_1 = \frac{a - jb}{2},\ A_2 = \frac{c}{4},\ A_3 = \frac{d}{4} \text{ in } D. \tag{3.18}$$

Due to $z = x + jy = \mu e_1 + \nu e_2$, $w = u_z = \xi e_1 + \eta e_2$, and

$$w_z = \frac{1}{2}(w_x - jw_y) = \xi_\nu e_1 + \eta_\mu e_2,$$

$$w_{\bar{z}} = \frac{1}{2}(w_x + jw_y) = \xi_\mu e_1 + \eta_\nu e_2,$$

the quasilinear hyperbolic equation (3.13) can be rewritten in the form

$$\xi_\mu e_1 + \eta_\nu e_2 = [A(z, u, w)\xi + B(z, u, w)\eta + C(z, u, w)u + D(z, u, w)]e_1$$

$$+[A(z, u, w)\xi + B(z, u, w)\eta + C(z, u, w)u + D(z, u, w)]e_2, \text{ i.e.}$$

$$\begin{cases} \xi_\mu = A(z, u, w)\xi + B(z, u, w)\eta + C(z, u, w)u + D(z, u, w) \\ \eta_\nu = A(z, u, w)\xi + A(z, u, w)\eta + C(z, u, w)u + D(z, u, w) \end{cases} \text{in } D, \tag{3.19}$$

in which

$$A = \frac{a - b}{4},\ B = \frac{a + b}{4},\ C = \frac{c}{4},\ D = \frac{d}{4}. \tag{3.20}$$

In the following, we mainly discuss oblique derivative problem for linear hyperbolic equation (3.12) and quasilinear hyperbolic equation (3.13) in the simply connected domain. We first prove the boundedness of solutions of the boundary value problem and give a priori estimates of its solutions, and then prove the solvability of the boundary value problem for the hyperbolic equations.

3.2 Formulation of oblique derivative problem and representations of solutions

The oblique derivative problem for equation (3.13) may be formulated as follows:

Problem P Find a continuously differentiable solution $u(z)$ of (3.13) in $\bar{D}$ satisfying the boundary conditions

$$\begin{gathered}\frac{1}{2}\frac{\partial u}{\partial \nu} = \mathrm{Re}[\overline{\lambda(z)}u_z] = r(z),\ z\in L = L_3\cup L_4,\\ u(z_3) = b_0,\ \mathrm{Im}[\overline{\lambda(z)}u_z]|_{z=z_3} = b_1,\end{gathered} \tag{3.21}$$

where ν is a given vector at every point on L, $\lambda(z) = a(z) + jb(z) = \cos(\nu, x) + j\cos(\nu, y)$, $z \in L$, b_0, b_1 are real constants, and $\lambda(z)$, $r(z)$, b_0, b_1 satisfy the conditions

$$\begin{gathered}C_\alpha[\lambda(z), L] \le k_0, C_\alpha[r(z), L] \le k_2, |b_0|, |b_1| \le k_2,\\ \max_{z\in L_3}\frac{1}{|a(z)-b(z)|} \le k_0,\ \max_{z\in L_4}\frac{1}{|a(z)+b(z)|} \le k_0,\end{gathered} \tag{3.22}$$

in which $\alpha\,(0 < \alpha < 1), k_0, k_2$ are non-negative constants. The above boundary value problem for (3.13) with $A_3(z, u, w) = 0$, $z \in D, u \in \mathbf{R}$, $w \in \mathbf{C}$ and $r(z) = b_0 = b_1 = 0$, $z \in L$ will be called Problem P_0.

By $z = x+jy = \mu e_1+\nu e_2$, $w = u_z = \xi e_1+\eta e_2$, the boundary condition (3.21) can be reduced to

$$\begin{gathered}\mathrm{Re}[\overline{\lambda(z)}(\xi e_1 + \eta e_2)] = r(z),\ u(z_3) = b_0,\\ \mathrm{Im}[\overline{\lambda(z)}(\xi e_1 + \eta e_2)]|_{z=z_3} = b_1,\end{gathered} \tag{3.23}$$

where $\overline{\lambda(z)} = (a-b)e_1 + (a+b)e_2$. Moreover, the domain D is transformed into $Q = \{0 \le \mu \le 2R_1,\ 0 \le \nu \le 2R\}$, $R = R_2 - R_1$, which is a rectangle and A, B, C, D are known functions of (μ, ν) and unknown continuous functions u, w, and they satisfy the boundary conditions

$$\begin{cases}\overline{\lambda(z_3)}w(z_3) = \overline{\lambda(z)}[\xi e_1 + \eta e_2]|_{z=z_3} = r(z_3) + jb_1,\ u(z_3) = b_0,\\ \mathrm{Re}[\overline{\lambda(z)}w(2Re_1+\nu e_2)] = r(z),\ \text{if } (x,y)\in L_3 = \{\mu=2R,\ 0\le\nu\le 2R_1\},\\ \mathrm{Re}[\overline{\lambda(z)}w(\mu e_1 + 0e_2)] = r(z),\ \text{if } (x,y)\in L_4 = \{0\le\mu\le 2R, \nu = 0\},\end{cases} \tag{3.24}$$

where $\lambda(z)$, $r(z)$, b_0, b_1 are as stated in (3.21). Moreover we can assume that $w(z_3) = 0$.

It is not difficult to see that the oblique derivative boundary value problem (Problem P) includes the Dirichlet boundary value problem (Problem

D) as a special case. In fact, the boundary condition of Dirichlet problem (Problem D) for equation (3.13) or (3.19) is as follows

$$u(z) = \phi(x) \ \text{on} \ L = L_3 \cup L_4. \tag{3.25}$$

We find the derivative with respect to the tangent direction $l = (x \mp jy)$ for (3.25), in which $\mp$ are determined by L_3 and L_4 respectively, and

$$ds = \sqrt{(dx)^2 + (dy)^2} = \sqrt{2}dx = -\sqrt{2}dy \ \text{on} \ L_3,$$
$$ds = \sqrt{(dx)^2 + (dy)^2} = \sqrt{2}dx = \sqrt{2}dy \ \text{on} \ L_4,$$

it is clear that the following equalities hold:

$$\begin{aligned}
&\frac{u_s}{2} = \frac{u_x x_s + u_y y_s}{2} = \frac{u_x - u_y}{2\sqrt{2}} = \text{Re}[\overline{\lambda(z)}u_z] = \frac{\phi_x}{2\sqrt{2}}, \ z \in L_3, \\
&\text{Im}[\overline{\lambda(z)}u_z]|_{z=z_3-0} = b_1^-, \\
&\frac{u_s}{2} = \frac{u_x x_s + u_y y_s}{2} = \frac{u_x + u_y}{2\sqrt{2}} = \text{Re}[\overline{\lambda(z)}u_z] = \frac{\phi_x}{2\sqrt{2}}, \ z \in L_4, \\
&\text{Im}[\overline{\lambda(z)}u_z]|_{z=z_3+0} = b_1^+,
\end{aligned} \tag{3.26}$$

in which

$$\begin{aligned}
&\lambda(z) = a + jb = \begin{cases} \dfrac{1-j}{\sqrt{2}} \ \text{on} \ L_3, \\ \dfrac{1+j}{\sqrt{2}} \ \text{on} \ L_4, \end{cases} \quad r(z) = \frac{\phi_x}{2\sqrt{2}} \ \text{on} \ L = L_3 \cup L_4, \\
&\begin{aligned} b_1^- &= \text{Im}[\frac{1+j}{\sqrt{2}} u_z(z_3)] = \frac{\phi_x}{2\sqrt{2}}\Big|_{z=z_3-0}, \\ b_1^+ &= \text{Im}[\frac{1-j}{\sqrt{2}} u_z(z_3)] = -\frac{\phi_x}{2\sqrt{2}}\Big|_{z=z_3+0}, \end{aligned} \quad b_0 = \phi(0),
\end{aligned} \tag{3.27}$$

in which $a = 1/\sqrt{2} \neq b = -1/\sqrt{2}$ on L_3 and $a = 1/\sqrt{2} \neq -b = -1/\sqrt{2}$ on L_4.

Noting that Problem P for (3.13) is equivalent to the Riemann-Hilbert problem (Problem A) for the complex equation of first order and boundary conditions:

$$w_{\bar{z}} = \text{Re}[A_1 w] + A_2 u + A_3 \ \text{in} \ D, \tag{3.28}$$

$$\text{Re}[\overline{\lambda(z)}w(z)] = r(z), \ z \in L, \ \text{Im}[\overline{\lambda(z)}w(z)]|_{z=z_3} = b_1, \tag{3.29}$$

and the relation

$$u(z) = 2\text{Re}\int_{z_3}^{z} \overline{w(z)}dz + b_0 \text{ in } D. \tag{3.30}$$

In addition, it is not difficult to derive that

$$\begin{aligned} 2\text{Re}\int_{\Gamma} \overline{w(z)}dz &= \int_{\Gamma} \overline{w(z)}dz + \int_{\Gamma} w(z)d\bar{z} \\ = 2j\iint_{D}[\bar{w}_z - w_{\bar{z}}]dxdy &= 4\iint_{D} \text{Im}[\bar{w}_z]dxdy = 0, \end{aligned}$$

the above equality for any subdomain in D is also true, hence the function determined by the integral in (3.30) is independent of integral paths in $\bar{D}$. In this case we may choose that the integral path is along two families of characteristic lines, namely first along one of characteristic line: $x + y = \mu\,(0 \le \mu \le 2R_1)$ and then along one of characteristic line: $x - y = \nu\,(0 \le \nu \le 2R)$, for instance, the value of $u(z^*)(z^* = x^* + jy^* \in D,\, y^* \le 0)$ can be obtained by the integral

$$u(z^*) = 2\text{Re}[\int_{s_1} \overline{w(z)}dz + \int_{s_2} \overline{w(z)}dz] + b_0,$$

in which $l_1 = \{x+y = 0, 0 \le x \le (x^* - y^*)/2\}$, $l_2 = \{x - y = x^* - y^*, (x^* - y^*)/2 \le x \le x^*\}$, in which $x^* - y^*$ is the intersection of the characteristic line: $\{x - y = x^* - y^*\}$ passing through the point z^* and real axis. In particular, when $A_j = 0, j = 1, 2, 3$, equation (3.13) becomes the simplest hyperbolic complex equation

$$u_{z\bar{z}} = 0. \tag{3.31}$$

Problem P for (3.31) is equivalent to Problem A for the simplest hyperbolic complex equation of first order

$$w_{\bar{z}} = 0 \text{ in } D \tag{3.32}$$

with the boundary condition (3.29) and the relation (3.30). Hence similarly to Theorem 1.2, we can derive the representation and existence theorem of solutions of Problem A for the simplest equation (3.32), namely

Theorem 3.1 *Any solution $u(z)$ of Problem P for the hyperbolic equation (3.31) can be expressed as (3.30), where $w(z)$ is as follows*

$$\begin{aligned} w(z) &= f(x-y)e_1 + g(x+y)e_2 \\ &= \frac{1}{2}\{f(x-y)+g(x+y)+j[f(x-y)-g(x+y)]\}, \end{aligned} \tag{3.33}$$

here and in the following, for convenience denote by the functions $a(x)$, $b(x), r(x)$ *of* x *the functions* $a(z), b(z), r(z)$ *of* z *in* (3.29), *and* $f(x-y)$, $g(x+y)$ *possess the forms*

$$\begin{aligned}
f(x-y) &= \frac{2r((x-y)/2+R)}{a((x-y)/2+R)-b((x-y)/2+R)}\\
&-\frac{[a((x-y)/2+R)+b((x-y)/2+R)]\,g(2R)}{a((x-y)/2+R)-b((x-y)/2+R)},\\
&\qquad 0\le x-y\le 2R,\\
(a(R)+b(R))g(2R) &= (a(R)+b(R))(U(z_3)-V(z_3))\\
&= r(R)-b_1 \text{ or } 0,\\
g(x+y) &= \frac{2r((x+y)/2)-[a((x+y)/2)-b((x+y)/2)]f(0)}{a((x+y)/2)+b((x+y)/2)},\\
&\qquad 0\le x+y\le 2R_1,\\
(a(R)-b(R))f(0) &= (a(R)-b(R))(U(z_3)+V(z_3))\\
&= r(R)+b_1 \text{ or } 0.
\end{aligned} \tag{3.34}$$

Moreover $u(z)$ *satisfies the estimate*

$$C^1_\delta[u(z),\overline{D}]\le M_1,\ C^1_\delta[u(z),\overline{D}]\le M_2k_2, \tag{3.35}$$

where $\delta=\delta(\alpha,k_0,k_2,D)\,(<1), M_1=M_1(\alpha,k_0,k_2,D), M_2=M_2(\alpha,k_0,D)$ *positive constants.*

Proof Let the general solution

$$w(z)=u_z=\frac{1}{2}\{f(x-y)+g(x+y)+j[f(x-y)-g(x+y)]\}$$

of (3.32) be substituted in the boundary condition (3.29), thus (3.29) can be rewritten as

$$\begin{aligned}
&a(z)U(z)-b(z)V(z)=r(z) \text{ on } L,\ \overline{\lambda(z_3)}w(z_3)=r(z_3)+jb_1, \text{ i.e.}\\
&[a(x)-b(x)]f(2x-2R)+[a(x)+b(x)]g(2R)=2r(x)\ \text{ on } L_3,\\
&[a(x)-b(x)]f(0)+[a(x)+b(x)]g(2x)=2r(x)\ \text{ on } L_4,
\end{aligned}$$

the above formulas can be rewritten as

$$
\begin{aligned}
&[a(\frac{t}{2}+R)-b(\frac{t}{2}+R)]f(t)+[a(\frac{t}{2}+R)+b(\frac{t}{2}+R)]g(2R)\\
&=2r(\frac{t}{2}+R),\ t\in[0,2R],\\
&(a(R)+b(R))g(2R)=(a(R)+b(R))(U(z_3)-V(z_3))=r(R)-b_1 \text{ or } 0,\\
&[a(\frac{t}{2})-b(\frac{t}{2})]f(0)+[a(\frac{t}{2})+b(\frac{t}{2})]g(t)=2r(\frac{t}{2}),\ t\in[0,2R_1],\\
&(a(R)-b(R))f(0)=(a(R)-b(R))(U(z_3)+V(z_3))=r(R)+b_1 \text{ or } 0,
\end{aligned}
$$

thus the solution $w(z)$ can be expressed as (3.33), (3.34). Here we mention that for the Dirichlet boundary condition, noting (3.27), we have $(a(R)+b(R))g(2R)=0$, $(a(R)-b(R))f(0)=0$. From the condition (3.22) and the relation (3.30), we see that the estimate (3.35) of $u(z)$ for (3.31) is obviously true.

Next we give the representation of Problem P for the quasilinear equation (3.13).

Theorem 3.2 *Under Condition C, any solution $u(z)$ of Problem P for the hyperbolic equation* (3.13) *can be expressed as*

$$
\begin{aligned}
&u(z)=2\text{Re}\int_{z_3}^{z}\overline{w(z)}dz+b_0 \text{ in } D,\\
&w(z)=w_0(z)+\Phi(z)+\Psi(z) \text{ in } D,\\
&w_0(z)=f(\nu)e_1+g(\mu)e_2,\ \Phi(z)=\tilde{f}(\nu)e_1+\tilde{g}(\mu)e_2,\\
&\Psi(z)=\int_{2R}^{\mu}[A\xi+B\eta+Cu+D]e_1d\mu+\int_{0}^{\nu}[A\xi+B\eta+Cu+D]e_2d\nu,
\end{aligned}
\tag{3.36}
$$

in which $f(\nu)$, $g(\mu)$ are as stated in (3.34) *and $\tilde{f}(\nu)$, $\tilde{g}(\mu)$ are similar to $f(\nu)$, $g(\mu)$ in* (3.34), *but $r(z)$, b_1 are replaced by $-\text{Re}[\overline{\lambda(z)}\Psi(z)]$, $-\text{Im}[\overline{\lambda(z_3)}\Psi(z_3)]$, namely*

$$
\begin{aligned}
&\text{Re}[\overline{\lambda(z)}\Phi(z)]=-\text{Re}[\overline{\lambda(z)}\Psi(z)] \text{ on } L,\\
&\text{Im}[\overline{\lambda(z_3)}\Phi(z_3)]=-\text{Im}[\overline{\lambda(z_3)}\Psi(z_3)].
\end{aligned}
\tag{3.37}
$$

Proof Let the solution $u(z)$ of Problem P be substituted into the coefficients of equation (3.13). Then the equation in this case can be seen as a linear hyperbolic equation (3.12). Due to Problem P is equivalent to the Problem A for the complex equation (3.28) with the relation (3.30), according to Theorem 1.3, it is not difficult to see that the function $\Psi(z)$ satisfies the complex equation

$$[\Psi]_{\bar{z}}=[A\xi+B\eta+Cu+D]e_1+[A\xi+B\eta+Cu+D]e_2 \text{ in } D, \tag{3.38}$$

and $\Phi(z) = w(z) - w_0(z) - \Psi(z)$ satisfies the complex equation and the boundary conditions

$$\xi_\mu e_1 + \eta_\nu e_2 = 0, \tag{3.39}$$

$$\begin{aligned} &\text{Re}[\overline{\lambda(z)}(\xi e_1 + \eta e_2)] = -\text{Re}[\overline{\lambda(z)}\Psi(z)] \text{ on } L, \\ &\text{Im}[\overline{\lambda(z)}(\xi e_1 + \eta e_2)]|_{z=z_3} = -\text{Im}[\overline{\lambda(z_3)}\Psi(z_3)]. \end{aligned} \tag{3.40}$$

According to the way to derive the representation (1.27) of solutions of Problem A for (1.22), we can obtain the representation (3.36) of Problem P for (3.13).

3.3 Existence and uniqueness of solutions of oblique derivative problems

Theorem 3.3 *If the complex equation* (3.13) *satisfies Condition C, then Problem P for* (3.13) *has a solution.*

Proof We consider the expression of $u(z)$ as in the form (3.36). In the following, by using the successive approximation we shall find a solution of Problem P for equation (3.13). Firstly, substitute

$$u_0(z) = 2\text{Re}\int_{z_3}^{z} \overline{w_0(z)}dz + b_0,\ w_0(z) = u_{0z} = \xi_0 e_1 + \eta_0 e_2, \tag{3.41}$$

into the position of $u, w = \xi e_1 + \eta e_2$ in the right-hand side of (3.13), where $w_0(z)$ is the same function in (3.36) and satisfies the estimate (3.35). Moreover, we have

$$\begin{aligned} &u_1(z)=2\text{Re}\int_{z_3}^{z}\overline{w_1(z)}dz+b_0,\ w_1(z)=w_0(z)+\Phi_1(z)+\Psi_1(z), \\ &\Psi_1(z) = \int_{2R}^{x+y} [A\xi_0 + B\eta_0 + Cu_0 + D]e_1 d(x+y) \\ &\qquad + \int_0^{x-y} [A\xi_0 + B\eta_0 + Cu_0 + D]e_2 d(x-y), \end{aligned} \tag{3.42}$$

from the first equality in (3.42), the estimate

$$C[u_1(z),\bar{D}] \le M_3 C[w_1(z),\overline{D}]R' + k_2 \tag{3.43}$$

can be derived, where $M_3 = M_3(D)$. From the second and third equalities in (3.42), we can obtain

$$\begin{aligned}
&C[\Psi_1(z),\bar{D}] \le 2M_4[(4+M_3R')m+k_2+1]R',\\
&C[\Phi_1(z),\bar{D}] \le 8M_4k_0^2(1+2k_0^2)[(4+M_3R')m+k_2+1]R',\\
&C[w_1(z)-w_0(z),\bar{D}] \le 2M_4M[(4+M_3R')m+1]R',
\end{aligned} \tag{3.44}$$

where $M_4=\max_{\bar{D}}(|A|,|B|,|C|,|D|)$, $M=1+4k_0^2(1+2k_0^2)$ are non-negative constants, $R'=\max(2R_1,2R)$, $m=||w_0(z)||_{C(\overline{D})}$. Thus we can find a sequence of functions $\{w_n(z)\}$ satisfying

$$\begin{aligned}
&u_{n+1}(z)=2\mathrm{Re}\int_{z_3}^{z}\overline{w_{n+1}(z)}dz+b_0,\ w_{n+1}(z)=w_0(z)+\Phi_n(z)\\
&+\int_{2R}^{\mu}[A\xi_n+B\eta_n+Cu_n+D]e_1d\mu+\int_0^{\nu}[B\eta_n+A\eta_n+Cu_n+D]e_2d\nu,
\end{aligned} \tag{3.45}$$

and then

$$\begin{aligned}
&||w_n-w_{n-1}|| \le \{2M_4M[(4+M_3R')m+1]\}^n\\
&\times\int_0^{R'}\frac{R'^{n-1}}{(n-1)!}dR' \le \frac{\{2M_4M[(4+M_3R')m+1]R'\}^n}{n!}.
\end{aligned} \tag{3.46}$$

From the above inequality, we can see that the sequence of functions $\{w_n(z)\}$, i.e.

$$w_n(z)=w_0(z)+[w_1(z)-w_0(z)]+\cdots+[w_n(z)-w_{n-1}(z)](n=1,2,...) \tag{3.47}$$

in $\bar{D}$ uniformly converges a function $w_*(z)$, and $w_*(z)$ satisfies the equality

$$\begin{aligned}
&w_*(z)=\xi_*e_1+\eta_*e_2=w_0(z)+\Phi_*(z)\\
&+\int_{2R}^{\mu}[A\xi_*+B\eta_*+Cu_*+D]e_1d\mu+\int_0^{\nu}[A\xi_*+B\eta_*+Cu_*+D]e_2d\nu,
\end{aligned} \tag{3.48}$$

and the function

$$u_*(z)=2\mathrm{Re}\int_{z_3}^{z}\overline{w_*(z)}dz+b_0, \tag{3.49}$$

is just a solution of Problem P for equation (3.13) in the closed domain $\bar{D}$.

Theorem 3.4 *Suppose that Condition C holds. Then Problem P for the complex equation* (3.13) *has at most one solution in* $\bar{D}$.

Proof Let $u_1(z), u_2(z)$ be any two solutions of Problem P for (3.13), we see that $u(z) = u_1(z) - u_2(z)$ and $w(z) = u_{1z}(z) - u_{2z}(z)$ satisfies the homogeneous complex equation and boundary conditions

$$w_{\bar{z}} = \text{Re}[\tilde{A}_1 w] + \tilde{A}_2 u \text{ in } D, \tag{3.50}$$

$$\text{Re}[\overline{\lambda(z)}w(z)] = 0 \text{ on } L, \ \text{Im}[\overline{\lambda(z_3)}w(z_3)] = 0, \tag{3.51}$$

and the relation

$$u(z) = 2\text{Re}\int_{z_3}^{z} \overline{w(z)}dz, \ z \in \bar{D}. \tag{3.52}$$

From Theorem 3.2, we see that the function $w(z)$ can be expressed in the form

$$\begin{gathered} w(z) = \Phi(z) + \Psi(z), \\ \Psi(z) - \int_{2R}^{\mu} [\tilde{A}\xi + \tilde{B}\eta + \tilde{C}u]e_1 d\mu + \int_0^{\nu} [\tilde{A}\xi + \tilde{B}\eta + \tilde{C}u]e_2 d\nu, \end{gathered} \tag{3.53}$$

moreover from (3.52),

$$C[u(z), \bar{D}] \leq M_3 C[w(z), \overline{D}]R'$$

can be obtained, in which $M_3 = M_3(D)$ is a non-negative constant. By using the successive procedure similar to the proof of Theorem 3.3, we can get

$$||w(z)|| = ||w_1 - w_2|| \leq \frac{[2M_5 M[(4 + M_3 R')m + 1]R']^n}{n!}, \tag{3.54}$$

where $M_5 = \max_{\bar{D}}(|\tilde{A}|, |\tilde{B}|, |\tilde{C}|)$. Let $n \to \infty$, thus we have $w(z) = w_1(z) - w_2(z) = 0$, $\Psi(z) = \Phi(z) = 0$ in $\bar{D}$. This proves the uniqueness of solutions of Problem P for (3.13) (see [86]33)).

4 The Oblique Derivative Problem for Degenerate Hyperbolic Equations of Second Order

This section deals with the discontinuous oblique derivative problem for the degenerate hyperbolic equations. We first give a representation theorem of solutions of the boundary value problem for the hyperbolic equations, and then prove the existence and uniqueness of solutions for the above oblique derivative problem. Finally we discuss the above boundary value problem in general domains.

4.1 Formulation of discontinuous oblique derivative problem for degenerate hyperbolic equations

We first consider the linear degenerate hyperbolic equation of second order

$$K(y)u_{xx} - u_{yy} = au_x + bu_y + cu + d \text{ in } D, \tag{4.1}$$

in which D is bounded by the segment $L_0 = [0, 2R_1]$ and two characteristic lines

$$\begin{aligned} L_3 : x &= 2R_1 - G(y) = 2R_1 - \int_0^y H(t)dt, \\ L_4 : x &= G(y) = \int_0^y H(t)dt, \end{aligned} \tag{4.2}$$

where $R_1 = G(y_3)$ is a positive number, $H(y) = \sqrt{K(y)}$, $l = 1, 2$, $K(y) = y^m h(y)$, m is a positive number, $h(y)$ is a continuously differentiable positive function, and $z_3 = R_1 + jy_3$ is the intersection of L_3 and L_4. Suppose that the coefficients of (4.1) satisfy **Condition** C, namely

$$\begin{aligned} &\hat{C}[\eta, \overline{D}] = C[\eta, \overline{D}] + C[\eta_x, \overline{D}] \le k_0, \eta = a, b, c, \\ &\hat{C}[d, \overline{D}] \le k_1, \ C[ya/H, \overline{D}] \le \varepsilon(y), \ m \ge 2, \end{aligned} \tag{4.3}$$

where k_0, k_1 are positive constants, and $\varepsilon(y) \to 0$ as $y \to 0$. It is clear that equation (4.1) is a degenerate hyperbolic equation, and the solution of equation (4.1) in D is a generalized solution. If the conditions in (4.3) are replaced by

$$\begin{aligned} &C_\alpha[ya/H, \overline{D}] \le \varepsilon(y), \ m \ge 2, \ C_\alpha[K(y), \overline{D}] \le k_0, \\ &\hat{C}_\alpha[\eta, \overline{D}] = C_\alpha[\eta, \overline{D}] + C_\alpha[\eta_x, \overline{D}] \le k_0, \eta = a, b, c, \hat{C}_\alpha[d, \overline{D}] \le k_1, \end{aligned} \tag{4.4}$$

in which $\alpha\,(0 < \alpha < 1)$, k_0, k_1 are positive constants, which will be called **Condition** C', it is clear that the solution of equation (4.1) in D is a classical solution of (4.1) in D.

The oblique derivative boundary value problem for equation (4.1) may be formulated as follows:

Problem P Find a continuous solution $u(z)$ of (4.1) in $\bar{D}\backslash L_0$, which satisfies the boundary conditions

$$\begin{aligned} &\frac{1}{2}\frac{\partial u}{\partial \nu} = \frac{1}{H(y)} \mathrm{Re}[\overline{\lambda(z)u_{\tilde{z}}}] = \mathrm{Re}[\overline{\Lambda(z)u_z}] = r(z), \ z \in L = L_3 \cup L_4, \\ &u(z_3) = b_0, \ \frac{1}{H(y)} \mathrm{Im}[\overline{\lambda(z)u_{\tilde{z}}}]|_{z=z_3} = \mathrm{Im}[\overline{\Lambda(z)u_z}]|_{z=z_3} = b_1, \end{aligned} \tag{4.5}$$

in which ν is a given vector at every point $z \in L$, $u_{\tilde{z}} = [H(y)u_x - ju_y]/2$, $u_{\bar{\tilde{z}}} = [H(y)u_x + ju_y]/2$, b_0, b_1 are real constants, $\lambda(z) = a(x) + jb(x)$, $\Lambda(z) = \cos(\nu, x) + j\cos(\nu, y)$, $R(z) = H(y)r(z)$, $z \in L$, $b_1' = H(y_3)b_1$ and $\lambda(z), \Lambda(z), r(z), b_0, b_1$ satisfy the conditions

$$\begin{gathered} C^1[\lambda(z), L_j] \le k_0,\ C^1[r(z), L_j] \le k_2,\ j = 3, 4, \\ |b_0|, |b_1| \le k_2,\ \max_{z \in L_3} \frac{1}{|a(z) - b(z)|},\ \max_{z \in L_4} \frac{1}{|a(z) + b(z)|} \le k_0, \end{gathered} \tag{4.6}$$

in which k_0, k_2 are positive constants.

For the Dirichlet problem (Problem D) with the boundary condition:

$$u(z) = \phi(x) \text{ on } L = L_3 \cup L_4, \tag{4.7}$$

where L_3, L_4 are as stated before, we find the derivative for (4.7) according to the parameter $s = x$ on L_3, L_4, and obtain

$$\begin{aligned} &u_s = u_x + u_y y_x = u_x - \frac{u_y}{H(y)} = \phi'(x) \text{ on } L_3, \\ &u_s = u_x + u_y y_x = u_x + \frac{u_y}{H(y)} = \phi'(x) \text{ on } L_4, \text{ i.e.} \\ &U(z) + V(z) = \frac{1}{2}H(y)\phi'(x) = R(z) \text{ on } L_3, \\ &U(z) - V(z) = \frac{1}{2}H(y)\phi'(x) = R(z) \text{ on } L_4, \text{ i.e.} \\ &\mathrm{Re}[(1+j)(U+jV)] = U(z) + V(z) = R(z) \text{ on } L_3, \\ &\mathrm{Im}[(1+j)(U+jV)] = [U(z) + V(z)]|_{z=z_3-0} = R(z_3 - 0), \\ &\mathrm{Re}[(1-j)(U+jV)] = U(z) - V(z) = R(z) \text{ on } L_4, \\ &\mathrm{Im}[(1-j)(U+jV)] = [-U(z) + V(z)]|_{z=z_3+0} = -R(z_3 + 0), \end{aligned}$$

where

$$\begin{aligned} &U(z) = \frac{1}{2}H(y)u_x,\ V(z) = -\frac{1}{2}u_y, \\ &a + jb = 1 - j,\ a = 1 \ne b = -1 \text{ on } L_3, \\ &a + jb = 1 + j,\ a = 1 \ne -b = -1 \text{ on } L_4. \end{aligned}$$

From the above formulas, we can write the complex forms of boundary conditions of $U+jV$:

$$\mathrm{Re}[\overline{\lambda(z)}(U+jV)]=R(z)\ \text{ on }\ L,$$

$$\mathrm{Im}[\overline{\lambda(z)}(U+jV)]|_{z=z_3-0}=R(z_3-0)=b_1',$$

$$\lambda(z)=\begin{cases}1-j=a+jb,\\ 1+j=a+jb,\end{cases}\quad R(z)=\begin{cases}\dfrac{1}{2}H(y)\phi'(x)\ \text{ on }\ L_3,\\ \dfrac{1}{2}H(y)\phi'(x)\ \text{ on }\ L_4,\end{cases}$$

and

$$u(z)=2\mathrm{Re}\int_{z_3}^{z}(\frac{U(z)}{H(y)}-jV)dz+\phi(R)\ \text{ in }\ D.$$

Hence Problem D is a special case of Problem P.

4.2 Representation and solvability of oblique derivative problem for degenerate hyperbolic equations

In this section, we first write the complex form of equation (4.1). Denote

$$\begin{aligned}
w=u_{\bar z}&=\frac{1}{2}[\sqrt{K}u_x-ju_y]=\frac{1}{2}[Hu_x-ju_y],\\
w_{\bar z}&=\frac{1}{2}[\sqrt{K}w_x+jw_y]=\frac{1}{2}[Hw_x+jw_y]\\
&=\frac{1}{4}[H^2u_{xx}-u_{yy}-jH(u_{yx}-u_{xy})+jH_yu_x]\\
&=\frac{1}{4}[(\frac{a}{H}+\frac{jH_y}{H})Hu_x+bu_y+cu+d]\\
&=\frac{1}{4}[\frac{a}{H}+\frac{jH_y}{H}](w+\overline{w})+\frac{1}{4}jb(\overline{w}-w)+\frac{1}{4}(cu+d)]\\
&=\frac{1}{4}[\frac{a}{H}+\frac{jH_y}{H}-jb]w+\frac{1}{4}[\frac{a}{H}+\frac{jH_y}{H}+jb]\overline{w}+\frac{1}{4}(cu+d)\\
&=A_1(z)w+A_2(z)\overline{w}+A_3(z)u+A_4(z)\\
&=\frac{e_1}{4}\{[\frac{a}{H}+\frac{H_y}{H}-b](\mathrm{Re}w+\mathrm{Im}w)\\
&\quad+[\frac{a}{H}+\frac{H_y}{H}+b](\mathrm{Re}w-\mathrm{Im}w)+cu+d\}
\end{aligned}$$

$$+\frac{e_2}{4}\{[\frac{a}{H}-\frac{H_y}{H}-b](\text{Re}w+\text{Im}w)$$
$$+[\frac{a}{H}-\frac{H_y}{H}+b](\text{Re}w-\text{Im}w)+cu+d\}, \tag{4.8}$$

where $e_1=(1+j)/2$, $e_2=(1-j)/2$. Denoting $w=U+jV$, $U=\text{Re}w$, $V=\text{Im}w$ and noting

$$\mu=x+\int_0^y\sqrt{K}dt=x+\int_0^y H(t)dt=x+G(y),$$
$$\nu=x-\int_0^y\sqrt{K}dt=x-\int_0^y H(t)dt=x-G(y),$$
$$\mu+\nu=2x, \mu-\nu=2G, \frac{\partial G}{\partial y}=H=\sqrt{K(y)},$$
$$\frac{\partial x}{\partial\mu}=\frac{\partial x}{\partial\nu}=\frac{1}{2}, \frac{\partial y}{\partial\mu}=-\frac{\partial y}{\partial\nu}=\frac{1}{2H}=\frac{1}{2\sqrt{K}},$$

we have

$$w_{\tilde{z}}=H(y)w_{\overline{Z}}=\tfrac{1}{2}[H(U+jV)_x+j(U+jV)_y]$$
$$=\frac{e_1}{2}[HU_x+V_y+HV_x+U_y]+\frac{e_2}{2}[HU_x+V_y-HV_x-U_y]$$
$$=\frac{e_1}{2}[H(U+V)_x+(U+V)_y]+\frac{e_2}{2}[H(U-V)_x-(U-V)_y]$$
$$=H(y)[e_1\left(\frac{(U+V)_x}{2}+\frac{(U+V)_y}{2H}\right)+e_2\left(\frac{(U-V)_x}{2}-\frac{(U-V)_y}{2H}\right)] \tag{4.9}$$
$$=He_1(U+V)_\mu+He_2(U-V)_\nu$$
$$=\frac{e_1}{4}\{[\frac{a}{H}+\frac{H_y}{H}-b](U+V)+[\frac{a}{H}+\frac{H_y}{H}+b](U-V)+cu+d\}$$
$$+\frac{e_2}{4}\{[\frac{a}{H}-\frac{H_y}{H}-b](U+V)+[\frac{a}{H}-\frac{H_y}{H}+b](U-V)+cu+d\}.$$

Hence the complex equation (4.9) can be reduced to

$$H(U+V)_\mu=\frac{1}{4}\{[\frac{a}{H}+\frac{H_y}{H}-b](U+V)$$
$$+[\frac{a}{H}+\frac{H_y}{H}+b](U-V)+cu+d\},$$
$$H(U-V)_\nu=\frac{1}{4}\{[\frac{a}{H}-\frac{H_y}{H}-b](U+V)$$

$$+[\frac{a}{H}-\frac{H_y}{H}+b](U-V)+cu+d\}, \text{ i.e.}$$

$$H(U+V)_x+(U+V)_y=[\frac{a}{H}+\frac{K_y}{2K}]U-bV+\frac{1}{2}(cu+d), \tag{4.10}$$

$$H(U-V)_x-(U-V)_y=[\frac{a}{H}-\frac{K_y}{2K}]U-bV+\frac{1}{2}(cu+d).$$

If $H(y)=\sqrt{y^m h(y)}=\sqrt{|K(y)|}$, then $w(z)=\sqrt{y^m h(y)}U+jV$ is a solution of the first order hyperbolic complex equation

$$w_{\bar{z}}=A_1(z)w+A_2(z)\overline{w}+A_3(z)u+A_4(z) \text{ in } D, \tag{4.11}$$

where

$$A_1=\frac{1}{4}[\frac{a}{H}+\frac{jH_y}{H}-jb]=\frac{a}{4(y^m h)^{1/2}}+j\left(\frac{h_y}{8h}+\frac{m}{8y}-\frac{b}{4}\right), \; A_3=\frac{c}{4},$$

$$A_2=\frac{1}{4}[\frac{a}{H}+\frac{jH_y}{H}+jb]=\frac{a}{4(y^m h)^{1/2}}+j\left(\frac{h_y}{8h}+\frac{m}{8y}+\frac{b}{4}\right), \; A_4=\frac{d}{4},$$

and

$$u(z)=2\text{Re}\int_{z_3}^{z} u_{\bar{z}}dz+b_0=2\text{Re}\int_{z_3}^{z}(\frac{U}{H}-jV)d(x+jy)+b_0$$

is a solution of equation (4.11) with $K(y)=|y|^m h(y)$. In particular, the complex equation

$$W_{\bar{z}}=0 \tag{4.12}$$

can be rewritten in the system

$$(U+V)_\mu=0, \;\; (U-V)_\nu=0 \text{ in } D_\tau, \tag{4.13}$$

where D_τ is the image domain of D through the mapping $\tau=\mu+j\nu=x+G(y)+j(x-G(y))$.

Now, we give the representation of solutions for the oblique derivative problem (Problem P) for system (4.13) in D. For this, we first discuss the Riemann-Hilbert problem(Problem A) for the system with the boundary conditions

$$\begin{gathered}\text{Re}[\overline{\lambda(z)}(U+jV)]=R(z) \text{ on } L=L_3\cup L_4,\\ u(z_3)=b_0, \; \text{Im}[\overline{\lambda(z)}(U+iV)]|_{z=z_3}=b_1',\end{gathered} \tag{4.14}$$

in which $\lambda(z) = a(z) + jb(z)$ on $L_3 \cup L_4$, and $b'_1 = H(\text{Im}z_3)b_1$. It is clear that the solution of Problem A for (4.13) can be expressed as

$$\begin{aligned} &\xi = U(z) + V(z) = f(\nu),\ \ \eta = U(z) - V(z) = g(\mu), \\ &U(z) = [f(\nu) + g(\mu)]/2,\ V(z) = [f(\nu) - g(\mu)]/2, \text{ i.e.} \\ &W(z) = U(z) + jV(z) = [(1+j)f(\nu) + (1-j)g(\mu)]/2, \end{aligned} \tag{4.15}$$

in which $f(t), g(t)$ are two arbitrary real continuous functions on $[0, 2R_1]$. For convenience, denote by the functions $a(x), b(x), R(x)$ of x the functions $a(z), b(z), R(z)$ of z in (4.14), thus (4.14) can be rewritten as

$$\begin{aligned} &a(x)U(z) - b(x)V(z) = H(y)r(x) = R(z) \text{ on } L, \text{ i.e.} \\ &[a(x)-b(x)]f(x-G(y)) + [a(x)+b(x)]g(x+G(y)) = 2R(x) \text{ on } L, \text{ i.e.} \\ &[a(x) - b(x)]f(0) + [a(x) + b(x)]g(2x) = 2R(x),\ x \in [0, R_1], \\ &[a(x)-b(x)]f(2x-2R_1) + [a(x)+b(x)]g(2R_1) = 2R(x), x \in [R_1, 2R_1], \text{ i.e.} \\ &[a(t/2) - b(t/2)]f(0) + [a(t/2) + b(t/2)]g(t) = 2R(t/2),\ t \in [0, 2R_1], \\ &[a(t/2+R_1) - b(t/2+R_1)]f(t) + [a(t/2+R_1) + b(t/2+R_1)]g(2R_1) \\ &= 2R(t/2 + R_1),\ t \in [0, 2R_1], \end{aligned}$$

where

$$\begin{aligned} &(a(R_1)-b(R_1))f(0) = (a(R_1)-b(R_1))(U(z_3)+V(z_3)) = R(R_1)+b_1 \text{ or } 0, \\ &(a(R_1)+b(R_1))g(2R_1) = (a(R_1)+b(R_1))(U(z_3)-V(z_3)) = R(R_1)-b_1 \text{ or } 0. \end{aligned}$$

From Subsection 4.1, for the Dirichlet boundary condition, we have $(a(R_1)-b(R_1))f(0)=0$ and $(a(R_1)+b(R_1))g(2R_1)=0$. Moreover we can derive

$$\begin{aligned} U(z) &= \frac{1}{2}\{\frac{2R(\nu/2+R_1) - [a(\nu/2+R_1)+b(\nu/2+R_1)]g(2R_1)}{a(\nu/2+R_1)-b(\nu/2+R_1)} + g(\mu)\} \\ &= \frac{1}{2}\{f(\nu) + \frac{2R(\mu/2) - [a(\mu/2) - b(\mu/2)]f(0)}{a(\mu/2) + b(\mu/2)}\}, \\ V(z) &= \frac{1}{2}\{\frac{2R(\nu/2+R_1) - [a(\nu/2+R_1)+b(\nu/2+R_1)]g(2R_1)}{a(\nu/2+R_1)-b(\nu/2+R_1)} - g(\mu)\} \\ &= \frac{1}{2}\{f(\nu) - \frac{2R(\mu/2) - [a(\mu/2) - b(\mu/2)]f(0)}{a(\mu/2) + b(\mu/2)}\}, \end{aligned} \tag{4.16}$$

if $a(x)+b(x)\neq 0$ on $[0,R_1]$ and $a(x)-b(x)\neq 0$ on $[R_1,2R_1]$ respectively. From the above formulas, it follows that

$$\begin{aligned}&\mathrm{Re}[(1+j)W(x)]=U(x)+V(x)=\frac{2R(x/2+R_1)-K(x)}{a(x/2+R_1)-b(x/2+R_1)},\\&K(x)=[a(x/2+R_1)+b(x/2+R_1)]g(2R_1),\\&\mathrm{Re}[(1-j)W(x)]=U(x)-V(x)=\frac{2R(x/2)-[a(x/2)-b(x/2)]f(0)}{a(x/2)+b(x/2)},\\&\qquad x\in[0,2R_1],\end{aligned}\tag{4.17}$$

if $a(x)+b(x)\neq 0$ on $[0,R_1]$ and $a(x)-b(x)\neq 0$ on $[R_1,2R_1]$ respectively. From (4.17), the solution

$$W(z)=\begin{cases}\frac{1}{2}\{(1+j)\frac{2R((x-G(y))/2+R_1)-K(z)}{a((x-G(y))/2+R_1)-b((x-G(y))/2+R_1)}\\ +(1-j)\frac{2R((x+G(y))/2)-N(z)}{a((x+G(y))/2)+b((x+G(y))/2)}\},\\ K(z)=[a((x-G(y))/2+R_1)+b((x-G(y))/2+R_1)]g(2R_1),\\ N(z)=[a((x+G(y))/2)-b((x+G(y))/2)]f(0)\end{cases}\tag{4.18}$$

is obtained. In the following we discuss the case of $f(x)=-g(x)$ on L_0, i.e. $2U(x)=f(x)+g(x)=0$ on L_0, because when we handle the complex equations of mixed type in Chapters IV-VI, the case will be appeared.

Theorem 4.1 *Problem A of equation* (4.12) *or system* (4.13) *in* $\bar D$ *has a unique solution, which can be expressed in the form* (4.18), *if* $\mathrm{Re}W(x)=U(x)=0$ *on* L_0, *then the solution* $W(z)$ *satisfies the estimates*

$$|\mathrm{Re}W(z)|\leq M_1,\ |\mathrm{Im}W(z)|\leq M_1\ \text{in}\ D,\ C_\delta[W(z),\overline{D}]\leq M_2,\tag{4.19}$$

where δ *is a sufficiently small positive constant, if* $H(y)=y^{m/2}$, *and* $M_l=M_l(\delta,k_0,k_2,D)\ (l=1,2)$ *are positive constants.*

Proof From Condition C and (4.18), the first two estimates in (4.19) are obtained. If $H(y)=y^{m/2}, m\geq 2$, $H[J((x+G(y))/2)]$, $H[J((x-G(y))/2+R_1)]\in C_\delta(\overline{D})$, where $y=J(x)$ and $y=J(2R_1-x)$ are the inverse functions of $x=G(y)=\int_0^y y^{m/2}dy=2y^{(m+2)/2}/(m+2)=Jy^{(m+2)/2}$ and $2R_1-x=G(y)$ respectively. Noting that the condition (4.6), and $a(x,y)$ is replaced by $a[(x+G(y))/2,((x+G(y))/2J)^{2/(m+2)}]$, or $a[(2R_1+x-G(y))/2,((2R_1-x+G(y))/2J)^{2/(m+2)}]$, we can get the third estimate in (4.19).

Next we state the representation and existence of solutions of Problem P for (4.1).

Theorem 4.2 *Under Condition C, any solution of Problem P for equation* (4.1) *can be expressed as*

$$\begin{aligned}&u(z)=2\mathrm{Re}\int_{z_3}^{z}[\frac{\mathrm{Re}w}{H}-j\mathrm{Im}w]dz+b_0,\, w(z)=W(z)+\Phi(z)+\Psi(z) \text{ in } D,\\ &W(z)=f(\nu)e_1+g(\mu)e_2,\ \Phi(z)=\tilde{f}(\nu)e_1+\tilde{g}(\mu)e_2,\\ &\Psi(z)=\int_{2R_1}^{\mu}g_1(z)e_1d\mu+\int_0^{\nu}g_2(z)e_2d\nu,\, g_l(z)=\hat{A}_l\xi+\hat{B}_l\eta+\hat{C}u+\hat{D},\, l=1,2,\end{aligned}\tag{4.20}$$

where

$$\begin{aligned}\hat{A}_1&=\frac{1}{4H}[\frac{a}{H}+\frac{H_y}{H}-b],\,\hat{B}_1=\frac{1}{4H}[\frac{a}{H}+\frac{H_y}{H}+b],\,\hat{C}=\frac{c}{4H},\\ \hat{A}_2&=\frac{1}{4H}[\frac{a}{H}-\frac{H_y}{H}-b],\,\hat{B}_2=\frac{1}{4H}[\frac{a}{H}-\frac{H_y}{H}+b],\,\hat{D}=\frac{d}{4H},\end{aligned}$$

and $f(\nu)$, $g(\mu)$ are as stated in (4.15), *and $\tilde{f}(\nu)$, $\tilde{g}(\mu)$ are similar to $f(\nu)$, $g(\mu)$ in* (4.15), *which satisfy the boundary condition*

$$\mathrm{Re}[\overline{\lambda(z)}(\Phi(z)+\Psi(z))]=0,\, z\in L,\, \mathrm{Im}[\overline{\lambda(z_3)}(\Phi(z_3)+\Psi(z_3))]=0.\tag{4.21}$$

Proof Due to Problem P is equivalent to the Problem A for the complex equation (4.8) with the first formula in (4.20), from Theorem 4.1 and (4.9), it is not difficult to see that the function $\Psi(z)$ satisfies the complex equation

$$[\Psi]_{\tilde{\bar{z}}}=H\{[\hat{A}_1\xi+\hat{B}_1\eta+\hat{C}u+\hat{D}]e_1+[\hat{A}_2\xi+\hat{B}_2\eta+\hat{C}u+\hat{D}]e_2\} \text{ in } D,\tag{4.22}$$

and $\Phi(z)=w(z)-W(z)-\Psi(z)$ satisfies equation (4.12) and the boundary conditions

$$\begin{aligned}&\mathrm{Re}[\overline{\lambda(z)}\Phi(z)]=-\mathrm{Re}[\overline{\lambda(z)}\Psi(z)] \text{ on } L,\\ &\mathrm{Im}[\overline{\lambda(z_3)}\Phi(z_3)]=-\mathrm{Im}[\overline{\lambda(z_3)}\Psi(z_3)].\end{aligned}\tag{4.23}$$

By the representation of solutions of Problem A for (4.11) as stated in the last three formulas of (4.20), we can obtain the representation of solutions of Problem P for (4.1) as stated in the first formula of (4.20).

Moreover by using the way in the proof of Theorem 3.3, namely we first find a solution $w(z)$ of Problem A for (4.9) in $D_\delta=D\cap\{y>\delta\}$ by using the successive approximation, where δ is a small positive number, and the function $u(z)$ in (4.20) is a solution of Problem P for equation (4.1) in D_δ. Moreover, let $\delta\to 0$, then we can obtain the solution $u(z)$ of Problem P for (4.1) in D. In addition, the uniqueness of solutions of Problem P for (4.1) in D is derived by the above same method. Thus the following theorem is proved.

Theorem 4.3 *If equation* (4.1) *in* D *satisfies Condition* C, *then Problem* P *for* (4.1) *in* D *has a unique solution.*

Finally we shall give the Hölder continuous estimates of solutions of Problem A for (4.1).

Theorem 4.4 *Suppose that the linear equation* (4.1) *satisfies Condition* C'. *Then the solution* $u(z), u_{\bar z}=w(z)=W(z)+\Phi(z)+\Psi(z)$ *in* $\bar D$ *satisfies the following estimates*

$$\begin{gathered}C_\delta[W(z),\overline{D}]\le M_3, C_\delta[\Psi(z),\overline{D}]\le M_3, C_\delta[\Phi(z),\overline{D}]\le M_3,\\ \tilde C_\delta[w(z),\overline{D}]=C_\delta[w(z),\overline{D}]+C_\delta[u(z),\overline{D}]\le M_3,\end{gathered}\tag{4.24}$$

where $W(z),\Phi(z),\Psi(z)$ *are as stated in* (4.20), δ *is a sufficiently small positive constant,* $M_3=M_3(\delta,k,D)$ *is a positive constant.*

Proof Due to the function $W(z)=w_0(z)$ in (4.18) satisfies the estimate (4.19), hence the first estimate in (4.24) can be derived. In order to prove that $\Psi(z)=\Psi_1(z)=\tilde\Psi^1(z)e_1+\tilde\Psi^2(z)e_2$ satisfies the second estimate in (4.24), from

$$\begin{gathered}\tilde\Psi^1(z)=\int_{2R_1}^{\mu}g_1(z)d\mu,\ \tilde\Psi^2(z)=\int_0^{\nu}g_2(z)d\nu,\\ g_l(z)=\hat A_l(z)\xi+\hat B_l(z)\eta+\hat C(z)u+\hat D,\ l=1,2.\end{gathered}\tag{4.25}$$

and (4.16), we see that $\tilde\Psi^1(z)=\tilde\Psi^1(\mu,\nu)$, $\tilde\Psi^2(z)=\tilde\Psi^2(\mu,\nu)$ in $\overline{D}$ with respect to $\mu=x+G(y)$, $\nu=x-G(y)$ satisfy the estimates

$$C_\delta[\tilde\Psi^1(\mu,\cdot),\overline{D}]\le M_4,\ C_\delta[\tilde\Psi^2(\cdot,\nu),\overline{D}]\le M_4,\tag{4.26}$$

respectively, where $M_4=M_4(\delta,k,D)$ is a positive constant. If we substitute the solution $w_0=w_0(z)=\xi_0e_1+\eta_0e_2$ of Problem A for (4.12) into the position of $w=\xi e_1+\eta e_2$ for (4.20), and $\xi_0=\mathrm{Re}w_0+\mathrm{Im}w_0$, $\eta_0=\mathrm{Re}w_0-\mathrm{Im}w_0$, from (4.4) and (4.19), we obtain

$$\begin{gathered}C_\delta[g_1(\cdot,\nu),\overline{D}]\le M_5,\ C_\delta[g_2(\mu,\cdot),\overline{D}]\le M_5,\\ C_\delta[\tilde\Psi^1(\cdot,\nu),\overline{D}]\le M_5,\ C_\delta[\tilde\Psi^2(\mu,\cdot),\overline{D}]\le M_5,\end{gathered}\tag{4.27}$$

in which $M_5 = M_5(\delta, k, D)$ is a positive constant. Due to $\Phi(z) = \Phi_1(z)$ satisfies the complex equation (4.12) and boundary condition (4.21), and $\Phi_1(z)$ possesses the representation similar to (4.18), it is easy to see that the estimate

$$C_\delta[\Phi_1(z), \overline{D}] \le M_6 = M_6(\delta, k, D) \tag{4.28}$$

is obtained. Thus setting $w_1(z) = w_0(z) + \Phi_1(z) + \Psi_1(z)$, $\tilde{w}_1(z) = w_1(z) - w_0(z)$, it is clear that the functions $\tilde{w}_1^1(z) = \mathrm{Re}\tilde{w}_1(z) + \mathrm{Im}\tilde{w}_1(z)$, $\tilde{w}_1^2(z) = \mathrm{Re}\tilde{w}_1(z) - \mathrm{Im}\tilde{w}_1(z)$ satisfy the estimates about $\mu = x + G(y)$, $\nu = x - G(y)$ respectively:

$$C_\delta[\tilde{w}_1^1(\cdot, \nu), \overline{D}] \le M_7,\ C_\delta[\tilde{w}_1^2(\mu, \cdot), \overline{D}] \le M_7, \tag{4.29}$$

where $M_7 = M_7(\delta, k, D)$. By using the successive approximation, we obtain the sequence of functions: $w_n(z)\,(n = 1, 2, ...)$, and the corresponding functions $\tilde{w}_n^1(z) = \mathrm{Re}\tilde{w}_n + \mathrm{Im}\tilde{w}_n$, $\tilde{w}_n^2(z) = \mathrm{Re}\tilde{w}_n - \mathrm{Im}\tilde{w}_n$ satisfy the estimates

$$C_\delta[\tilde{w}_n^1(\cdot, \nu), \overline{D}] \le \frac{(M_7)^n}{n!},\ C_\delta[\tilde{w}_n^2(\mu, \cdot), \overline{D}] \le \frac{(M_7)^n}{n!}, \tag{4.30}$$

and denote the limit function $w(z)$ of $w_n(z) = \sum_{m=0}^{n} \tilde{w}_n(z)$ in $\overline{D}$, the corresponding functions $\tilde{w}^1(z) = \mathrm{Re}w(z) + \mathrm{Im}w(z)$, $\tilde{w}^2(z) = \mathrm{Re}w(z) - \mathrm{Im}w(z)$ satisfy the estimates

$$C_\delta[\tilde{w}^1(\cdot, \nu), \overline{D}] \le e^{M_7},\ C_\delta[\tilde{w}^2(\mu, \cdot), \overline{D}] \le e^{M_7}.$$

Combining the first formula in (4.24), (4.26)-(4.30) and the above formulas, the second and third estimates in (4.24) is derived. Moreover by the first formula in (4.20) and the above estimates, we immediately the fourth formula in (4.24).

Theorem 4.5 *Suppose that the quasilinear complex equation* (4.1) *satisfy Condition C'. Then the solution $w(z)$ of Problem A in $\bar{D}$ for* (4.1) *satisfies the following estimate*

$$\tilde{C}_\delta[w(z), \overline{D}] = C_\delta[w(z), \overline{D}] + C^1[u(z), \overline{D}] \le M_8 k, \tag{4.31}$$

in which δ is a constant as stated in (4.24), *and $k = k_1 + k_2$, $M_8 = M_8(\delta, k_0, D)$ are positive constants.*

Proof If $k = k_1 + k_2 > 0$, then the system of functions $[w^*(z), u^*(z)] = [w(z)/k, u(z)/k]$ is a solution of Problem A for the complex equation

$$w^*_{\bar{\bar{z}}} = A_1 \mathrm{Re}w^* + A_2 \mathrm{Im}w^* + A_3 u^* + A_4/k \ \text{ in } D,$$

with the boundary conditions

$$\mathrm{Re}[\overline{\lambda(z)}w^*(z)] = H(y)r(z)/k,\ z \in L = L_3 \cup L_4,$$

$$u(z_3) = b_0/k, \mathrm{Im}[\overline{\lambda(z)}w^*(z)]|_{z=z_3} = H(\mathrm{Im}z_3)b_1/k.$$

Noting that $H(y)r(z)/k$, b_0/k, b_1/k satisfy the conditions

$$C^1[r(z)/k, L_l] \le 1,\ l = 3, 4,\ |b_0/k|, |b_1/k| \le 1,$$

by means of the result of Theorem 4.4, we can get the estimate of the solution $[w^*(z), u^*(z)]$:

$$\tilde{C}_\delta[w^*(z), \overline{D}] \le M_8 = M_8(\delta, k, D).$$

From the above estimate it follows that the estimate (4.31) holds with $k > 0$. When $k = 0$, by using the uniqueness of solutions of Problem P for equation (4.1) as stated in Theorem 4.3, it is clear that $u(z) = 0$, hence the estimate (4.31) with $k = 0$ obviously holds.

After obtaining the estimates in (4.24), then similarly to the method of Section 4, Chapter I, we can use the fixed-point theorem to prove the existence of solutions of Problem P for equation (4.1).

4.3 Oblique derivative problem for degenerate hyperbolic equations in general domains

Now we consider some general domains with non-characteristic boundary and prove the unique solvability of Problem P for equation (4.8). Let D be a simply connected bounded domain D in the hyperbolic complex plane $\mathbf{C}$ with the boundary $\partial D = L_0 \cup L$, where $L_0 = [0, 2R_1]$, $L = L_3 \cup L_4$ is as follows:

$$L_4 = \{x = G(y), 0 \le x \le R_1\}, L_3 = \{x = 2R_1 - G(y), R_1 \le x \le 2R_1\}.$$

1. We consider the domain D' with the boundary $L_0 \cup L'$, $L' = L'_3 \cup L'_4$, where the parameter equations of the curves L'_3, L'_4 are as follows:

$$L'_3 = \{x + G(y) = 2R_1, l \le x \le 2R_1\}, L'_4 = \{y = \gamma_1(s), 0 \le s \le s_0\}, \tag{4.32}$$

in which $Y = G(y) = \int_0^y \sqrt{K(y)}dy$, s is the parameter of are length of L'_4, $\gamma_1(s)$ on $\{0 \le s \le s_0\}$ is continuously differentiable, $\gamma_1(0) = 0, \gamma_1(s) > 0$

on $\{0 < s \le s_0\}$, and the slope of curve L_4' at a point z^* is not equal to $dy/dx = -1/H(y)$ of the characteristic curve $s_2 : dy/dx = -1/H(y)$ at the point, where z^* is an intersection point of L_4' and the characteristic curve of s_2, and $z_3' = l + j\gamma_1(s_0)$ is the intersection point of L_3' and L_4'.

We consider the oblique derivative boundary value problem (Problem P') for equation (4.1) in D' with the boundary conditions

$$\begin{gathered}\frac{1}{2}\frac{\partial u}{\partial \nu} = \frac{1}{H(y)}\mathrm{Re}[\overline{\lambda(z)}u_{\tilde{z}}] = r(z),\ z \in L' = L_3' \cup L_4',\\ u(z_3') = b_0,\ \frac{1}{H(y)}\mathrm{Im}[\overline{\lambda(z)}u_{\tilde{z}}]|_{z=z_3'} = b_1,\end{gathered} \tag{4.33}$$

where $\lambda(z) = a(x) + ib(x)$, $R(z) = H(y)r(z)$ on L', $b_1' = H(\mathrm{Im} z_3')b_1$, and $\lambda(z)$, $r(z)$, b_1' satisfy the conditions

$$\begin{gathered}C^1[\lambda(z), L'] \le k_0,\ C^1[r(z), L'] \le k_2,\ |b_0|, |b_1| \le k_2,\\ \max_{z\in L_3'} \frac{1}{|a(x) - b(x)|} \le k_0,\ \max_{z\in L_4'} \frac{1}{|a(x) + b(x)|} \le k_0,\end{gathered} \tag{4.34}$$

in which k_0, k_2 are positive constants.

Setting $Y = G(y) = \int_0^y \sqrt{K(t)}dt$. By the conditions in (4.32), the inverse function $x = \tau(\mu) = (\mu + \nu)/2$ of $\mu = x + G(y)$ can be found, and then $\nu = 2\tau(\mu) - \mu$, $0 \le \mu \le l + \gamma_1(s_0)$. We make a transformation

$$\tilde{\mu} = \mu,\ \tilde{\nu} = \frac{2R_1[\nu - 2\tau(\mu) + \mu]}{2R_1 - 2\tau(\mu) + \mu}, 0 \le \mu \le 2R_1, 2\tau(\mu) - \mu \le \nu \le 2R_1, \tag{4.35}$$

where μ, ν are real variables, its inverse transformation is

$$\mu = \tilde{\mu}, \nu = \frac{[2R_1 - 2\tau(\mu) + \mu]\tilde{\nu}}{2R_1} + 2\tau(\mu) - \mu, 0 \le \tilde{\mu} \le 2R_1, 0 \le \tilde{\nu} \le 2R_1. \tag{4.36}$$

It is not difficult to see that the transformation in (4.35) maps the domain D' onto D. The transformation (4.35) and its inverse transformation (4.36) can be rewritten as

$$\begin{cases}\tilde{x} = \frac{1}{2}(\tilde{\mu} + \tilde{\nu}) = \dfrac{4R_1x - (2R_1 + x + Y)[2\tau(x + \gamma_1(s)) - x - \gamma_1(s)]}{4R_1 - 4\tau(x + \gamma_1(s)) + 2x + 2\gamma_1(s)},\\ \tilde{Y} = \frac{1}{2}(\tilde{\mu} - \tilde{\nu}) = \dfrac{4R_1Y + (2R_1 - x - Y)[2\tau(x + \gamma_1(s)) - x - \gamma_1(s)]}{4R_1 - 4\tau(x + \gamma_1(s)) + 2x + 2\gamma_1(s)},\end{cases} \tag{4.37}$$

and

$$\begin{cases} x=\dfrac{1}{2}(\mu+\nu)=\dfrac{4R_1\tilde{x}-[2\tau(x+\gamma_1(s))-x-\gamma_1(s)](\tilde{x}-\tilde{Y}-2R_1)}{4R_1}, \\ Y=\dfrac{1}{2}(\mu-\nu)=\dfrac{4R_1\tilde{Y}+[2\tau(x+\gamma_1(s))-x-\gamma_1(s)](\tilde{x}-\tilde{Y}-2R_1)}{4R_1}. \end{cases} \tag{4.38}$$

Denote by $\tilde{Z}=\tilde{x}+j\tilde{Y}=f(Z)$, $Z=x+jY=f^{-1}(\tilde{Z})$ the transformation (4.37) and the inverse transformation (4.38) respectively. In this case, the system of equations is

$$\begin{aligned} \xi_\nu &= \hat{A}_1\xi+\hat{B}_1\eta+\hat{C}u+\hat{D}, \\ & \qquad\qquad\qquad\qquad z\in D', \\ \eta_\mu &= \hat{A}_2\xi+\hat{B}_2\eta+\hat{C}u+\hat{D}, \end{aligned} \tag{4.39}$$

which is another form of (4.10) in D'. Suppose that (4.1) in D' satisfies Condition C, through the transformation (4.35), we obtain $\xi_{\tilde{\mu}}=\xi_\mu$, $\eta_{\tilde{\nu}}=[2R_1-2\tau(\mu)+\mu]\xi_\nu/2R_1$ in D', where $\xi=U+V,\eta=U-V$, and then

$$\begin{aligned} \xi_{\tilde{\mu}} &= \hat{A}_1\xi+\hat{B}_1\eta+\hat{C}u+\hat{D}, \\ & \qquad\qquad\qquad\qquad\qquad \text{in } D, \\ \eta_{\tilde{\nu}} &= \frac{2R_1-2\tau(\mu)+\mu}{2R_1}[\hat{A}_2\xi+\hat{B}_2\eta+\hat{C}u+\hat{D}], \end{aligned} \tag{4.40}$$

and through the transformation (4.37), the boundary condition (4.33) is reduced to

$$\begin{aligned} &\mathrm{Re}[\overline{\lambda(f^{-1}(\tilde{Z}))}w(f^{-1}(\tilde{Z}))]=R(f^{-1}(\tilde{Z})),\ \tilde{Z}=\tilde{x}+j\tilde{Y}\in L, \\ &\mathrm{Im}[\overline{\lambda(f^{-1}(\tilde{Z}_3))}w(f^{-1}(\tilde{Z}_3))]=b_1', \end{aligned} \tag{4.41}$$

where $Z=f^{-1}(\tilde{Z}),\tilde{Z}_3=f(Z_3'),Z_3'=l+jG[\gamma_1(s_0)]$. Therefore the boundary value problem (4.1),(4.33) is transformed into the boundary value problem (4.40),(4.41), i.e. the corresponding Problem A in D. According to Theorem 4.3, we can prove that the boundary value problem (4.40),(4.41) has a unique solution $w(f^{-1}(\tilde{Z}))$, and the corresponding function

$$u(z)=2R\int_{z_3}^{z}[\mathrm{Re}w/H-j\mathrm{Im}w]dz+b_0 \text{ in } D$$

is just a solution of Problem P' for (4.1) in D' with the boundary conditions (4.33).

Theorem 4.6 *If equation* (4.1) *in D' satisfies Condition C in the domain D' with the boundary $L_0\cup L_3'\cup L_4'$, where L_3', L_4' are as stated in* (4.32), *then Problem P' for* (4.1) *with the boundary conditions* (4.33) *has a unique solution* $u(z)$.

2. Next let the domain D'' be a simply connected domain with the boundary $L_0\cup L_3''\cup L_4''$, where L_0 is as stated before and

$$L_3''=\{y=\gamma_2(s),0\le s\le s_0'\},\ L_4''=\{y=\gamma_1(s),0\le x\le s_0\}, \tag{4.42}$$

in which s is the parameter of are length of L_3'' or L_4'', $\gamma_1(0)=0,\gamma_2(0)=0$, $\gamma_1(s)>0, 0<s\le s_0$, $\gamma_2(s)>0, 0<x\le s_0'$, $\gamma_1(s)$ on $0\le x\le s_0$ and $\gamma_2(s)$ on $0\le s\le s_0'$ are continuously differentiable, $z_3''=l+j\gamma_1(s_0)=l+j\gamma_2(s_0')$. Denote by two points z_1^*, z_2^* the intersection points of L_4'', L_3'' and the characteristic curves $s_2: dy/dx=-1/H(y)$, $s_1: dy/dx=1/H(y)$ respectively, we require that the slopes of curves L_4'', L_3'' at z_1^*, z_2^* arc not equal to those at the characteristic curves s_2, s_1 at the corresponding points, hence $\gamma_1(s),\gamma_2(s)$ can be expressed by $\gamma_1[s(\mu)]$ $(0\le\mu\le 2R_1)$, $\gamma_2[s(\nu)]$ $(0\le\nu\le 2R_1)$. We consider the oblique derivative problem (Problem P'') for equation (4.1) in D'' with the boundary conditions

$$\begin{aligned}&\mathrm{Re}[\overline{\lambda(z)}u_{\bar z}]=R(z),\ z\in L''=L_3''\cup L_4'',\\ &u(z_3'')=b_0,\ \ \mathrm{Im}[\overline{\lambda(z)}u_{\bar z}]|_{z=z_3''}=b_1,\end{aligned} \tag{4.43}$$

where $\lambda(z)$, $r(z)$ satisfy the corresponding conditions

$$\begin{aligned}&C^1[\lambda(z),L'']\le k_0, C^1[r(z),L'']\le k_2, |b_0|,|b_1|\le k_2,\\ &\max_{z\in L_3''}\frac{1}{|a(x)-b(x)|},\ \max_{z\in L_4''}\frac{1}{|a(x)+b(x)|}\le k_0,\end{aligned} \tag{4.44}$$

in which k_0, k_2 are positive constants. By the conditions in (4.42), the inverse function $x=(\mu+\nu)/2=\tau(\mu), x=(\mu+\nu)/2=\sigma(\nu)$ of $\mu=x+G(y), \nu=x-G(y)$ can be found, namely

$$\begin{aligned}&\nu=2\tau(\mu)-\mu,\ \mu=2\sigma(\nu)-\nu,\\ &0\le\mu=x+\gamma_1(s)\le l+\gamma_1(s_0),\ 0\le\nu=x-\gamma_2(s)\le l-\gamma_2(s_0').\end{aligned} \tag{4.45}$$

We make a transformation

$$\tilde\mu=\frac{2R_1\mu}{2\sigma(\nu)-\nu},\ \tilde\nu=\nu,\ 0\le\mu\le 2\sigma(\nu)-\nu,\ 0\le\nu\le 2R_1 \tag{4.46}$$

where μ,ν are real variables, its inverse transformation is

$$\mu=\frac{[2\sigma(\nu)-\nu]\tilde\mu}{2R_1},\ \nu=\tilde\nu,\ 0\le\tilde\mu\le 2R_1, 0\le\tilde\nu\le 2R_1. \tag{4.47}$$

Hence we have

$$\begin{aligned}\tilde{x}=\frac{1}{2}(\tilde{\mu}+\tilde{\nu})&=\frac{2R_1(x+Y)+(x-Y)[2\sigma(x-\gamma_2(s))-x+\gamma_2(s)]}{2[2\sigma(x-\gamma_2(s))-x+\gamma_2(s)]},\\ \tilde{Y}=\frac{1}{2}(\tilde{\mu}-\tilde{\nu})&=\frac{2R_1(x+Y)-(x-Y)[2\sigma(x-\gamma_2(x))-x+\gamma_2(s)]}{2[2\sigma(x-\gamma_2(s))-x+\gamma_2(s)]},\\ x=\frac{1}{2}(\mu+\nu)&=\frac{1}{4R_1}[(2\sigma(x-\gamma_2(s))-x+\gamma_2(s))(\tilde{x}+\tilde{Y})+2R_1(\tilde{x}-\tilde{Y})],\\ Y=\frac{1}{2}(\mu-\nu)&=\frac{1}{4R_1}[(2\sigma(x-\gamma_2(s))+x-\gamma_2(s))(\tilde{x}+\tilde{Y})-2R_1(\tilde{x}-\tilde{Y})].\end{aligned} \tag{4.48}$$

Denote by $\tilde{Z}=\tilde{x}+j\tilde{Y}=g(Z)$, $Z=x+jY=g^{-1}(\tilde{Z})$ the transformation and its inverse transformation in (4.48) respectively. Through the transformation (4.46), we obtain

$$\left.\begin{aligned}(U+V)_{\tilde{\mu}}&=\frac{2\sigma(\nu)-\nu}{2R_1}(U+V)_{\mu},\\ (U-V)_{\tilde{\nu}}&=(U-V)_{\nu},\end{aligned}\right\}\ \text{in } D'', \tag{4.49}$$

and system (4.39) in D'' is reduced to

$$\left.\begin{aligned}\xi_{\tilde{\mu}}&=\frac{2\sigma(\nu)-\nu}{2R_1}[\hat{A}_1\xi+\hat{B}_1\eta+\hat{C}u+\hat{D}],\\ \eta_{\tilde{\nu}}&=\hat{A}_2\xi+\hat{B}_2\eta+\hat{C}u+\hat{D},\end{aligned}\right\}\ \text{in } D'. \tag{4.50}$$

Moreover, through the transformation (4.48), the boundary condition (4.43) on L'' is reduced to

$$\begin{aligned}\operatorname{Re}[\overline{\lambda(g^{-1}(\tilde{Z}))}w(g^{-1}(\tilde{Z}))]&=R[g^{-1}(\tilde{Z})]\ \text{on } L',\\ \operatorname{Im}[\overline{\lambda(g^{-1}(\tilde{Z}_1'))}w(g^{-1}(\tilde{Z}_3')]&=b_1',\end{aligned} \tag{4.51}$$

in which $Z=g^{-1}(\tilde{Z})$, $\tilde{Z}_3'=g(Z_3'')$, $Z_3''=l+jG[\gamma_2(s_0')]$. Therefore the boundary value problem (4.1),(4.43) is transformed into the boundary value problem (4.50),(4.51). According to the method in the proof of Theorem 4.6, we can see that the boundary value problem (4.50),(4.51) has a unique solution $w(g^{-1}(\tilde{Z}))$, and then the corresponding function $u=u(z)$ is a solution of the boundary value problem (Problem P''), i.e. equation (4.1) in D'' and the boundary condition (4.43) on L''. But we mention that through the transformation (4.46) or (4.48), the boundaries L_3'', L_4'' are reduced to L_3', L_4' respectively, such that L_3', L_4' satisfy the condition as stated in (4.32).

Theorem 4.7 *If equation* (4.1) *satisfies Condition* C *in the domain* D'' *bounded by the boundary* $L_0 \cup L_3'' \cup L_4''$, *where* L_3'', L_4'' *are as stated in* (4.42), *then Problem* P'' *for* (4.1) *in* D'' *with the boundary condition* (4.43) *on* L'' *has a unique solution* $u(z)$.

5 The Oblique Derivative Problem for Second Order Hyperbolic Equations with Degenerate Rank 0

This section deals with the oblique derivative problem for the hyperbolic equations with degenerate rank 0. We first give the formulation and representation of solutions of the oblique derivative problem for hyperbolic equations, and then prove the existence of its solutions.

5.1 Formulation of oblique derivative problem for second order hyperbolic equations

Let D be a domain bounded by the segment $L_0 = [0, 2]$ on $x-$axis and two characteristic lines

$$L_1 : x = -G(y) = -\int_0^y H(t)dt,\ L_2 : x = G(y)+2 = \int_0^y H(t)dt+2,$$

in which $H(y) = H_1(y)/H_2(y)$, $H_l(y) = \sqrt{|K_l(y)|}$, $l = 1, 2$, $K_l(y) = -|y|^{m_l} h_l(y)$ $(l = 1, 2)$, m_l $(l = 1, 2,\ m_2 < 1)$ are positive numbers, $h_l(y)$ $(l = 1, 2)$ are continuously differentiable positive functions in $\overline{D}$, and $z_1 = 1+jy_1$ is the intersection of L_1 and L_2. We consider the linear hyperbolic equation of second order with the parabolic degeneracy

$$K_1(y)u_{xx} - K_2(y)u_{yy} + a(x, y)u_x + b(x, y)u_y + c(x, y)u = -d(x, y) \tag{5.1}$$

satisfying **Condition** C in D, namely

$$\begin{gathered} C[ya/H_1H_2, \overline{D}] \le \varepsilon_1(y),\ m_1 + m_2 \ge 2, \\ \hat{C}[\eta, \overline{D}] = C[\eta, \overline{D}] + C[\eta_x, \overline{D}] \le k_0, \eta = a, b, c,\ \hat{C}[d, \overline{D}] \le k_1, \end{gathered} \tag{5.2}$$

where k_0, k_1 are positive constants, and $\varepsilon_1(y) \to 0$ as $y \to 0$. If the above conditions are replaced by

$$\begin{gathered} C_\alpha[ya/H_1H_2, \overline{D}] \le \varepsilon_1(y),\ m_1 + m_2 \ge 2, \\ \hat{C}_\alpha[\eta, \overline{D}] = C_\alpha[\eta, \overline{D}] + C_\alpha[\eta_x, \overline{D}] \le k_0, \eta = a, b, c, \hat{C}_\alpha[d, \overline{D}] \le k_1, \end{gathered} \tag{5.3}$$

in which $\alpha\,(0<\alpha<1)$, k_0, k_1 are positive constants, then the conditions will be called **Condition** C'. In this section, we can only consider $H_l(y)=|y|^{m_l/2}$, $l=1,2, H(y)=H_1(y)/H_2(y)=y^m, m=m_1-m_2$, here $m_1(>0), m_2(0<m_2<1)$ are two real constants, then

$$G_l(y)=\int_0^y H_l(t)dy=-\frac{2}{m_l+2}|y|^{(m_l+2)/2},\ l=1,2,$$

$$Y=G(y)=\int_0^y H(t)dy=-\frac{2}{m+2}|y|^{(m+2)/2}, m=m_1-m_2>-1.$$

The oblique derivative boundary value problem for equation (5.1) may be formulated as follows:

Problem P The another oblique derivative boundary value problem for equation (5.1) is to find a continuously differentiable solution $u(z)$ of (5.1) in $D^*=\overline{D}\backslash\{0,2\}$, which is continuous in $\bar{D}\backslash L_0$ and satisfies the boundary conditions

$$\begin{aligned}&\frac{1}{2}\frac{\partial u}{\partial\nu}=\frac{1}{H_1(y)}\mathrm{Re}[\overline{\lambda(z)}u_{\tilde{z}}]=\mathrm{Re}[\overline{\Lambda(z)}u_z]=r(z)\ \text{ on } L_1\cup L_2,\\ &u(z_1)=b_0,\ \frac{1}{H(y)}\mathrm{Im}[\overline{\lambda(z)}u_{\tilde{z}}]|_{z=z_1}=\mathrm{Im}[\overline{\Lambda(z)}u_z]|_{z=z_1}=b_1,\end{aligned}\tag{5.4}$$

where ν is a given vector at every point $z\in L_1\cup L_2$, $\lambda(z), r(z), b_0, b_1$ satisfy the conditions

$$\begin{aligned}&C^1_\alpha[\lambda(z),L_l]\le k_0,\ C^1_\alpha[r(z),L_l]\le k_2, l=1,2,\ |b_j|\le k_2, j=0,1,\\ &\max_{z\in L_1}\frac{1}{|\mathrm{Re}\lambda(z)-\mathrm{Im}\lambda(z)|}\le k_0,\ \max_{z\in L_2}\frac{1}{|\mathrm{Re}\lambda(z)+\mathrm{Im}\lambda(z)|}\le k_0,\end{aligned}\tag{5.5}$$

in which n is the outward normal vector at every point on $L_1\cup L_2$, k_0, k_2 are positive constants.

5.2 Representations of solutions of oblique derivative problems for second order hyperbolic equations

Denote

$$\begin{aligned}W(z)&=U+jV=[H_1(y)u_x-jH_2(y)u_y]/2\\ &=u_{\tilde{z}}=H_1(y)[u_x-ju_Y]/2=H_1(y)u_Z,\end{aligned}$$

$$H_1(y)W_{\overline{Z}} = H_1(y)[W_x + jW_Y]/2$$
$$=[H_1(y)W_x+jH_2(y)W_y]/2=H_1(y)W_{\bar{\tilde{z}}} \text{ in } \overline{D},$$

we get

$$\begin{aligned}
W_{\bar{\tilde{z}}} &= \frac{1}{4}\{H_1^2u_{xx} - H_2^2u_{yy} - jH_1H_2(u_{yx} - u_{xy}) + H_2(jH_{1y}u_x \\
&-H_{2y}u_y)\} = \frac{1}{4}[(\frac{a}{H_1} + \frac{jH_2H_{1y}}{H_1})H_1u_x + (\frac{b}{H_2} - H_{2y})H_2u_y + cu + d] \\
&= \frac{1}{4}[(\frac{a}{H_1} + \frac{jH_2H_{1y}}{H_1})(W+\overline{W}) + j(\frac{b}{H_2} - H_{2y})(\overline{W} - W) + cu + d] \\
&= \frac{1}{4}[\frac{a}{H_1} + \frac{jH_2H_{1y}}{H_1} - \frac{jb}{H_2} + jH_{2y}]W \\
&+ \frac{1}{4}[\frac{a}{H_1} + \frac{jH_2H_{1y}}{H_1} + \frac{jb}{H_2} - jH_{2y}]\overline{W} + \frac{1}{4}(cu + d) \\
&= \frac{e_1}{4}\{[\frac{a}{H_1} + \frac{H_2H_{1y}}{H_1} - \frac{b}{H_2} + H_{2y}](\text{Re}W + \text{Im}W) \\
&+ [\frac{a}{H_1} + \frac{H_2H_{1y}}{H_1} + \frac{b}{H_2} - H_{2y}](\text{Re}W - \text{Im}W) + cu + d\} \\
&+ \frac{e_2}{4}\{[\frac{a}{H_1} - \frac{H_2H_{1y}}{H_1} - \frac{b}{H_2} + H_{2y}](\text{Re}W + \text{Im}W) \\
&+ [\frac{a}{H_1} - \frac{H_2H_{1y}}{H_1} + \frac{b}{H_2} - H_{2y}](\text{Re}W - \text{Im}W) + cu + d\} \text{ in } \overline{D},
\end{aligned} \tag{5.6}$$

where $e_1 = (1 + j)/2, e_2 = (1 - j)/2$. Noting that

$$\mu = x + \int_0^y H(t)dt = x + G(y),\ \nu = x - \int_0^y H(t)dt = x - G(y),$$
$$\mu + \nu = 2x,\ \mu - \nu = 2G(y),\ \frac{\partial G(y)}{\partial y} = H(y),$$
$$\frac{\partial x}{\partial \mu} = \frac{\partial x}{\partial \nu} = \frac{1}{2},\ \frac{\partial y}{\partial \mu} = -\frac{\partial y}{\partial \nu} = \frac{1}{2H(y)} \text{ in } \overline{D},$$

we can obtain

$$\begin{aligned}
W_{\bar{\bar{z}}} &= \frac{1}{2}[H_1(U+jV)_x + jH_2(U+jV)_y]\\
&=\frac{e_1}{2}[H_1U_x+H_2V_y+H_1V_x+H_2U_y]+\frac{e_2}{2}[H_1U_x+H_2V_y-H_1V_x-H_2U_y]\\
&=\frac{e_1}{2}[H_1(U+V)_x+H_2(U+V)_y]+\frac{e_2}{2}[H_1(U-V)_x-H_2(U-V)_y]\\
&= H_1[e_1(U+V)_\mu + e_2(U-V)_\nu]\\
&=\frac{e_1}{4}\{[\frac{a}{H_1}+\frac{H_2H_{1y}}{H_1}-\frac{b}{H_2}+H_{2y}](U+V)\\
&+[\frac{a}{H_1}+\frac{H_2H_{1y}}{H_1}+\frac{b}{H_2}-H_{2y}](U-V)+cu+d\}\\
&+\frac{e_2}{4}\{[\frac{a}{H_1}-\frac{H_2H_{1y}}{H_1}-\frac{b}{H_2}+H_{2y}](U+V)\\
&+[\frac{a}{H_1}-\frac{H_2H_{1y}}{H_1}+\frac{b}{H_2}-H_{2y}](U-V)+cu+d\} \text{ in } \overline{D}.
\end{aligned} \tag{5.7}$$

It is clear that Problem P for (5.1) is equivalent to the Riemann-Hilbert problem (Problem A) for the complex equation of first order

$$\begin{aligned}
&W_{\bar{z}} = A_1(z)W + A_2(z)\overline{W} + A_3(z)u + A_4(z) \text{ in } \overline{D},\\
&A_1 = \frac{1}{4}[\frac{a}{H_1}+\frac{jH_2H_{1y}}{H_1}-\frac{jb}{H_2}+jH_{2y}],\ A_3=\frac{c}{4},\\
&A_2 = \frac{1}{4}[\frac{a}{H_1}+\frac{jH_2H_{1y}}{H_1}+\frac{jb}{H_2}-jH_{2y}],\ A_4=\frac{d}{4},
\end{aligned} \tag{5.8}$$

with the boundary conditions

$$\begin{aligned}
&\operatorname{Re}[\overline{\lambda(z)}W(z)]=H_1(y)r(z) = R(z),\ z\in L=L_1\cup L_2,\\
&\operatorname{Im}[\overline{\lambda(z)}u_{\bar{z}}]|_{z=z_1} = H_1(y_1)b_1 = b_1',
\end{aligned} \tag{5.9}$$

and the relation

$$u(z)=2\operatorname{Re}\int_{z_1}^{z}[\frac{\operatorname{Re}W}{H_1(y)}-j\frac{\operatorname{Im}W}{H_2(y)}]dz+b_0 \text{ in } \overline{D}. \tag{5.10}$$

Moreover we see the complex equation of first order

$$W_{\bar{\bar{z}}} = 0 \text{ in } \overline{D}, \tag{5.11}$$

or the system of first order equations

$$(U+V)_\mu=0,\ \ (U-V)_\nu=0 \text{ in } \overline{D_\tau}. \tag{5.12}$$

is a special case of equation (5.7). It is not difficult to see that the general solution of Problem A for (5.12) in $\overline{D}$ can be expressed as

$$\begin{aligned}&\xi=U(z)+V(z)=f(\nu),\ \eta=U(z)-V(z)=g(\mu),\\&U(z)=[f(\nu)+g(\mu)]/2,\ V(z)=[f(\nu)-g(\mu)]/2, \text{ i.e.}\\&W(z)=U(z)+jV(z)=[(1+j)f(\nu)+(1-j)g(\mu)]/2,\end{aligned} \tag{5.13}$$

in which $f(t),g(t)$ are two arbitrary real continuous functions on $L_0=[0,2]$. For convenience, denote by the functions $a(x),b(x),R(x)$ of x the functions $a(z),b(z),R(z)$ of z in (5.9), thus (5.9) can be rewritten as

$$\begin{aligned}&a(x)U(z)-b(x)V(z)=R(z) \text{ on } L, \text{ i.e.}\\&[a(x)-b(x)]f(x-G(y))+[a(x)+b(x)]g(x+G(y))=2R(x) \text{ on } L, \text{ i.e.}\\&[a(x)-b(x)]f(2x)+[a(x)+b(x)]g(0)=2R(x),\ x\in[0,1],\\&[a(x)-b(x)]f(2)+[a(x)+b(x)]g(2x-2)=2R(2x-2),\ x\in[1,2], \text{ and}\\&[a(t/2)-b(t/2)]f(t)+[a(t/2)+b(t/2)]g(0)=2R(t/2),\ t\in[0,2],\\&[a(t/2+1)-b(t/2+1)]f(2)+[a(t/2+1)+b(t/2+1)]g(t)\\&=2R(t/2+1),\ t\in[0,2],\end{aligned}$$

where

$$\begin{aligned}&(a(1)-b(1))f(2)=(a(1)-b(1))(U(z_1)+V(z_1))=r(1)+b_1' \text{ or } 0,\\&(a(1)+b(1))g(0)=(a(1)+b(1))(U(z_1)-V(z_1))=r(1)-b_1' \text{ or } 0.\end{aligned}$$

Moreover we can derive

$$\begin{aligned}&U(z)=\frac{1}{2}\{\frac{2R(\nu/2)-[a(\nu/2)+b(\nu/2)]g(0)}{a(\nu/2)-b(\nu/2)}+g(\mu)\},\\&V(z)=\frac{1}{2}\{\frac{2R(\nu/2)-[a(\nu/2)+b(\nu/2)]g(0)}{a(\nu/2)-b(\nu/2)}-g(\mu)\},\\&U(z)=\frac{1}{2}\{f(\nu)+\frac{2R(\mu/2+1)-[a(\mu/2+1)-b(\mu/2+1)]f(2)}{a(\mu/2+1)+b(\mu/2+1)}\},\\&V(z)=\frac{1}{2}\{f(\nu)-\frac{2R(\mu/2+1)-[a(\mu/2+1)-b(\mu/2+1)]f(2)}{a(\mu/2+1)+b(\mu/2+1)}\},\end{aligned}$$

if $a(x)-b(x)\neq 0$ on $[0,1]$ and $a(x)+b(x)\neq 0$ on $[1,2]$ respectively. From the above formulas, the solution

$$W(z)=\begin{cases}\dfrac{1}{2}\{(1+j)\dfrac{2R((x-G(y))/2)-K(z)}{a((x-G(y))/2)-b((x-G(y))/2)}\\[2ex] +(1-j)\dfrac{2R((x+G(y))/2+1)-N(z)}{a((x+G(y))/2+1)+b((x+G(y))/2+1)}\},\\[2ex] K(z)=[a((x-G(y))/2)+b((x-G(y))/2)]g(0),\\[1ex] N(z)=[a((x+G(y))/2+1)-b((x+G(y))/2+1)]f(2)\end{cases}\qquad(5.14)$$

of Problem A for (5.11) is obtained. Similarly to Theorem 4.1, we have the following theorem.

Theorem 5.1 *Problem A of equation* (5.11) *or* (5.12) *has a unique solution, which can be expressed in the form* (5.14), *and the solution $W(z)=W_0(z)$ satisfying the estimate*

$$|\mathrm{Re}W_0(z)|\le M_1,\ |\mathrm{Im}W_0(z)|\le M_1\ \text{ in } D,\ C_\delta[W_0(z),\overline{D}]\le M_2,\qquad(5.15)$$

where δ is a sufficiently small positive constant, and $M_l=M_l(\delta,k_0,k_2,D)$ $(l=1,2)$ are positive constants.

Now we give two representations of solutions for Problem P for equation (5.1).

Theorem 5.2 *Under Condition C, any solution $u(z)$ of Problem P for equation* (5.1) *can be can be expressed as in* (5.10), *where $W(z)=u_{\bar z}$ possesses the form*

$$\begin{aligned}&W(z)=W_0(z)+\Phi(z)+\Psi(z)\ \text{ in } D,\\ &W_0(z)=f(\nu)e_1+g(\mu)e_2,\ \Phi(z)=\tilde f(\nu)e_1+\tilde g(\mu)e_2,\\ &\Psi(z)=\int_0^\mu \hat g_1(z)e_1d\mu+\int_2^\nu \hat g_2(z)e_2d\nu,\\ &\hat g_l(z)=\hat A_l\xi+\hat B_l\eta+\hat C_lu+\hat D_l,\ l=1,2,\end{aligned}\qquad(5.16)$$

in which $e_1=(1+j)/2$, $e_2=(1-j)/2$, $\mu=x+G(y)$, $\nu=x-G(y)$, and

$$\hat A_1=\frac{1}{4H_1}[\frac{a}{H_1}+\frac{H_2H_{1y}}{H_1}-\frac{b}{H_2}+H_{2y}],\ \hat C_1=\frac{c}{4H_1},$$
$$\hat B_1=\frac{1}{4H_1}[\frac{a}{H_1}+\frac{H_2H_{1y}}{H_1}+\frac{b}{H_2}-H_{2y}],\ \hat D_1=\frac{d}{4H_1},$$

$$\hat{A}_2 = \frac{1}{4H_1}[\frac{a}{H_1} - \frac{H_2H_{1y}}{H_1} - \frac{b}{H_2} + H_{2y}],\ \hat{C}_2 = \frac{c}{4H_1},$$
$$\hat{B}_2 = \frac{1}{4H_1}[\frac{a}{H_1} - \frac{H_2H_{1y}}{H_1} + \frac{b}{H_2} - H_{2y}],\ \hat{D}_2 = \frac{d}{4H_1},$$

where $W_0(z)$ *is a solution of Problem A for equation* (5.11) *in* D, $f(\nu)$, $g(\mu)$ *are as stated in* (5.13), *and* $\tilde{f}(\nu)$, $\tilde{g}(\mu)$ *are similar to* $f(\nu)$, $g(\mu)$ *in* (5.13), *which satisfy the boundary conditions*

$$\begin{aligned}&\mathrm{Re}[\overline{\lambda(z)}(\Phi(z)+\Psi(z))]=0, z\in L_1\cup L_2,\\&\mathrm{Im}[\overline{\lambda(z_1)}(\Phi(z_1)+\Psi(z_1))]=0.\end{aligned} \tag{5.17}$$

Proof Let $u(z)$ be a solution of Problem P for equation (5.1), and denote $\xi = \mathrm{Re}W(z) + \mathrm{Im}W(z), \eta = \mathrm{Re}W(z) - \mathrm{Im}W(z)$. Substitute u, ξ, η in the positions of u, ξ, η in (5.16), thus the function $\Psi(z)$ in $\overline{D}$ can be determined. Moreover from Theorem 5.1, we can find the solution $\Phi(z)$ in $\overline{D}$ of (5.11) with the boundary condition (5.17), thus the function $W(z)$ in the first formula in (5.16), i.e.

$$W(z) = W_0(z) + \Phi(z) + \Psi(z) \text{ in } D \tag{5.18}$$

is the solution of Problem A in D for equation (5.8), and then the function $u(z)$ in (5.10) is just a solution of Problem P for (5.1).

Theorem 5.3 *Under Condition C, any solution* $u(z)$ *of Problem P for equation* (5.1) *in D can be expressed as follows*

$$\begin{aligned}&u(z)=u(x)-2\int_0^y \frac{V(z)}{H_2(y)}dy=\int_0^z \left[\frac{\mathrm{Re}W}{H_1(y)}-j\frac{\mathrm{Im}W}{H_2(y)}\right]dz+b_0 \text{ in } D,\\&W(z) = W_0(z) + \phi(z) + \psi(z),\ \phi(z) + \psi(z) = \xi(z)e_1 + \eta(z)e_2,\\&\xi(z)=\zeta(z)+\int_0^y g_1(z)dy, z\in s_1, \eta(z)=\theta(z)+\int_0^y g_2(z)dy, z\in s_2,\\&g_l(z)=\tilde{A}_l(U+V)+\tilde{B}_l(U-V)+2\tilde{C}_lU+2\tilde{D}_lV+\tilde{E}_lu+\tilde{F}_l, l=1,2,\end{aligned} \tag{5.19}$$

where $u(0) = b_0$, $W_0(z)$ *is as stated in Theorem* 5.2, $U = H_1u_x/2, V = -H_2u_y/2$, $\phi(z) = \int_{S_1} g_1(z)e_1 + \int_{S_2} g_2(z)e_2$ *is a solution of* (5.11) *in* $\overline{D}$, *and* s_1, s_2 *are two characteristics of families in* D:

$$s_1: \frac{dx}{dy}=\sqrt{K(y)}=H(y),\ s_2: \frac{dx}{dy}=-\sqrt{K(y)}=-H(y) \tag{5.20}$$

passing through $z = x + jy \in D$ from two points on L_0 respectively, and

$$\begin{aligned}
&W(z) = U(z) + jV(z) = \frac{1}{2}H_1 u_x - \frac{j}{2}H_2 u_y,\\
&\xi(z) = \mathrm{Re}\psi(z) + \mathrm{Im}\psi(z), \eta(z) = \mathrm{Re}\psi(z) - \mathrm{Im}\psi(z),\\
&\tilde{A}_1 = \frac{1}{4}[\frac{h_{1y}}{h_1} + \frac{h_{2y}}{h_2}],\ \tilde{A}_2 = \frac{1}{4}[\frac{h_{1y}}{h_1} - \frac{h_{2y}}{h_2}],\\
&\tilde{B}_1 = \frac{1}{4}[\frac{h_{1y}}{h_1} - \frac{h_{2y}}{h_2}],\ \tilde{B}_2 = \frac{1}{4}[\frac{h_{1y}}{h_1} + \frac{h_{2y}}{h_2}],\\
&\tilde{C}_1 = \frac{a}{4H_1H_2} + \frac{m_1}{8y},\ \tilde{C}_2 = -\frac{a}{4H_1H_2} + \frac{m_1}{8y},\\
&\tilde{D}_1 = -\frac{b}{4H_2^2} + \frac{m_2}{8y},\ \tilde{D}_2 = \frac{b}{4H_2^2} - \frac{m_2}{8y},\\
&\tilde{E}_1 = -\tilde{E}_2 = \frac{c}{2H_2},\ \tilde{F}_1 = -\tilde{F}_2 = \frac{d}{2H_2},
\end{aligned} \tag{5.21}$$

in which we choose $H_l(y) = [|y|^{m_l} h_l(y)]^{1/2}$, $l = 1, 2$, m_l, $h_l(y)\,(l = 1, 2)$ are as stated before, and

$$d\mu = d[x + G(y)] = 2H(y)dy \ \text{ on } s_1,$$

$$d\nu = d[x - G(y)] = -2H(y)dy \ \text{ on } s_2.$$

Proof It is not difficult to derive the expression (5.19) by (5.7). In the following we shall derive (5.19) by the expression (5.16). In fact any solution of Problem P for equation (5.1) in $\overline{D}$ can be expressed as

$$\begin{aligned}
&u(z) = u(x) - 2\int_0^y \frac{V(z)}{[|y|^{m_2}h_2]^{1/2}}dy = 2\mathrm{Re}\int_0^z \left[\frac{\mathrm{Re}W}{H_1(y)} - j\frac{\mathrm{Im}W}{H_2(y)}\right]dz + b_0,\\
&W(z) = W_0(z) + \phi(z) + \psi(z),\ \phi(z) + \psi(z) = \xi(z)e_1 + \eta(z)e_2,\\
&\xi(z) = \int_0^\mu \xi_\mu d\mu = \int_0^\mu \frac{1}{2H(y)}[H(y)\xi_x + \xi_y]d\mu\\
&= \int_{S_1} g_1(z)dy + \int_0^y g_1(z)dy = \zeta(z) + \int_0^y g_1(z)dy,\ z \in s_1,\\
&\eta(z) = \int_2^\nu \eta_\nu d\nu = \int_2^\nu \frac{1}{2H(y)}[H(y)\xi_x - \xi_y]d\nu
\end{aligned}$$

$$= \int_{S_2} g_2(z)dy + \int_0^y g_2(z)dy = \theta(z) + \int_0^y g_2(z)dy, \ z \in s_2, \tag{5.22}$$

where $\phi(z) = \zeta(z)e_1 + \theta(z)e_2 = \int_{S_1} g_1(z)e_1 + \int_{S_2} g_2(z)e_2$, s_1, s_2 are as stated in (5.20), and S_1, S_2 are two characteristics as in (5.20) form the corresponding points of L_1, L_2 to a point on L_0 respectively. Noting (5.20) and

$$ds_1 = \sqrt{(dx)^2+(dy)^2} = -\sqrt{1+(dx/dy)^2}dy = -\sqrt{1+K}dy = -\frac{\sqrt{1+K}}{\sqrt{K}}dx,$$

$$ds_2 = \sqrt{(dx)^2+(dy)^2} = -\sqrt{1+(dx/dy)^2}dy = -\sqrt{1+K}dy = \frac{\sqrt{1+K}}{\sqrt{K}}dx,$$

the system (5.22) is just (5.19), where the coefficients are as stated in (5.21).

From Theorem 5.3, we see that the function $w(z) = \phi(z) + \psi(z)$ is a solution of the homogeneous problem (Problem A_0) for the complex equation

$$w_{\bar{\bar{z}}} = A_1(z)w + A_2(z)\overline{w} + A_3(z)u + A(z), \ A(z) = A_1 W_0 + A_2 \overline{W_0} + A_3 u_0 \tag{5.23}$$

with the homogeneous boundary conditions

$$\mathrm{Re}[\overline{\lambda(z)}w(z)] = 0 \ \text{ on } L_1 \cup L_2, \ u(z_1) = 0, \ \mathrm{Im}[\overline{\lambda(z_1)}w(z_1)] = 0. \tag{5.24}$$

Due the solution $W_0(z)$ of Problem A for equation (5.11) satisfies the estimate (5.15), it is clear that the coefficient $A(z)$ of equation (5.23) satisfy the condition

$$|y|^\tau |A(z)| \le M_3 \ \text{ in } D, \tag{5.25}$$

where $\tau = \max(0, 1 - m_1/2 - m_2/2, 1 - m_2)$ and $M_3 = M_3(\alpha, k_0, k_2, D)$ are positive constants.

Besides, Theorem 5.3 can be used to prove the existence of Darboux type problem for equation (5.1) with the boundary conditions

$$u(x) = \phi(x) \ \text{ on } L_1, \ u(x) = \psi(x) \ \text{ on } L_0,$$

in which $\phi(x)$, $\psi(x)$ satisfy the conditions $\phi(0) = \psi(0)$, $C^2_\alpha[\phi(x), L_1] \le k_2$, $C^2_\alpha[\psi(x), L_0] \le k_2$, here $\alpha\,(0 < \alpha < 1$, k_2 are positive constants, which is the same as in the proof of Theorem 2.3, Chapter V below.

5.3 Unique solvability of oblique derivative problem for second order hyperbolic equations

We first prove the uniqueness of Problem P for equation (5.1).

Theorem 5.4 *Let D be given as above and equation* (5.1) *satisfy Condition C. Then the oblique derivative problem* (*Problem P*) *for* (5.1) *in D at most has a solution.*

Proof We assume that m_1, m_2 are positive numbers as stated before, and $m_1 - m_2 > -1$, denote by $u_1(z), u_2(z)$ two solutions of Problem P for (5.1), by Theorem 5.2, we see that the function $u_{\bar{z}}(z) = u_{1\bar{z}}(z) - u_{2\bar{z}}(z) = U(z) + jV(z)$ in $\overline{D}$ is a solution of the homogeneous system of integral equations

$$\begin{aligned} &u(z) = 2\text{Re}\int_{z_1}^{z}\left[\frac{\text{Re}w}{H_1(y)} - j\frac{\text{Im}w}{H_2(y)}\right]dz \text{ in } \overline{D},\\ &w(z)=\phi(z) + \psi(z) = \xi(z)e_1 + \eta(z)e_2,\\ &\xi(z) = \int_0^{\mu} g_1(z)d\mu = \int_0^{\mu}[\hat{A}_1\xi+\hat{B}_1\eta+\hat{C}_1u]d\mu, z\in s_1,\\ &\eta(z) = \int_2^{\nu} g_2(z)d\nu = \int_2^{\nu}[\hat{A}_2\xi+\hat{B}_2\eta+\hat{C}_2u]d\nu, z\in s_2, \end{aligned} \tag{5.26}$$

where

$$\xi = U(z) + V(z) = [H_1(y)u_x - H_2(y)u_y]/2,$$

$$\eta = U(z) - V(z) = [H_1(y)u_x + H_2(y)u_y]/2.$$

We first consider the closed domain $D_0 = \overline{D} \cap \{0 \le \mu \le \delta_0, 2 \le \nu \le 2-\delta_0\}$, in which δ, δ_0 are sufficiently small positive constants, such that the following requirements hold. Noting that u_x, u_y are continuous in $\overline{D}\backslash\{0,2\}$, we see that there exist positive numbers $N(>1), \gamma < 1$ dependent on $u(z), D_0$, such that

$$|u(z)| \le N\gamma,\ |\xi(z)| \le N\gamma,\ |\eta(z)| \le N\gamma. \tag{5.27}$$

From (5.26) and (5.27), we can obtain

$$|\xi(z)| = |\int_0^{\mu}[\hat{A}_1\xi+\hat{B}_1\eta+\hat{C}_1u]d\mu| \le |\int_0^{\mu} 3k_3N\gamma d\mu| \le k_3N\gamma^2,$$

where $|\hat{A}_l|, |\hat{B}_l|, |\hat{C}_l| \le k_3, l = 1, 2$, k_3 is a positive constant, and we require $3\delta_0 \le \gamma$. Similarly we have

$$|\eta(z)|=|\int_2^{\nu}[\hat{A}_2\xi+\hat{B}_2\eta+\hat{C}_2u]d\nu| \le k_3N\gamma^2,$$

$$|u(z)| = 2|\text{Re}\int_{z_1}^{z}\left[\frac{\text{Re}w}{H_1(y)}-j\frac{\text{Im}w}{H_2(y)}\right]dz| \le k_3N\gamma^2.$$

Applying the repeated insertion, the inequalities

$$|u(z)|\le k_3N\gamma^k,\ |\xi(z)|\le k_3N\gamma^k,\ |\eta(z)|\le k_3N\gamma^k,\ k=2,3,... \tag{5.28}$$

can be obtained. This shows that $u(z)=0,\xi(z)=0,\eta(z)=0$ in D_0. By the similar way, we may derive $u(z)=0,\ \xi(z)=0,\ \eta(z)=0\ \in\ D_1=\{0\le\mu\le\delta_0,2-2\delta_0\le\nu\le2-\delta_0\}\cup\{\delta_0\le\mu\le2\delta_0,2-\delta_0\le\nu\le2\}$. Moreover we can handle the above estimates along the positive direction of $\mu=x+G(y),\ \nu=x-G(y)$ succesively, and finally derive $\xi(z)=0,\ \eta(z)=0,\ u(z)=0$ in D^*. Taking into account the arbitrariness of δ, we can obtain $\xi(z)=0,\ \eta(z)=0,\ u(z)=0$ in D.

Theorem 5.5 *Problem P for equation* (5.1) *is solvable.*

Proof On the basis of the solution $W_0(z)$ of Problem A for equation (5.11) or (5.12), and the corresponding function of $u_0(z)$ determined by (5.10), we see that the solution $W(z)=w(z)+W_0(z)$ of equation (5.8) can be reduced to the solution $w(z)$ of Problem A_0 for equation (5.23). By using the way as stated in the proof of Theorem 4.3, we can also prove the result.

5.4 Oblique derivative problems for quasilinear hyperbolic equations with degenerate rank 0

In this section, we consider the second order quasilinear hyperbolic equation

$$K_1(y)u_{xx}-K_2(y)u_{yy}+au_x+bu_y+cu+d=0 \ \text{ in } \overline{D}, \tag{5.29}$$

where a,b,c,d are functions of $z(\in\overline{D}),u,u_x,u_y\ (\in\mathbf{R})$, its complex form is the following complex equation of second order

$$W_{\overline{\overline{z}}}=F(z,u,W),F=A_1W+A_2\overline{W}+A_3u+A_4 \ \text{ in } \overline{D}, \tag{5.30}$$

where $A_l=A_l(z,u,W)$, $j=1,2,3,4$, and

$$\begin{aligned}&W(z)=u_{\overline{z}}=\frac{1}{2}[H_1u_x+jH_2u_y],W_{\overline{\overline{z}}}=\frac{1}{2}[H_1W_x+jH_2W_y],\\&A_1=\frac{1}{4}[\frac{a}{H_1}+\frac{jH_{1y}H_2}{H_1}-\frac{jb}{H_2}+jH_{2y}],\ A_3=\frac{c}{4},\\&A_2=\frac{1}{4}[\frac{a}{H_1}+\frac{jH_{1y}H_2}{H_1}+\frac{jb}{H_2}-jH_{2y}],\ A_4=\frac{d}{4},\end{aligned} \tag{5.31}$$

where $H_1=\sqrt{|K_1|}$, $H_2=\sqrt{|K_2|}$.

Suppose that the equation (5.29) satisfies the following conditions, namely **Condition C:**

1) $A_l(z,u,u_z)\,(l=1,2,3)$ are continuous in $z\in\overline{D}$ for any continuously differentiable function $u(z)$ in $D^*=\overline{D}\backslash\{0,2\}$, and satisfy

$$C[y\mathrm{Re}A_1/H_2,\overline{D}]\le\varepsilon_1(y),\ m_1+m_2\ge 2,$$
$$\hat{C}[H_1\mathrm{Re}A_1,\overline{D}]\le k_0,\hat{C}[|y|^\tau\mathrm{Im}(A_1\pm A_2),\overline{D}]\le k_0, \tag{5.32}$$
$$\hat{C}[A_3,\overline{D}]\le k_0,\ \hat{C}[A_4,\overline{D}]\le k_1.$$

2) For any continuously differentiable functions $u_1(z),u_2(z)$ in D^*, the equality

$$F(z,u_1,u_{1\bar{z}})-F(z,u_2,u_{2\bar{z}})=\tilde{A}_1u_{\bar{z}}+\tilde{A}_2u_{\overline{\bar{z}}}+\tilde{A}_3u \tag{5.33}$$

in $\overline{D}$ holds, where $u=u_1-u_2$, $\tilde{A}_j=\tilde{A}_j(z,u_1,u_2)\,(j=1,2)$ satisfy the conditions

$$C[\mathrm{Re}\tilde{A}_1|y|/H_2,\overline{D}]\le\varepsilon_1(y),m_1+m_2\ge 2,\tilde{C}[H_1\mathrm{Re}\tilde{A}_1,\overline{D}]\le k_0,$$
$$\hat{C}[|y|^\tau\mathrm{Im}(\tilde{A}_1\pm\tilde{A}_2),\overline{D}]\le k_0,\ \hat{C}[\tilde{A}_3,\overline{D}]\le k_0, \tag{5.34}$$

in (5.32),(5.34), $\tau=\max(1-m_2/2,0)$, k_0,k_1 are positive constants, and $\varepsilon_1(y)\to 0$ as $y\to 0$. In particular, when (5.29) is a linear equation, the condition (5.34) obviously holds.

For equation (5.29) with Condition C, we can similarly give the representation of solutions of Problem P, and prove the unique solvability of Problem P in $\bar{D}\backslash L_0$ for equation (5.29).

6 The Cauchy Problem for Hyperbolic Equations of Second Order with Degenerate Rank 0

This section deals with the Cauchy problem for hyperbolic equations of second order with degenerate rank 0. We first transform the Cauchy problem into the equivalent problem for system of integral equations by the complex method, which is simpler and clearer than other methods, and then prove the existence and uniqueness of solutions for the equations by using the successive approximation. The result in this section includes the result in [67]1) as a special case. Here we mention that the advantage of the method in [67]1) is absorbed, and some mistakes in [67]1) are corrected, which can be seen in (6.15) and (6.20) below.

6.1 Formulation of Cauchy problem for second order hyperbolic equations

We first consider the linear hyperbolic equation of second order with the degenerate rank 0: (5.1), namely

$$\begin{gathered}K_1(y)u_{xx}-K_2(y)u_{yy}+a(x,y)u_x+b(x,y)u_y\\ +c(x,y)u+d(x,y)=0 \ \text{ in } D,\end{gathered}\tag{6.1}$$

and the coefficients of (6.1) satisfy Condition C as stated in Section 5.

The Cauchy problem for equation (6.1) may be formulated as follows:

Problem C Find a continuously differentiable solution $u(z)$ of (6.1) in $\overline{D}$ satisfying the initial conditions

$$u(x,0)=\tau(x),\ u_y(x,0)=\gamma(x),\ x\in L_0=[0,2],\tag{6.2}$$

in which D is a domain as stated before, and the functions $\tau(x),\gamma(x)$ in (6.2) possess the third order continuous derivatives on $0\le x\le 2$, i.e.

$$C^3[\tau(x),L_0]\le k_2,\ C^3[\gamma(x),L_0]\le k_2,\tag{6.3}$$

in which $\alpha\,(0<\alpha<1), k_2$ are positive constants. The above Cauchy problem is called **Problem** C. Problem C with the conditions $\tau(z)=0,\ \gamma(z)=0,\ z\in L$ will be called Problem C_0.

Denote

$$U(z)=u_x/2,\ V(z)=-u_y/2,\ \mu=x+G(y),\ \nu=x-G(y),$$

where $G(y)=\int_0^y H(t)dt$, $H(y)=H_1(y)/H_2(y)$, $H_l(y)=\sqrt{|K_l(y)|}, l=1,2$. It is not difficult to see that Problem C for equation (6.1) is equivalent to the boundary value problem (Problem P), i.e. the system of first order equations

$$\begin{cases}(U+V)_\mu &=\hat{A}_1\xi+\hat{B}_1\eta+\hat{C}_1u+\hat{D}_1,\\ (U-V)_\nu &=\hat{A}_2\xi+\hat{B}_2\eta+\hat{C}_2u+\hat{D}_2,\end{cases}\quad \text{in } \overline{D},$$

$$\hat{A}_1=\frac{1}{4H_1}[\frac{a}{H_1}+\frac{H_2H_{1y}}{H_1}-\frac{b}{H_2}+H_{2y}],$$

$$\hat{B}_1=\frac{1}{4H_1}[\frac{a}{H_1}+\frac{H_2H_{1y}}{H_1}+\frac{b}{H_2}-H_{2y}],$$

$$\hat{A}_2=\frac{1}{4H_1}[\frac{a}{H_1}-\frac{H_2H_{1y}}{H_1}-\frac{b}{H_2}+H_{2y}],$$
$$\hat{B}_2=\frac{1}{4H_1}[\frac{a}{H_1}-\frac{H_2H_{1y}}{H_1}+\frac{b}{H_2}-H_{2y}], \tag{6.4}$$
$$\hat{C}_1=\hat{C}_2=\frac{c}{4H_1},\ \hat{D}_1=\hat{D}_2=\frac{d}{4H_1},$$

and the boundary conditions

$$U(x)=u_x/2=\tau'(x)/2,\ V(x)=-u_y/2=-\gamma(x)/2,\ x\in L_0, \tag{6.5}$$

with the relation

$$u(z)=\tau(x)-2\int_0^y\frac{V(z)}{H_2(y)}dy=2\mathrm{Re}\int_0^z\left[\frac{\mathrm{Re}W}{H_1(y)}-j\frac{\mathrm{Im}W}{H_2(y)}\right]dz+b_0 \text{ in } \overline{D}, \tag{6.6}$$

in which $W(z)=U(z)+jV(z)$, $b_0=\tau(0)$. The boundary condition (6.5) can be rewritten as

$$\begin{cases}\mathrm{Re}[(1+j)W(x)]=U(x)+V(x)=[H_1(0)\tau'(x)-H_2(0)\gamma(x)]/2=0,\\ \mathrm{Re}[(1-j)W(x)]=U(x)-V(x)=[H_1(0)\tau'(x)+H_2(0)\gamma(x)]/2=0\end{cases} \tag{6.7}$$

on L_0.

6.2 Reduction of Cauchy problem for degenerate hyperbolic equations to integral equations

In this subsection, we transform Problem C for equation (6.1) into a system of integral equations. In the next subsection, we shall prove that the problem has a unique continuously differentiable solution by using the approximation method. Equation (6.1) is hyperbolic in domain D and parabolic on the partial boundary L_0. The characteristics of (6.1) are given by the two families of curves

$$s_1:\ \frac{dx}{dy}=\sqrt{K(y)}=H(y),\ s_2:\ \frac{dx}{dy}=-\sqrt{K(y)}=-H(y), \tag{6.8}$$

in which $K(y)=K_1(y)/K_2(y)$. Construct a new unknown function

$$v(x,y)=u(x,y)-y\gamma(x)-\tau(x),$$

then we have

$$\begin{cases} u_x(x,y) = v_x(x,y) + y\gamma'(x) + \tau'(x), \\ u_{xx}(x,y) = v_{xx} + y\gamma''(x) + \tau''(x), \\ u_y(x,y) = v_y(x,y) + \gamma(x),\ v_{yy}(x,y) = v_{yy}(x,y), \end{cases}$$

and equation (6.1) is reduced to the form

$$K_1(y)v_{xx} - K_2(y)v_{yy} + a(x,y)v_x + b(x,y)v_y + c(x,y)v + F(x,y) = 0, \quad (6.9)$$

here

$$F(x,y) = K_1(y)[y\gamma''(x) + \tau''(x)] + a(x,y)[y\gamma'(x) + \tau'(x)]$$
$$+ b(x,y)\gamma(x) + c(x,y)[y\gamma(x) + \tau(x)] + d(x,y)$$

is a differentiable function of x and y. The initial condition (6.2) becomes

$$v(x,0) = v_y(x,0) = 0,\ 0 \le x \le 2. \quad (6.10)$$

The Cauchy problem will be called Problem C_0 for (6.9). In the following there is no harm in assuming the case: $H_l(y) = \sqrt{|K_l(y)|} = \sqrt{|y|^{m_l} h_l(y)}$, $l = 1, 2$, and first prove the existence of solutions of Problem C_0 for (6.9). From (6.4), equation (6.9) can be written as the system of integral equations

$$\begin{aligned} v(z) &= -2\int_0^y \frac{V(z)}{|y|^{m_2/2} h_2^{1/2}} dy = \int_0^y \frac{\eta - \xi}{(|y|^{m_2} h_2)^{1/2}} dy, \\ U + V &= \int_0^y [H_1(U+V)_x + H_2(U+V)_y]/H_2 dy = \int_0^y g_1(z) dy \text{ on } s_1, \\ U - V &= -\int_0^y [H_1(U-V)_x - H_2(U-V)_y]/H_2 dy = \int_0^y g_2(z) dy \text{ on } s_2, \\ g_l(z) &= \tilde{A}_l(U+V) + \tilde{B}_l(U-V) + 2\tilde{C}_l U + 2\tilde{D}_l V + \tilde{E}_l u + \tilde{F}_l,\ l = 1, 2, \end{aligned} \quad (6.11)$$

in which s_1, s_2 are two characteristics from two points on L_0 to a point $z = x + jy \in D$ respectively, and

$$\tilde{A}_1 = \frac{1}{4}[\frac{h_{1y}}{h_1} + \frac{h_{2y}}{h_2}],\quad \tilde{A}_2 = \frac{1}{4}[\frac{h_{1y}}{h_1} - \frac{h_{2y}}{h_2}],$$
$$\tilde{B}_1 = \frac{1}{4}[\frac{h_{1y}}{h_1} - \frac{h_{2y}}{h_2}],\quad \tilde{B}_2 = \frac{1}{4}[\frac{h_{1y}}{h_1} + \frac{h_{2y}}{h_2}],$$

$$\begin{aligned}
&\tilde{C}_1 = \frac{a}{2H_1H_2} + \frac{m_1}{4y}, \quad \tilde{C}_2 = -\frac{a}{2H_1H_2} + \frac{m_1}{4y},\\
&\tilde{D}_1 = -\frac{b}{2H_2^2} + \frac{m_2}{4y}, \quad \tilde{D}_2 = \frac{b}{2H_2^2} - \frac{m_2}{4y},\\
&\tilde{E}_1 = -\tilde{E}_2 = \frac{c}{2H_2}, \quad \tilde{F}_1 = -\tilde{F}_2 = \frac{d}{2H_2},
\end{aligned} \tag{6.12}$$

where we use

$$\begin{aligned}
&d\mu = d[x + G(y)] = 2H(y)dy \ \text{ on } s_1,\\
&d\nu = d[x - G(y)] = -2H(y)dy \ \text{ on } s_2.
\end{aligned} \tag{6.13}$$

This is a system of integral equations, here the integrals of $\xi = U + V, \eta = U - V$ are along the direction of the characteristics s_1, s_2 respectively. Thus the system of integral equations can be rewritten as

$$\begin{aligned}
&v(z) = -2\int_0^y \frac{V(z)}{y^{m_2/2}h_2^{1/2}}dy = \int_0^y \frac{\eta - \xi}{|y|^{m_2/2}h_2^{1/2}}dy,\\
&\xi(z) = \int_0^y g_1(z)dy, z \in s_1, \ \eta(z) = \int_0^y g_2(z)dy, z \in s_2,\\
&g_l(z) = \tilde{A}_l\xi + \tilde{B}_l\eta + \tilde{C}_l(\xi + \eta) + \tilde{D}_l(\xi - \eta) + \tilde{E}_l v + \tilde{F}_l, l = 1, 2.
\end{aligned} \tag{6.14}$$

It is clear that for any two points $\tilde{z}_1 = \tilde{x}_1 + j\tilde{y} \in s_1, \tilde{z}_2 = \tilde{x}_2 + j\tilde{y} \in s_2$, where s_1, s_2 are two characteristic curves passing through a point $z = x + jy \in \overline{D}$ and x_1, x_2 are the intersection points with the axis $y = 0$ respectively, we have

$$|\tilde{x}_1 - \tilde{x}_2| \le |x_1 - x_2| = 2\left|\int_0^y \sqrt{|K(t)|}dt\right| \le \frac{4\max_{\overline{D}}\sqrt{h}}{m+2}|y|^{m/2+1} \le M|y|^{m/2+1} \tag{6.15}$$

for $-\delta < y < 0$, herein $m = m_1 - m_2 > -1$, $M(>\max[2\sqrt{h(x,y)}, 1, k_0, k_1])$ is a positive constant, δ is a sufficiently small positive constant. From (5.2), we can assume that the coefficients of (6.12) possess continuously differentiable with respect to $x \in L_0$ and satisfy the conditions

$$\begin{aligned}
&|\tilde{A}_l|, |\tilde{A}_{lx}|, |\tilde{B}_l|, |\tilde{B}_{lx}|, |H_2\tilde{E}_l|, |H_2\tilde{E}_{lx}|, |H_2\tilde{F}_l|, |H_2\tilde{F}_{lx}|,\\
&|1/\sqrt{h_l}|, |h_{ly}/h_l|, |h_{lx}/h_l|, |a|, |a_x|, |b|, |b_x| \le M \ \text{ in } D, \ l = 1, 2,\\
&|ya/H_1H_2| \le \varepsilon_1(y), \ m_1 + m_2 \ge 2,
\end{aligned} \tag{6.16}$$

in which $\varepsilon_1(y)$ is a positive function, $|b/H_2^2| = |by/H_2^2 y| \le \varepsilon_2(y)/|y|, \varepsilon_2(y) = |by|/H_2^2$, $\varepsilon_l(y) \to 0\,(l = 1, 2)$ as $y \to 0$. We can find a solution of Problem C_0

on the segment $-\delta \le y \le 0$, it is not difficult to see that if δ is a sufficiently small positive number, then there exists a positive constant $\gamma\,(<1)$, such that the following conditions are easily satisfied:

$$\begin{gathered}
\frac{4M^3}{2-m_2}|y|^{\beta_1}<\gamma,\ \frac{(4\varepsilon_1(y)+2\varepsilon_2(y))M+m_1+m_2}{2\beta'}<\gamma,\\
3M|y|^{\beta}+\frac{2\varepsilon_2(y)M+m_2}{4\beta}+\frac{2\varepsilon_1(y)M+m_1}{4\beta'}|y|^{m_1/2}<\gamma,\\
\frac{16M^2}{1-m_2+\beta}|y|^{1-m_2}+\frac{2\varepsilon_2(y)M+m_2}{2\beta}<\gamma,\ (\frac{7M}{1-m_2+\beta}\\
+\frac{4\varepsilon_2(y)M+m_2}{2\beta})|y|^{1-m_2}+\frac{8M|y|+4(\varepsilon_1(y)+\varepsilon_2(y))M+m_1+m_2}{2\beta'}<\gamma,
\end{gathered} \tag{6.17}$$

where $0 \le |y| \le \delta$, $\beta = 1 - m_2/2 - \beta_1, \beta' = 1 + m_1/2 - m_2/2 - \beta_1, \beta_1$ is a sufficiently small positive constant. In the next section we shall find a solution of Problem C_0 for (6.9) with the condition $0 \le m_2 < 1$.

6.3 Existence of solutions of Cauchy problem for degenerate hyperbolic equations

Let D_0 be the domain bounded by a segment $(0\le)a_0\le x\le a_1(\le 2)$ of the x-axis and the characteristics s_1 and s_2 of the families (6.8) emanating from $(a_0, 0)$ and $(a_1, 0)$ respectively, which intersect at a point $z = x + jy \in \overline{D}$. We choose $v_0 = v_0(z) = 0, \xi_0 = \xi_0(z) = 0, \eta_0 = \eta_0(z) = 0$ and substitute them into the corresponding positions of v, ξ, η in the right-hand sides of (6.14), and obtain

$$\begin{aligned}
&v_1(z)=-2\int_0^y\frac{V_0(z)}{(|y|^{m_2}h_2)^{1/2}}dy=\int_0^y\frac{\eta_0-\xi_0}{(|y|^{m_2}h_2)^{1/2}}dy,\\
&\xi_1(z)=\int_0^y[\tilde{A}_1\xi_0+\tilde{B}_1\eta_0+\tilde{C}_1(\xi_0+\eta_0)\\
&+\tilde{D}_1(\xi_0-\eta_0)+\tilde{E}_1v_0+\tilde{F}_1]dy=\int_0^y\tilde{F}_1dy,\ z\in s_1,\\
&\eta_1(z)=\int_0^y[\tilde{A}_2\xi_0+\tilde{B}_2\eta_0+\tilde{C}_2(\xi_0+\eta_0)\\
&+\tilde{D}_2(\xi_0-\eta_0)+\tilde{E}_2v_0+\tilde{F}_2]dy=\int_0^y\tilde{F}_2dy,\ z\in s_2.
\end{aligned} \tag{6.18}$$

By the successive approximation, we find the sequences of functions $\{v_k\}, \{\xi_k\}, \{\eta_k\}$, which satisfy the relations

$$\begin{aligned}
&v_{k+1}(z) = -2\int_0^y \frac{V_k(z)}{|y|^{m_2/2}h_2^{1/2}}dy = \int_0^y \frac{\eta_k-\xi_k}{|y|^{m_2/2}h_2^{1/2}}dy,\\
&\xi_{k+1}(z)=\int_0^y[\tilde{A}_1\xi_k+\tilde{B}_1\eta_k+\tilde{C}_1(\xi_k+\eta_k)+\tilde{D}_1(\xi_k-\eta_k)+\tilde{E}_1v_k+\tilde{F}_1]dy, z\in s_1,\\
&\eta_{k+1}(z)=\int_0^y[\tilde{A}_2\xi_k+\tilde{B}_2\eta_k+\tilde{C}_2(\xi_k+\eta_k)+\tilde{D}_2(\xi_k-\eta_k)+\tilde{E}_2v_k+\tilde{F}_2]dy, z\in s_2,\\
&k=0,1,2,....
\end{aligned}\tag{6.19}$$

We can prove that $\{v_k\}, \{\xi_k\}, \{\eta_k\}$ in D_0 satisfy the estimates

$$\begin{aligned}
&|v_k(z)|, |\xi_k(z)|, |\eta_k(z)|, |\xi_k(z)-\eta_k(z)|\le M\sum_{j=0}^k\gamma^j|y|^\beta,\\
&|v_{k+1}(z)-v_k(z)|, |\xi_{k+1}(z)-\xi_k(z)|, |\eta_{k+1}(z)-\eta_k(z)|\le M\gamma^k|y|^\beta,\\
&|\xi_{k+1}(z)-\eta_{k+1}(z)-\xi_k(z)+\eta_k(z)|\le M\gamma^k|y|^\beta,\\
&|\xi_k(z_1)-\xi_k(z_2)|\le M\sum_{j=0}^k\gamma^j|y|^\beta(|x_1-x_2|+|y|^{m_1/2}),\\
&|v_k(z_1)-v_k(z_2)|, |\eta_k(z_1)-\eta_k(z_2)|\le M\sum_{j=0}^k\gamma^j|y|^\beta(|x_1-x_2|+|y|^{m_1/2}),\\
&|v_{k+1}(z_1)-v_{k+1}(z_2)-v_k(z_1)+v_k(z_2)|\le M\gamma^k|y|^\beta(|x_1-x_2|+|y|^{m_1/2}),\\
&|\xi_{k+1}(z_1)-\xi_{k+1}(z_2)-\xi_k(z_1)+\xi_k(z_2)|\le M\gamma^k|y|^\beta(|x_1-x_2|+|y|^{m_1/2}),\\
&|\eta_{k+1}(z_1)-\eta_{k+1}(z_2)-\eta_k(z_1)+\eta_k(z_2)|\le M\gamma^k|y|^\beta(|x_1-x_2|+|y|^{m_1/2}),\\
&|\xi_k(z)+\eta_k(z)|\le M\sum_{j=0}^k\gamma^j|y|^{\beta'},\\
&|\xi_{k+1}(z)+\eta_{k+1}(z)-\xi_k(z)-\eta_k(z)|\le M\gamma^k|y|^{\beta'},
\end{aligned}\tag{6.20}$$

in which $z_1 = x_1 + jy$, $z_2 = x_2 + jy$ are the same as $\tilde{z}_1, \tilde{z}_2$ in (6.15), $|x_1 - x_2| < 1$, $\beta = 1 - m_2/2 - \beta_1, \beta' = 1 + m_1/2 - m_2/2 - \beta_1$, β_1 is a sufficiently small positive constant. In fact, from (6.18), it follows that the first formulas with $k = 1$ hold, namely

$$|\xi_1(z)|=|\int_0^y\tilde{F}_1dy|=|\int_0^yM|y|^{-m_2/2}dy|\le\frac{2M}{2-m_2}|y|^{\beta+\beta_1}$$

$$\le M|y|^\beta \le M\gamma^0|y|^\beta \le M\sum_{j=0}^{1}\gamma^j|y|^\beta, |\eta_1(z)|, |v_1(z)| \le M\sum_{j=0}^{1}\gamma^j|y|^\beta.$$

Next from (6.18), we can get

$$|\xi_1(z)+\eta_1(z)|=2|U_1(z)|=|H_1(y)v_{1x}|\le|\int_0^y[\tilde{F}_1(z_1)+\tilde{F}_2(z_2)]dy|$$
$$\le|\int_0^y\frac{d(z_1)-d(z_2)}{2H_2(y)}dy|\le|\int_0^y\frac{d_x|x_1-x_2|}{2H_2(y)}dy|\le\frac{M^2}{2}|\int_0^y|y|^{-m_2/2}|x_1-x_2|dy|$$
$$\le\frac{M^3}{2-m_2}|y|^{1-m_2/2}|y|^{1+m/2}\le M\gamma|y|^\beta|y|^{1+m/2}\le M\sum_{j=0}^{1}\gamma^j|y|^{\beta'},$$
$$|\xi_1(z)-\eta_1(z)|=2|V_1(z)|=|H_2(y)v_{1y}|\le|\int_0^y[\tilde{F}_1(z_1)-\tilde{F}_2(z_2)]dy|$$
$$\le 2M|\int_0^y|y|^{-m_2/2}dy|\le\frac{4M}{2-m_2}|y|^{1-m_2/2}=\frac{4M}{2-m_2}|y|^{\beta+\beta_1}\le M\sum_{j=0}^{1}\gamma^j|y|^\beta,$$
$$|\xi_1(z_1)-\xi_1(z_2)|\le|\int_0^y[\tilde{F}_1(x_1+jt)-\tilde{F}_1(x_2+jt)]dt|$$
$$\le|\int_0^y|\tilde{F}_{1x}||x_1-x_2|dy|\le M|\int_0^y|y|^{-m_2/2}|x_1-x_2|dy|$$
$$\le\frac{2M}{2-m_2}|y|^{1-m_2/2}|x_1-x_2|\le M\gamma|y|^\beta|x_1-x_2|\le M\sum_{j=0}^{1}\gamma^j|y|^\beta|x_1-x_2|,$$
$$|\eta_1(z_1)-\eta_1(z_2)|=|\int_0^y[\tilde{F}_2(x_1+jt)-\tilde{F}_2(x_2+jt)]dt|\le\sum_{j=0}^{1}\gamma^j|y|^\beta|x_1-x_2|,$$
$$|v_1(z_1)-v_1(z_2)|\le M\sum_{j=0}^{1}\gamma^j|y|^\beta|x_1-x_2|,$$
$$|\xi_1(z)-\xi_0(z)|=|\xi_1(z)|\le M\gamma^0|y|^\beta,$$
$$|\eta_1(z)-\eta_0(z)|=|\eta_1(z)|\le M\gamma^0|y|^\beta, |v_1(z)-v_0(z)|=|v_1(z)|\le M\gamma^0|y|^\beta,$$
$$|\xi_1(z)-\eta_1(z)-\xi_0(z)+\eta_0(z)|=|\xi_1(z)-\eta_1(z)|\le M\gamma^0|y|^\beta,$$
$$|\xi_1(z)+\eta_1(z)-\xi_0(z)-\eta_0(z)|=|\xi_1(z)+\eta_1(z)|\le M\gamma^0|y|^{\beta'},$$
$$|\xi_1(z_1)-\xi_1(z_2)-\xi_0(z_1)+\xi_0(z_2)|=|\xi_1(z_1)-\xi_1(z_2)|\le M\gamma^0|y|^\beta|x_1-x_2|,$$
$$|\eta_1(z_1)-\eta_1(z_2)-\eta_0(z_1)+\eta_0(z_2)|=|\eta_1(z_1)-\eta_1(z_2)|\le M\gamma^0|y|^\beta|x_1-x_2|,$$
$$|v_1(z_1)-v_1(z_2)-v_0(z_1)+v_0(z_2)|=|v_1(z_1)-v_1(z_2)|\le M\gamma^0|y|^\beta|x_1-x_2|.$$

In addition, we use the inductive method, namely suppose the estimates in (6.20) for $k = n$ are valid, then they are also valid for $k = n + 1$. In the following, we only give the estimates of $|\xi_{n+1}(z)|, |\xi_{n+1}(z_1) - \xi_{n+1}(z_2)|$ and $|\xi_{n+1}(z) + \eta_{n+1}(z)|$, the other estimates can be similarly given. From (6.19), we have

$$
\begin{aligned}
&|\xi_{n+1}(z)| \le |\int_0^y \{[|\tilde{A}_1|+|\tilde{B}_1|+|\tilde{E}_1|]|y|^\beta+|\tilde{C}_1||y|^{\beta'}+|\tilde{D}_1||y|^\beta]\\
&\times M\sum_{j=0}^n \gamma^j+|\tilde{F}_1|\}dy| \le M|\int_0^y \{[3M|y|^{\beta-m_2/2}+\left(\frac{\varepsilon_1(y)}{2|y|}+\frac{m_1}{4|y|}\right)|y|^{\beta'}\\
&+\left(\frac{\varepsilon_2(y)}{|y|}+\frac{m_2}{4|y|}\right)|y|^\beta]\sum_{j=0}^n \gamma^j+|y|^{-m_2/2}\}dy|\\
&\le M\{[3M|y|^\beta+(2\varepsilon_1(y)+m_1)\frac{|y|^{m_1/2}}{4\beta'}\\
&+(2\varepsilon_2(y)+m_2)\frac{1}{4\beta}]\sum_{j=0}^n \gamma^j+1\}|y|^\beta \le M\sum_{j=0}^{n+1}\gamma^j|y|^\beta.
\end{aligned}
$$

Moreover we have

$$
\begin{aligned}
&|\xi_{n+1}(z_1)-\xi_{n+1}(z_2)| \le |\int_0^y [\tilde{A}_1(z_1)\xi_n(z_1)+\tilde{B}_1(z_1)\eta_n(z_1)\\
&+\tilde{C}_1(z_1)(\xi_n(z_1)+\eta_n(z_1))+\tilde{D}_1(z_1)(\xi_n(z_1)-\eta_n(z_1))+\tilde{E}_1(z_1)v_n(z_1)\\
&+\tilde{F}_1(z_1)-\tilde{A}_1(z_2)\xi_n(z_2)-\tilde{B}_1(z_2)\eta_n(z_2)-\tilde{C}_1(z_2)(\xi_n(z_2)+\eta_n(z_2))\\
&-\tilde{D}_1(z_2)(\xi_n(z_2)-\eta_n(z_2))-\tilde{E}_1(z_2)v_n(z_2)-\tilde{F}_1(z_2)]dy|\\
&\le |\int_0^y \{M\sum_{k=0}^n \gamma^k|y|^\beta[(8M|y|^{-m_2}+\frac{2\varepsilon_2(y)M+m_2}{2|y|})|x_1-x_2|\\
&+(\frac{2\varepsilon_2(y)M+m_2}{2|y|}+\frac{4\varepsilon_1(y)M+m_1}{2|y|})|y|^{\beta'}]+M^2|y|^{-m_2/2}|x_1-x_2|\}dy|\\
&\le M[(\frac{16M^2}{1-m_2+\beta}|y|^{1-m_2}+\frac{2\varepsilon_2(y)M+m_2}{2\beta})|y|^\beta|x_1-x_2|\\
&+(\frac{2\varepsilon_2(y)M+m_2}{2\beta'}+\frac{4\varepsilon_1(y)M+m_1}{2\beta'})|y|^{\beta'}]\sum_{k=0}^n \gamma^k\\
&+\frac{2M^2}{2-m_2}|y|^{1-m_2/2}|x_1-x_2|| \le M(\sum_{k=0}^n \gamma^k+1)|y|^\beta(|x_1-x_2|+|y|^{m_1/2})
\end{aligned}
$$

$$\leq M\sum_{k=0}^{n+1}\gamma^k|y|^\beta(|x_1-x_2|+|y|^{m_1/2}),$$

in which we use

$$|\tilde{E}_1(z_1)v_n(z_1)-\tilde{E}_1(z_2)v_n(z_2)|\leq|(\tilde{E}_1(z_1)-\tilde{E}_1(z_2))v_n(z_1)$$
$$+\tilde{E}_1(z_2)(v_n(z_1)-v_n(z_2))|\leq M^2\sum_{k=0}^{n}\gamma^k|y|^\beta|x_1-x_2|[2|y|^{-m_2/2}$$
$$+|y|^{m_1/2-m_2/2}]\leq M^2\sum_{k=0}^{n}\gamma^k|y|^{\beta-m_2/2}[2|x_1-x_2|+|y|^{m_1/2}],$$
$$|\tilde{C}_1(z_1)(\xi_n(z_1)+\eta_n(z_1))-\tilde{C}_1(z_2)(\xi_n(z_2)+\eta_n(z_2))$$
$$+\tilde{D}_1(z_1)(\xi_n(z_1)-\eta_n(z_1))-\tilde{D}_1(z_2)(\xi_n(z_2)-\eta_n(z_2))|$$
$$\leq\frac{1}{2H_1H_2}|(a(z_1)-a(z_2))(\xi_n(z_1)+\eta_n(z_1))+a(z_2)(\xi_n(z_1)+\eta_n(z_1)$$
$$-\xi_n(z_2)-\eta_n(z_2))|+\frac{1}{2H_2^2}|(b(z_1)-b(z_2))(\xi_n(z_1)-\eta_n(z_1))$$
$$+b(z_2)(\xi_n(z_1)-\eta_n(z_1)-\xi_n(z_2)+\eta_n(z_2))|$$
$$+|\frac{m_1}{4y}(\xi_n(z_1)+\eta(z_1)-\xi_n(z_2)-\eta_n(z_2))|$$
$$+|\frac{m_2}{4y}(\xi_n(z_1)-\eta_n(z_1)-\xi_n(z_2)+\eta_n(z_2))|$$
$$\leq M\sum_{k=0}^{n}\gamma^k[(2M^2|y|^{-m_2}+\frac{\varepsilon_2(y)M}{|y|}+\frac{m_2}{2|y|})|y|^\beta|x_1-x_2|$$
$$+(\frac{2\varepsilon_2(y)M+m_2}{2|y|}+\frac{4\varepsilon_1(y)M+m_1}{2|y|})|y|^{\beta'}].$$

In addition we consider

$$|\xi_{n+1}(z)+\eta_{n+1}(z)|\leq|\int_0^y[\tilde{A}_1(z_1)\xi_n(z_1)+\tilde{A}_2(z_2)\xi_n(z_2)+\tilde{B}_1(z_1)\eta_n(z_1)$$
$$+\tilde{B}_2(z_2)\eta_n(z_2)+\tilde{C}_1(z_1)(\xi_n(z_1)+\eta_n(z_1))+\tilde{C}_2(z_2)(\xi_n(z_2)+\eta_n(z_2))$$
$$+\tilde{D}_1(z_1)(\xi_n(z_1)-\eta_n(z_1))+\tilde{D}_2(z_2)(\xi_n(z_2)-\eta_n(z_2))$$
$$+\tilde{E}_1(z_1)v_n(z_1)+\tilde{E}_2(z_2)v_n(z_2)+\tilde{F}_1(z_1)+\tilde{F}_2(z_2)]dy|,$$

taking into account $\tilde{A}_1(z_2)+\tilde{A}_2(z_2)=\tilde{B}_1(z_2)+\tilde{B}_2(z_2)$ and

$$\begin{aligned}
&|\tilde{A}_1(z_1)\xi_n(z_1)+\tilde{A}_2(z_2)\xi_n(z_2)+\tilde{B}_1(z_1)\eta_n(z_1)\\
&+\tilde{B}_2(z_2)\eta_n(z_2)+\tilde{E}_1(z_1)v_n(z_1)+\tilde{E}_2(z_2)v_n(z_2)|\\
&\leq|[\tilde{A}_1(z_1)-\tilde{A}_1(z_2)]\xi_n(z_1)+[\tilde{A}_1(z_2)+\tilde{A}_2(z_2)]\xi_n(z_1)\\
&+\tilde{A}_2(z_2)[\xi_n(z_2)-\xi_n(z_1)]+[\tilde{B}_1(z_1)-\tilde{B}_1(z_2)]\eta_n(z_1)\\
&+[\tilde{B}_1(z_2)+\tilde{B}_2(z_2)]\eta_n(z_1)+\tilde{B}_2(z_2)[\eta_n(z_2)-\eta_n(z_1)]\\
&+[\tilde{E}_1(z_1)-\tilde{E}_2(z_2)]v_n(z_1)+\tilde{E}_2(z_2)[v_n(z_2)-v_n(z_1)]|\\
&\leq M^2[(3|y|^{-m_2/2}+3|y|^{-m_2/2})|x_1-x_2|+4|y|^{m_1/2}]\sum_{j=0}^{n}\gamma^j|y|^\beta\\
&\leq M^2\sum_{j=0}^{n}\gamma^j[6|y|^{\beta-m_2/2}|x_1-x_2|+4|y|^{\beta+m_1/2}],\\
&|\tilde{C}_1(z_1)(\xi_n(z_1)+\eta_n(z_1))+\tilde{C}_2(z_2)(\xi_n(z_2)+\eta_n(z_2))|\\
&=|[\tilde{C}_1(z_1)-\tilde{C}_2(z_2)](\xi_n(z_1)+\eta_n(z_1))+\tilde{C}_2(z_2)[\xi_n(z_2)\\
&+\eta_n(z_2)+\xi_n(z_1)+\eta_n(z_1)]|\leq M[\frac{2\varepsilon_1(y)M}{|y|}+\frac{m_1}{2|y|})|y|^{\beta'}]\sum_{j=0}^{k}\gamma^j,\\
&|\tilde{D}_1(z_1)(\xi_n(z_1)-\eta_n(z_1))+\tilde{D}_2(z_2)(\xi_n(z_2)-\eta_n(z_2))|\\
&=|[\tilde{D}_1(z_1)+\tilde{D}_2(z_2)](\xi_n(z_1)-\eta_n(z_1))\\
&+\tilde{D}_2(z_2)[\xi_n(z_2)-\eta_n(z_2)-\xi_n(z_1)+\eta_n(z_1)]|\\
&\leq M[\frac{|b(z_1)-b(z_2)|}{2H_2^2}|y|^\beta+(\frac{|b|}{H_2^2}+\frac{m_2}{2|y|})|y|^\beta(|x_1-x_2|\\
&+|y|^{m_1/2})]\sum_{j=0}^{k}\gamma^j\leq\frac{M^2}{2}|y|^{-m_2}|x_1-x_2|\\
&+M(\frac{2\varepsilon_2(y)M}{|y|}+\frac{m_2}{2|y|})\sum_{j=0}^{k}\gamma^j|y|^\beta(|x_1-x_2|+|y|^{m_1/2}),
\end{aligned}$$

where $\varepsilon_2(y)=|by|/H^2(y)$, thus provided that δ is small enough, the inequality

$$|\xi_{n+1}(z)+\eta_{n+1}(z)|\leq|\int_0^y\{[7M^2|y|^{-m_2}+M(\frac{2\varepsilon_2(y)M}{|y|}+\frac{m_2}{2|y|})]$$

$$\times\sum_{j=0}^{n}\gamma^j|y|^\beta|x_1-x_2|+(4M^2+\frac{2(\varepsilon_1(y)+\varepsilon_2(y))M}{|y|}$$

$$+\frac{m_1+m_2}{2|y|})\sum_{j=0}^{n}\gamma^j|y|^{\beta'}+M^2|y|^{-m_2/2}|x_1-x_2|\}dy|$$

$$\leq M\{[\frac{7M|y|^{1-m_2}}{1-m_2+\beta}+\frac{4\varepsilon_2(y)M+m_2}{2\beta}]\sum_{j=0}^{n}\gamma^j|y|^\beta|x_1-x_2|$$

$$+\frac{8M^2|y|+4\varepsilon_1(y)M+m_1}{2\beta'}\sum_{j=0}^{n}\gamma^j|y|^{\beta'}+\frac{4\varepsilon_2(y)M+m_2}{2\beta'}\sum_{j=0}^{k}\gamma^j|y|^{\beta'}$$

$$+\frac{2M}{2-m_2}|y|^{1-m_2/2}|x_1-x_2|\}\leq M\{\sum_{j=0}^{n}\gamma^j[\frac{7M}{1-m_2+\beta}|y|^{1-m_2}$$

$$+\frac{4\varepsilon_2(y)M+m_2}{2\beta}|y|^\beta+\frac{4\varepsilon_2(y)M+m_2}{2\beta'}+\frac{8M|y|+4\varepsilon_1(y)M+m_1}{2\beta'}]$$

$$+1\}|y|^{\beta'}\leq M[\sum_{j=1}^{n+1}\gamma^j+1]|y|^{\beta'}\leq M\sum_{j=0}^{n+1}\gamma^j|y|^{\beta'}$$

is derived. Similarly we can verify

$$|\xi_{n+1}(z)-\xi_n(z)|\leq M\gamma^n|y|^\beta,$$

$$|\xi_{n+1}(z)+\eta_{n+1}(z)-\xi_n(z)-\eta_n(z)|\leq M\gamma^n|x_1-x_2|^{\beta_1}|y|^{\beta'},$$

$$|\xi_{n+1}(z_1)-\xi_{n+1}(z_2)-\xi_n(z_1)+\xi_n(z_2)|\leq M\gamma^n|y|^\beta[|x_1-x_2|+|y|^{m_1/2}].$$

From the estimate (6.20), the convergence of $\{v_n(z)\}, \{\xi_n(z)\}, \{\eta_n(z)\}$, and the comparison test, we can derive that $\{v_n(z)\}, \{\xi_n(z)\}, \{\eta_n(z)\}$ in D_0 uniformly converge to $v_*(z), \xi_*(z), \eta_*(z)$ satisfying

$$v_*(z)=-2\int_0^y\frac{V_*(z)}{(|y|^{m_2}h_2)^{1/2}}dy=\int_0^y\frac{\eta_*-\xi_*}{(|y|^{m_2}h_2)^{1/2}}dy,$$

$$\xi_*(z)=\int_0^y[\tilde{A}_1\xi_*+\tilde{B}_1\eta_*+\tilde{C}_1(\xi_*+\eta_*)+\tilde{D}_1(\xi_*-\eta_*)+\tilde{E}_1v_*+\tilde{F}_1]dy, z\in s_1,$$

$$\eta_*(z)=\int_0^y[\tilde{A}_2\xi_*+\tilde{B}_2\eta_*+\tilde{C}_2(\xi_*+\eta_*)+\tilde{D}_2(\xi_*-\eta_*)+\tilde{E}_2v_*+\tilde{F}_2]dy, z\in s_2,$$

and the function $v(z)=v_*(z)$ satisfies equation (6.9) and boundary condition (6.10), hence $u(z)=v(z)+y\gamma(x)+\tau(x)$ is a solution of Problem C

for (6.1). Thus the existence of solutions of Problem C for equation (6.1) is proved. Moreover it is easy to prove the uniqueness of solutions of Problem C for (6.1).

Theorem 6.1 *Under Condition C, the Cauchy problem (Problem C) for equation* (6.1) *has a unique solution.*

Remark 6.1 The condition $m_2 < 1$ in $K_2(y) = -|y|^{m_2}h_2(y)$ can be cancelled, provided that the last four conditions in (5.2) are replaced by the conditions $\hat{C}[a/|y|^{[m_2]},\overline{D}] \le k_0$, $\hat{C}[b/|y|^{[m_2]},\overline{D}] \le k_0, \hat{C}[c/|y|^{[m_2]},\overline{D}] \le k_0$, $\hat{C}[d/|y|^{[m_2]},\overline{D}] \le k_1$, where $[m_2]$ is the integer part of m_2, then Theorem 6.1 is also valid.

From Section 3, Chapter III, [12]3), we know that in general Problem C for equation (6.1) is ill-posed, and (6.16) is a sufficient condition of unique solvability of Problem C for (6.1). Besides, if the coefficients $K_l(y)(j=1,2)$ of equation (6.1) are replaced by the functions $K_l(x,y)(l=1,2)$ with some conditions, namely we consider the quasilinear equation of second order

$$K_1(x,y)u_{xx}-K_2(x,y)u_{yy}+au_x+bu_y+cu+d=0 \text{ in } D, \tag{6.21}$$

satisfying the conditions as stated in Section 5, in which

$$-K_l(x,y) = [H_l(x,y)]^2 = |y|^{m_l}h_l(x,y),\ l=1,2,$$

herein $m_l\,(l=1,2,m_2<1)$ are positive numbers, $h_l(x,y)\,(l=1,2)$ are continuously differentiable positive functions. Similarly to (5.30), $W(z)=[H_1(x,y)u_x-jH_2(x,y)u_y]/2$ in $\overline{D}$ can be written in the complex form

$$W_{\bar{\bar{z}}}=F(z,u,W),\ F=A_1W+A_2\overline{W}+A_3u+A_4 \text{ in } \overline{D},$$

where $A_j = A_j(z,u,W)$, $j=1,2,3,4$, and the system corresponding to (6.4) and its coefficients are replaced by

$$\begin{aligned}&(U+V)_\mu=\hat{A}_1(U+V)+\hat{B}_1(U-V)+\hat{C}_1u+\hat{D}_1,\\&(U-V)_\nu=\hat{A}_2(U+V)+\hat{B}_2(U-V)+\hat{C}_2u+\hat{D}_2,\end{aligned} \tag{6.22}$$

and

$$\hat{A}_1=\frac{1}{4H_1}\Big[\frac{a}{H_1}+\frac{H_1H_{2x}}{H_2}+\frac{H_2H_{1y}}{H_1}-\frac{b}{H_2}+H_{1x}+H_{2y}\Big],$$
$$\hat{B}_1=\frac{1}{4H_1}\Big[\frac{a}{H_1}-\frac{H_1H_{2x}}{H_2}+\frac{H_2H_{1y}}{H_1}+\frac{b}{H_2}+H_{1x}-H_{2y}\Big],$$

$$\hat{A}_2 = \frac{1}{4H_1}[\frac{a}{H_1} - \frac{H_1H_{2x}}{H_2} - \frac{H_2H_{1y}}{H_1} - \frac{b}{H_2} + H_{1x} + H_{2y}],$$

$$\hat{B}_2 = \frac{1}{4H_1}[\frac{a}{H_1} + \frac{H_1H_{2x}}{H_2} - \frac{H_2H_{1y}}{H_1} + \frac{b}{H_2} + H_{1x} + H_{2y}],$$

$$\hat{C}_1 = \hat{C}_2 = \frac{c}{4H_1}, \ \hat{D}_1 = \hat{D}_2 = \frac{d}{4H_1}.$$

Applying the similar method as stated before, the following theorem can be obtained.

Theorem 6.2 *Suppose that equation* (6.21) *satisfies Condition C. Then Problem C for* (6.21) *has a unique solution in D.*

CHAPTER IV

FIRST ORDER COMPLEX EQUATIONS OF MIXED TYPE

In this chapter, we mainly discuss the discontinuous Riemann-Hilbert boundary value problem for first order complex equations of mixed (elliptic-hyperbolic) type with parabolic degeneracy. In Section 1, we first introduce the corresponding results of the Riemann-Hilbert boundary value problem for first order complex equations of mixed type without parabolic degeneracy.

1 The Riemann-Hilbert Problem for First Order Complex Equations of Mixed Type

In this section we discuss the Riemann-Hilbert boundary value problem for first order linear complex equations of mixed (elliptic-hyperbolic) type in a simply connected domain. Firstly, we give the representation theorem and prove the uniqueness of solutions for the above boundary value problem. Secondly by using the method of successive approximation, the existence of solutions for the above problem is proved.

1.1 Formulation of Riemann-Hilbert problem of complex equations of mixed type

Let D be a simply connected bounded domain in the complex plane $\mathbf{C}$ with the boundary $\partial D = \Gamma \cup L$, where $\Gamma(\subset \{y > 0\}) \in C^1_\mu(0 < \mu < 1)$ with the end points $z = 0, 2$ and $L = L_1 \cup L_2, L_1 = \{x = -y, 0 \le x \le 1\}, L_2 = \{x = y + 2, 1 \le x \le 2\}$. Denote $D^+ = D \cap \{y > 0\}$, $D^- = D \cap \{y < 0\}$ and $z_0 = 1 - j$. Without loss of generality, we may assume that $\Gamma = \{|z - 1| = 1, y \ge 0\}$, otherwise through a conformal mapping, this requirement can be realized. In the following, we use the complex number $x + iy$ and complex function $w(z) = u(z) + iv(z)$ in D^+, and apply the hyperbolic unit j with the condition $j^2 = 1$ in D^-, and $x + jy$, $w(z) = u(z) + jv(z)$ are called the hyperbolic number and hyperbolic function in D^- respectively. We discuss the first order linear system of mixed (elliptic-hyperbolic) type equations

$$\begin{cases} u_x - \text{sgny}\, v_y = au + bv + f, \\ v_x + u_y = cu + dv + g, \end{cases} \text{ in } D, \tag{1.1}$$

in which a, b, c, d, f, g are real functions of $(x, y)\,(\in D)$. Denote

$$w(z) = u + iv,\ w_z = \frac{1}{2}[w_x - iw_y],\ w_{\bar{z}} = \frac{1}{2}[w_x + iw_y] \text{ in } D^+,$$
$$w(z) = u + jv,\ w_z = \frac{1}{2}[w_x - jw_y],\ w_{\bar{z}} = \frac{1}{2}[w_x + jw_y] \text{ in } D^-,$$

then system (1.1) in D can be reduced to the complex form

$$w_{\bar{z}} = A_1(z)w + A_2(z)\bar{w} + A_3(z) \text{ in } D,$$
$$A_1 = \begin{cases} \frac{1}{4}[a - ib + ic + d], \\ \frac{1}{4}[a + jb + jc + d], \end{cases} \quad A_2 = \begin{cases} \frac{1}{4}[a + ib + ic - d], \\ \frac{1}{4}[a - jb + jc - d], \end{cases} \tag{1.2}$$
$$A_3 = \begin{cases} \frac{1}{2}[f + ig] \\ \frac{1}{2}[f + jg] \end{cases} \text{ in } \begin{Bmatrix} D^+ \\ D^- \end{Bmatrix}.$$

In particular, if $A_l(z) = 0$, $l = 1, 2, 3$, then equation (1.2) becomes

$$w_{\bar{z}} = 0 \text{ in } D. \tag{1.3}$$

Suppose that the complex equation (1.2) satisfies the following conditions, namely **Condition** C.

The coefficients $A_l(z)\,(l = 1, 2, 3)$ are measurable in $z \in D^+$ and continuous in $\overline{D^-}$ in $D^* = \bar{D}\backslash\{0, 2\}$, and satisfy

$$L_p[A_l, \overline{D^+}] \le k_0,\ l = 1, 2,\ L_p[A_3, \overline{D^+}] \le k_1,$$
$$\hat{C}[A_l, \overline{D^-}] = C[A_l, \overline{D^-}] + C[A_{lx}, \overline{D^-}] \le k_0, l = 1, 2, \hat{C}[A_3, \overline{D^-}] \le k_1, \tag{1.4}$$

where $p(> 2)$, k_0, k_1 are positive constants.

The Riemann-Hilbert boundary value problem for the complex equation (1.2) may be formulated as follows:

Problem A Find a continuous solution $w(z)$ of (1.2) in $D^* = \bar{D}\backslash\{0,2\}$ satisfying the boundary conditions

$$\mathrm{Re}[\overline{\lambda(z)}w(z)] = r(z), \ \ z \in \Gamma, \tag{1.5}$$

$$\mathrm{Re}[\overline{\lambda(z)}w(z)]=r(z), \ z\in L_l \ (l=1 \text{ or } 2), \ \mathrm{Im}[\overline{\lambda(z_0)}w(z_0)]=b_0, \tag{1.6}$$

where $\lambda(z) = a(z) + ib(z) \neq 0$, $z \in \Gamma \cup L_l$ $(l = 1$ or $2)$, b_1 is a real constant, and $\lambda(z)$, $r(z)$, b_0 satisfy the conditions

$$\begin{gathered} C_\alpha[\lambda(z),\Gamma] \leq k_0, \ C_\alpha[r(z),\Gamma] \leq k_2, \ |b_0| \leq k_2, \\ C_\alpha[\lambda(z),L_l]\leq k_0, \ C_\alpha[r(z),L_l]\leq k_2, \ l=1 \text{ or } 2, \\ \max_{z\in L_1}\frac{1}{|a(z)-b(z)|}\leq k_0, \text{ or } \max_{z\in L_2}\frac{1}{|a(z)+b(z)|}\leq k_0, \end{gathered} \tag{1.7}$$

in which $\alpha(0 < \alpha < 1), k_0, k_2$ are positive constants. For convenience, later on sometimes we assume that $w(z_0) = 0$.

This Riemann-Hilbert problem (Problem A) for (1.2) with $r(z) = 0$, $z \in \Gamma \cup L_1$ (or L_2) and $b_1 = 0$ will be called Problem A_0. The number

$$K = \frac{1}{2}(K_1 + K_2), \tag{1.8}$$

is called the index of Problem A and Problem A_0 on the boundary ∂D^+ of D^+, where

$$K_l=\left[\frac{\phi_l}{\pi}\right]+J_l, \ J_l=0 \text{ or } 1, e^{i\phi_l}=\frac{\lambda(t_l-0)}{\lambda(t_l+0)}, \gamma_l=\frac{\phi_l}{\pi}-K_l, l=1,2, \tag{1.9}$$

in which $t_1 = 0$, $t_2 = 2$, $\lambda(t) = 1 + i$ on $L_0 = [0, 2]$ or $\lambda(t) = 1 - i$ on $L_0 = [0, 2]$ and $\lambda(t_1 + 0) = \lambda(t_2 - 0) = \sqrt{2}\exp(\pi i/4)$ or $\sqrt{2}\exp(7\pi i/4)$. Here we only discuss the case of $K = (K_1 + K_2)/2 = -1/2$ on ∂D^+.

1.2 Representation of Riemann-Hilbert problem for mixed complex equations

We first introduce a lemma, which is a special case of Theorem 1.1, Chapter I.

Lemma 1.1 *Suppose that the complex equation* (1.2) *satisfies Condition* C. *Then any solution of Problem* A *for* (1.2) *in* D^+ *with the boundary conditions* (1.5) *and*

$$\mathrm{Re}[\overline{\lambda(x)}w(x)]=s(x), \lambda(x)=1+i \text{ or } 1-i, x\in L_0, C_\alpha[s(x),L_0]\leq k_3, \tag{1.10}$$

can be expressed as

$$w(z) = \Phi(z)e^{\phi(z)} + \psi(z), \ \ z \in \overline{D^+}, \tag{1.11}$$

where $\mathrm{Im}[\phi(z)] = 0$, $z \in L_0 = [0, 2]$, *and* $\phi(z)$, $\psi(z)$ *satisfies the estimates*

$$C_\delta[\phi, \overline{D^+}] + L_{p_0}[\phi_{\bar z}, \overline{D^+}] \le M_1, C_\delta[\psi, \overline{D^+}] + L_{p_0}[\psi_{\bar z}, \overline{D^+}] \le M_2, \tag{1.12}$$

in which k_3, $\delta\,(0 < \delta \le \alpha)$, $p_0\,(2 < p_0 \le 2)$, $M_j = M_j(p_0, \delta, k, D^+)\,(j = 1, 2)$ *are positive constants,* $k = (k_0, k_1, k_2, k_3)$, $\Phi(z)$ *is analytic in* D^+ *and* $w(z)$ *satisfies the estimate*

$$C_\delta[X(z)w(z), \overline{D^+}] \le M_3(k_1 + k_2 + k_3), \tag{1.13}$$

in which

$$X(z) = |z - t_1|^{\eta_1}|z - t_2|^{\eta_2}, \eta_l = \max(-2\gamma_l, 0) + 4\delta, l = 1, 2, \tag{1.14}$$

here $\gamma_j\,(j = 1, 2)$ *are real constants as stated in* (1.9) *and* δ *is a sufficiently small positive constant, and* $M_3 = M_3(p_0, \delta, k_0, D^+)$ *is a positive constant.*

Theorem 1.2 *If the complex equation* (1.2) *satisfies Condition C in D, then any solution of Problem A with the boundary conditions* (1.5), (1.6) *for* (1.2) *can be expressed as*

$$w(z) = w_0(z) + W(z), \tag{1.15}$$

where $w_0(z)$ *is a solution of Problem A for the complex equation* (1.3) *and* $W(z)$ *possesses the form*

$$\begin{aligned}
&W(z) = w(z) - w_0(z), w(z) = \tilde\Phi(z)e^{\tilde\phi(z)} + \tilde\psi(z) \text{ in } D^+, \\
&\tilde\phi(z) = \tilde\phi_0(z) + Tg = \tilde\phi_0(z) - \frac{1}{\pi}\iint_{D^+}\frac{g(\zeta)}{\zeta - z}d\sigma_\zeta, \tilde\psi(z) = Tf \text{ in } D^+, \\
&W(z) = \Phi(z) + \Psi(z), \Psi(z) = \int_0^\mu g_1(z)d\mu e_1 + \int_2^\nu g_2(z)d\nu e_2 \text{ in } D^-,
\end{aligned} \tag{1.16}$$

in which $\mathrm{Im}\tilde\phi(z) = 0$ *on* L_0, $\tilde\phi_0(z)$ *is an analytic function in* D^+,

$$e_1 = (1 + j)/2, \ e_2 = (1 - j)/2, \ \mu = x + y, \ \nu = x - y,$$

$$g(z) = \begin{cases} A_1 + A_2\dfrac{\overline{\tilde W(z)}}{\tilde W(z)}, \tilde W(z) \ne 0, \\ 0, \ \tilde W(z) = 0, \end{cases} f = A_1\tilde\psi + A_2\overline{\tilde\psi} + A_3 \text{ in } D^+, \tag{1.17}$$

$$g_1(z) = A\xi + B\eta + E, g_2(z) = C\xi + D\eta + F \text{ in } D^-,$$

where

$$\tilde{W}(z) = w(z) - \tilde{\psi}(z),\ \xi = \mathrm{Re}w + \mathrm{Im}w,\ \eta = \mathrm{Re}w - \mathrm{Im}w,$$

$$A = \mathrm{Re}A_1 + \mathrm{Im}A_1, B = \mathrm{Re}A_2 + \mathrm{Im}A_2, E = \mathrm{Re}A_3 + \mathrm{Im}A_3,$$

$$C = \mathrm{Re}A_2 - \mathrm{Im}A_2, D = \mathrm{Re}A_1 - \mathrm{Im}A_1, F = \mathrm{Re}A_3 - \mathrm{Im}A_3,$$

and $\tilde{\phi}(z)$, $\tilde{\psi}(z)$ *satisfy the estimates*

$$C_\delta[\tilde{\phi}(z), \overline{D^+}] + L_{p_0}[\tilde{\phi}_{\bar{z}}, \overline{D^+}] \le M_4, C_\delta[\tilde{\psi}(z), \overline{D^+}] + L_{p_0}[\tilde{\psi}_{\bar{z}}, \overline{D^+}] \le M_4, \quad (1.18)$$

herein $M_4 = M_4(p_0, \delta, k, D^+)$ *is a positive constant,* $\tilde{\Phi}(z)$ *in* D^+ *and* $\Phi(z)$ *in* D^- *are the solutions of* (1.3) *satisfying the boundary conditions*

$$\begin{aligned}
&\mathrm{Re}[\overline{\lambda(z)}(e^{\tilde{\phi}(z)}\tilde{\Phi}(z) + \tilde{\psi}(z))] = r(z),\ z \in \Gamma,\\
&\mathrm{Re}[\overline{\lambda(x)}(\tilde{\Phi}(x)e^{\tilde{\phi}(x)} + \tilde{\psi}(x))] = s(x),\ x \in L_0,\\
&\mathrm{Re}[\overline{\lambda(x)}\Phi(x)] = \mathrm{Re}[\overline{\lambda(x)}(W(x) - \Psi(x))],\ x \in L_0,\\
&\mathrm{Re}[\overline{\lambda(z)}\Phi(z)] = -\mathrm{Re}[\overline{\lambda(z)}\Psi(z)],\ z \in L_1 \text{ or } L_2,\\
&\mathrm{Im}[\overline{\lambda(z_0)}\Phi(z_0)] = -\mathrm{Im}[\overline{\lambda(z_0)}\Psi(z_0)],
\end{aligned} \quad (1.19)$$

where $\lambda(x) = 1 + i$ *or* $1 - i$, $x \in L_0$ *in the second formula of* (1.19), *and* $\lambda(x) = 1 + j$ *or* $1 - j$, $x \in L_0$ *in the third formula of* (1, 19), $s(x)$ *is a known continuous function. Moreover the solution* $w_0(z)$ *of Problem A for* (1.3) *satisfies the estimate*

$$C_\delta[X(z)w_0(z), \overline{D^+}] + C_\delta[Y^\pm(z)w_0^\pm(z), \overline{D^-}] \le M_5(k_1 + k_2), \quad (1.20)$$

in which $w_0^\pm(z) = \mathrm{Re}w_0 \pm \mathrm{Im}w_0$ *and*

$$\begin{aligned}
&X(z) = \prod_{l=1}^{2} |z - t_l|^{\eta_l}, Y^\pm(z) = \prod_{l=1}^{2} |x \pm y - t_l|^{\eta_l},\\
&\eta_l = \max(-2\gamma_l, 0) + 4\delta,\ l = 1, 2,
\end{aligned} \quad (1.21)$$

herein $\delta\,(\le \alpha)$ *is a sufficiently small positive constant, and* $M_5 = M_5(p_0, \delta, k_0, D)$ *is a positive constant* (*see*[86]21)).

Proof Let the solution $w(z)$ be substituted into the position of w in the complex equation (1.2) and (1.17), thus the functions $g_1(z)$, $g_2(z)$ and

$\Psi(z)$ in $\overline{D^-}$ in (1.16), (1.17) can be determined. Moreover we can find the solution $\Phi(z)$ of (1.3) with the boundary condition (1.19), in which

$$s(x)=\begin{cases}\dfrac{2r((1-i)x/2)-2R((1-i)x/2)}{a((1-i)x/2)-b((1-i)x/2)}\\ +\mathrm{Re}[(1-i)\Psi(x)], \quad \text{or}\\ \dfrac{2r((1+i)x/2+1-i)-2R((1+i)x/2+1-i)}{a((1+i)x/2+1-i)+b((1+i)x/2+1-i)}\\ +\mathrm{Re}[(1+i)\Psi(x)],\end{cases} \tag{1.22}$$

herein $x\in L_0$, $R(z)=\mathrm{Re}[\overline{\lambda(z)}\Psi(z)]$, $\lambda(z)$ on L_1 or L_2, thus

$$w(z)=w_0(z)+W(z)=\begin{cases}\tilde{\Phi}(z)e^{\tilde{\phi}(z)}+\tilde{\psi}(z) \ \text{ in } \ D^+,\\ w_0(z)+\Phi(z)+\Psi(z) \ \text{in } D^-,\end{cases} \tag{1.23}$$

is the solution of Problem A for the complex equation

$$w_{\bar{z}}=A_1w+A_2\bar{w}+A_3 \ \text{ in } \ D, \tag{1.24}$$

which can be expressed as in (1.15) and (1.16).

1.3 Unique solvability of Riemann-Hilbert problem for complex equations of mixed type

Theorem 1.3 *Let the mixed complex equation* (1.2) *satisfy Condition* C. *Then Problem* A *for* (1.2) *has a solution in* D.

Proof In order to find a solution $w(z)$ of Problem A in D, we consider the representation: $w(z)$ in the form (1.15) – (1.17). In the following, we shall find a solution of Problem A by using the successive approximation. First of all, denoting the solution $w_0(z)\,(=\xi_0e_1+\eta_0e_2)$ of Problem A for (1.3), and substituting it into the position of $w\,(=\xi e_1+\eta e_2)$ in the right-hand side of (1.2), we can find the functions

$$W_1(z)=w_1(z)-w_0(z),\ w_1(z)=\tilde{\Phi}_1(z)e^{\tilde{\phi}_1(z)}+\tilde{\psi}_1(z),$$

$$\tilde{\phi}_1(z)=\tilde{\phi}_0(z)-\frac{1}{\pi}\iint_{D^+}\frac{g_0(\zeta)}{\zeta-z}d\sigma_\zeta,\ \tilde{\psi}_1(z)=Tf_0 \ \text{ in } \ D^+,$$

$$
\begin{aligned}
&w_1(z) = w_0(z) + W_1(z),\ W_1(z) = \Phi_1(z) + \Psi_1(z),\\
&\Psi_1(z) = \int_0^\mu [A\xi_0 + B\eta_0 + E]e_1 d\mu + \int_2^\nu [C\xi_0 + D\eta_0 + F]e_2 d\nu \text{ in } \overline{D^-},
\end{aligned}
\tag{1.25}
$$

where $f_0(z), g_0(z), \Psi_1(z)$ are known, $\mu = x+y, \nu = x-y$, and the solution $w_0(z)$ satisfies the estimate (1.20). Next we find the solutions $\tilde{\phi}(z)$ in $\overline{D^+}$ satisfying the boundary conditions

$$
\begin{aligned}
&\mathrm{Re}[\overline{\lambda(z)}(e^{\tilde{\phi}_1(z)}\tilde{\Phi}_1(z) + \tilde{\psi}_1(z))] = r(z),\ z \in \Gamma,\\
&\mathrm{Re}[\overline{\lambda(x)}(\tilde{\Phi}_1(x)e^{\tilde{\phi}_1(x)} + \tilde{\psi}_1(x))] = s_1(x),\ x \in L_0,
\end{aligned}
\tag{1.26}
$$

where $\lambda(x) = 1+i$ or $1-i$ on L_0 is as stated in (1.19) and

$$
s_1(x) = \begin{cases} \dfrac{2r((1-i)x/2) - 2R_1((1-i)x/2)}{a((1-i)x/2) - b((1-i)x/2)} + \mathrm{Re}[(1-i)\Psi_1(x)], \text{ or}\\[2ex] \dfrac{2r((1+i)x/2+1-i) - 2R_1((1+i)x/2+1-i)}{a((1+i)x/2+1-i) + b((1+i)x/2+1-i)} + \mathrm{Re}[(1+i)\Psi_1(x)]. \end{cases}
$$

Moreover we find the function $\Phi_1(z)$ in D^- of (1.3) satisfying the boundary conditions

$$
\begin{aligned}
&\mathrm{Re}[\overline{\lambda(x)}\Phi_1(x)] = \mathrm{Re}[\overline{\lambda(x)}(W_1(x) - \Psi_1(x))], x \in L_0,\\
&\mathrm{Re}[\overline{\lambda(z)}\Phi_1(z)] = -\mathrm{Re}[\overline{\lambda(z)}\Psi_1(z)],\ z \in L_1 \text{ or } L_2,\\
&\mathrm{Im}[\overline{\lambda(z_0)}\Phi_1(z_0)] = -\mathrm{Im}[\overline{\lambda(z_0)}\Psi_1(z_0)],
\end{aligned}
$$

where $\lambda(x) = 1+j$ or $1-j$ on L_0. Thus we obtain the function

$$
w_1(z) = w_0(z) + W_1(z) = \begin{cases} \tilde{\Phi}_1(z)e^{\tilde{\phi}_1(z)} + \tilde{\psi}_1(z) \text{ in } D^+,\\ w_0(z) + \Phi_1(z) + \Psi_1(z) \text{ in } \overline{D^-}, \end{cases}
\tag{1.27}
$$

which satisfies the estimate

$$
C_\delta[X(z)w_1(z), \overline{D^+}] + C[Y^\pm(z)w_1^\pm(z), \overline{D^-}] \le M_6 = M_6(p_0, \delta, k, D), \tag{1.28}
$$

where $\tilde{\phi}_1(z)$, $\tilde{\psi}_1(z)$, $\tilde{\Phi}_1(z)$ are similar to the functions $\tilde{\phi}(z)$, $\tilde{\psi}(z)$, $\tilde{\Phi}(z)$ in Theorem 1.2. In addition we substitute $w_1(z) = w_0(z) + W_1(z)$ and the corresponding functions $w_1^+(z) = \xi_1(z) = \mathrm{Re}w_1(z) + \mathrm{Im}w(z), w_1^-(z) = \eta_1(z) = \mathrm{Re}w_1(z) - \mathrm{Im}w(z)$ into the positions of w, ξ, η in (1.16), (1.17), and similarly to (1.25)–(1.27), we can find the corresponding functions $\tilde{\phi}_2(z), \tilde{\psi}_2(z), \tilde{\Phi}_2(z)$

in D^+ and $\Psi_2(z)$, $\Phi_2(z)$ and $W_2(z)=\Phi_2(z)+\Psi_2(z)$ in $\overline{D^-}$, and the function

$$w_2(z)=w_0(z)+W_2(z)=\begin{cases}\tilde{\Phi}_2(z)e^{\tilde{\phi}_2(z)}+\tilde{\psi}_2(z) \text{ in } D^+,\\ w_0(z)+\Phi_2(z)+\Psi_2(z) \text{ in } \overline{D^-}\end{cases}\tag{1.29}$$

satisfies a similar estimate in the form (1.28). Thus there exists a sequence of functions $\{w_n(z)\}$ as follows

$$w_n(z)=w_0(z)+W_n(z)=\begin{cases}\tilde{\Phi}_n(z)e^{\tilde{\phi}_n(z)}+\tilde{\psi}_n(z) \text{ in } D^+,\\ w_0(z)+\Phi_n(z)+\Psi_n(z) \text{ in } D^-,\\ \Psi_n(z)=\int_0^\mu [A\xi_{n-1}+B\eta_{n-1}+E]e_1d\mu\\ +\int_2^\nu [C\xi_{n-1}+D\eta_{n-1}+F]e_2d\nu \text{ in } \overline{D^-},\end{cases}\tag{1.30}$$

and then

$$\begin{aligned}&|Y^\pm[w_1^\pm-w_0^\pm]|\le|Y^\pm\Phi_1^\pm|+\sqrt{2}[|Y^+\int_0^\mu[A\xi_0+B\eta_0+E]e_1d\mu|\\&+|Y^-\int_2^\nu[C\xi_0+D\eta_0+F]e_2d\nu|]\le 2M_7M(4m+1)R' \text{ in } \overline{D^-},\end{aligned}\tag{1.31}$$

where $M_7=\max_{z\in\overline{D^-}}(|A|,|B|,|C|,D||E|,|F|)$, $m=\max\{C[Y^+(z)w_0^+(z),\overline{D^-}]+C[Y^-(z)w_0^-(z),\overline{D^-}]\}$, $R'=2$, $M=1+4k_0^2(1+2k_0^2)$. It is clear that $w_n(z)-w_{n-1}(z)$ satisfies

$$\begin{aligned}&w_n(z)-w_{n-1}(z)=\Phi_n(z)-\Phi_{n-1}(z)\\&+\int_0^\mu[A(\xi_n-\xi_{n-1})+B(\eta_n-\eta_{n-1})]e_1d\mu\\&+\int_2^\nu[C(\xi_n-\xi_{n-1})+D(\eta_n-\eta_{n-1})]e_2d\nu \text{ in } \overline{D^-},\end{aligned}\tag{1.32}$$

where $n=1,2,\ldots$. From the above equality, we can obtain

$$\begin{aligned}&|Y^\pm(z)[w_n^\pm-w_{n-1}^\pm]|\le[2M_7M(4m+1)]^n\\&\times\int_0^{R'}\frac{R'^{n-1}}{(n-1)!}dR'\le\frac{[2M_7M(4m+1)R']^n}{n!} \text{ in } \overline{D^-},\end{aligned}\tag{1.33}$$

and then we can see that the sequence of functions $\{Y^{\pm}(z)w_n^{\pm}(z)\}$, i.e.

$$Y^{\pm}(z)w_n^{\pm}(z)=Y^{\pm}(z)\{w_0^{\pm}(z)+[w_1^{\pm}(z)-w_0^{\pm}(z)]+\cdots+[w_n^{\pm}(z)-w_{n-1}^{\pm}(z)]\} \tag{1.34}$$

$(n=1,2,...)$ in $\overline{D^-}$ uniformly converge to functions $Y^{\pm}(z)w_*^{\pm}(z)$, and $w_*(z)$ satisfies the equality

$$\begin{gathered}w_*(z)=w_0(z)+\Phi_*(z)+\Psi_*(z),\\ \Psi_*(z)=\int_0^{\mu}[A\xi_*+B\eta_*+E]e_1d\mu+\int_2^{\nu}[C\xi_*+D\eta_*+F]e_2d\nu \text{ in } \overline{D^-},\end{gathered} \tag{1.35}$$

in which $\xi^*=\text{Re}w^*+\text{Im}w^*$, $\eta=\text{Re}w^*-\text{Im}w^*$, and $w_*(z)$ satisfies the estimate

$$C[Y^{\pm}(z)w_*^{\pm}(z),\overline{D^-}]\le e^{2M_7M(4m+1)R'}. \tag{1.36}$$

In addition, we can find a sequence of functions $\{w_n(z)\}$ $(w_n(z)=\tilde\Phi_n(z)e^{\tilde\phi_n(z)}+\tilde\psi_n(z))$ in D^+ and $\tilde\Phi_n(z)$ is an analytic function in D^+ satisfying the boundary conditions

$$\begin{gathered}\text{Re}[\overline{\lambda(z)}(\tilde\Phi_n(z)e^{\tilde\phi_n(z)}+\tilde\psi_n(z))]=r(z),\ z\in\Gamma,\\ \text{Re}[\overline{\lambda(x)}(\tilde\Phi_n(x)e^{\tilde\phi_n(x)}+\tilde\psi_n(x))]=s(x),\ x\in L_0,\end{gathered} \tag{1.37}$$

in which

$$s_n(x)=\begin{cases}\dfrac{2r((1-i)x/2)-2R_n((1-i)x/2)}{a((1-i)x/2)-b((1-i)x/2)}+\text{Re}[(1-i)\Psi_n(x)], \text{ or}\\ \dfrac{2r((1+i)x/2+1-i)-2R_n((1-i)x/2+1-i)}{a((1+i)x/2+1-i)+b((1+i)x/2+1-i)}+\text{Re}[(1+i)\Psi_n(x)],\end{cases} \tag{1.38}$$

herein $x\in L_0$, $R_n(z)=\text{Re}[\overline{\lambda(z)}\Psi_n(z)]$, $\lambda(z)$ on L_1 or L_2. From (1.33), it follows that

$$C_\delta[X(x)s_n(x),L_0]\le 2k_2k_0+(1+2k_0)\frac{[2M_7M(4m+1)R']^n}{n!}=M_8, \tag{1.39}$$

and the estimate

$$C_\delta[X(z)w_n(z),\overline{D^+}]\le M_3(k_1+k_2+M_8), \tag{1.40}$$

thus from $\{X(z)w_n(z)\}$, we can choose a subsequence which uniformly converges a function $X(z)w_*(z)$ in $\overline{D^+}$. Combining (1.36) and (1.40), it is obvious that the solution $w_*(z)$ of Problem A for (1.2) in $\bar D$ satisfies the estimate

$$C_\delta[X(z)w_*(z),\overline{D^+}]+C[Y^{\pm}(z)w_*^{\pm}(z),\overline{D^-}]\le M_9=M_9(p_0,\delta,k,D), \tag{1.41}$$

where M_9 is a positive constant.

Theorem 1.4 *Suppose that the complex equation* (1.2) *satisfies Condition* C. *Then Problem* A *for* (1.2) *has at most one solution in* D.

Proof Let $w_1(z), w_2(z)$ be any two solutions of Problem A for (1.2). By Condition C, we see that $w(z) = w_1(z) - w_2(z)$ satisfies the homogeneous complex equation and boundary conditions

$$Lw = \tilde{A}_1 w + \tilde{A}_2 \bar{w} \text{ in } D, \tag{1.42}$$

$$\begin{aligned} &\operatorname{Re}[\overline{\lambda(z)}w(z)] = 0, z \in \Gamma, \operatorname{Re}[\overline{\lambda(x)}w(x)] = s(x), x \in L_0, \\ &\operatorname{Re}[\overline{\lambda(z)}w(z)] = 0, \ z \in L_1 \text{ or } L_2, \ \operatorname{Re}[\overline{\lambda(z_0)}w(z_0)] = 0. \end{aligned} \tag{1.43}$$

From Theorem 1.2, the solution $w(z)$ can be expressed in the form

$$w(z) = \begin{cases} \tilde{\Phi}(z)e^{\tilde{\phi}(z)}, \ \tilde{\phi}(z) = \tilde{T}g \text{ in } D^+, \\ g(z) = \begin{cases} A_1 + A_2\overline{w}/w, w(z) \neq 0, z \in D^+, \\ 0, \ w(z) = 0, \ z \in D^+, \end{cases} \\ \Phi(z) + \Psi(z) \text{ in } \overline{D^-}, \\ \Psi(z) = \displaystyle\int_0^\mu [A\xi + B\eta]e_1 d\mu + \int_2^\nu [C\xi + D\eta]e_2 d\nu \text{ in } \overline{D^-}, \end{cases} \tag{1.44}$$

where $\tilde{\Phi}(z)$ is analytic in D^+ and $\Phi(z)$ is a solution of (1.3) in D^- satisfying the boundary condition (1.19), but $\tilde{\psi}(z) = 0, z \in D^+$, $r(z) = 0, z \in \Gamma$, and

$$s(x) = \begin{cases} \dfrac{-2R[(1-i)x/2]}{a[(1-i)x/2] - b[(1-i)x/2]} + \operatorname{Re}[(1-i)\Psi(x)], \text{ or} \\ \dfrac{-2R[(1-i)x/2+1-i]}{a[(1+i)x/2+1-i] + b[(1+i)x/2+1-i]} + \operatorname{Re}[(1+i)\Psi(x)], \end{cases}$$

herein $x \in L_0$, $R(z) = \operatorname{Re}[\overline{\lambda(z)}\Psi(z)]$, $\lambda(z)$ on L_1 or L_2. By using the method in the proof of Theorem 1.3, we can derive that

$$|Y^{\pm}(z)w^{\pm}(z)| \leq \frac{[2M_7M(4m+1)R']^n}{n!} \text{ in } \overline{D^-}. \tag{1.45}$$

Let $n \to \infty$, we get $w^{\pm}(z) = 0$, i.e. $w(z) = w_1(z) - w_2(z) = 0$, $\Psi(z) = \Phi(z) = 0$ in D^-. Noting that $w(z) = \tilde{\Phi}(z)e^{\tilde{\phi}(z)}$ satisfies the boundary conditions in (1.43), we see that the analytic function $\tilde{\Phi}(z)$ in D^+ satisfies the boundary conditions

$$\mathrm{Re}[\overline{\lambda(z)}e^{\tilde{\phi}(z)}\tilde{\Phi}(z)]=0,\ z\in\Gamma,\ \mathrm{Re}[\overline{\lambda(x)}\tilde{\Phi}(x)]=0,\ x\in L_0, \tag{1.46}$$

and the index of the boundary value problem (1.46) is $K=-1/2$, hence $\tilde{\Phi}(z)=0$ in D^+, and then $w(z)=\tilde{\Phi}(z)e^{\tilde{\phi}(z)}=0$ in D^+, namely $w(z)=w_1(z)-w_2(z)=0$ in D^+. This proves the uniqueness of solutions of Problem A for (1.2).

From Theorems 1.3 and 1.4, we see that under Condition C, Problem A for equation (1.2) has a unique solution $w(z)$, which can be found by using successive approximation, and $w(z)$ of Problem A satisfies the estimates

$$C_\delta[X(z)w(z),\overline{D^+}]\le M_{10},\ C[Y^\pm(z)w^\pm(z),\overline{D^-}]\le M_{11}, \tag{1.47}$$

where $w^\pm(z)=\mathrm{Re}w(z)\pm\mathrm{Im}w(z)$, $X(z)$, $Y^\pm(z)$ are as stated in (1.21), and $\delta(0<\delta\le\alpha)$, $M_j=M_j(p_0,\delta,k,D)\,(j=10,11)$ are positive constants, $k=(k_0,k_1,k_2)$. Moreover, we can derive the following theorem.

Theorem 1.5 *Suppose that equation* (1.2) *satisfies Condition C. Then any solution $w(z)$ of Problem A for* (1.2) *satisfies the estimates*

$$\begin{aligned}C_\delta[X(z)w(z),\overline{D^+}]&\le M_{12}(k_1+k_2),\\ C[Y^\pm(z)w^\pm(z),\overline{D^-}]&\le M_{13}(k_1+k_2),\end{aligned} \tag{1.48}$$

in which $M_j=M_j(p_0,\delta,k_0,D)(j=12,13)$ are positive constants.

Proof When $k=k_1+k_2=0$, from Theorem 1.4, it is easy to see that (1.48) holds. If $k=k_1+k_2>0$, then we see that the function $W(z)=w(z)/k$ is a solution of the homogeneous boundary value problem

$$Lw=F(z,w)/k,\ F/k=A_1W+A_2\overline{W}+A_3/k\ \text{ in } D,$$

$$\mathrm{Re}[\overline{\lambda(z)}W(z)]=r(z)/k,\ z\in\Gamma,\ \mathrm{Im}[\overline{\lambda(z_0)}W(z_0)]=b_0/k,$$

$$\mathrm{Re}[\overline{\lambda(z)}W(z)]=r(z)/k,\ z\in L_j,\ j=1\text{ or }2,$$

where

$$L_p[A_3/k,\overline{D^+}]\le 1,\ \hat{C}[A_3/k,\overline{D^-}]\le 1,\ C_\alpha[r(z)/k,\Gamma]\le 1,$$

$$C_\alpha[r(z)/k,L_l]\le 1,\ l=1\ \text{or}\ 2,\ |b_0/k|\le 1.$$

On the basic of the estimate (1.47), we can obtain the estimates

$$C_\delta[X(z)w(z),\overline{D^+}]\le M_{12},\ C[Y^\pm(z)w^\pm(z),\overline{D^-}]\le M_{13}, \tag{1.49}$$

where $M_l = M_l(p_0, \delta, k_0, D)\,(l = 12, 13)$ are positive constants. From (1.49), it follows the estimate (1.48).

From the estimates (1.48), (1.49), we can see the regularity of solutions of Problem A for (1.2). Moreover, we can give the Hölder continuous estimate of solutions of Problem A for first order quasilinear complex equation of mixed type with the more restrictive conditions than Condition C, which includes the linear complex equation (1.2) as a special case (see [86]21)).

Finally, we mention that if the index K is an arbitrary integer or $2K$ is an arbitrary odd integer, the above Riemann-Hilbert problem for (1.2) can be considered, but in general the boundary value problem for $K < -1$ have some solvability conditions or its solution for $K \geq 0$ is not unique.

2 The Riemann-Hilbert Problem for First Order Linear Complex Equations of Mixed Type with Parabolic Degeneracy

This section deals with the Riemann-Hilbert problem for linear mixed (elliptic-hyperbolic) complex equations of first order in a simply connected domain. Firstly, we give the representation theorem and prove the uniqueness of solutions for the above boundary value problem. Afterwards by using the method of successive approximation, the existence of solutions for the above problem is proved.

2.1 Formulation of Riemann-Hilbert problem for linear degenerate mixed complex equations

Let D be a simply connected bounded domain in the complex plane $\mathbf{C}$ with the boundary $\partial D = \Gamma \cup L$, where $\Gamma(\subset \{y > 0\}) \in C^1_\mu(0 < \mu < 1)$ with the end points $z = 0, 2$, $L = L_1 \cup L_2$, $L_1 = \{x + G(y) = 0, 0 \leq x \leq 1\}$, $L_2 = \{x - G(y) = 2, 1 \leq x \leq 2\}$ are two characteristic curves, and z_0 is the intersection point of L_1 and L_2, in which $G(y) = \int_0^y \sqrt{|K(t)|}dt$, $K(y) = \text{sgn}y|y|^m h(y)$, m is a positive number, $h(y)$ is a continuously differentiable positive function in $\overline{D}$. Similarly to Section 2, Chapter II, we may assume that the boundary Γ is a smooth curve with the form $x = \tilde{G}(y)$ and $x = 2 - \tilde{G}(y)$ near to the points $z = 0, 2$, where $x = \tilde{G}(y)$ and $x = 2 - \tilde{G}(y)$ are vertical to the axis $\text{Im}z = 0$ at $z = 0, 2$ respectively. Denote $D^+ = D \cap \{y > 0\}$ and $D^- = D \cap \{y < 0\}$. We consider the linear degenerate mixed system of first order equations

$$\begin{cases} H(y)u_x - \operatorname{sgn} y\, v_y = a_1 u + b_1 v + c_1 \\ H(y)v_x + u_y = a_2 u + b_2 v + c_2 \end{cases} \text{ in } D, \tag{2.1}$$

where $H(y) = G'(y) = \sqrt{|K(y)|}$, and $a_l, b_l, c_l (l = 1, 2)$ are real functions of $z(\in D)$. The following degenerate mixed system is a special case of system (2.1) with $H(y) = |y|^{m/2}$:

$$\begin{cases} |y|^{m/2} u_x - \operatorname{sgn} y\, v_y = a_1 u + b_1 v + c_1 \\ |y|^{m/2} v_x + u_y = a_2 u + b_2 v + c_2 \end{cases} \text{ in } D, \tag{2.2}$$

where m is a positive constant. We denote

$$\begin{aligned} &w(z) = u + iv,\ w_{\tilde{z}} = \frac{1}{2}[H(y)w_x - iw_y],\ w_{\bar{\tilde{z}}} = \frac{1}{2}[H(y)w_x + iw_y] \text{ in } D^+, \\ &w(z) = u + jv,\ w_{\tilde{z}} = \frac{1}{2}[H(y)w_x - jw_y],\ w_{\bar{\tilde{z}}} = \frac{1}{2}[H(y)w_x + jw_y] \text{ in } D^-. \end{aligned} \tag{2.3}$$

then system (2.2) in D can be reduced to the form

$$w_{\bar{\tilde{z}}} = A_1(z)w + A_2(z)\bar{w} + A_3(z) = g(Z) \text{ in } D, \tag{2.4}$$

where

$$Z = x + iY = x + iG(y) \text{ in } D^+,\ Z = x + jY = x + jG(y) \text{ in } \overline{D^-},$$

the above coefficients are as follows

$$\begin{aligned} &A_1 = \begin{cases} \frac{1}{4}[a_1 + ia_2 - ib_1 + b_2], \\ \frac{1}{4}[a_1 + ja_2 + jb_1 + b_2], \end{cases} \quad A_2 = \begin{cases} \frac{1}{4}[a_1 + ia_2 + ib_1 - b_2], \\ \frac{1}{4}[a_1 + ja_2 - jb_1 - b_2], \end{cases} \\ &A_3 = \begin{cases} \frac{1}{2}[c_1 + ic_2] \\ \frac{1}{2}[c_1 + jc_2] \end{cases} \text{ in } \begin{Bmatrix} D^+ \\ D^- \end{Bmatrix}. \end{aligned} \tag{2.5}$$

Suppose that equation (2.4) satisfies the following conditions: **Condition** C

The coefficients $A_j(z)\,(l = 1,2,3)$ in (2.4) are measurable in D^+ and continuous in $\overline{D^-}$, and satisfy

$$\begin{gathered}L_\infty[A_l, D^+] \le k_0, l = 1,2,\ L_\infty[A_3, D^+] \le k_1,\\ \hat{C}[A_l,\overline{D^-}]=C[A_l,\overline{D^-}]+C[A_{lx},\overline{D^-}]\le k_0, l=1,2, \hat{C}[A_3,\overline{D^-}]\le k_1,\end{gathered} \tag{2.6}$$

where k_0, k_1 ($\ge \max[1, 6k_0]$) are positive constants.

Now we formulate the Riemann-Hilbert problem as follows:

Problem A Find a continuous solution $w(z)$ of (2.4) in $D^* = \bar{D}\backslash\{0,2\}$, which satisfies the boundary conditions

$$\text{Re}[\overline{\lambda(z)}w(z)] = r(z), z\in\Gamma\cup L_1,\ \text{Im}[\overline{\lambda(z_0)}w(z_0)] = b_0, \tag{2.7}$$

where $\lambda(z) = a(x) + ib(x)$, b_0 is a real constants, and $\lambda(z) \ne 0$, $r(z)$, b_0 satisfy the conditions

$$\begin{gathered}C_\alpha[\lambda(z),\Gamma]\le k_0, C_\alpha[\lambda(z),L_1]\le k_0, C_\alpha[r(z),\Gamma]\le k_2,\\ C_\alpha[r(z),L_1]\le k_2, |b_0| \le k_2,\ \max_{z\in L_1}\frac{1}{|a(z)-b(z)|}\le k_0,\end{gathered} \tag{2.8}$$

in which $\alpha\,(0 < \alpha < 1), k_0, k_2$ are positive constants. In particular, if $\lambda(z) = a(x) + ib(x) = 1$, then Problem A is the Dirichlet boundary value problem, which will be called Problem D. For convenience, we may assume that $w(z_0) = 0$. Problem A with the conditions $A_3 = 0$ in D, $r(z) = 0$ on $\Gamma\cup L_1$ and $b_0 = 0$ is called Problem A_0.

The number

$$K = \frac{1}{2}(K_1 + K_2) \tag{2.9}$$

is called the index of Problem A and Problem A_0, where

$$K_l=\left[\frac{\phi_l}{\pi}\right]+J_l, J_l=0 \text{ or } 1, e^{i\phi_l}=\frac{\lambda(t_l-0)}{\lambda(t_l+0)}, \gamma_l=\frac{\phi_l}{\pi}-K_l, l=1,2, \tag{2.10}$$

in which $t_1 = 0$, $t_2 = 2$ on $L_0 = [0,2]$. Later on we shall only discuss the case: $K = -1/2$, and the other cases can be similarly discussed.

2.2 Representation and uniqueness of solutions of Riemann-Hilbert problem

Now we give the representation theorems of solutions for equation (2.4).

Theorem 2.1 *Suppose that the equation (2.4) satisfies Condition C. Then any solution of Problem A for (2.4) in D^+ can be expressed as*

$$w(z)=w_0(z)+\Phi[Z(z)]+\Psi[Z(z)]=\Phi_0[Z(z)]e^{\phi[Z(z)]}+\psi[Z(z)] \text{ in } D^+, \quad (2.11)$$

where $w_0(z)$ is a solution of Problem A for the complex equation

$$w_{\bar{z}} = 0 \text{ in } D^+, \ \text{ i.e. } w_{\overline{Z}} = 0 \text{ in } D_Z^+ \quad (2.12)$$

with the boundary condition (2.7), and $W(z)$ in D^+ possesses the form

$$W[z(Z)]=\Phi[Z(z)]+\Psi[Z(z)],\ \Psi(Z)=\tilde{T}F=-\frac{1}{\pi}\iint_{D_t^+}\frac{F(t)}{t-Z}d\sigma_t \text{ in } D_Z^+, \quad (2.13)$$

in which $Z = x + iG(y)$, and

$$F(Z)=\frac{g(Z)}{H(y)}=\frac{1}{H(y)}[A_1(z)w+A_2(z)\overline{w}+A_3(z)] \text{ in } D_Z^+,$$

$$\psi(Z)=-\frac{1}{\pi}\iint_{D_t}\frac{f(t)}{t-Z}d\sigma_t \text{ in } D_Z^+,\ H(y)f(Z)\in L_\infty(D_Z^+),$$

$$\phi(Z)=-\frac{1}{\pi}\iint_{D_t^+}\frac{h(t)}{t-Z}d\sigma_t \text{ in } D_Z^+,$$

$$h(Z)=\begin{cases}\dfrac{1}{H(y)}\{A_1[z(Z)]+A[z(Z)]\dfrac{\overline{W[z(Z)]}}{W[z(Z)]}\} \text{ if } W[z(Z)]\neq 0, Z\in D_Z^+,\\ 0 \text{ in if } W[z(Z)]=0,\ Z\in D_Z^+,\end{cases} \quad (2.14)$$

where $W(z) = w(z) - \psi[Z(z)]$, $z(Z)$ is the mapping from $Z \in D_Z^+$ to $z \in D^+$, $\Phi(Z), \Phi_0(Z)$ are solutions of equation (2.12), $\psi(Z)$ satisfies the estimate as that of the function $\psi(Z)$ in (2.13), Chapter I, and $\Phi(Z)+\Psi(Z)$ in ∂D^+ satisfy the homogeneous boundary conditions

$$\begin{aligned}&\mathrm{Re}[\overline{\lambda(z)}(\Phi[Z(z)]+\Psi[Z(z)])]=0 \text{ in } \Gamma,\\ &\mathrm{Re}[(1-i)(\Phi(Z)+\Psi(Z))]=0 \text{ on } L_0.\end{aligned} \quad (2.15)$$

Proof Let $w(z)$ be a solution of Problem A for equation (2.4), and be substituted in the positions of w in (2.13),(2.14), thus the functions $F(Z)$ and $\Psi(Z)$ in $\overline{D_Z^+}$ are determined. Moreover we can find the solutions $w_0(z), \Phi(Z)$ in $\overline{D^+}$ of (2.12) with the boundary conditions (2.7) and (2.15) respectively, thus

$$w(z) = w_0(z) + W(z) \text{ in } D^+ \quad (2.16)$$

is the solution of Problem A in D^+ for equation (2.4), where $W(z) = \Phi(z)+\Psi(z)$ satisfies the homogeneous boundary condition of Problem A.

Theorem 2.2 *Suppose that the equation* (2.4) *satisfies Condition* C. *Then any solution of Problem A for* (2.4) *in* D *can be expressed as*

$$w(z) = u(z) + jv(z) = w_0(z) + W(z) \text{ in } D, \tag{2.17}$$

where $w_0(z)$ *is a solution of Problem A for the complex equation*

$$w_{\bar{z}} = 0 \text{ in } D \tag{2.18}$$

with the boundary condition (2.7), *and* $W(z)$ *in* D *possesses the form*

$$\begin{aligned}
&W(z) = \Phi[Z(z)] + \Psi[Z(z)] \text{ in } D^+,\\
&\Psi(Z)=\tilde{T}f=-\frac{1}{\pi}\iint_{D_t^+}\frac{F(t)}{t-Z}d\sigma_t \text{ in } D_Z^+,\\
&W(z)=\phi(z)+\psi(z)=\xi(z)e_1+\eta(z)e_2 \text{ in } \overline{D^-},\\
&\xi(z)=\int_0^{\mu}[g_1(z)/2H(y)]d\mu=\zeta(z)+\int_0^y g_1(z)dy, z\in s_1,\\
&\eta(z)=\int_2^{\nu}[g_2(z)/2H(y)]d\nu=\theta(z)+\int_0^y g_2(z)dy, z\in s_2,\\
&g_l(z) = \hat{A}_l(U+V) + \hat{B}_l(U-V) + \hat{C}_l,\ l=1,2,\\
&\frac{g_1(z)}{2H} = A(U+V) + B(U-V) + E,\\
&-\frac{g_2(z)}{2H}=C(U+V)+D(U-V)+F,
\end{aligned} \tag{2.19}$$

where $F(z)$ *is as stated in* (2.13), $\Phi(z)$ *in* D^+ *and* $\phi(z) = \zeta(z)e_1 + \theta(z)e_2$ *in* D^- *are the solutions of* (2.18), $\zeta(z) = \int_{S_1} g_1(z)dy\ D^-$, *and* s_1, s_2 *are two families of characteristics in* D^-:

$$s_1 : \frac{dx}{dy}=\sqrt{|K(y)|}=H(y),\ s_2 : \frac{dx}{dy}=-\sqrt{|K(y)|}=-H(y) \tag{2.20}$$

passing through $z = x + jy \in \overline{D^-}$, S_1 *is the characteristic curve from a point on* L_1 *to a point on* L_0, *and*

$$W(z) = \xi(z)e_1 + \eta(z)e_2,$$

$$\hat{A}_1=\frac{a_1+a_2+b_1+b_2}{2},\ \hat{B}_1=\frac{a_1+a_2-b_1-b_2}{2},$$

$$\hat{A}_2=\frac{a_2-a_1+b_2-b_1}{2},\ \hat{B}_2=\frac{a_2-a_1-b_2+b_1}{2},$$
$$\hat{C}_1=c_1+c_2,\ \hat{C}_2=c_2-c_1\ \text{ in } D^-.$$

Moreover $\Phi(z)$ *and* $\phi(z)$ *satisfy the boundary conditions*

$$\begin{aligned}
&\mathrm{Re}[\overline{\lambda(z)}(\Phi(z)+\Psi(z))]=0,\ z\in\Gamma,\\
&\mathrm{Re}[\overline{\lambda(x)}(\Phi(x)+\Psi(x))]=s(x),\ \ x\in L_0,\\
&\mathrm{Re}[\overline{\lambda(x)}\phi(x)]=\mathrm{Re}[\overline{\lambda(x)}(W(x)-\psi(x))],\ x\in L_0,\\
&\mathrm{Re}[\overline{\lambda(z)}\phi(z)]=-\mathrm{Re}[\overline{\lambda(z)}\psi(z)],\ z\in L_1 \text{ or } L_2,\\
&\mathrm{Im}[\overline{\lambda(z_0)}\phi(z_0)]=-\mathrm{Im}[\overline{\lambda(z_0)}\psi(z_0)],
\end{aligned}\tag{2.21}$$

in which $\lambda(x)=1+i$, $x\in L_0$ *in the second formula of* (2.21), *and* $\lambda(x)=1+j$, $x\in L_0$ *in the third formula of* (2.21), $s(x)$ *is a known continuous function. Here we choose* $H(y)=[|y|^m h(y)]^{1/2}$, m, $h(y)$ *are as stated before, and*

$$\begin{aligned}
&d\mu=d[x+G(y)]=2H(y)\mathrm{d}y\ \text{ on } s_1,\\
&d\nu=d[x-G(y)]=-2H(y)\mathrm{d}y\ \text{ on } s_2.
\end{aligned}\tag{2.22}$$

Proof It is clear that the formulas in D^+ are true. As to the formulas in D^-, from (2.4), we have

$$\begin{aligned}
w_{\bar{z}}&=[H(y)w_x+jw_y]/2=[H(y)(u+jv)_x+j(u+jv)_y]/2\\
&=\frac{e_1}{2}[H(y)u_x+v_y+H(y)v_x+u_y]+\frac{e_2}{2}[H(y)u_x+v_y-H(y)v_x-u_y]\\
&=H(y)[e_1\left(\frac{u_x}{2}+\frac{v_y}{2H(y)}+\frac{v_x}{2}+\frac{u_y}{2H(y)}\right)+e_2\left(\frac{u_x}{2}+\frac{v_y}{2H(y)}-\frac{v_x}{2}-\frac{u_y}{2H(y)}\right)]\\
&=H(y)e_1(u+v)_\mu+H(y)e_2(u-v)_\nu\ \text{ in } D^-,
\end{aligned}\tag{2.23}$$

where $e_1=(1+j)/2, e_2=(1-j)/2$, and

$$\begin{aligned}
&\mu=x+G(y)=x+\int_0^y\sqrt{-K(t)}dt,\nu=x-G(y)=x-\int_0^y\sqrt{-K(t)}dt,\\
&\mu+\nu=2x,\mu-\nu=2G(y),\ \frac{\partial x}{\partial\mu}=\frac{\partial x}{\partial\nu}=\frac{1}{2},\frac{\partial y}{\partial\mu}=-\frac{\partial y}{\partial\nu}=\frac{1}{2H(y)}.
\end{aligned}\tag{2.24}$$

Similarly to (2.12), Chapter III, that equation (2.3) in $\overline{D^-}$ can be reduced to the system of integral equations

$$\begin{aligned}&W(z) = \phi(z) + \psi(z) = \xi(z)\mathrm{e}_1 + \theta(z)\mathrm{e}_2,\\&\xi(z)=\int_0^{\mu}\xi_\mu d\mu=\zeta(z)+\int_0^y[H\xi_x+\xi_y]dy=\zeta(z)+\int_0^y g_1(z)\mathrm{d}y,\ z\in s_1,\\&\eta(z)=\int_2^{\nu}\xi_\nu d\nu=\theta(z)+\int_0^y[-H\eta_x+\eta_y]dy=\theta(z)+\int_0^y g_2(z)dy,\ z\in s_2,\\&g_l(z) = \hat{A}_l(U+V)+\hat{B}_l(U-V)+\hat{C}_l, l = 1,2.\end{aligned}\tag{2.25}$$

Noting (2.20) and

$$\begin{aligned}&ds_1=\sqrt{(dx)^2+(dy)^2}=-\sqrt{1+(dx/dy)^2}dy=-\sqrt{1-K}dy=-\frac{\sqrt{1-K}}{\sqrt{-K}}dx,\\&ds_2=\sqrt{(dx)^2+(dy)^2}=-\sqrt{1+(dx/dy)^2}dy=-\sqrt{1-K}dy=\frac{\sqrt{1-K}}{\sqrt{-K}}dx,\end{aligned}\tag{2.26}$$

it is clear that the system (2.25) is just (2.19) in D^-.

Theorem 2.3 *Suppose that equation* (2.4) *satisfies Condition C. Then Problem A for* (2.4) *has at most one solution in D.*

Proof Let $w_1(z), w_2(z)$ be any two solutions of Problem A for (2.4). It is easy to see that $w(z) = w_1(z) - w_2(z)$ satisfies the homogeneous equation and boundary conditions

$$w_{\bar{\bar{z}}} = A_1 w + A_2\overline{w} \text{ in } D,\tag{2.27}$$

$$\operatorname{Re}[\overline{\lambda(z)}w(z)] = 0,\ \ z\in\Gamma\cup L_1,\ \operatorname{Im}[\overline{\lambda(z_0)}w(z_0)] = 0.\tag{2.28}$$

On the basis of Theorem 2.2, the solution $w(z)$ in the domain D can be expressed in the form

$$\begin{aligned}&w(z) = \Phi(Z)+\Psi(Z) \text{ in } \overline{D^+}, \Psi(z) = -2i\operatorname{Im}Tf,\\&Tf=-\frac{1}{\pi}\iint_{D_t^+}\frac{f(t)}{t-Z}d\sigma_t, f(Z)=\frac{g(Z)}{H(y)} \text{ in } \overline{D^+},\\&w(z)=\phi(z)+\psi(z)=\xi(z)e_1+\eta(z)e_2 \text{ in } \overline{D^-},\\&\xi(z) = \zeta(z) + \int_0^y[\hat{A}_1(U + V) + \hat{B}_1(U - V)]dy,\ z\in s_1,\end{aligned}$$

$$\eta(z) = \theta(z) + \int_0^y [\hat{A}_2(U+V) + \hat{B}_2(U-V)]dy,\ z \in s_2, \tag{2.29}$$

where $\Phi(Z)$ is an analytic function in D^+.

Firstly we can prove the solution $w(z) = 0$ in $\overline{D^-}$. In fact, choose any closed set $D_0 = \overline{D^-} \cap \{0 < d_0 < x < d_1 < 2\}$, where d_0, d_1 are positive numbers. According to the proof of Theorem 1.4, and noting that the continuity of $U(z), V(z)$ in D_0, there exists a positive number N dependent on $w(z), D_0$, such that

$$|\xi(z)| \le N,\ |\eta(z)| \le N \ \text{ in } \overline{D_0}.$$

From (2.19) and similar to (1.45), we can obtain

$$|\xi(z)| = |\int_{y_1}^y [\hat{A}_1\xi + \hat{B}_1\eta]dy| \le |\int_{y_1}^y N[|\hat{A}_1| + |\hat{B}_1|]dy|$$

$$\le |\int_{y_1}^y Nk_3 dy| \le Nk_3|y - y_1| = N\frac{(k_3|y-y_1|)^k}{k!} \quad \text{on } s_1,\ k = 1.$$

Similarly we have

$$|\eta(z) - \theta(z)| = |\int_{y_1'}^y [\hat{A}_2\xi + \hat{B}_2\eta]dy| \le N\frac{(k_3|y-y_1'|)^k}{k!} \quad \text{on } s_2,\ k = 1,$$

where $\max_{D_0}[|\hat{A}_1|, |\hat{B}_1|] \le k_3/2$, k_3 is a positive constant, where $z_1 = x_1 + jy_1$, $z_1' = x_1' + jy_1'$ are two intersection points of L_1, L_2 and two families of characteristics lines: s_1, s_2 as stated in (2.20) passing through $z = x + jy \in \overline{D^-}$ respectively. Applying the repeated insertion, the inequalities

$$|\xi(z)| \le N\frac{(k_3|y-y_1|)^k}{k!},$$

$$|\eta(z) - \theta(z)| \le N\frac{(k_3|y-y_1'|)^k}{k!},\ k = 2, 3, \ldots$$

can be obtained. This shows that $\xi(z) = 0, \eta(z) = 0$ in D_0. Taking into account the arbitrariness of d_0, d_1, we can derive $\xi(z) = 0, \eta(z) = 0$ in $\overline{D^-}$, thus $w(z) = 0$ in D^-. Moreover noting that $w(z)$ satisfies the homogeneous equation (2.27), the homogeneous boundary condition (2.28), and the index $K = -1/2$ of Problem A on ∂D^+, we can derive the solution $w(z) = 0$ in $\overline{D^+}$.

2.3 Solvability of Riemann-Hilbert problem for degenerate mixed equations

We first prove the existence and representation of solutions for Problem A for equation

$$\begin{gathered} w_{\bar z}=A_3(z) \text{ i.e. } w_{\overline{Z}}=A_3/H(y) \text{ in } D, \text{ or}\\ (U+iV)_{\overline{Z}}=A_3/H \text{ in } D^+, \\ (U+V)_\mu=g_1^0, (U-V)_\nu=g_2^0 \text{ in } \overline{D^-}. \end{gathered} \tag{2.30}$$

For this, we first find the solution of the system of first order equations (2.30) in D^- with the boundary condition

$$\begin{gathered} \mathrm{Re}[\overline{\lambda(z)}(U+jV)]=r(z)-R(z),\ z\in L_1,\\ \mathrm{Re}[\overline{\lambda(z)}(U+jV)]=R_0(x)-R(x), x\in L_0, \end{gathered} \tag{2.31}$$

where $\lambda(z)=a(z)+jb(z)$ on L_1, and $\lambda(x)=1+j$ on L_0, $R_0(x)$ on L_0 is an undetermined real function, and $R(z)=\mathrm{Re}[\overline{\lambda(z)}\psi(z)]$, $\psi(z)=\int_0^\mu g_1^0 d\mu e_1+\int_2^\nu g_2^0 d\nu e_2=\psi_1(z)e_1+\psi_2(z)e_2$. In fact the solution of Problem A for (2.30) in D^- can be expressed as

$$\begin{gathered} \xi=U+V+\psi_1(z)=f(\nu)+\psi_1(z), \nu=x-G(y),\\ \eta=U-V+\psi_2(z)=g(\mu)+\psi_2(z), \mu=x+G(y),\\ U(x,y)=[f(\nu)+g(\mu)]/2,\ V(x,y)=[f(\nu)-g(\mu)]/2, \text{ and}\\ w(z)=[(1+j)(f(\nu)+\psi_1(z))+(1-j)(g(\mu)+\psi_2(z)]/2, \end{gathered} \tag{2.32}$$

in which $f(t), g(t)$ are two arbitrary real continuous functions on $L_0=[0,2]$, thus the formulas in (2.31) can be rewritten as

$$\begin{gathered} a(z)U(z)-b(z)V(z)=r(z)-R(z) \text{ on } L_1,\\ U(x)-V(x)=R_0(x)-R(x) \text{ on } L_0, \text{ i.e.}\\ [a(z)-b(z)]f(x-G(y))+[a(z)+b(z]g(x+G(y))\\ =2[r(z)-R(z)] \text{ on } L_1,\\ U(x)-V(x)=R_0(x)-R(x) \text{ on } L_0, \text{ i.e.}\\ [a(h(x))-b(h(x))]f(2x)+[a(h(x))+b(h(x))]g(0) \end{gathered}$$

$$= 2[r(h(x)) - R(h(x))] \text{ on } L_1,$$

$$U(x) - V(x) = R_0(x) - R(x) \text{ on } L_0, \text{ i.e.}$$

$$[a(h(t/2)) - b(h(t/2))]f(t) + [a(h(t/2)) + b(h(t/2))]g(0) \tag{2.33}$$

$$= 2[r(t/2) - R(h(t/2))],$$

$$U(t) - V(t) = R_0(t) - R(t), \ t \in [0, 2],$$

where

$$(a(h(1)/2) + b(h(1/2)))g(0) = (a(h(1/2)) + b(h(1/2)))(U(z_0) - V(z_0))$$

$$= r(h(1/2)) - R(h(1/2)) - b_0 + b_0' \text{ or } 0,$$

and $h(x) = x + j(-G)^{-1}(x)$, $y = (-G)^{-1}(x)$ is the inverse function of $x = -G(y)$, $b_0' = \operatorname{Im}[\overline{\lambda(h(1)}\psi(h(1))]$. Noting the boundary conditions in (2.31), we can derive

$$U = \frac{1}{2}\{\frac{2[r(h(\nu/2)) - R(h(\nu/2))] - (a(h(\nu/2)) + b(h(\nu/2)))g(0)}{a(h(\nu/2)) - b(h(\nu/2))}$$

$$+R_0(\mu) - R(\mu)\},$$

$$V = \frac{1}{2}\{\frac{2[r(h(\nu/2)) - R(h(\nu/2))] - (a(h(\nu/2)) + b(h(\nu/2)))g(0)}{a(h(\nu/2)) - b(h(\nu/2))}$$

$$-R_0(\mu) + R(\mu))\},$$

if $a(z) - b(z) \neq 0$ on L_1. From the above formulas, it follows that

$$\operatorname{Re}[(1 + j)W(x)] = U(x) + V(x)$$

$$= \frac{2[r(h(x/2)) - R(h(x/2))] - (a(h(x/2)) + b(h(x/2)))g(0)}{a(h(x/2)) - b(h(x/2))}, \text{ i.e.}$$

$$\operatorname{Re}[(1 - i)W(x)] = U(x) + V(x)$$

$$= \frac{2[r(h(x/2)) - R(h(x/2))] - (a(h(x/2)) + b(h(x/2)))g(0)}{a(h(x/2)) - b(h(x/2))}, x \in [0, 2]. \tag{2.34}$$

In addition, from the above condition and the first boundary condition in (2.7), noting that the index $K = -1/2$, there exists a unique solution $w(z) = U + iV$ of the equation (2.12) in D^+, and then the function $R_0(x) = U(x) - V(x)$ on L_0 and the solution

$$
\begin{aligned}
w(z) &= W(z) + \Psi(z) = U(z) + iV(z) + \Psi(z) \text{ in } D^+,\\
w(z) &= \frac{1}{2}[(1+j)\frac{2[r(h(\nu/2))-R(h(\nu/2))]-(a(h(\nu/2))+b(h(\nu/2)))g(0)}{a(h(\nu/2))-b(h(\nu/2))}\\
&\qquad +(1-j)(R_0(\mu)-\psi_2(\mu))]\\
&= \frac{2[r(h(\nu/2))-R(h(\nu/2))]-(a(h(\nu/2))+b(h(\nu/2)))g(0)}{a(h(\nu/2))-b(h(\nu/2))}\\
&\qquad +R_0(\mu)-\psi_2(\mu)\\
&\quad +j\{\frac{2[r(h(\nu/2))-R(h(\nu/2))]-(a(h(\nu/2))+b(h(\nu/2)))g(0)}{a(h(\nu/2))-b(h(\nu/2))}\\
&\qquad -(R_0(\mu)-\psi_2(\mu))\} \text{ in } D^-
\end{aligned}
\tag{2.35}
$$

are obtained. Hence we have the following theorem.

Theorem 2.4 *Problem A for* (2.12) *in D has a unique solution in the form* (2.32) *and* (2.35), *which satisfies the estimates*

$$
\begin{aligned}
&|w(z)| = |U(z) + iV(z) + \Psi(z)| \le M_1 \text{ in } D^+_\varepsilon,\\
&|f(\nu) + \psi_1(z)| \le M_1,\ |g(\mu) + \psi_2(z)| \le M_1 \text{ in } D^-_\varepsilon,
\end{aligned}
\tag{2.36}
$$

where $\nu = x - G(y), \mu = x + G(y)$, $D^\pm_\varepsilon = \overline{D^\pm} \cap \{|z-t_1| \ge \varepsilon\} \cap \{|z-t_2| \ge \varepsilon\}$, ε is a sufficiently small positive number, and $M_1 = M_1(\alpha, k_0, k_1, D^\pm_\varepsilon)$ is a positive constant.

Moreover we shall give the estimates of solutions of the Riemann-Hilbert problem (Problem A) for mixed complex equation (2.4).

Theorem 2.5 *Suppose that complex equation* (2.4) *satisfies Condition C. Then any solution $w(z)$ of Problem A for* (2.4) *satisfies the estimates*

$$
C_\delta[X(z)w(z), \overline{D^+}] + C[Y^\pm(z)w^\pm(z), \overline{D^-}] \le M_2, \tag{2.37}
$$

where

$$
\begin{aligned}
X(z) &= \prod_{l=1}^{2} |z-t_l|^{\eta_l},\ Y^\pm(z) = \prod_{l=1}^{2} |x \pm y - t_l|^{\eta_l},\\
\eta_l &= \max(-2\gamma_l, 0) + 4\delta,\ l = 1, 2,
\end{aligned}
$$

herein $\delta\,(\le \alpha)$ is a sufficiently small positive constant, and $M_2 = M_2(p_0, \delta, k, D)$ is a positive constant.

Proof By using the method in Section 1 and the result as in Theorem 2.2, the above solution $w(z) = w_0(z) + W(z)$ possesses the expressions (2.17) and (2.19), where $W(z) = \Phi(z) + \Psi(z)$ in D^+ and $W(z) = \phi(z) + \psi(z)$ in D^- satisfy the boundary condition (2.21). From the boundary condition (2.21) on $\Gamma \cup L_0$, namely

$$\begin{aligned}
&\mathrm{Re}[\overline{\lambda(z)}(\Phi(z)+\Psi(z))]=0, \quad z\in\Gamma,\\
&\mathrm{Re}[\overline{\lambda(x)}(\Phi(x)+\Psi(x))]=s(x), \quad x\in L_0,\\
&s(x)=\frac{2r(h(x/2))-2R(h(x/2))}{a(h(x/2))-b(h(x/2))}+\mathrm{Re}[\overline{\lambda(x)}\psi(x)],
\end{aligned}$$

in which $\lambda(x) = 1+i$ on L_0, $h(x)$ is as stated in (2.33) below, and the result in Theorem 2.4, Chapter I, we can obtain the estimate of the solution $w(z)$ as follows

$$\hat{C}_\delta[w(z),\overline{D^+}]=C_\delta[X(Z)w(z(Z)),\overline{D^+_Z}]\le M_3,\ \hat{C}_\delta[w(z),\overline{D^+}]\le M_4(k_1+k_2),$$

where $M_3 = M_3(\delta, k, H, D)$, $M_4 = M_4(\delta, k_0, H, D)$ are non-negative constants. Moreover from the boundary condition (2.21) on $L_1 \cup L_0$, namely

$$\begin{aligned}
&\mathrm{Re}[\overline{\lambda(x)}\phi(x)]=\mathrm{Re}[\overline{\lambda(x)}(W(x)-\psi(x))], \ x\in L_0,\\
&\mathrm{Re}[\overline{\lambda(z)}\phi(z)]=-\mathrm{Re}[\overline{\lambda(z)}\psi(z)], \ z\in L_1,\\
&\mathrm{Im}[\overline{\lambda(z_0)}\phi(z_0)]=-\mathrm{Im}[\overline{\lambda(z_0)}\psi(z_0)],
\end{aligned}$$

where $\lambda(x) = 1 + j$ on L_0, and by using the way in the proof of Theorem 1.5, the following estimate can be derived

$$C[Y^\pm(z)w^\pm(z),\overline{D^-}]\le M_5,$$

where $M_5 = M_5(p_0, \delta, k_0, D)$ is a positive constant. Hence the estimate (2.37) is derived.

Finally we find a solution of the Riemann-Hilbert problem (Problem A) for general degenerate mixed complex equation (2.4) with the boundary condition (2.7), i.e. the following theorem.

Theorem 2.6 *If the complex equation* (2.4) *satisfies Condition C, then Problem A for* (2.4) *has a solution.*

Proof According to the proof of Theorem 1.3, and noting the results in Section 2, Chapter I and Section 2, Chapter III, we know that it is sufficient to find the solutions of two boundary value problems, i.e. Problem A^+: (2.4) in D^+ and the boundary conditions

$$\text{Re}[\overline{\lambda(z)}w(z)] = R(z) \text{ on } \Gamma \cup L_0, \tag{2.38}$$

and Problem A^-: (2.4) in D^- and the boundary conditions

$$\text{Re}[\overline{\lambda(z)}w(z)] = R(z) \text{ on } L_1 \cup L_0, \text{Im}[\overline{\lambda(z_0)}w(z_0)] = b_0, \tag{2.39}$$

where $R(z) = r(z)$ on $\Gamma \cup L_1$ in (2.7) and $\lambda(x) = 1 + i$, $R(x) = \hat{R}_0(x)$ and $\lambda(x) = 1+j$, $R(x) = \tilde{R}_0(x)$ on $L_0 = [0, 2]$. In fact according to Theorem 2.5, Chapter I, we see that Problem A^+ for equation (2.4) in D^+ has a unique solution $w(z)$, thus the function $\tilde{R}_0(x)$ on $L_0 = [0, 2]$ is obtained. Moreover by using the way of successive approximation, a solution of Problem A^- for equation (2.4) in D^- can be found. Hence the solvability of Problem A for (2.4) is proved.

In order to find a solution $w(z)$ of Problem A in D, we consider the representation: $w(z)$ in the form (2.17), (2.19). In the following, we shall find a solution of Problem A. First of all, denoting the solution $w_0(z)$ $(=\xi_0 e_1 + \eta_0 e_2 = f_0 e_1 + g_0 e_2)$ of Problem A for (2.18), and substituting it into the position of w $(= \xi e_1 + \eta e_2)$ in the right-hand side of (2.4) and (2.19), we can find the functions

$$\begin{aligned}
&W_1(z) = w_1(z) - w_0(z),\ w_1(z) = \Phi_1(z) + \Psi_1(z),\\
&\Psi_1(z) = TF_0(z) = -\frac{1}{\pi}\iint_{D^+}\frac{F_0(\zeta)}{\zeta - z}d\sigma_\zeta,\\
&F_0 = \frac{1}{H(y)}[A_1 w_0 + A_2\overline{w_0} + A_3] \text{ in } D^+,\\
&w_1(z) = w_0(z) + W_1(z),\ W_1(z) = \phi_1(z) + \psi_1(z),\\
&\psi_1(z) = \int_0^\mu [A\xi_0 + B\eta_0 + E]e_1 d\mu + \int_2^\nu [C\xi_0 + D\eta_0 + F]e_2 d\nu \text{ in } \overline{D^-},
\end{aligned} \tag{2.40}$$

where $f_0(z), g_0(z), \psi_1(z)$ are known, $\mu = x + y, \nu = x - y$, and the solution $w_0(z)$ satisfies the estimate (2.36). Next we find the solutions $\Phi(z)$ in $\overline{D^+}$ satisfying the boundary conditions

$$\begin{aligned}
&\text{Re}[\overline{\lambda(z)}(\Phi_1(z) + \Psi_1(z))] = 0,\ z \in \Gamma,\\
&\text{Re}[\overline{\lambda(x)}(\Phi_1(x) + \Psi_1(x))] = s_1(x),\ x \in L_0,
\end{aligned} \tag{2.41}$$

where $\lambda(x) = 1 + i$ on L_0 is as stated in (2.21) and

$$s_1(x) = \frac{2r(h(x/2)) - 2R_1(h(x/2))}{a(h(x/2)) - b(h(x/2))} + \text{Re}[\overline{\lambda(x)}\psi_1(x)],$$

in which $\lambda(x) = 1 + i$ on L_0, and $h(x)$ is as stated in (2.33). Moreover we find the function $\Phi_1(z)$ in D^- of (2.18) satisfying the boundary conditions

$$\begin{aligned}
&\mathrm{Re}[\overline{\lambda(x)}\phi_1(x)] = \mathrm{Re}[\overline{\lambda(x)}(W_1(x) - \psi_1(x))],\ x \in L_0,\\
&\mathrm{Re}[\overline{\lambda(z)}\phi_1(z)] = -\mathrm{Re}[\overline{\lambda(z)}\psi_1(z)],\ z \in L_1,\\
&\mathrm{Im}[\overline{\lambda(z_0)}\phi_1(z_0)] = -\mathrm{Im}[\overline{\lambda(z_0)}\psi_1(z_0)].
\end{aligned}$$

where $\lambda(x) = 1 + j$ on L_0. From Theorem 2.5, we can obtain the estimate

$$C_\delta[X(z)w_1(z), \overline{D^+}] + C[Y^\pm(z)w_1^\mp(z), \overline{D^-}] \le M_6 = M_6(p_0, \delta, k, D), \quad (2.42)$$

where $\Phi_1(z)$, $\Psi_1(z)$ are similar to the functions $\Phi(z)$, $\Psi(z)$ in Theorem 2.2. In addition we substitute $w_1(z) = w_0(z) + W_1(z)$ and the corresponding functions $w_1^+(z) = \xi_1(z) = \mathrm{Re}w_1(z) + \mathrm{Im}w_1(z), w_1^-(z) = \eta_1(z) = \mathrm{Re}w_1(z) - \mathrm{Im}w(z)$ into the positions of w, ξ, η in (2.4), (2.19), and similarly to (2.38)-(2.41), we can find the corresponding functions $\Phi_2(z)$, $\Psi_2(z)$, $W_2(z) = \Phi_2(z) + \Psi_2(z)$ in D^+ and $\phi_2(z), \psi_2(z), W_2(z) = \phi_2(z) + \psi_2(z)$ in $\overline{D^-}$, and then the function

$$w_2(z) = w_0(z) + W_2(z) = \begin{cases} \Phi_2(z) + \Psi_2(z) \text{ in } D^+, \\ w_0(z) + \phi_2(z) + \psi_2(z) \text{ in } \overline{D^-}, \end{cases} \quad (2.43)$$

satisfies a similar estimate as in the form (2.42). Thus there exists a sequence of functions $\{w_n(z)\}$ as follows

$$w_n(z) = w_0(z) + W_n(z) = \begin{cases} \Phi_n(z) + \Psi_n(z) \text{ in } D^+, \\ w_0(z) + \phi_n(z) + \psi_n(z) \text{ in } D^-, \\ \psi_n(z) = \int_0^\mu [A\xi_{n-1} + B\eta_{n-1} + E]e_1 d\mu \\ + \int_2^\nu [C\xi_{n-1} + D\eta_{n-1} + F]e_2 d\nu \text{ in } \overline{D^-}, \end{cases} \quad (2.44)$$

and then

$$\begin{aligned}
&|Y^\pm[w_1^\mp - w_0^\mp]| \le |Y^\pm\phi_1^\mp| + \sqrt{2}[|Y^+\int_0^\mu [A\xi_0 + B\eta_0 + E]e_1 d\mu| \\
&+ |Y^-\int_2^\nu [C\xi_0 + D\eta_0 + F]e_2 d\nu|] \le 2M_3 M(4m+1)R' \text{ in } \overline{D^-},
\end{aligned} \quad (2.45)$$

where

$$M_7=\max_{z\in\overline{D^-}}(|A|,|B|,|C|,|D|,|E|,|F|),\ M=1+4k_0^2(1+2k_0^2),$$
$$m=\max\{C[Y^+(z)w_0^+(z),\overline{D^-}]+C[Y^-(z)w_0^-(z),\overline{D^-}]\},\ R'=2.$$

It is easy to see that $w_n(z)-w_{n-1}(z)$ satisfies

$$\begin{aligned}&w_n(z)-w_{n-1}(z)=\phi_n(z)-\phi_{n-1}(z)\\&+\int_0^\mu[A(\xi_n-\xi_{n-1})+B(\eta_n-\eta_{n-1})]e_1d\mu\\&+\int_2^\nu[C(\xi_n-\xi_{n-1})+D(\eta_n-\eta_{n-1})]e_2d\nu\ \text{ in }\overline{D^-},\end{aligned}\tag{2.46}$$

where $n=1,2,\ldots$. From the above equality, we can obtain

$$\begin{aligned}&|Y^\pm(z)[w_n^\pm-w_{n-1}^\pm]|\le[2M_3M(4m+1)]^n\\&\times\int_0^{R'}\frac{R'^{n-1}}{(n-1)!}dR'\le\frac{[2M_3M(4m+1)R']^n}{n!}\ \text{ in }\overline{D^-},\end{aligned}\tag{2.47}$$

and then we can see that the sequence of functions $\{Y^\pm(z)w_n^\pm(z)\}$, i.e.

$$\begin{aligned}&Y^\pm(z)w_n^\pm(z)=Y^\pm(z)\{w_0^\pm(z)+[w_1^\pm(z)-w_0^\pm(z)]\\&+\cdots+[w_n^\pm(z)-w_{n-1}^\pm(z)]\}\ \text{ in }\overline{D^-},\ n=1,2,\ldots\end{aligned}\tag{2.48}$$

uniformly converge to functions $Y^\pm(z)w_*^\pm(z)$, and $w_*(z)$ satisfies the equality

$$\begin{aligned}&w_*(z)=w_0(z)+\phi_*(z)+\psi_*(z),\ \psi_*(z)\\&=\int_0^\mu[A\xi_*+B\eta_*+E]e_1d\mu+\int_2^\nu[C\xi_*+D\eta_*+F]e_2d\nu\text{ in }\overline{D^-},\end{aligned}\tag{2.49}$$

in which $\xi^*=\mathrm{Re}w^*+\mathrm{Im}w^*$, $\eta=\mathrm{Re}w^*-\mathrm{Im}w^*$, and $w_*(z)$ satisfies the estimate

$$C[Y^\pm(z)w_*^\pm(z),\overline{D^-}]\le e^{2M_3M(4m+1)R'}.\tag{2.50}$$

In the meantime, we find a sequence of functions $\{w_n(z)\}$ $(w_n(z)=\Phi_n(z)+\Psi_n(z))$ in D^+ and $\Phi_n(z)$ is an analytic function in D^+ satisfying the boundary conditions

$$\begin{aligned}&\mathrm{Re}[\overline{\lambda(z)}(\Phi_n(z)+\Psi_n(z))]=0,\ z\in\Gamma,\\&\mathrm{Re}[\overline{\lambda(x)}(\Phi_n(x)+\Psi_n(x))]=s(x),\ x\in L_0,(2.51)\\&s_n(x)=\frac{2r(h(x/2))-2R_n(h(x/2))}{a(h(x/2))-b(h(x/2))}+\mathrm{Re}[\overline{\lambda(x)}\psi_n(x)],\end{aligned}\tag{2.51}$$

herein $x \in L_0, R_n(z) = \text{Re}[\overline{\lambda(z)}\psi_n(z)]$, $\lambda(z)$ on L_1 or L_2. From (2.47), it follows that

$$\begin{gathered}C_\delta[X(x)s_n(x), L_0] \le 2k_2k_0 + (1+2k_0)\frac{[2M_3M(4m+1)R']^n}{n!} = M_7,\\ C_\delta[X(z)w_n(z), \overline{D^+}] \le M_8(k_1 + k_2 + M_7),\end{gathered} \tag{2.52}$$

where $M_8 = M_8(p_0, \delta, k_0, D)$ is a positive constant. Finally from $\{X(z)w_n(z)\}$, we can choose a subsequence which uniformly converges a function $X(z)w_*(z)$ in $\overline{D^+}$. Combining (2.50) and (2.52), it is obvious that the solution $w_*(z)$ of Problem A for (2.4) in $\bar{D}$ satisfies the estimate (2.36).

3 The Discontinuous Riemann-Hilbert Problem for First Order Quasilinear Complex Equations of Mixed Type with Degenerate Line

This section deals with the Riemann-Hilbert boundary value problem for first order quasilinear complex equations of mixed (elliptic-hyperbolic) type with parabolic degeneracy, we first discuss the problem in a special domains, and then consider the problem in the general domains.

3.1 Representation and uniqueness of solutions of discontinuous Riemann-Hilbert problem

Let D be a simply connected bounded domain in the complex plane $\mathbf{C}$ with the boundary $\partial D = \Gamma \cup L$, where Γ, $L = L_1 \cup L_2$ are as stated before, and $H(y) = G'(y) = \sqrt{|K(y)|}$, $G(y)$ are as stated before. We consider the first order quasilinear degenerate system of mixed type

$$\left\{\begin{aligned} &H(y)u_x - \text{sgn}y\, v_y = a_1u + b_1v + c_1\\ &H(y)v_x + u_y = a_2u + b_2v + c_2\end{aligned}\right. \quad \text{in } D, \tag{3.1}$$

where $a_l, b_l, c_l\,(l = 1, 2)$ are real functions of $(x, y)\,(\in D)$, $u, v\,(\in \mathbf{R})$. Similarly to (2.1), that system (3.1) in D can be reduced to the complex form

$$w_{\bar{\bar{z}}} = F(z, w),\ F(z, w) = A_1(z, w)w + A_2(z, w)\bar{w} + A_3(z, w) \text{ in } D, \tag{3.2}$$

where the coefficients are in the form

$$A_1=\begin{cases}\dfrac{1}{4}[a_1+ia_2-ib_1+b_2],\\ \dfrac{1}{4}[a_1+ja_2+jb_1+b_2],\end{cases}\quad A_2=\begin{cases}\dfrac{1}{4}[a_1+ia_2+ib_1-b_2],\\ \dfrac{1}{4}[a_1+ja_2-jb_1-b_2],\end{cases}$$

$$A_3=\begin{cases}\dfrac{1}{2}[c_1+ic_2]\\ \dfrac{1}{2}[c_1+jc_2]\end{cases}\ \text{in}\ \begin{Bmatrix}D^+\\ D^-\end{Bmatrix}.$$

Suppose that equation (3.2) satisfies the following conditions: **Condition** C

1) $A_l(z,w)\,(l=1,2,3)$ are measurable in D^+ and continuous in $\overline{D^-}$ for all continuous functions $w(z)$ in $D^*=\bar{D}\backslash\{0,2\}$, and satisfy

$$\begin{gathered}L_\infty[A_l,\overline{D^+}]\le k_0,\ l=1,2,\ \ L_\infty[A_3,\overline{D^+}]\le k_1,\\ \hat{C}[A_l,\overline{D^-}]=C[A_l,\overline{D^-}]+C[A_{lx},\overline{D^-}]\le k_0, l=1,2, \hat{C}[A_3,\overline{D^-}]\le k_1.\end{gathered}\tag{3.3}$$

2) For any continuous functions $w_1(z),w_2(z)$ on D^*, the following equality holds:

$$\begin{aligned}F(z,w_1)-F(z,w_2)=&\tilde{A}_1(z,w_1,w_2)(w_1-w_2)\\ &+\tilde{A}_2(z,w_1,w_2)\,(\overline{w_1}-\overline{w_2})\ \text{ in } D,\end{aligned}\tag{3.4}$$

where

$$L_\infty[\tilde{A}_l,D^+]\le k_0,\ \hat{C}[\tilde{A}_l,\overline{D^-}]\le k_0,\ l=1,2,\tag{3.5}$$

in (3.3), (3.5), k_0, k_1 are positive constants. In particular, when (3.2) is a linear equation (2.4), the condition (3.4) is obviously valid.

The boundary conditions of discontinuous Riemann-Hilbert problem (**Problem** B) for the complex equation (3.2) are as follows.

$$\text{Re}[\overline{\lambda(z)}w(z)]=r(z), z\in\Gamma\cup L_1, \text{Im}[\overline{\lambda(z_l)}w(z_l)]=b_l, l=0,1,...,J,\tag{3.6}$$

where $J=2K+1$, $\lambda(z)=a(x)+ib(x)$, $z_l\,(l=1,...,2K+1=J)$ are distinct points on Γ, $b_l\,(l=0,1,...,2K+1)$ are real constants, $2K+1$ is a positive integer and K is called the index of Problem B, and $\lambda(z)\ne 0$, $r(z)$, $b_l\,(l=$

$0, 1, ..., 2K+1)$ satisfy the conditions

$$
\begin{gathered}
C_\alpha[\lambda(z), \Gamma] \le k_0,\ C_\alpha[\lambda(z), L_1] \le k_0, \\
C_\alpha[r(z), \Gamma] \le k_2,\ C_\alpha[r(z), L_1] \le k_2, \\
|b_l| \le k_2, l=0, 1, ..., J,\ \max_{z\in L_1} \frac{1}{|a(z)-b(z)|} \le k_0,
\end{gathered} \tag{3.7}
$$

in which $\alpha\,(0 < \alpha < 1), k_0, k_2$ are positive constants. Problem B with the conditions $A_3 = 0$ in D, $r(z) = 0$ on $\Gamma \cup L_1$, $b_l = 0\,(l = 0, 1, ..., J)$ will be called Problem B_0. Besides we mention that if we choose the index $K = -1/2$ of Problem B in D^+, then the last $2K+1$ point conditions should be cancelled.

Let the solution $w(z)$ of Problem B be substituted in the coefficients of (3.2). Then the equation can be viewed as a linear equation (2.4). Hence we have the results as Theorems 2.3 and 2.5. Moreover we can prove the following theorem.

Theorem 3.1 *Suppose that the quasilinear complex equation* (3.2) *satisfies Condition* C. *Then Problem* B *for* (3.2) *has a unique solution in* D.

Proof We first prove the uniqueness of the solution of Problem B for (3.2). Let $w_1(z), w_2(z)$ be any two solutions of Problem B for (3.2). By Condition C, we see that $w(z) = w_1(z) - w_2(z)$ is a solution of Problem B_0, which satisfies the homogeneous complex equation and boundary conditions

$$
\begin{gathered}
w_{\bar z} = \tilde A_1 w + \tilde A_2 \bar w \text{ in } D, \\
\operatorname{Re}[\overline{\lambda(z)} w(z)] = 0, z \in L_1,\ \operatorname{Re}[\overline{\lambda(z_l)} w(z_l)] = 0, l = 0, 1, ..., J,
\end{gathered}
$$

where the conditions about the coefficients $\tilde A_l (l = 1, 2)$ are similar as those in the proof of Theorem 2.3 for the linear equation (2.4). Besides the remained proof is the same in the proof of Theorems 2.3 and 2.5, or by Theorem 2.5, Chapter I and Theorem 2.3, Chapter III. Next noting the conditions (3.3), (3.4), by using the same way as in the proof of Theorem 2.5, the existence of solutions of Problem B for (3.2) can be proved.

In order to give the Hölder continuous estimate of solutions for (3.2), we need to add the following condition.

3) For any complex numbers z_1, $z_2 (\in \bar D)$, w_1, w_2, the above functions

satisfy

$$\begin{aligned}&|A_l(z_1,w_1)-A_l(z_2,w_2)|\leq k_0[|z_1-z_2|^\alpha+|w_1-w_2|],\ l=1,2,\\&|A_3(z_1,w_1)-A_3(z_2,w_2)|\leq k_1[|z_1-z_2|^\alpha+|w_1-w_2|],\ z\in\overline{D^-},\end{aligned}\tag{3.8}$$

in which $\alpha(0<\alpha<1), k_0,\ k_1$ are positive constants.

On the basis of the results of Theorem 3.3 in Chapter I and the proof of Theorem 1.3, we can derive the following theorem.

Theorem 3.2 *Let the quasilinear complex equation* (3.2) *satisfy Condition C and* (3.8). *Then any solution $w(z)$ of Problem B for* (3.2) *satisfies the following estimates*

$$C_\delta[X(z)w(z),\overline{D^+}]\leq M_1,\ C[Y^\pm(z)w^\pm(z),\overline{D^-}]\leq M_2,\tag{3.9}$$

in which $w^\pm(z)=\text{Re}w(z)\pm\text{Im}w(z)$ and $X(z)$, $Y^\pm(z)$ are as stated in (1.21), *$\gamma_l\ (l=1,2)$ are real constants as stated in* (2.10) *and δ is a sufficiently small positive constant, and $M_j=M_j(\delta,k,D)\ (l=4,5)$ are positive constants, $k=(k_0,k_1,k_2)$.*

3.2 Riemann-Hilbert problem for quasilinear mixed equations in general domains

Let D be a simply connected bounded domain D in the complex plane $\mathbf{C}$ with the boundary $\partial D=\Gamma\cup L$, where Γ,L are as stated in Section 2. Now, we consider the domain D' with the boundary $\Gamma\cup L_1'\cup L_2'$, where Γ can be replaced by another smooth curve Γ' as stated in Section 2, Chapter II, here we we only consider the curves L_1', L_2', whose parameter equations are as follows:

$$L_1'=\{\gamma_1(s)+y=0,0\leq s\leq s_0\},L_2'=\{x-G(y)=2,l\leq x\leq 2\},\tag{3.10}$$

in which $Y=G(y)=\int_0^y\sqrt{|K(y)|}dy$, $\gamma_1(s)$ on $0\leq s\leq s_0$ is continuously differentiable, s is the parameter of arc length of L_1', and $\gamma_1(0)=0$, $\gamma_1(s)>0$ on $0<s\leq s_0$, and the slope of the curve L_1' at the intersection point z^* of L_1' and the characteristic curve $s_1: dy/dx=1/H(y)$ is not equal to that of the characteristic curve at the point, this shows that $\gamma_1(s)$ can be expressed by $\gamma_1[s(\nu)]\ (0\leq\nu\leq 2)$. Denote $D'^+=D'\cap\{y>0\}=D^+$, $D'^-=D'\cap\{y<0\}$ and $z_0'=l-j\gamma_1(s_0)$ is the intersection point of L_1' and L_2'.

We consider the first order quasilinear complex equation of mixed type as stated in (3.2) in D', and assume that (3.2) satisfies Condition C in $\overline{D'}$.

The Riemann-Hilbert boundary value problem for equation (3.2) may be formulated as follows:

Problem A' Find a continuous solution $w(z)$ of (3.2) in $D_* = \bar{D}\backslash\{0,2\}$, which satisfies the boundary conditions

$$\mathrm{Re}[\overline{\lambda(z)}w(z)] = r(z),\ z \in \Gamma, \tag{3.11}$$

$$\mathrm{Re}[\overline{\lambda(z)}w(z)]=r(z), z\in L_1',\ \mathrm{Im}[\overline{\lambda(z_0')}w(z_0')]=b_0, \tag{3.12}$$

in which the function $\lambda(z) = a(x)+ib(x) \neq 0$ on Γ, $\lambda(z) = a(x)+jb(x) \neq 0$ on L_1', and b_0 is a real constant, and $\lambda(z)$, $r(z)$, b_0 satisfy the conditions

$$\begin{aligned} &C_\alpha[\lambda(z),\Gamma]\le k_0, C_\alpha[r(z),\Gamma]\le k_2, C_\alpha[\lambda(z),L_1']\le k_0,\\ &C_\alpha[r(z),L_1']\le k_2, |b_0| \le k_2,\ \max_{z\in L_1'}\frac{1}{|a(z)-b(z)|}\le k_0, \end{aligned} \tag{3.13}$$

in which $\alpha\,(0<\alpha<1), k_0, k_2$ are all positive constants. The number

$$K = \frac{1}{2}(K_1 + K_2)$$

is called the index of Problem A' as stated in Section 1. Similarly we only discuss the case of $K = -1/2$ on ∂D^+, because in this case the solution of Problem A' is unique. In the following, we first discuss the domain D' and then discuss another general domain D''.

1. Setting $Y = G(y) = \int_0^y \sqrt{|K(t)|}dt$, the system (3.1) can be rewritten in the form

$$\begin{cases} H(y)u_x - \mathrm{sgn} y\, v_y = a_1u+b_1v+c_1 \\ H(y)v_x + u_y = a_2u + b_2v + c_2 \end{cases} \text{in } D'. \tag{3.14}$$

By the conditions in (3.10), the inverse function $x = \sigma(\nu) = (\mu+\nu)/2$ of $\nu = x - G(y)$ can be found, i.e. $\mu = 2\sigma(\nu) - \nu$, $0 \le \nu \le l + \gamma_1(l)$, and the curve L_1' can be expressed by $\mu = 2\sigma(\nu) - \nu = 2\sigma(x+\gamma_1(s)) - x - \gamma_1(s)$, $0 \le s \le s_0$. We make a transformation

$$\begin{aligned} &\tilde{\mu} = 2[\mu - 2\sigma(\nu) + \nu]/[2 - 2\sigma(\nu) + \nu], \tilde{\nu} = \nu,\\ &2\sigma(\nu) - \nu \le \mu \le 2,\ 0 \le \nu \le 2, \end{aligned} \tag{3.15}$$

where μ, ν are real variables, its inverse transformation is

$$\begin{gathered}\mu=[2-2\sigma(\nu)+\nu]\tilde{\mu}/2+2\sigma(\nu)-\nu,\ \nu=\tilde{\nu},\\ 0\le\tilde{\mu}\le 2,\ 0\le\tilde{\nu}\le 2.\end{gathered}\tag{3.16}$$

It is not difficult to see that the transformation in (3.15) maps the domain D'^- onto D^-. The transformation (3.15) and its inverse transformation (3.16) can be rewritten as

$$\begin{cases}\tilde{x}=\dfrac{1}{2}(\tilde{\mu}+\tilde{\nu})=\dfrac{4x-(2+x-Y)[2\sigma(x+\gamma_1(s))-x-\gamma_1(s)]}{4-4\sigma(x+\gamma_1(x))+2x+2\gamma_1(s)},\\ \tilde{Y}=\dfrac{1}{2}(\tilde{\mu}-\tilde{\nu})=\dfrac{4Y-(2-x+Y)[2\sigma(x+\gamma_1(s))-x-\gamma_1(s)]}{4-4\sigma(x+\gamma_1(x))+2x+2\gamma_1(s)},\end{cases}\tag{3.17}$$

and

$$\begin{cases}x=\dfrac{1}{2}(\mu+\nu)=\dfrac{[2-2\sigma(x+\gamma_1(s))+x+\gamma_1(s)](\tilde{x}+\tilde{Y})}{4}\\ +\sigma(x+\gamma_1(x))-\dfrac{x+\gamma_1(x)-\tilde{x}+\tilde{Y}}{2},\\ Y=\dfrac{1}{2}(\mu-\nu)=\dfrac{[2-2\sigma(x+\gamma_1(s))+x+\gamma_1(s)](\tilde{x}+\tilde{Y})}{4}\\ +\sigma(x+\gamma_1(x))-\dfrac{x+\gamma_1(x)+\tilde{x}-\tilde{Y}}{2}.\end{cases}\tag{3.18}$$

Denote by $\tilde{Z}=\tilde{x}+j\tilde{Y}=f(Z)$, $Z=x+jY=f^{-1}(\tilde{Z})$ the transformation (3.17) and the inverse transformation (3.18) respectively. In this case, the system of equations is

$$\xi_\mu=\hat{A}_1\xi+\hat{B}_1\eta+\hat{C}_1,\ \eta_\nu=\hat{A}_2\xi+\hat{B}_2\eta+\hat{C}_2,\ z\in D'^-,\tag{3.19}$$

which is another form of (3.14) in D'^-. Suppose that (3.14) in D' satisfies Condition C, through the transformation (3.13), we obtain $\xi_{\tilde{\mu}}=[2-2\sigma(\nu)+\nu]\xi_\mu/2$, $\eta_{\tilde{\nu}}=\eta_\nu$ in D'^-, where $\xi=u+v,\eta=u-v$, and then

$$\left.\begin{aligned}\xi_{\tilde{\mu}}&=[2-2\sigma(\nu)+\nu][\hat{A}_1\xi+\hat{B}_2\eta+\hat{C}_1]/2\\ \eta_{\tilde{\nu}}&=\hat{A}_2\xi+\hat{B}_2\eta+\hat{C}_2\end{aligned}\right\}\ \text{in } D^-,\tag{3.20}$$

and through the transformation (3.17), the boundary condition (3.12) is reduced to

$$\begin{gathered}\mathrm{Re}[\overline{\lambda(f^{-1}(\tilde{Z}))}w(f^{-1}(\tilde{Z}))]=r(f^{-1}(\tilde{Z})),\ \tilde{Z}=\tilde{x}+j\tilde{Y}\in L_1,\\ \mathrm{Im}[\overline{\lambda(f^{-1}(\tilde{Z}_0))}w(f^{-1}(\tilde{Z}_0)]=b_0,\end{gathered}\tag{3.21}$$

in which $\tilde{Z} = f(Z)$, $\tilde{Z}_0 = f(Z_0')$, $Z_0' = l + jG[-\gamma_1(s_0)]$. Therefore the boundary value problem (3.19),(3.12) is transformed into the boundary value problem (3.20), (3.21), i.e. the corresponding Problem A in D. On the basis of Theorem 3.1, we see that the boundary value problem (3.14) (in D^+), (3.20), (3.21) has a unique solution $w(\tilde{Z})$, and

$$w(z) = \begin{cases} w[\tilde{Z}(x+iy)] \text{ in } D^+, \\ w[\tilde{Z}(x+jy)] \text{ in } D^- \end{cases}$$

is just a solution of Problem A' for (3.14) in D', herein $Z = x + jG(y)$, $z = x + jy = x + jG^{-1}(Y)$.

Theorem 3.3 *If the mixed equation* (3.14) *in* D' *satisfies Condition* C' *in the domain* D' *with the boundary* $\Gamma \cup L_1' \cup L_2'$, *where* L_1', L_2' *are as stated in* (3.21), *then Problem* A' *for* (3.14) *with the boundary conditions* (3.11), (3.12) *has a unique solution* $w(z)$.

2. Next let the domain D'' be a simply connected domain with the boundary $\Gamma \cup L_1'' \cup L_2''$, where Γ is as stated before and the parameter expression of arc length of L_1'', L_2'' are as follows:

$$L_1'' = \{\gamma_1(s) + y = 0, 0 \le s \le s_0\}, L_2'' = \{\gamma_2(s) + y = 0, 0 \le s \le s_0'\}, \tag{3.22}$$

in which $\gamma_1(0) = 0, \gamma_2(0) = 0, \gamma_1(s) > 0, 0 < s \le s_0$; $\gamma_2(s) > 0$, $0 < s \le s_0'$; $\gamma_1(s)$ on $0 \le x \le s_0$, $\gamma_2(x)$ on $0 \le x \le s_0'$ are continuously differentiable, and the slope of the curves L_1', L_2' at z_1^*, z_2^* are not equal to those of the characteristic curves $s_1 : dy/dx = 1/H(y)$, $s_2 : dy/dx = -1/H(y)$ at the points respectively, where $z_k^*\,(k=1,2)$ are the intersection points of $L_k'\,(k = 1, 2)$ and $s_k\,(k = 1, 2)$ respectively, thus $\gamma_1(s), \gamma_2(s)$ can be expressed by $\gamma_1[s_1(\nu)], \gamma_2[s_2(\mu)]$. Denote $D''^+ = D'' \cap \{y > 0\} = D^+$ and $D''^- = D'' \cap \{y < 0\}$ and $z_0'' = l - i\gamma_1(s_0) = l - i\gamma_2(s_0')$. We consider the Riemann-Hilbert problem (Problem A'') for equation (3.14) in D'' with the boundary conditions (3.11) and

$$\text{Re}[\overline{\lambda(z)}w(z)] = r(z), z \in L_1'', \text{Im}[\overline{\lambda(z_0'')}w(z_0'')] = b_0, \tag{3.23}$$

where $\lambda(z)$, $r(z)$ satisfy the corresponding condition

$$\begin{gathered} C_\alpha[\lambda(z), \Gamma] \le k_0, \ \ C_\alpha[r(z), \Gamma] \le k_2, \\ C_\alpha[\lambda(z), L_1''] \le k_0, C_\alpha[r(z), L_1''] \le k_2, \\ |b_0| \le k_2, \ \max_{z \in L_1''} \frac{1}{|a(z) - b(z)|} \le k_0, \end{gathered} \tag{3.24}$$

in which $\alpha\,(0<\alpha<1), k_0, k_2$ are positive constants. By the conditions in (3.22), the inverse function $x=\tau(\mu)=(\mu+\nu)/2$ of $\mu=x+G(y)$ can be found, i.e. $\nu=2\tau(\mu)-\mu$, $0\le\mu\le 2$, and the curve L_2'' can be expressed as $x=\tau(\mu)=(\mu+\nu)/2$, namely

$$\nu=2\tau(\mu)-\mu=2\tau(x-\gamma_2(s))-x+\gamma_2(s),\ 0\le s\le s_0'. \tag{3.25}$$

We make a transformation

$$\tilde\mu=\mu,\ \tilde\nu=\frac{2\nu}{2\tau(\mu)-\mu},\ 0\le\mu\le 2,\ 0\le\nu\le 2\tau(\mu)-\mu, \tag{3.26}$$

where μ,ν are real variables, its inverse transformation is

$$\begin{gathered}\mu=\tilde\mu=\tilde x+\tilde Y,\ \nu=\frac{[2\tau(\mu)-\mu]\tilde\nu}{2}\\ =\frac{[2\tau(x-\gamma_2(s))-x+\gamma_2(s)](\tilde x-\tilde Y)}{2},\ 0\le\tilde\mu\le 2, 0\le\tilde\nu\le 2.\end{gathered} \tag{3.27}$$

Hence we have

$$\begin{aligned}&\tilde x=\frac{1}{2}(\tilde\mu+\tilde\nu)=\frac{2(x-Y)+(x+Y)[2\tau(x-\gamma_2(s))-x+\gamma_2(s)]}{2[2\tau(x-\gamma_2(s))-x+\gamma_2(s)]},\\ &\tilde Y=\frac{1}{2}(\tilde\mu-\tilde\nu)=\frac{-2(x-Y)+(x+Y)[2\tau(x-\gamma_2(s))-x+\gamma_2(s)]}{2[2\tau(x-\gamma_2(s))-x+\gamma_2(s)]},\\ &x=\frac{1}{2}(\mu+\nu)=\frac{1}{4}[(2\tau(x-\gamma_2(s))-x+\gamma_2(s))(\tilde x-\tilde Y)+2(\tilde x+\tilde Y)],\\ &Y=\frac{1}{2}(\mu-\nu)=\frac{1}{4}[(-2\tau(x-\gamma_2(s))+x-\gamma_2(s))(\tilde x-\tilde Y)+2(\tilde x+\tilde Y)].\end{aligned} \tag{3.28}$$

Denote by $\tilde Z=\tilde x+j\tilde Y=g(Z)$, $Z=x+jY=g^{-1}(\tilde Z)$ the transformation (3.28) and its inverse transformation in (3.28) respectively. Through the transformation (3.26), we obtain

$$(u+v)_{\tilde\mu}=(u+v)_\mu,\ (u-v)_{\tilde\nu}=[\tau(\mu)-\mu/2](u-v)_\nu \text{ in } D'^-. \tag{3.29}$$

System (3.30) in D''^- is reduced to

$$\left.\begin{aligned}&\xi_{\tilde\mu}=\hat A_1\xi+\hat B_1\eta+\hat C_1\\ &\eta_{\tilde\nu}=[\tau(\mu)-\mu/2][\hat A_2\xi+\hat B_2\eta+\hat C_2]\end{aligned}\right\} \text{ in } D'^-. \tag{3.30}$$

Moreover, through the transformation (3.28), the boundary condition (3.23) on L''_2 is reduced to

$$\begin{gathered}\operatorname{Re}[\overline{\lambda(g^{-1}(\tilde{Z}))w(g^{-1}(\tilde{Z}))}] = r[g^{-1}(\tilde{Z})], z = x + jy \in L'_1,\\ \operatorname{Im}[\overline{\lambda(g^{-1}(Z'_0))}w(g^{-1}(Z'_0))] = b_0,\end{gathered} \tag{3.31}$$

in which $\tilde{Z} = g(Z)$, $\tilde{Z}'_0 = g(Z''_0), Z''_0 = l + jG[-\gamma_2(s'_0)]$. Therefore the boundary value problem (3.19), (3.23) in D'' is transformed into the boundary value problem (3.30), (3.31). According to the method in the proof of Theorem 3.3, we can see that the boundary value problem (3.14) (in D^+), (3.30), (3.31) has a unique solution $w(\tilde{Z})$, and then $w = w(z)$ is a solution of the boundary value problem (3.14), (3.11), (3.23). But we mention that through the transformation (3.26) or (3.28), the boundaries L''_1, L''_2 are reduced to L'_1, L'_2 respectively, such that L'_1, L'_2 satisfy the condition as stated in (3.10).

Theorem 3.4 *If the mixed equation* (3.14) *satisfies Condition C' in the domain D'' with the boundary $\Gamma \cup L''_1 \cup L''_2$, where L''_1, L''_2 are as stated in* (3.22)*, then Problem A'' for* (3.14), (3.11), (3.23) *in D'' has a unique solution $w(z)$.*

4 The General Boundary Value Problem for Quasilinear Mixed Complex Equations with Degenerate Line

This section deals with the general boundary value problem for quasilinear mixed (elliptic-hyperbolic) complex equations of first order in a simply connected domain. Firstly, we give the representation theorem and prove the uniqueness of solutions for the above boundary value problem. Afterwards by using the method of successive approximation, the existence of solutions for the above problem is proved.

4.1 Formulation of general boundary value problem for complex equations of mixed type

Let D be a simply connected domain with the boundary $\Gamma \cup L_1 \cup L_2$, and the boundary of Γ possess the form $x + \tilde{G}(y) = 0$ near $z = 0$ and $x - \tilde{G}(y) = 2$ near $z = 2$ as stated in Section 2, and $L = L_1 \cup L_2, L_1 = \{x = -G(y), 0 \le x \le 1\}$, $L_2 = \{x = G(y) + 2, 1 \le x \le 2\}$, and $D^+ = D \cap \{y > 0\}$, $D^- =$

$D \cap \{y < 0\}$. Here, there are n points $E_1 = a_1$, $E_2 = a_2$, ..., $E_n = a_n$ on the segment $AB = [0, 2] = L_0$, where $a_0 = 0 < a_1 < a_2 < ... < a_n < a_{n+1} = 2$, and denote by $A = A_0 = 0, A_1 = a_1/2 - j|(-G)^{-1}(a_1/2)|$, $A_2 = a_2/2 - j|(-G)^{-1}(a_2/2)|$, ..., $A_n = a_n/2 - j|(-G)^{-1}(a_n/2)|$, $A_{n+1} = C = 1 - j|(-G)^{-1}(1)|$ and $B_1 = 1 - j|(-G)^{-1}(1)| + a_1/2 + j|(-G)^{-1}(a_1/2)|$, $B_2 = 1 - j|(-G)^{-1}(1)| + a_2/2 + j|(-G)^{-1}(a_2/2)|$, ..., $B_n = 1 - j|(-G)^{-1}(1)| + a_n/2 + j|(-G)^{-1}(a_n/2)|$, $B_{n+1} = B = 2$ on the segments AC, CB respectively. Moreover, we denote $D_1^- = \overline{D^-} \cap \{\cup_{l=0}^{[n/2]}(a_{2l} \le x - G(y) \le a_{2l+1})\}$, $D_2^- = \overline{D^-} \cap \{\cup_{l=1}^{[(n+1)/2]}(a_{2l-1} \le x + G(y) \le a_{2l})\}$ and $\tilde{D}_{2l+1}^- = \overline{D^-} \cap \{a_{2l} \le x - G(y) \le a_{2l+1}\}$, $l = 0, 1, ..., [n/2]$, $\tilde{D}_{2l}^- = \overline{D^-} \cap \{a_{2l-1} \le x + G(y) \le a_{2l}\}$, $l = 1, ..., [(n+1)/2]$, and $D_*^+ = \overline{D^+} \backslash Z_0, Z_0 = \{a_0, a_1, ..., a_{n+1}\}$, $D_*^- = \overline{D^-} \backslash Z_0$, $D_* = D_*^+ \cup D_*^-$.

We discuss the quasilinear equations of mixed type (3.2), i.e.

$$w_{\bar{\bar{z}}} = A_1(z, w)w + A_2(z, w)\overline{w} + A_3(z, w) \text{ in } D, \tag{4.1}$$

and suppose that (4.1) satisfies Condition C as stated in Section 3. In the following, we introduce the general Riemann-Hilbert boundary value problem for the complex equation (4.1) as follows.

Problem B Find a continuous solution $w(z)$ of (4.1) in $D^* = \bar{D} \backslash Z_0$ and $w(z)$ satisfies the boundary conditions

$$\text{Re}[\overline{\lambda(z)}w(z)] = r(z), \ z \in \Gamma, \tag{4.2}$$

$$\begin{aligned} &\text{Re}[\overline{\lambda(z)}w(z)] = r(z), z \in L_3 = \sum_{l=0}^{[n/2]} A_{2l}A_{2l+1}, \\ &\text{Re}[\overline{\lambda(z)}w(z)] = r(z), z \in L_4 = \sum_{l=1}^{[(n+1)/2]} B_{2l-1}B_{2l}, \end{aligned} \tag{4.3}$$

$$\text{Im}[\overline{\lambda(z)}w(z)]|_{z=a_0} = b_0,$$

$$\begin{aligned} &\text{Im}[\overline{\lambda(z)}w(z)]|_{z=A_{2l+1}} = c_{2l+1}, \ l \in Z', \\ &\text{Im}[\overline{\lambda(z)}w(z)]|_{z=B_{2l-1}} = c_{2l}, \ l \in Z'', \end{aligned} \tag{4.4}$$

where $Z' = \{0, 1, ..., [n/2]\}$, $Z'' = \{1, ..., [(n+1)/2]\}$, c_l $(l = 1, ..., n+1)$ are real constants, $\lambda(z) = a(x) + ib(x) \ne 0$, $z \in \Gamma$, and $\lambda(z)$, $r(z)$, c_l $(l = 1, ..., n+1)$

satisfy the conditions

$$\begin{aligned}&C_\alpha[\lambda(z),\Gamma]\le k_0, C_\alpha[r(z),\Gamma]\le k_2, |c_l|\le k_2, l=1,...,n+1,\\&|b_0|\le k_2,\ C_\alpha[\lambda(z),L_l]\le k_0,\ C_\alpha[r(z),L_l]\le k_2, l=3,4,\\&\max_{z\in L_3}\frac{1}{|a(z)-b(z)|}\le k_0,\ \max_{z\in L_4}\frac{1}{|a(z)+b(z)|}\le k_0,\end{aligned}\tag{4.5}$$

where $\alpha(0<\alpha<1)$, k_0, k_2 are positive constants. The above discontinuous Riemann-Hilbert boundary value problem for (4.1) is called Problem B.

Denote by $\lambda(t_l-0)$ and $\lambda(t_l+0)$ the left limit and right limit of $\lambda(z)$ as $z\to t_l=a_l\ (l=0,1,...,n+1)$, and

$$\begin{aligned}&e^{i\phi_l}=\frac{\lambda(t_l-0)}{\lambda(t_l+0)},\ \gamma_l=\frac{1}{\pi i}\ln\frac{\lambda(t_l-0)}{\lambda(t_l+0)}=\frac{\phi_l}{\pi}-K_l,\\&K_l=\left[\frac{\phi_l}{\pi}\right]+J_l,\ J_l=0\ \text{or}\ 1,\ l=0,1,...,n+1,\end{aligned}\tag{4.6}$$

in which $[a]$ is the largest integer not exceeding the real number a, $\lambda(z)=\exp(i\pi/2)$ on $L_1'=L_0\cap\overline{D_1^-}$ and $\lambda(a_{2l}+0)=\lambda(a_{2l+1}-0)=\exp(i\pi/2)$, $l=0,1,...,[n/2]$, and $\lambda(z)=\exp(-i\pi/2)$ on $L_2'=L_0\cap\overline{D_2^-}$ and $\lambda(a_{2l-1}+0)=\lambda(a_{2l}-0)=\exp(-i\pi/2)$, $l=1,...,[(n+1)/2]$, and $0\le\gamma_l<1$ when $J_l=0$, and $-1<\gamma_l<0$ when $J_l=1$, $l=0,1,...,n+1$, and

$$K=\frac{1}{2}(K_0+K_1+\cdots+K_{n+1})=\sum_{l=0}^{n+1}\left(\frac{\phi_l}{2\pi}-\frac{\gamma_l}{2}\right)\tag{4.7}$$

is called the index of Problem B. We can require that the solution $w(z)$ in D^+ satisfy the conditions

$$w[z(Z)]=O(|Z-a_l|^{-\eta_l}),\ l=0,1,...,n+1,\tag{4.8}$$

in the neighborhood of $a_l(0\le l\le n+1)$ in $Z(D^+)$, where $Z=Z(z)=x+iG(y)$, $z=z(Z)$ is the inverse function of $Z=Z(z)$, $\eta_l=\max(0,-\gamma_l)+2\delta$ $(l=1,...,n),\eta_l=\max(0,-2\gamma_l)+4\delta\ (l=0,n+1)$ and $\gamma_l\ (l=0,1,...,n+1)$ are real constants in (4.6), δ is a sufficiently small positive number, and choose the index $K=0$ or $-1/2$, if we select $K=-1/2$, then the first point condition in (4.4) should be cancelled.

4.2 Representation of solutions for general boundary value problem

Now, we give the representation of solutions for the discontinuous Riemann-Hilbert problem (Problem B) for system (4.1) in $\overline{D}$. For this, we first discuss the Riemann-Hilbert problem (Problem B) for system of first order equations:

$$w_{\overline{z}} = 0, \text{ in } D, \tag{4.9}$$

i.e.

$$(\xi + i\eta)_{\mu - i\nu} = 0 \text{ in } D^+, \ \xi_\mu = 0, \eta_\nu = 0 \text{ in } D^-, \tag{4.10}$$

with the boundary conditions

$$\begin{aligned} &\text{Re}[\overline{\lambda(z)}(u+jv)] = \begin{cases} r(z)=R(z), \ z \in L_3 \cup L_4, \\ R_0(x), x \in \{L_0 \cap D_1^-\} \cup \{L_0 \cap D_2^-\}, \end{cases} \\ &\text{Im}[\overline{\lambda(z)}(u+jv)]|_{z=A_{2l+1}} = c_{2l+1}, \ l = 0, 1, ..., [n/2], \\ &\text{Im}[\overline{\lambda(z)}(u+jv)]|_{z=B_{2l-1}} = c_{2l}, \ l = 1, ..., [(n+1)/2], \end{aligned} \tag{4.11}$$

in which $\lambda(z) = a(z)+jb(z)$ on L_1 and $\lambda(z) = 1+j$ on $L_0 \cap D_1^-$ $\lambda(z) = 1-j$ on $L_0 \cap D_2^-$, and $R_0(x)$ on $\{L_0 \cap D_1^-\} \cup \{L_0 \cap D_2^-\}$ is an undetermined real function. It is clear that the solution of Problem B for the second system of (4.10) in $\overline{D^-}$ can be expressed as

$$\begin{aligned} &\xi = u(z) + v(z) = f(\nu), \ \eta = u(z) - v(z) = g(\mu), \\ &u(z) = [f(\nu) + g(\mu)]/2, \ v(z) = [f(\nu) - g(\mu)]/2, \text{ i.e.} \\ &W(z) = u(z) + jv(z) = [(1+j)f(\nu) + (1-j)g(\mu)]/2, \end{aligned} \tag{4.12}$$

where $f(t), g(t)$ are two arbitrary real continuous functions on $L_0 = [0, 2]$. For convenience, sometimes we denote by the functions $a(x)$, $b(x)$, $R(x)$ of x the functions $a(z), b(z), R(z)$ of z in (4.11), thus (4.11) can be rewritten as

$$\begin{aligned} &a(x)u(z) - b(x)v(z) = R(x) \text{ on } L_3 \cup L_4, \\ &u(x) - v(x) = R_0(x) \text{ on } L_0 \cap D_1^-, \ u(x) + v(x) = R_0(x) \text{ on } L_0 \cap D_2^-, \\ &[(a(z) - jb(z))(u(z) + jv(z))]|_{z=A_{2l+1}} = R(A_{2l+1}) + jc_{2l+1}, \ l \in Z', \\ &[(a(z) - jb(z))(u(z) + jv(z))|_{z=B_{2l-1}} = R(B_{2l-1}) + jc_{2l}, \ l \in Z''. \end{aligned} \tag{4.13}$$

From (4.13), we have

$$[a(x)-b(x)]f(2x)+[a(x)+b(x)]h_{2l}=2R(x) \text{ on } L_3,$$

$$[a(x)-b(x)]h_{2l-1}+[a(x)+b(x)]g(2x-2)=2R(x) \text{ on } L_4,$$

$$\mathrm{Im}[\overline{\lambda(z)}u_{\tilde{z}}]|_{z=A_{2l+1}}=R(\mathrm{Re}A_{2l+1})+jc_{2l+1}, l=0,1,..,[n/2],$$

$$\mathrm{Im}[\overline{\lambda(z)}u_{\tilde{z}}]|_{z=B_{2l-1}}=R(\mathrm{Re}B_{2l-1})+jc_{2l}, l=1,..,[(n+1)/2],$$

$$h_{2l}=u(A_{2l+1})-v(A_{2l+1})=\frac{R(\mathrm{Re}A_{2l+1})-c_{2l+1}}{a(\mathrm{Re}A_{2l+1})+b(\mathrm{Re}A_{2l+1})}, l\in Z',$$

$$h_{2l-1}=u(B_{2l-1})+v(B_{2l-1})=\frac{R(\mathrm{Re}B_{2l-1})+c_{2l}}{a(\mathrm{Re}B_{2l-1})-b(\mathrm{Re}B_{2l-1})}, l\in Z''.$$

The above formulas can be rewritten as

$$\begin{aligned}
f(x-G(y))&=\frac{2R((x-G(y))/2)}{a((x-G(y))/2)-b((x-G(y))/2)}\\
&-\frac{[a((x-G(y))/2)+b((x-G(y))/2)]h_{2l}}{a((x-G(y))/2)-b((x-G(y))/2)} \text{ in } D_1^-,\\
g(x+G(y))&=\frac{2R((x+G(y))/2+1)}{a((x+G(y))/2+1)+b((x+G(y))/2+1)}\\
&-\frac{[a((x+G(y))/2+1)-b((x+G(y))/2+1)]h_{2l-1}}{a((x+G(y))/2+1)+b((x+G(y))/2+1)} \text{ in } D_2^-.
\end{aligned} \tag{4.14}$$

Thus the solution $w(z)$ of (4.9) can be expressed as

$$\begin{aligned}
w(z)&=w_0(z)=f(x-G(y))e_1+g(x+G(y))e_2\\
&=\frac{1}{2}\{f(x-G(y))+g(x+G(y))+j[f(x-G(y))-g(x+G(y))]\},
\end{aligned} \tag{4.15}$$

where $f(x-G(y))$, $g(x+G(y))$ are as stated in (4.14) and h_{2l-1}, h_{2l} are as stated before.

Next we find a solution of the Riemann-Hilbert boundary value problem for equation (4.9) in D^+ with the boundary conditions

$$\begin{aligned}
&\mathrm{Re}[\overline{\lambda(z)}(u(x)+iv(x))]=R(z) \text{ on } \Gamma,\\
&\mathrm{Re}[\overline{\lambda(z)}(u(x)+iv(x))]=R_0(x) \text{ on } \{L_0\cap D_1^-\}\cup\{L_0\cap D_2^-\},\\
&\mathrm{Im}[\overline{\lambda(z)}(u(z)+iv(z))]|_{z=a_0}=b_0,
\end{aligned} \tag{4.16}$$

where $\lambda(z)=1+i$ on $L_0\cap D_1^-$, $\lambda(z)=1-i$ on $L_0\cap D_2^-$, $R_0(x)=f(x)$ on $L_0\cap D_1^-$ and $R_0(x)=g(x)$ on $L_0\cap D_2^-$, where $f(x-G(y))$, $g(x+G(y))$ are functions as stated in (4.14). Noting that the index of the above boundary condition is $K=0$, by the result in Section 1, Chapter I, we know that the above Riemann-Hilbert problem has a unique solution $w(z)$ in D^+, and then

$$\begin{aligned} u(x)-v(x)&=\mathrm{Re}[(1-j)(u(x)+jv(x)]=R_0(x) \text{ on } L_0\cap D_1^-,\\ u(x)+v(x)&=\mathrm{Re}[(1+j)(u(x)+jv(x)]=R_0(x) \text{ on } L_0\cap D_2^-,\end{aligned} \tag{4.17}$$

is determined. This shows that Problem B for equation (4.9) is uniquely solvable, namely

Theorem 4.1 *Problem B of equation (4.9) or system (4.10) in $\overline{D}$ has a unique solution $w(z)=w_0(z)$ as stated in (4.15), which satisfies the estimates*

$$|f(\nu)|\le M_1,\ |g(\mu)|\le M_1 \text{ in } D_\varepsilon^-, \tag{4.18}$$

in which $\nu=x-G(y),\mu=x+G(y)$, $D_\varepsilon^-=\overline{D^-}\cap\{|z-t_0|\ge\varepsilon\}\cap\cdots\cap\{|z-t_{n+1}|\ge\varepsilon\}$, the positive number ε is small enough, and $M_1=M_1(\alpha,k_0,k_1,D_\varepsilon^-)$ is a positive constant.

Now we state and verify the representation of solutions of Problem B for equation (4.1).

Theorem 4.2 *Under Condition C, any solution $w(z)$ of Problem B for equation (4.1) in D^- can be expressed as follows*

$$\begin{aligned} &w(z)=w_0(z)+\Phi[Z(z)]+\Psi[Z(z)], \Psi(Z)=-\frac{1}{\pi}\iint_{D_t^+}\frac{f(t)}{t-Z}d\sigma_t \text{ in } \overline{D_Z^+},\\ &w(z)=w_0(z)+\phi(z)+\psi(z)=\xi(z)e_1+\eta(z)e_2 \text{ in } \overline{D^-},\\ &\xi(z)=\zeta(z)+\int_0^y g_1(z)dy=\xi_0(z)+\int_{S_1}g_1(z)dy+\int_0^y g_1(z)dy,\\ &g_1(z)=\hat{A}_1(U+V)+\hat{B}_1(U-V)+\hat{C}_1,\ z\in s_1,\\ &\eta(z)=\theta(z)+\int_0^y[\hat{A}_2(U+V)+\hat{B}_2(U-V)+\hat{C}_2]dy, z\in s_2,\end{aligned} \tag{4.19}$$

where $\phi(z)$ is a solution of (4.9) in D^-, $\xi_0=\mathrm{Re}w_0(z)+\mathrm{Im}w_0(z),\eta_0=\mathrm{Re}w_0(z)-\mathrm{Im}w_0(z)$, $w_0(z)$ is as stated in (4.15), and s_1,s_2 are two families *of characteristics in* D^-:

$$s_1 : \frac{dx}{dy} = \sqrt{|K(y)|} = H(y),\; s_2 : \frac{dx}{dy} = -\sqrt{|K(y)|} = -H(y) \tag{4.20}$$

passing through $z = x + jy \in D^-$, S_1 *is the characteristic curve from a point on* L_1 *to a point on* L_0, *and*

$$w(z) = \xi(z)e_1 + \eta(z)e_2,\; \xi(z) = \mathrm{Re}\psi(z) + \mathrm{Im}\psi(z),\; \eta(z) = \mathrm{Re}\psi(z) - \mathrm{Im}\psi(z),$$

$$\hat{A}_1 = \frac{a_1 + a_2 + b_1 + b_2}{2},\; \hat{B}_1 = \frac{a_1 + a_2 - b_1 - b_2}{2},\; \hat{C}_1 = c_1 + c_2,$$

$$\hat{A}_2 = \frac{a_2 - a_1 + b_2 - b_1}{2},\; \hat{B}_2 = \frac{a_2 - a_1 - b_2 + b_1}{2},\; \hat{C}_2 = c_2 - c_1,$$

in which we choose $H(y) = [|y|^m h(y)]^{1/2}, h(y)$ *is a positive continuously differentiable function and*

$$\begin{aligned} d\mu &= d[x + G(y)] = 2H(y)dy \text{ on } s_1, \\ d\nu &= d[x - G(y)] = -2H(y)dy \text{ on } s_2. \end{aligned} \tag{4.21}$$

Proof From Section 2, Chapter III, we see that equation (4.1) in $\overline{D^-}$ can be reduced to the system of integral equations

$$\begin{aligned} &w(z) = w_0(z) + \phi(z) + \psi(z) = \xi(z)e_1 + \theta(z)e_2, \\ &\xi(z) = \mathrm{Re}w_0(z) + \mathrm{Im}w_0(z) + \int_0^\mu \xi_\mu d\mu = \zeta(z) + \int_0^y [H\xi_x + \xi_y]dy \\ &= \zeta(z) + \int_0^y [\hat{A}_1(U+V) + \hat{B}_1(U-V) + \hat{C}_1]dy,\; z \in s_1, \\ &\eta(z) = \mathrm{Re}w_0(z) - \mathrm{Im}w_0(z) + \int_2^\nu \eta_\nu d\nu = \theta(z) - \int_0^y [H\eta_x - \eta_y]dy \\ &= \theta(z) + \int_0^y [\hat{A}_2(U+V) + \hat{B}_2(U-V) + \hat{C}_2]dy,\; z \in s_2. \end{aligned} \tag{4.22}$$

Noting (4.20) and

$$\begin{aligned} ds_1 &= \sqrt{(dx)^2 + (dy)^2} = -\sqrt{1 + (dx/dy)^2}dy = -\sqrt{1-K}dy = -\frac{\sqrt{1-K}}{\sqrt{-K}}dx, \\ ds_2 &= \sqrt{(dx)^2 + (dy)^2} = -\sqrt{1 + (dx/dy)^2}dy = -\sqrt{1-K}dy = \frac{\sqrt{1-K}}{\sqrt{-K}}dx, \end{aligned} \tag{4.23}$$

it is clear that the system (4.22) is just (4.19).

4.3 Solvability of general boundary value problem for degenerate mixed equations

In this section, we prove the existence of solutions of Problem B for the complex equation (4.1) with the boundary conditions (4.2)-(4.4). We can see that Problem B can be divided into two problems, i.e. Problem B_1: equation (4.1) in D^+ with the boundary conditions

$$\begin{aligned} &\mathrm{Re}[\overline{\lambda(z)}w(z)] = R(z), z \in \Gamma \cup L_0, \\ &\mathrm{Im}[\overline{\lambda(z)}w(z)]|_{z=a_0} = b_0, \end{aligned} \tag{4.24}$$

and Problem B_2: equation (4.1) in D^- with the boundary conditions

$$\begin{aligned} &\mathrm{Re}[\overline{\lambda(z)}w(z)] = R(z), z \in L_3 \cup L_4 \cup L_0, \\ &\mathrm{Im}[\overline{\lambda(z)}w(z)]|_{z=A_{2l+1}} = c_{2l+1}, l = 0, 1, ..., [n/2], \\ &\mathrm{Im}[\overline{\lambda(z)}w(z)]|_{z=B_{2l-1}} = c_{2l}, l = 1, ..., [(n+1)/2], \end{aligned} \tag{4.25}$$

where $\lambda(x) = 1 + i$ or $1 + j$ on $L_0 \cap D_1^-$ and $\lambda(z) = 1 - i$ or $1 - j$ on $L_0 \cap D_2^-$, and the function $R(x) = R_0(x)$ on L_0 can be determined by the solvability of Problem B_2. The solvability of Problem B_1 can be obtained by the result in Section 3, Chapter I, and the unique solvability of Problem B_2 will be proved as follows.

We first prove the existence of solutions of Problem B_2 for equation (4.1). Denote $D_0 = \{\delta_0 \le x \le 2 - \delta_0, -\delta \le y \le 0\}$, and s_1, s_2 are the characteristics of families (4.20) emanating from any two points $(b_0, 0), (b_1, 0) (\delta_0 \le b_0 < b_1 \le 2 - \delta_0)$ respectively, which intersect at a point $z = x + jy \in D$, $[b_0, b_1] \cap \{a_0, a_1, ..., a_{n+1}\} = \emptyset$, and δ_0, δ are sufficiently small positive numbers. We choose the solution $w_0(z)$ of Problem B_2 for equation (4.9), and substitute the corresponding functions $\xi_0 = \mathrm{Re}w_0 + \mathrm{Im}w_0, \eta_0 = \mathrm{Re}w_0 - \mathrm{Im}w_0$ into the positions of ξ, η in the right sides of (4.19), and by the successive approximation, we can find the sequences of functions $\{\xi_k\}$, $\{\eta_k\}$, which satisfy the relations

$$\begin{aligned} &\xi_{k+1}(z) = \zeta_{k+1}(x) + \int_0^y [\hat{A}_1\xi_k + \hat{B}_1\eta_k + \hat{C}_1]dy, \ z \in s_1, \\ &\eta_{k+1}(z) = \theta_{k+1}(x) + \int_0^y [\hat{A}_2\xi_k + \hat{B}_2\eta_k + \hat{C}_2]dy, \ z \in s_2, \end{aligned} \tag{4.26}$$

We may only discuss the case of $K(y) = -|y|^m h(y)$, because otherwise we can similarly discuss. It is clear that for two characteristics s_1, s_2 passing through a point $z=x+jy\in D$ and x_1,x_2 are the intersection points with the axis $y=0$ respectively, for any two points $\tilde z_1=\tilde x_1+j\tilde y\in s_1,\tilde z_2=\tilde x_2+j\tilde y\in s_2$, we have

$$|\tilde x_1-\tilde x_2|\le|x_1-x_2|=2|\int_0^y\sqrt{-K(t)}dt|\le M|y|^{m/2+1}, \tag{4.27}$$

in which $M(>\max[2\sqrt{h(y)},1])$ is a positive constant. From Condition C, we can assume that the coefficients of (4.19) are continuously differentiable with respect to $x\in L_0$ and satisfy the condition

$$|\hat A_l|,|\hat A_{lx}|,|\hat B_l|,|\hat B_{lx}|,|\hat C_l|,|\hat C_{lx}|,\left|\frac{1}{\sqrt{h}}\right|,\left|\frac{h_y}{h}\right|\le M, z\in\bar D, l=1,2. \tag{4.28}$$

According to the proof of Theorem 2.5, we can find a solution $w_*(z)=\xi_*(z)e_1+\eta_*(z)e_2$ of Problem B_2 in D^- satisfying the system of integral equations

$$\xi_*(z)=\zeta_*(z)+\int_0^y[\hat A_1\xi_*+\hat B_1\eta_*+\hat C_1]dy, z\in s_1,$$

$$\eta_*(z)=\theta_*(z)+\int_0^y[\hat A\xi_*+\hat B_2\eta_*+\hat C_2]dy,\ z\in s_2,$$

and the function $w_*(z)=e_1\xi_*(x)+e_2\eta_*(z)$ satisfies equation (4.1) and boundary condition (4.25), this shows that Problem B_2 has a solution for equation (4.1) in $\overline{D^-}\cap\{-\delta\le y\le 0\}$. Next apply the similar method in Section 1, the result in $\overline{D^-}\cap\{y<-\delta\}$ can be obtained, hence the function $w_*(z)$ in D^- is a solution of Problem B_2 for (4.1) in D^-. Thus the existence of solutions of Problem B for equation (4.1) is proved. In addition, we can verify that the solution of Problem B for (4.1) in D is unique.

From the above discussion, we have the following theorem.

Theorem 4.3 *Let equation* (4.1) *satisfies Condition C. Then the discontinuous Riemann-Hilbert problem* (*Problem B*) *for* (4.1) *has a unique solution.*

Finally we mention that the coefficients $K(y)$ in equation (3.1) can be replaced by functions $K(x,y)$ with some conditions, for instance $K(x,y)=\text{sgn}y|y|^m h(x,y)$, m is as stated before, and $h(x,y)$ is a continuously differentiable positive function in $\overline{D}$. In this case, the first order quasilinear degenerate system of mixed type corresponding to (3.1) is in the form

$$\begin{cases} H(x,y)u_x-\text{sgn}y\,v_y=a_1u+b_1v+c_1 \\ H(x,y)v_x+u_y=a_2u+b_2v+c_2 \end{cases} \text{in } D, \tag{4.29}$$

where $H(x,y)=\sqrt{|K(x,y)|}$, $a_l,b_l,c_l\,(l=1,2)$ are real functions of $(x,y)\,(\in D)$, $u,v\,(\in \mathbf{R})$, and the complex form of system (4.29) in D is as follows

$$w_{\overline{\tilde{z}}}=A_1(z,w)w+A_2(z,w)\bar{w}+A_3(z,w) \text{ in } D, \tag{4.30}$$

where the coefficients (4.30) are the same as those in (3.2), but

$$w_{\overline{\tilde{z}}}=[H(x,y)u+iv]/2 \text{ in } D^+,\ w_{\overline{\tilde{z}}}=[H(x,y)u+jv]/2 \text{ in } \overline{D^-}.$$

For the complex equation (4.30), under Condition C as stated in Section 3, we can discuss the solvability of Problem B by the similar method.

CHAPTER V

SECOND ORDER LINEAR EQUATIONS OF MIXED TYPE

In this chapter, we mainly discuss some boundary value problems including the Tricomi problem, oblique derivative problem, exterior Tricomi-Rassias problem and the Frankl problem for second order linear equations of mixed type with parabolic degeneracy, in which we establish the representation of solutions for the boundary value problems, and prove the uniqueness and existence of solutions for the above problems.

1 The Oblique Derivative Problem for Second Order Equations of Uniformly Mixed Type

In this section, we first introduce the uniqueness and existence of solutions of oblique derivative problem for second order linear equations of mixed type without parabolic degeneracy.

1.1 Formulation of oblique derivative problem for second order equations of mixed type

Let D be a simply connected bounded domain D in the complex plane $\mathbf{C}$ with the boundary $\partial D = \Gamma \cup L$ as stated in Section 1, Chapter IV. We consider the second order linear equation of mixed type

$$u_{xx} + \operatorname{sgn} y\, u_{yy} = au_x + bu_y + cu + d \text{ in } \overline{D}, \tag{1.1}$$

where a, b, c, d are real functions of $z\,(\in D)$, its complex form is as follows

$$u_{z\bar{z}} = \operatorname{Re}[A_1(z)u_z] + A_2(z)u + A_3(z) \text{ in } \overline{D}, \tag{1.2}$$

where

$$\begin{aligned} &z=x+iy, u_z = \frac{1}{2}[u_x - iu_y] = \overline{u_{\bar{z}}}, u_{z\bar{z}} = \frac{1}{4}[u_{xx} + u_{yy}] \text{ in } \overline{D^+},\\ &z=x+jy, u_z = \frac{1}{2}[u_x - ju_y] = \overline{u_{\bar{z}}}, u_{z\bar{z}} = \frac{1}{4}[u_{xx} - u_{yy}] \text{ in } \overline{D^-},\\ &A_1 = \frac{a+ib}{2} \text{ in } \overline{D^+}, A_1 = \frac{a-jb}{2} \text{ in } \overline{D^-}, A_2 = \frac{c}{4}, A_3 = \frac{d}{4} \text{ in } \overline{D}. \end{aligned} \tag{1.3}$$

Suppose that equation (1.2) satisfies the following conditions: **Condition** C

The coefficients $A_l(z)\,(l=1,2,3)$ in (1.2) are measurable in $z\in D^+$ and continuous in $\overline{D^-}$ and satisfy

$$L_p[A_l,\overline{D^+}]\le k_0,\ l=1,2,\ L_p[A_3,\overline{D^+}]\le k_1,\ A_2\ge 0 \text{ in } D^+, \tag{1.4}$$

$$\hat{C}[A_l,\overline{D^-}]=C[A_l,\overline{D^-}]+C[A_{lx},\overline{D^-}]\le k_0, l=1,2, \hat{C}[A_3,\overline{D^-}]\le k_1, \tag{1.5}$$

where $p\,(>2)$, k_0, k_1 are positive constants. If the condition (1.5) is replaced by

$$C_\alpha[A_l,\overline{D^\pm}]\le k_0,\ l=1,2,\ C_\alpha[A_3,\overline{D^\pm}]\le k_1, \tag{1.6}$$

in which $\alpha(0<\alpha<1)$ is a real constant, then thc conditions will be called **Condition** C'.

Problem P Find a continuously differentiable solution $u(z)$ of (1.2) in $D^*=\bar{D}\backslash\{0,2\}$, which is continuous in $\bar{D}$ and satisfies the boundary conditions

$$\frac{1}{2}\frac{\partial u}{\partial \nu}=\mathrm{Re}[\overline{\lambda(z)}u_z]=r(z),\ \ z\in\Gamma,\ \ u(0)=b_0,\ u(2)=b_2, \tag{1.7}$$

$$\frac{1}{2}\frac{\partial u}{\partial \nu}=\mathrm{Re}[\overline{\lambda(z)}u_z]=r(z), z\in L_l\,(l=1 \text{ or } 2), \mathrm{Im}[\overline{\lambda(z)}u_z]|_{z=z_0}=b_1, \tag{1.8}$$

where ν is a given vector at every point on $\Gamma\cup L_l\,(l=1$ or $2)$, $\lambda(z)=a(x)+ib(x)=\cos(\nu,x)-i\cos(\nu,y)$, if $z\in\Gamma$, and $\lambda(z)=a(z)+jb(z)=\cos(\nu,x)+j\cos(\nu,y)$, if $z\in L_l\,(l=1$ or $2)$, b_0,b_1,b_2 are real constants, and $\lambda(z)$, $r(z)$, b_0, b_1, b_2 satisfy the conditions

$$C_\alpha[\lambda(z),\Gamma\cup L_l]\le k_0,\ C_\alpha[r(z),\Gamma\cup L_l]\le k_2, l=1 \text{ or } 2, \cos(\nu,n)\ge 0 \ \text{ on } \Gamma,$$

$$|b_l|\le k_2, l=0,1,2,\ \max_{z\in L_1}\frac{1}{|a(z)-b(z)|}\le k_0 \text{ or } \max_{z\in L_2}\frac{1}{|a(z)+b(z)|}\le k_0, \tag{1.9}$$

in which n is the outward normal vector at every point on Γ, $\alpha\,(0<\alpha<1)$, k_0,k_2 are positive constants. For convenience, we may assume that $u_z(z_0)=0$, otherwise through a transformation of function, the requirement can be realized.

The boundary value problem for (1.2) with $A_3(z)=0$, $r(z)=0$, and $b_0=b_1=b_2=0$ will be called Problem P_0. The number

$$K=\frac{1}{2}(K_1+K_2) \tag{1.10}$$

is called the index of Problem P and Problem P_0, where

$$K_l=\left[\frac{\phi_l}{\pi}\right]+J_l,\, J_l=0 \text{ or } 1,\, e^{i\phi_l}=\frac{\lambda(t_l-0)}{\lambda(t_l+0)},\, \gamma_l=\frac{\phi_l}{\pi}-K_l,\, l=1,2, \tag{1.11}$$

in which $t_1=0$, $t_2=2$, $\lambda(t)=e^{i\pi/4}$ on $L_0=(0,2)$ on $x-$axis and $\lambda(t_1+0)=\lambda(t_2-0)=\exp(i\pi/4)$, or $\lambda(t)=e^{i7\pi/4}$ on L_0 and $\lambda(t_1+0)=\lambda(t_2-0)=\exp(i7\pi/4)$. Here we choose $K=0$, or $K=-1/2$ on the boundary ∂D^+ of D^+ if $\cos(\nu,n)\equiv 0$ on Γ and the condition $u(2)=b_2$ can be cancelled, because in this case from the boundary condition (1.7), we can determine the value $u(2)$ by the value $u(0)$, namely

$$u(2)=2\mathrm{Re}\int_0^2 u_z dz+u(0)=2\int_\pi^0 \mathrm{Re}[\overline{\lambda(z)}u_z]d\theta+b_0=2\int_\pi^0 r(z)d\theta+b_0, \tag{1.12}$$

in which $\overline{\lambda(z)}=i(z-1)$, $\theta=\arg(z-1)$ on $\Gamma=\{|z-1|=1, \mathrm{Im}z>0\}$. Problem P in this case still includes the Dirichlet problem as a special case. In brief, we choose that

$$K=\begin{cases}0,\\ -\dfrac{1}{2},\end{cases} \text{the point condition is} \begin{Bmatrix} u(0)=b_0,\, u(2)=b_2\\ u(0)=b_0\end{Bmatrix}, \quad \text{if} \begin{cases}\cos(\nu,n)\not\equiv 0\\ \cos(\nu,n)\equiv 0\end{cases} \text{on } \Gamma. \tag{1.13}$$

Later on we shall only discuss the case: $K=0$, and the case: $K=-1/2$ can be similarly discussed.

Setting that $u_z=w(z)$, it is clear that Problem P for (1.2) is equivalent to the Riemann-Hilbert boundary value problem (Problem A) for the first order complex equation of mixed type

$$w_{\bar z}=F,\ F=\mathrm{Re}[A_1(z)w]+A_2(z)u+A_3(z) \text{ in } D \tag{1.14}$$

with the boundary conditions

$$\begin{gathered}\mathrm{Re}[\overline{\lambda(z)}w(z)]=r(z),\ z\in\Gamma,\ u(2)=b_2,\\ \mathrm{Re}[\overline{\lambda(z)}w(z)]=r(z), z\in L_l\, (l=1 \text{ or } 2), \mathrm{Im}[\overline{\lambda(z_0)}w(z_0)]=b_1,\end{gathered} \tag{1.15}$$

and the relation

$$u(z)=2\mathrm{Re}\int_0^z \hat w(z)dz+b_0=\begin{cases}2\mathrm{Re}\displaystyle\int_0^z w(z)dz+b_0 \text{ in } \overline{D^+},\\ 2\mathrm{Re}\displaystyle\int_0^z \overline{w(z)}dz+b_0 \text{ in } \overline{D^-},\end{cases} \tag{1.16}$$

here and later on we denote $\hat{w}(z)=w(z)$ in $\overline{D^+}$ and $\hat{w}(z)=\overline{w(z)}$ in $\overline{D^-}$. On the basis of the result in Section 1, Chapter IV, we can find a solution $w(z)$ of Problem A for the mixed complex equation (1.14) as stated in (1.15), Chapter IV.

1.2 Representation and uniqueness of solutions for oblique derivative problem

Now we give the representation theorems of solutions for equation (1.2).

Theorem 1.1 *Let equation (1.2) satisfy Condition C in D^+, $u(z)$ be a continuous solution of (1.2) in $\overline{D^+}$ and continuously differentiable in $D^+_*=\overline{D^+}\backslash\{0,2\}$. Then $u(z)$ can be expressed as*

$$\begin{aligned}&u(z)=U(z)\Psi(z)+\psi(z) \text{ in } D^+,\\ &U(z)=2\mathrm{Re}\int_0^z w(z)dz+b_0,\ w(z)=\Phi(z)e^{\phi(z)} \text{ in } D^+,\end{aligned}\tag{1.17}$$

where $\psi(z)$, $\Psi(z)$ are the solutions of equation (1.2) in D^+ and

$$u_{z\bar z}-\mathrm{Re}[A_1u_z]-A_2u=0 \text{ in } D^+ \tag{1.18}$$

respectively and satisfy the boundary conditions

$$\psi(z)=0,\ \ \Psi(z)=1 \text{ on } \Gamma\cup L_0, \tag{1.19}$$

where $\psi(z)$, $\Psi(z)$ satisfies the estimates

$$\begin{aligned}&C^1_\delta[\psi,\overline{D^+}]\le M_1,\ \|\psi\|_{W^2_{p_0}(D^+)}\le M_2,\\ &C^1_\delta[\Psi,\overline{D^+}]\le M_3,\ \|\Psi\|_{W^2_{p_0}(D^+)}\le M_3,\ \Psi(z)\ge M_4>0,\ z\in\overline{D^+},\end{aligned}\tag{1.20}$$

in which $\delta\,(0<\delta\le\alpha)$, $p_0\,(2<p_0\le p)$, $M_l=M_l(p_0,\delta,k,D)\,(l=1,2,3,4)$ are positive constants, $k=(k_0,k_1,k_2)$. Moreover $U(z)$ is a solution of the equation

$$U_{z\bar z}-\mathrm{Re}[AU_z]=0,\ A=-2(\ln\Psi)_{\bar z}+A_1 \text{ in } D^+, \tag{1.21}$$

where $\mathrm{Im}[\phi(z)]=0$, $z\in L_0=(0,2)$ and $\phi(z)$ satisfies the estimate

$$C_\delta[\phi,\overline{D^+}]+L_{p_0}[\phi_{\bar z},\overline{D^+}]\le M_5, \tag{1.22}$$

in which $\delta(0 < \delta \le \alpha)$, $M_5 = M_5(p_0, \delta, k_0, D)$ *are two positive constants,* $\Phi(z)$ *is analytic in* D^+. *If* $u(z)$ *is a solution of* (1.2) *in* D^+ *satisfying the boundary conditions* (1.7) *and*

$$\mathrm{Re}[\overline{\lambda(z)}u_z]|_{z=x} = s(x), \lambda(x) = 1+i \text{ or } 1-i, \ x \in L_0, C_\delta[s(x), L_0] \le k_3, \quad (1.23)$$

then the following estimate holds:

$$C_\delta[u(z), \overline{D^+}] + C_\delta[X(z)u_z, \overline{D^+}] \le M_6(k_1 + k_2 + k_3), \quad (1.24)$$

in which k_3 *is a positive constant,* $s(x)$ *can be seen as stated in the form* (1.33) *below, and*

$$X(z) = \prod_{l=1}^{2} |z - t_l|^{\eta_l}, \eta_l = \max[-2\gamma_l, 0] + 8\delta, \ l = 1, 2, \quad (1.25)$$

here $\gamma_l\ (l = 1, 2)$ *are real constants as stated in* (1.11) *and* δ *is a sufficiently small positive constant, and* $M_6 = M_6(p_0, \delta, k_0, D^+)$ *is a positive constant.*

Proof According to the method in the proof of Theorem 1.1, Chapter II, the equations (1.2), (1.18) in D^+ have the solutions $\psi(z)$, $\Psi(z)$ respectively, which satisfy the boundary condition (1.19) and the estimate (1.20). Setting that

$$U(z) = [u(z) - \psi(z)]/\Psi(z),$$

it is clear that $U(z)$ is a solution of equation (1.21) and $w(z)$ can be expressed the second formula in (1.17), where $\phi(z)$ satisfies the estimate as in (1.22) and $\Phi(z)$ is an analytic function in D^+. If $s(x)$ in (1.23) is a known function, then the boundary value problem (1.2), (1.7), (1.23) has a unique solution $u(z)$ as stated in the form (1.17), which satisfies the estimate (1.24).

Theorem 1.2 *Suppose that the equation* (1.2) *satisfies Condition* C. *Then any solution of Problem* P *for* (1.2) *can be expressed as*

$$u(z) = 2\mathrm{Re}\int_0^z \hat{w}(z)dz + b_0, \ w(z) = w_0(z) + W(z), \quad (1.26)$$

where $\hat{w}(z)$ *is as stated in* (1.16), *and* $w_0(z)$ *is a solution of Problem* A *for the complex equation*

$$w_{\bar{z}} = 0 \text{ in } D \quad (1.27)$$

with the boundary conditions (1.7), (1.8) ($w_0(z) = u_{0z}$), *and* $W(z)$ *possesses the form*

$$\begin{aligned}
&W(z) = w(z) - w_0(z) \text{ in } D,\ w(z) = \tilde{\Phi}(z)e^{\tilde{\phi}(z)} + \tilde{\psi}(z) \text{ in } D^+,\\
&\tilde{\phi}(z)=\tilde{\phi}_0(z)+Tg=\tilde{\phi}_0(z)-\frac{1}{\pi}\int\!\!\int_{D^+}\frac{g(\zeta)}{\zeta-z}d\sigma_\zeta,\ \tilde{\psi}(z)=Tf \text{ in } D^+,\\
&W(z)=\Phi(z)+\Psi(z),\ \Psi(z)=\int_0^\mu g^1(z)d\mu e_1+\int_2^\nu g^2(z)d\nu e_2 \text{ in } D^-,
\end{aligned} \tag{1.28}$$

in which $e_1 = (1+j)/2$, $e_2 = (1-j)/2$, $\mu = x+y$, $\nu = x-y$, $\tilde{\phi}_0(z)$ *is an analytic function in* D^+, *such that* $\mathrm{Im}[\tilde{\phi}(x)] = 0$ *on* L_0, *and*

$$\begin{aligned}
&g(z)=\begin{cases}\dfrac{A_1}{2}+\dfrac{\overline{A_1\tilde{W}}}{2\tilde{W}},\ \tilde{W}(z)\neq 0,\\ 0,\ \tilde{W}(z)=0, z\in D^+,\end{cases}\quad f(z)=\mathrm{Re}[A_1\tilde{\phi}_z]+A_2u+A_3,\\
&g^1(z)=g^2(z)=A\xi+B\eta+Cu+D, \xi=\mathrm{Re}w+\mathrm{Im}w, \eta=\mathrm{Re}w-\mathrm{Im}w,\\
&A=\frac{\mathrm{Re}A_1+\mathrm{Im}A_1}{2}, B=\frac{\mathrm{Re}A_1-\mathrm{Im}A_1}{2}, C=A_2, D=A_3 \text{ in } D^-,
\end{aligned} \tag{1.29}$$

where $\tilde{W}(z) = w(z) - \tilde{\psi}(z)$, $\tilde{\Phi}(z)$ *and* $\Phi(z)$ *are the solutions of equation* (1.27) *in* D^+ *and* D^- *respectively satisfying the boundary conditions*

$$\begin{aligned}
&\mathrm{Re}[\overline{\lambda(z)}(\tilde{\Phi}(z)e^{\tilde{\phi}(z)} + \tilde{\psi}(z))] = r(z),\ z\in\Gamma,\\
&\mathrm{Re}[\overline{\lambda(x)}(\tilde{\Phi}(x)e^{\tilde{\phi}(x)} + \tilde{\psi}(x))] = s(x),\ x\in L_0,\\
&\mathrm{Re}[\overline{\lambda(x)}\Phi(x)]=\mathrm{Re}[\overline{\lambda(x)}(W(x)-\Psi(x))], z\in L_0,\\
&\mathrm{Re}[\overline{\lambda(z)}\Phi(z)] = -\mathrm{Re}[\overline{\lambda(z)}\Psi(z)],\ z\in L_1 \text{ or } L_2,\\
&\mathrm{Im}[\overline{\lambda(z_0)}\Phi(z_0)] = -\mathrm{Im}[\overline{\lambda(z_0)}\Psi(z_0)],
\end{aligned} \tag{1.30}$$

where $\lambda(x) = 1+i$ or $1-i$, $x\in L_0$ *in second formulas of* (1.30) *and* $\lambda(x) = 1+j$ or $1-j$, $x \in L_0$ *in third formula of* (1.30). *Moreover by Theorem* 1.2, *Chapter* IV, *the solution* $w_0(z)$ *of Problem A for* (1.27) *and* $u_0(z)$ *satisfy the estimate in the form*

$$C_\delta[u_0(z),\overline{D}]+C_\delta[X(z)w_0(z),\overline{D^+}]+C_\delta[Y^\pm(z)w_0^\pm(z),\overline{D^-}]\le M_7(k_1+k_2), \tag{1.31}$$

where $w^\pm(z) = \mathrm{Re}w(z) \pm \mathrm{Im}w(z)$, $X(z)$, $Y^\pm(z)$ *are as stated in* (1.21), *Chapter* IV, *and*

$$u_0(z) = 2\mathrm{Re}\int_0^z \hat{w}_0(z)dz + b_0, \tag{1.32}$$

in which $\hat{w}_0(z) = w_0(z)$ *in* D^+ *and* $\hat{w}_0(z) = \overline{w_0(z)}$ *in* $\overline{D^-}$, *and* $M_7 = M_7(p_0, \delta, k_0, D)$ *is a positive constant. From* (1.32), *it follows that*

$$C_\delta[u_0(z), \bar{D}] \leq M_8\{C_\delta[X(z)w_0(z), \overline{D^+}] + C_\delta[Y^\pm(z)w_0^\pm(z), \overline{D^-}]\} + k_2,$$

where $M_8 = M_8(D)$ *is a positive constant.*

Proof Let $u(z)$ be a solution of Problem P for equation (1.2), and $w(z) = u_z$, $u(z)$ be substituted in the positions of w, u in (1.29), thus the functions $g(z)$, $f(z)$, $g^1(z)$, $g^2(z)$, and $\tilde{\psi}(z), \tilde{\phi}(z)$ in $\overline{D^+}$ and $\Psi(z)$ in $\overline{D^-}$ in (1.28), (1.29) can be determined. Moreover we can find the solution $\tilde{\Phi}(z)$ in D^+ and $\Phi(z)$ in $\overline{D^-}$ of (1.27) with the boundary conditions (1.30), where

$$s(x) = \begin{cases} \dfrac{2r(x/2) - 2R(x/2)}{a(x/2) - b(x/2)} + \text{Re}[(1-i)\Psi(x))] \text{ or} \\ \dfrac{2r(x/2+1) - 2R(x/2+1)}{a(x/2+1) + b(x/2+1)} + \text{Re}[(1+i)\Psi(x)] \text{ on } L_0, \end{cases} \tag{1.33}$$

here and later on $R(z) = \text{Re}[\overline{\lambda(z)}\Psi(z)]$ on L_1 or L_2, and thus

$$w(z) = w_0(z) + W(z) = \begin{cases} \tilde{\Phi}(z)^{\tilde{\phi}(z)} + \tilde{\psi}(z) \text{ in } D^+, \\ w_0(z) + \Phi(z) + \Psi(z) \text{ in } D^-, \end{cases}$$

is the solution of Problem A for the complex equation

$$w_{\bar{z}} = \text{Re}[A_1 w] + A_2 u + A_3 \text{ in } D, \tag{1.34}$$

which can be expressed as in (1.28), and $u(z)$ is a solution of Problem P for (1.2) as stated in the formula in (1.26).

Theorem 1.3 *If equation* (1.2) *satisfies Condition* C, *then Problem* P *for* (1.2) *has at most one solution in* D.

Proof Let $u_1(z), u_2(z)$ be any two solutions of Problem P for (1.2). By Condition C, we see that $u(z) = u_1(z) - u_2(z)$ and $w(z) = u_z$ satisfies the homogeneous equation and boundary condition

$$w_{\bar{z}} = \text{Re}[A_1 w] + A_2 u \text{ in } D, \tag{1.35}$$

$$\begin{gathered} \text{Re}[\overline{\lambda(z)}w(z)] = 0, \ z \in \Gamma, \ u(0) = 0, \ u(2) = 0, \\ \text{Re}[\overline{\lambda(z)}w(z)] = 0, \ z \in L_l (l = 1 \text{ or } 2), \ \text{Im}[\overline{\lambda(z_0)}w(z_0)] = 0. \end{gathered} \tag{1.36}$$

From Theorem 1.2, the solution $w(z)$ can be expressed in the form

$$w(z)=\begin{cases}\tilde{\Phi}(z)e^{\tilde{\phi}(z)}+\tilde{\psi}(z),\tilde{\psi}(z)=Tf,\tilde{\phi}(z)=\tilde{\phi}_0(z)+\tilde{T}g \text{ in } D^+,\\ w_0(z)+\Phi(z)+\Psi(z) \text{ in } D^-,\end{cases}$$
$$\Psi(z)=\int_0^\mu[A\xi+B\eta+Cu]e_1d\mu+\int_2^\nu[A\xi+B\eta+Cu]e_2d\nu \text{ in } D^-, \tag{1.37}$$

where $g(z)$ is as stated in (1.29), $\tilde{\Phi}(z)$ in D^+ is an analytic function and $\Phi(z)$ is a solution of (1.27) in $\overline{D^-}$ satisfying the boundary condition (1.30), $\tilde{\phi}(z)$, $\tilde{\psi}(z)$ possess the similar properties as $\phi(z)$, $\psi_z(z)$ in Theorem 1.1. If $A_2=0$ in D^+, then $\tilde{\psi}(z)=0$. Besides the functions $\tilde{\Phi}(z)$, $\Phi(z)$ satisfy the boundary conditions

$$\begin{cases}\operatorname{Re}[\overline{\lambda(x)}\tilde{\Phi}(x)]=s(x),\\ \operatorname{Re}[\overline{\lambda(x)}\Phi(x)]=\operatorname{Re}[\overline{\lambda(x)}(W(x)-\Psi(x))]\end{cases} \text{ on } L_0, \tag{1.38}$$

where $s(x)$ is as stated in (1.33), but $r(x)=0$. From (1.17) with $b_0=0$, we can obtain

$$\begin{aligned}C[u(z),\bar{D}]\le M_8\{C[X(z)w(z),\overline{D^+}]+\\ +C[Y^+(z)w^+(z),\overline{D^-}]+C[Y^-(z)w^-(z),\overline{D^-}]\}.\end{aligned} \tag{1.39}$$

Using the method of successive approximation, and noting that

$$C[u(z),\overline{D^-}]\le M_0\{C[Y^+(z)w^+(z),\overline{D^-}]+C[Y^-(z)w^-(z),\overline{D^-}]\}R',$$

where $M_0=M_0(D)$ is a positive constant, the estimate

$$C[Y^\pm(z)w^\pm(z),\overline{D^-}]\le\frac{2M_9M(4+M_0R')mR']^n}{n!} \tag{1.40}$$

can be derived, where $M_9=\max\{C[A,\overline{D^-}],C[B,\overline{D^-}],C[C,\overline{D^-}]\}$, $M=1+4k_0^2(1+2k_0^2)$, and $m=C[Y^+(z)w^+(z),\overline{D^-}]+C[Y^-(z)w^-(z),\overline{D^-}]$. Let $n\to\infty$, from (1.28) and (1.29), it follows that $\Psi(z)=0,\Phi(z)=0$, $z\in\overline{D^-}$, and then $w(z)=0$ in $\overline{D^-}$. Thus the solution $u(z)=2\operatorname{Re}\int_0^z\hat{w}(z)dz$ is the solution of equation (1.18) with the boundary conditions

$$\operatorname{Re}[\overline{\lambda(z)}u_z(z)]=0 \text{ on } \Gamma, \operatorname{Re}[\overline{\lambda(x)}u_z(x)]=0 \text{ on } L_0, u(0)=0, u(2)=0, \tag{1.41}$$

in which $\lambda(x)=1+i$, or $1-i$, $x\in L_0$. Similarly to the proof of Theorem 1.4, Chapter II, we can obtain $u(z)=0$ on $\overline{D^+}$. This shows the uniqueness of solutions of Problem P for (1.2).

1.3 The solvability of oblique derivative problem for second order equations of mixed type

Theorem 1.4 *Suppose that the mixed equation* (1.2) *satisfies Condition C. Then Problem P for* (1.2) *has a solution in D.*

Proof It is clear that Problem P for (1.2) is equivalent to Problem A for the complex equation of first order and boundary conditions:

$$w_{\bar{z}} = F, \quad F = \mathrm{Re}[A_1 w] + A_2 u + A_3 \ \text{ in } D, \tag{1.42}$$

$$\begin{gathered}\mathrm{Re}[\overline{\lambda(z)}w(z)] = r(z), \quad z \in \Gamma,\\ \mathrm{Re}[\overline{\lambda(z)}w(z)]=r(z), z\in L_l(l=1 \text{ or } 2), \mathrm{Im}[\overline{\lambda(z_0)}w(z_0)]=b_1,\end{gathered} \tag{1.43}$$

and the relation (1.16). From (1.16), it follows that

$$C[u(z), \bar{D}] \leq M_8[C(X(z)w(z), \overline{D^+}) + C(Y^{\pm}(z)w^{\pm}(z), \overline{D^-})] + k_2, \tag{1.44}$$

where $X(z), Y^{\pm}(z), w^{\pm}(z)$ are as stated in (1.31), $M_8 = M_8(D)$ is a positive constant. In the following, by using successive approximation, we shall find a solution of Problem A for the complex equation (1.42) in D. Firstly denoting the solution $w_0(z)\,(= \xi_0 e_1 + \eta_0 e_2)$ of Problem A for (1.27) and $u_0(z)$ in (1.32), and substituting them into the position of $w\,(= \xi e_1 + \eta e_2)$, $u(z)$ in the right-hand side of (1.42), similarly to the proof of Theorem 1.3, Chapter IV, we have the corresponding functions $f_1(z), g_1(z), g_2^1(z), g_2^1(z)$ and

$$\begin{aligned}&w_1(z)=\tilde{\Phi}_1(z)e^{\tilde{\phi}_1(z)}+\tilde{\psi}_1(z), W_1(z)=w_1(z)-w_0(z) \ \text{ in } D^+,\\ &\tilde{\phi}_1(z)=\tilde{\phi}_0(z)+Tg_1=\tilde{\phi}_0(z)-\frac{1}{\pi}\iint_{D^+}\frac{g_1(\zeta)}{\zeta-z}d\sigma_\zeta, \tilde{\psi}_1(z)=Tf_1 \text{ in } D^+,\\ &W_1(z)=\Phi(z)+\Psi(z), \Psi(z)=\int_0^\mu g_1^1(z)d\mu e_1+\int_2^\nu g_1^2(z)d\nu e_2 \text{ in } D^-,\end{aligned} \tag{1.45}$$

where $\mu = x + y$, $\nu = x - y$, where $\Phi_1(z)$ is a solution of (1.27) in D^- satisfying the boundary conditions

$$\begin{aligned}&\mathrm{Re}[\overline{\lambda(x)}\Phi_1(x)] = \mathrm{Re}[\overline{\lambda(x)}(W_1(z) - \Psi_1(x))], \ z \in L_0,\\ &\mathrm{Re}[\overline{\lambda(z)}\Phi_1(z)] = -\mathrm{Re}[\overline{\lambda(z)}\Psi_1(z)], \ z \in L_1 \text{ or } L_2,\\ &\mathrm{Im}[\overline{\lambda(z_0)}\Phi_1(z_0)] = -\mathrm{Im}[\overline{\lambda(z_0)}\Psi_1(z_0)],\end{aligned} \tag{1.46}$$

and

$$w_1(z) = w_0(z) + W_1(z) = w_0(z) + \Phi_1(z) + \Psi_1(z) \ \text{ in } \overline{D^-} \tag{1.47}$$

satisfies the estimate

$$C_\delta[X(z)w_1(z),\overline{D^+}]+C[Y^\pm(z)w_1^\pm(z),\overline{D^-}]\le M_{10}=M_{10}(p_0,\delta,k,D^-). \quad (1.48)$$

Furthermore we substitute $w_1(z)=w_0(z)+W_1(z)$ and corresponding functions $w_1(z)$, $\xi_1(z)=w^+(z)=\text{Re}w_1(z)+\text{Im}w_1(z)$, $\eta_1(z)=w^-(z)=\text{Re}w_1(z)-\text{Im}w_1(z)$, $u_1(z)$ into the positions $w(z)$, $\xi(z)$, $\eta(z)$, $u(z)$ in (1.28), (1.29), and similarly to (1.45)–(1.47), we can find the corresponding functions $\tilde\psi_2(z),\tilde\phi_2(z),\tilde\Phi_2(z)$ in $\overline{D^+}$, $\Psi_2(z),\Phi_2(z)$, $W_2(z)=\Phi_2(z)+\Psi_2(z)$ in $\overline{D^-}$, and the function

$$\begin{aligned}&w_2(z)=\tilde\Phi_2(z)e^{\tilde\phi_2(z)}+\tilde\psi_2(z) \text{ in } D^+,\\&w_2(z)=w_0(z)+W_2(z)=w_0(z)+\Phi_2(z)+\Psi_2(z) \text{ in } \overline{D^-}\end{aligned} \quad (1.49)$$

satisfies the similar estimate in the form (1.48). Thus there exists a sequence of functions: $\{w_n(z)\}$, i.e.

$$\begin{aligned}&w_n(z)=\tilde\Phi_n(z)e^{\tilde\phi_n(z)}+\tilde\psi_n(z) \text{ in } D^+,\\&w_n(z)=w_0(z)+W_n(z)=w_0(z)+\Phi_n(z)+\Psi_n(z),\\&\Psi_n(z)=\int_0^\mu g_n^1(z)e_1d\mu+\int_2^\nu g_n^2(z)e_2d\nu \text{ in } \overline{D^-},\end{aligned} \quad (1.50)$$

and then

$$\begin{aligned}&|Y^\pm(z)[w_1^-(z)-w_0^-(z)]|\le|Y^\pm(z)\Phi_1^\pm(z)|\\&+\sqrt2[|Y^+(z)\int_0^\mu[A\xi_0+B\eta_0+Cu_0+D]e_1d\mu|\\&+|Y^-(z)\int_2^\nu[A\xi_0+B\eta_0+Cu_0+D]e_2d\nu|]\\&\le 2M_{11}M(4m+M_0mR'+1)R' \text{ in } \overline{D^-},\end{aligned} \quad (1.51)$$

where $R'=2$, $M=1+4k_0^2(1+2k_0^2)$, $M_{11}=\max_{z\in\overline{D^-}}(|A|,|B|,|C|,|D|)$, and $m=C[Y^+(z)w_0^+(z),\overline{D^-}]+C[Y^-(z)w_0^-(z),\overline{D^-}]$. It is clear that $w_n(z)-w_{n-1}(z)$ satisfies

$$\begin{aligned}&w_n^-(z)-w_{n-1}^-(z)=\Phi_n(z)-\Phi_{n-1}(z)\\&+\int_0^\mu[A(\xi_n-\xi_{n-1})+B(\eta_n-\eta_{n-1})+C(u_n-u_{n-1})]e_1d\mu\\&+\int_2^\nu[A(\xi_n-\xi_{n-1})+B(\eta_n-\eta_{n-1})+C(u_n-u_{n-1})]e_2d\nu \text{ in } \overline{D^-},\end{aligned} \quad (1.52)$$

in which $n = 1, 2, \ldots$. From the above equality, and

$$C[u_n(z), \overline{D^-}] \le M_0\{C[Y^+(z)w_n^+(z), \overline{D^-}] + C[Y^-(z)w_n^-(z), \overline{D^-}]\}R' + k_2.$$

where $M_0 = M_0(D)$ is a positive constant, we can obtain the estimate

$$\begin{aligned} &|Y^\pm(z)[w_n^\pm - w_{n-1}^\pm]| \le [2M_{11}M(4m + M_0 mR' + 1)]^n \\ &\times \int_0^{R'} \frac{R'^{n-1}}{(n-1)!} dR' \le \frac{[2M_{11}M(4m + M_0 mR' + 1)R']^n}{n!} \text{ in } \overline{D^-}, \end{aligned} \tag{1.53}$$

we can see that the sequence of functions: $\{Y^\pm(z)w_n^\pm(z)\}$, i.e.

$$\begin{aligned} Y^\pm(z)w_n^\pm(z) &= Y^\pm(z)\{w_0^\pm(z) + [w_1^\pm(z) - w_0^\pm(z)] \\ &+ \cdots + [w_n^\pm(z) - w_{n-1}^\pm(z)]\} \ (n = 1, 2 \ldots) \end{aligned} \tag{1.54}$$

in $\overline{D^-}$ uniformly converge to $Y^\pm(z)w_*^\pm(z)$, and $w_*(z) = [w_*^+(z) + w_*^-(z) - i(w_*^+(z) - w_*^-(z))]/2$ satisfies the equality

$$\begin{aligned} &w_*(z) = w_0(z) + \Phi_*(z) + \Psi_*(z), \\ &\Psi_*(z) = \int_0^\mu [A\xi_* + B\eta_* + Cu_* + D]e_1 d\mu \\ &+ \int_2^\nu [A\xi_* + B\eta_* + Cu_* + D]e_2 d\nu \text{ in } \overline{D^-}, \end{aligned} \tag{1.55}$$

thus the corresponding function $u_*(z)$ is just a solution of Problem P for equation (1.2) in the domain D^-, and $w_*(z)$ satisfies the estimate

$$C[Y^\pm(z)w_*^\pm(z), \overline{D^-}] \le e^{2M_{11}M(4m + M_0 mR' + k_2 + 1)R'}. \tag{1.56}$$

In the meantime we can obtain the estimate

$$C_\delta[X(z)w_n(z), \overline{D^+}] \le M_{12} = M_{12}(p_0, \delta, k, D), \tag{1.57}$$

hence from the sequence $\{X(z)w_n(z)\}$, we can choose a subsequence, which uniformly converges to $X(z)w_*(z)$ in $\overline{D^+}$, and $w_*(z)$ satisfies the same estimate (1.57). Combining (1.56) and (1.57), it is obvious that the solution $w_*(z) = u_z$ of Problem A for (1.2) in $\bar{D}$ satisfies the estimate

$$C_\delta[X(z)w_*(z), \overline{D^+}] + C[Y^\pm(z)w_*^\pm(z), \overline{D^-}] \le M_{13} = M_{13}(p_0, \delta, k, D),$$

where M_{13} is a positive constant. Moreover the function $u(z)$ in (1.17) is a solution of Problem P for (1.2), where $w(z) = w^*(z)$.

From Theorems 1.3 and 1.4, we see that under Condition C, Problem A for equation (1.42) has a unique solution $w(z)$, which can be found by using successive approximation and the corresponding solution $u(z)$ of Problem P satisfies the estimates

$$\begin{aligned} &C_\delta[u(z),\overline{D^+}]+C_\delta[X(z)u_z,\overline{D^+}]\leq M_{14},\\ &C[u(z),\bar{D}]+C[Y^\pm(z)u_z^\pm,\bar{D}]\leq M_{15}, \end{aligned} \tag{1.58}$$

where $X(z), Y^\pm(z)$ is as stated in (1.31), and $M_l = M_l(p_0,\delta,k,D)\,(l = 14,15)$ are positive constants, $k=(k_0,k_1,k_2)$. Moreover we can derive the following theorem.

Theorem 1.5 *Suppose that equation* (1.2) *satisfies Condition* C. *Then any solution* $u(z)$ *of Problem* P *for* (1.2) *satisfies the estimates*

$$\begin{aligned} &C_\delta[u(z),\overline{D^+}]+C_\delta[X(z)u_z,\overline{D^+}]\leq M_{16}(k_1+k_2),\\ &C[u(z),\overline{D^-}]+C[Y^\pm(z)u_z^\pm,\overline{D^-}]\leq M_{17}(k_1+k_2), \end{aligned} \tag{1.59}$$

in which $M_l=M_l(p_0,\delta,k_0,D)\,(l=16,17)$ *are positive constants.*

From the estimates (1.58), (1.59), we can see that the regularity of solutions of Problem P for (1.2) (see [86]17),33)).

2 The Tricomi problem for Second Order Degenerate Equations of Mixed Type

In [12],[48],[71],[72],[74],[76] and so on, the authors posed and discussed the Tricomi problem of some second order equations of mixed type with parabolic degeneracy, which possesses important applications to gas dynamics. The section deals with the Tricomi problem for general mixed equations with parabolic degeneracy, which include the Tricomi problem for the Chaplygin equation as a special case. Firstly the formulation of the problem for the equations is given, next the representations and estimates of solutions for the above problem are obtained, finally the existence of solutions for the problem is proved by the successive approximation and the method of parameter extension. By using the similar method, we can discuss the oblique derivative problem, which is a special case of the result in next section.

2.1 The Tricomi problem of second order degenerate equations of mixed type

We first introduce the general second order linear equations of mixed type

$$Au_{xx} + 2Bu_{xy} + Cu_{yy} + Du_x + Eu_y + Fu + G = 0 \text{ in } D,$$

where A, B, C are sufficiently smooth in a bounded domain D including $L_0 = (0, 2)$ on $x-$axis, and

$$\Delta(z) = AC - B^2 = 0 \text{ on } L_0 = (0, 2),$$

$$\Delta(z)<0 \text{ in } D^+=D\cap\{y>0\}, \Delta(z)>0 \text{ in } D^-=D\cap\{y<0\}.$$

It is clear that the above equation is elliptic in D^+ and is hyperbolic in D^-, and L_0 is the parabolic degenerate line. On the basis of the theory of M. Cibrario, there exists a non-singular transformation of the independent variables $\xi = \xi(x, y)$, $\eta = \eta(x, y)$, the above general equation can be reduced to the form

$$\eta^m K(\xi, \eta)u_{\xi\xi} + u_{\eta\eta} + au_\xi + bu_\eta + cu + d = 0,$$

where m is a positive number (see [12]1)). Hence we mainly discuss the above standard equation of mixed type later on.

Let D be a simply connected bounded domain in the complex plane $\mathbf{C}$ with the boundary $\partial D = \Gamma \cup L$, where $\Gamma(\subset \{y > 0\}) \in C^2_\mu\,(0 < \mu < 1)$ is a curve with the end points $z = 0, 2$. Denote $D^+ = D \cap \{y > 0\}$, $D^- = D \cap \{y < 0\}$. There is no harm in assuming that the boundary Γ of the domain D^+ is a smooth curve with the form $x - \tilde{G}(y) = 0$ and $x + \tilde{G}(y) = 2$ including the line segments of $\text{Re}z = 0, 2$ near the points $z = 0$ and 2 respectively, and $L = L_1 \cup L_2$, where $\tilde{G}(y)$ is the same as stated in Section 2, Chapter II, and

$$L_1 = \{x + \int_0^y \sqrt{-K(t)}dt = 0,\ x \in [0, 1]\},$$

$$L_2 = \{x - \int_0^y \sqrt{-K(t)}dt = 2,\ x \in [1, 2]\},$$

where $K(y) = \text{sgn}y|y|^m h(y)$, m is a positive number, $h(y)$ is a continuously differentiable positive function in $\overline{D}$, and $z_0 = x_0 + jy_0 = 1 + jy_0$ is the intersection point of L_1 and L_2. Consider second order linear equation of mixed type with parabolic degeneracy

$$Lu=K(y)u_{xx}+u_{yy}+a(x,y)u_x+b(x,y)u_y+c(x,y)u=-d(x,y) \text{ in } D. \quad (2.1)$$

Denote $H(y)=\sqrt{|K(y)|}$, and suppose that the coefficients of (2.1) satisfy **Condition** C, namely a,b,c,d are measurable in D^+ and continuous in D^-, and satisfy

$$\begin{gathered}L_\infty[\eta,\overline{D^+}]\le k_0,\ \eta=a,b,c,\ \ L_\infty[d,\overline{D^+}]\le k_1,\ \ c\le 0\ \text{in}\ D^+\\ \hat{C}[d,\overline{D^-}]=C[d,\overline{D^-}]+C[d_x,\overline{D^-}]\le k_1,\hat{C}[\eta,\overline{D^-}]\le k_0,\eta=a,b,c,\end{gathered}\tag{2.2}$$

in the above conditions, the solution of equation (2.1) in D is a generalized solution of (2.1). If the above conditions are replaced by

$$\begin{gathered}L_\alpha[\eta,\overline{D}]\le k_0,\eta=a,b,c,\ L_\alpha[d,\overline{D}]\le k_1,\ c\le 0\ \text{in}\ \overline{D^+},\\ \hat{C}_\alpha[\eta,D^-]=C_\alpha[\eta,D^-]+C_\alpha[\eta_x,D^-]\le k_0,\eta=a,b,c,\hat{C}_\alpha[d,D^-]\le k_1,\end{gathered}\tag{2.3}$$

where $\alpha(0<\alpha<1),k_0(\ge\max[2\sqrt{h(y)},1/\sqrt{h(y)}]),\ k_1(>6\max[1,k_0])$ are positive constants, then the solution of equation (2.1) in D is a classical solution. If $H(y)=[|y|^m h(y)]^{1/2}$, here m is a positive number, $h(y)$ is a continuously differentiable positive function, then

$$Y=G(y)=\int_0^y H(t)dy,\ |Y|\le\frac{k_0}{m+2}|y|^{(m+2)/2}\ \text{in}\ \overline{D^\pm},\tag{2.4}$$

and its inverse function of $Y=G(y)$ is

$$\begin{aligned}y=\pm|G^{-1}(Y)|,\ |y|\le&\left(\frac{k_0(m+2)}{2}\right)^{2/(m+2)}|Y|^{2/(m+2)}\\ &=J(k_0|Y|)^{2/(m+2)}\ \text{in}\ \overline{D^\pm}.\end{aligned}$$

In particular, when the coefficients $a=b=c=d=0$ in equation (2.1), then (2.1) becomes the famous Chaplygin equation in gas dynamics

$$K(y)u_{xx}+u_{yy}=0\ \text{in}\ D.\tag{2.5}$$

From [9]2), [28] and [71]2), we can see the mechanical background of equations (2.5) and (2.1).

The Tricomi boundary value problem for equation (2.1) may be formulated as follows:

Problem T Find a continuous solution $u(z)$ of (2.1) in $\overline{D}$, where u_x,u_y are continuous in $D^*=\overline{D}\backslash\{0,2\}$, and $u(z)$ satisfies the boundary conditions

$$u(z)=\phi(z)\ \text{on}\ \Gamma,\ u(z)=\psi(x)\ \text{on}\ L_1,\tag{2.6}$$

where $\phi(z)$, $\psi(z)$ satisfy the conditions

$$C^2_\alpha[\phi(z),\Gamma]\le k_2, C^2_\alpha[\psi(x),L_1]\le k_2, \phi(0)=\psi(0), \tag{2.7}$$

in which $\alpha\,(0<\alpha<1), k_2$ are positive constants.

If the boundary Γ near $z=0,2$ possesses the form $x=\tilde{G}(y)$ or $x=2-\tilde{G}(y)$ as stated before, we find the derivative for (2.6) according to the parameters $s=\operatorname{Im} z=y$ on Γ and $s=\operatorname{Re} z=x$ on L_1, and obtain

$$\begin{aligned}
&u_s=u_xx_y+u_y=\tilde{H}(y)u_x+u_y=\phi'(y), \text{ i.e.}\\
&\tilde{H}(y)H(y)u_x/H(y)+u_y=\phi'(y) \text{ on } \Gamma \text{ near } x=0,\\
&u_s=u_xx_y+u_y=-\tilde{H}(y)u_x+u_y=\phi'(y), \text{ i.e.}\\
&H(y)\tilde{H}(y)u_x/H(y)-u_y=-\phi'(y) \text{ on } \Gamma \text{ near } x=2,\\
&u_s=u_x+u_yy_x=u_x-u_y/H(y)=\psi'(x), \text{ i.e.}\\
&H(y)u_x-u_y=H(y)\psi'(x)=H[-Jx^{2/(m+2)}]\psi'(x)\\
&=J^{m/2}x^{m/(m+2)}\psi'(x) \text{ on } L_1,
\end{aligned}$$

where $\tilde{G}'(y)=\tilde{H}(y)$, and $H(y)=y^{m/2}$. It is clear that the complex form of above conditions is as follows

$$\begin{aligned}
&\operatorname{Re}[\overline{\lambda(z)}(U+iV)]=\operatorname{Re}[\overline{\lambda(z)}(H(y)u_x-iu_y)]/2=R(z) \text{ on } \Gamma,\\
&\operatorname{Re}[\overline{\lambda(z)}(U+jV)]=\operatorname{Re}[\overline{\lambda(z)}(H(y)u_x-ju_y)]/2=R(z) \text{ on } L_1, \qquad (2.8)\\
&u(0)=b_0,\ \operatorname{Im}[\overline{\lambda(z)}(U+jV)]_{z=z_0}=H(y_0)\psi'(x_0)/2\sqrt{2}=b_1,
\end{aligned}$$

where $b_0=\phi(0)$, $\lambda(z)=a+ib$ on Γ, $\lambda(z)=a+ib$ on L_1, and

$$\lambda(z)=\begin{cases}(1-j)/\sqrt{2},\\ \tilde{H}(y)/H(y)-i,\\ -\tilde{H}(y)/H(y)+i,\\ i\end{cases}\quad R(z)=\begin{cases}R_1(x) \text{ on } L_1\\ R_2(x) \text{ on } \Gamma \text{ at } z=0,\\ R_3(x) \text{ on } \Gamma \text{ at } z=2,\\ R_4(x) \text{ on } L_0=(0,2),\end{cases}$$

in which $R_1(x)=J^{m/2}x^{m/(m+2)}\psi'(x)/2\sqrt{2}$, $R_2(x)=\phi'(y)/2$, $R_3(z)=$

$-\phi'(y)/2$, and $R_4(x)$ is an undetermined function. We have

$$e^{i\phi_1}=\frac{\lambda(t_1-0)}{\lambda(t_1+0)}=e^{-\pi i/2-\pi i/2}=e^{-\pi i},\ \gamma_1=\frac{-\pi}{\pi}-K_1=0, K_1=-1,$$
$$e^{i\phi_2}=\frac{\lambda(t_2-0)}{\lambda(t_2+0)}=e^{\pi i/2-\pi i/2}=e^{0\pi i},\ \gamma_2=\frac{i0\pi}{\pi}-K_2=0, K_2=0,$$

in which $t_1=0, t_2=2$, $\lambda(t)=e^{\pi i/2}$ on L_0, and $\lambda(t_1+0)=\lambda(t_2-0)=\exp(i\pi/2)$, and K_1, K_2 are chosen. Thus the index of Problem T in D^+ is

$$K=(K_1+K_2)/2=-1/2.$$

If we consider $\mathrm{Re}[\overline{\lambda(z)}(U+jV)]=0$ on L_0, where $\lambda(z)=1$, $\gamma_1=1/2$, $\gamma_2=-1/2$, then $K=-1/2$; and if $\gamma_1=\gamma_2=-1/2$, $K_1=K_2=0$, then $K=0$, in this case we can add one point condition $u(2)=\phi(2)=b_1$ in (2.8), such that Problem T is well-posed. Obviously the Tricomi problem for Chaplygin equation (2.5) is a special case of Problem T for equation (2.1).

Noting that $\phi(z)\in C^2_\alpha(\Gamma), \psi(x)\in C^2(L_1)\ (0<\alpha<1)$, we can find two twice continuously differentiable functions $u_0^\pm(z)$ in $\overline{D}^\pm$, for instance, which are the solutions of the Dirichlet problem with the boundary condition on $\Gamma\cup L_1$ in (2.6) for harmonic equations in $D^\pm$, thus the functions $v(z)=v^\pm(z)=u(z)-u_0^\pm(z)$ in $\overline{D}^\pm$ is the solution of the following equation

$$K(y)v_{xx}+v_{yy}+a(x,y)v_x+b(x,y)v_y+c(x,y)v+\tilde{d}(x,y)=0 \text{ in } D \qquad (2.9)$$

satisfying the corresponding boundary conditions

$$\begin{aligned}&v(z)=0 \text{ on } \Gamma\cup L_1, \text{ i.e.}\\&\mathrm{Re}[\overline{\lambda(z)}W(z)]=R(z) \text{ on } \Gamma\cup L_1,\\&v(0)=b_0,\ \mathrm{Im}[\overline{\lambda(z_0)}W(z_0)]_{z=z_0}=b_1,\end{aligned}\qquad (2.10)$$

where $\tilde{d}=d+Lu_0^\pm(z)$ in $D^\pm$, $W(z)=U+iV=v_{\bar z}^+$ in $\overline{D^+}$, $W(z)=U+jV=v_{\bar z}^-$ in $\overline{D^-}$, $R(z)=0$ on $\Gamma\cup L_1$ and $b_0=b_1=0$, and the coefficients of (2.9) satisfy the conditions similar to Condition C, hence later on we only discuss the homogeneous Tricomi problem (Problem $\tilde{T}$) for equation (2.1) with the boundary condition (2.10) and the case of index $K=-1/2$, the other case can be similarly discussed. From $v(z)=v^\pm(z)=u(z)-u_0^\pm(z)$ in $\overline{D}^\pm$, we have $u(z)=v^-(z)+u_0^-(z)$ in $\overline{D}^-$, $u(z)=v^+(z)+u_0^+(z)$ in $\overline{D}^+$, $v^+(z)=v^-(z)-u_0^+(z)+u_0^-(z)$, $u_y=v_y^\pm+u_{0y}^\pm$, $v_y^+=v_y^--u_{0y}^++u_{0y}^-=2\hat{R}_0(x)$, and $v_y^-=2\tilde{R}_0(x)$ on $L_0=D\cap\{0<x<2, y=0\}$.

2.2 Representation and uniqueness of solutions of Tricomi problem for degenerate mixed equations

In this section, we first write the complex form of equation (2.1). We use the complex function in the elliptic domain $\overline{D^+}$ and the hyperbolic function in $\overline{D^-}$, denote

$$\begin{aligned}
&W(z)=U+jV=[H(y)u_x-ju_y]/2=H(y)[u_x-ju_Y]/2=H(y)u_Z,\\
&W_{\bar z}=\frac{1}{2}[H(y)W_x+jW_y]=H(y)[W_x+jW_Y]/2=H(y)W_{\overline{Z}} \text{ in } \overline{D^-},\\
&W(z)=U+iV=[H(y)u_x-iu_y]/2=H(y)[u_x-iu_Y]/2=H(y)u_Z,\\
&W_{\bar z}=\frac{1}{2}[H(y)W_x+iW_y]=H(y)[W_x+iW_Y]/2=H(y)W_{\overline{Z}} \text{ in } \overline{D^+},
\end{aligned} \tag{2.11}$$

where $Z = x + jY$ in $\overline{D^-}$ and $Z = x + iY$ in D^+, we have

$$\begin{aligned}
W_{\bar z}&=\frac{1}{4}\{H^2u_{xx}-u_{yy}-jH[u_{yx}-u_{xy}]+jH_yu_x\}\\
&=\frac{1}{4}[(\frac{a}{H}+\frac{jH_y}{H})Hu_x+bu_y+cu+d]\\
&=\frac{1}{4}[(\frac{a}{H}+\frac{jH_y}{H})(W+\overline{W})+jb(\overline{W}-W)+cu+d]\\
&=\frac{1}{4}[\frac{a}{H}+\frac{jH_y}{H}-jb]W+\frac{1}{4}[\frac{a}{H}+\frac{jH_y}{H}+jb]\overline{W}+\frac{1}{4}(cu+d)\\
&=\frac{e_1}{4}\{[\frac{a}{H}+\frac{H_y}{H}-b](\mathrm{Re}W+\mathrm{Im}W)+[\frac{a}{H}+\frac{H_y}{H}+b]\\
&\times(\mathrm{Re}W-\mathrm{Im}W)+cu+d\}+\frac{e_2}{4}\{[\frac{a}{H}-\frac{H_y}{H}-b](\mathrm{Re}W+\mathrm{Im}W)\\
&+[\frac{a}{H}-\frac{H_y}{H}+b](\mathrm{Re}W-\mathrm{Im}W)+cu+d\} \text{ in } \overline{D^-},
\end{aligned} \tag{2.12}$$

where $e_1 = (1 + j)/2, e_2 = (1 - j)/2$. Noting that

$$\begin{aligned}
&\mu=x+\int_0^y H(t)dt=x+G(y),\ \nu=x-G(y),\\
&\mu+\nu=2x,\ \mu-\nu=2G(y),\ \frac{\partial G(y)}{\partial y}=H(y),\\
&\frac{\partial x}{\partial \mu}=\frac{\partial x}{\partial \nu}=\frac{1}{2},\ \frac{\partial y}{\partial \mu}=-\frac{\partial y}{\partial \nu}=\frac{1}{2H(y)} \text{ in } \overline{D^-},
\end{aligned} \tag{2.13}$$

we can obtain

$$
\begin{aligned}
W_{\bar{\bar{z}}} &= \frac{1}{2}[H(U+jV)_x + j(U+jV)_y] \\
&= \frac{e_1}{2}[HU_x+V_y+HV_x+U_y]+\frac{e_2}{2}[HU_x+V_y-HV_x-U_y] \\
&= \frac{e_1}{2}[H(U+V)_x+(U+V)_y]+\frac{e_2}{2}[H(U-V)_x-(U-V)_y] \\
&= H[e_1(U+V)_\mu + e_2(U-V)_\nu] = \frac{e_1}{4}\{[\frac{a}{H}+\frac{H_y}{H}-b] \\
&\times(U+V)+[\frac{a}{H}+\frac{H_y}{H}+b](U-V)+cu+d\}+\frac{e_2}{4}\{[\frac{a}{H}-\frac{H_y}{H}-b] \\
&\times(U+V)+[\frac{a}{H}-\frac{H_y}{H}+b](U-V)+cu+d\} \\
&= \frac{e_1}{4}\{(\frac{h_y}{2h}-b)(U+V)+(\frac{h_y}{2h}+b)(U-V) \\
&+[\frac{2a}{H}+\frac{m}{y}]U+cu+d\}+\frac{e_2}{4}\{(-\frac{h_y}{2h}-b)(U+V) \\
&-(\frac{h_y}{2h}-b)(U-V)+[\frac{2a}{H}-\frac{m}{y}]U+cu+d\} \text{ in } \overline{D^-},
\end{aligned} \tag{2.14}
$$

if $H(y) = \sqrt{|y|^m h(y)}$, where m, $h(y)$ are as stated before. In addition we have

$$
\begin{aligned}
W_{\bar{\bar{z}}} &= \frac{1}{2}[H(y)W_x+iW_y]=\frac{1}{4}\{H[Hu_x-iu_y]_x+i[Hu_x-iu_y]_y\} \\
&= \frac{1}{4}\{H^2u_{xx}+u_{yy}-iH[u_{yx}-u_{xy}]+iH_yu_x\} \\
&= \frac{1}{4}\{H^2u_{xx}+u_{yy}+iH_yu_x\}=\frac{1}{4}\{H^2u_{xx}+u_{yy}+i\frac{H_y}{H}(Hu_x)\} \\
&= \frac{1}{4}\{[\frac{iH_y}{H}-\frac{a}{H}](Hu_x)-bu_y-cu-d\} \\
&= \frac{1}{4}\{[\frac{iH_y}{H}-\frac{a}{H}](W+\overline{W})-ib(W-\overline{W})-cu-d\} \\
&= \frac{1}{4}\{[\frac{iH_y}{H}-\frac{a}{H}-ib]W+[\frac{iH_y}{H}-\frac{a}{H}+ib]\overline{W}-cu-d\} \text{ in } \overline{D^+}.
\end{aligned} \tag{2.15}
$$

Thus if $u(z)$ is a solution of (2.1), then the function

$$
W(z) = \begin{cases} U+iV \text{ in } \overline{D^+} \\ U+jV \text{ in } \overline{D^-} \end{cases}
$$

is a solution of the first order complex equation of mixed type

$$W_{\bar{\bar{z}}} = A_1(z)W + A_2(z)\overline{W} + A_3(z)u + A_4(z) = g(z) \text{ in } D,$$

$$A_1 = \begin{cases} \frac{1}{4}[-\frac{a}{H} + \frac{iH_y}{H} - ib], \\ \frac{1}{4}[\frac{a}{H} + \frac{jH_y}{H} - jb], \end{cases} A_2 = \begin{cases} \frac{1}{4}[-\frac{a}{H} + \frac{iH_y}{H} + ib], \\ \frac{1}{4}[\frac{a}{H} + \frac{jH_y}{H} + jb], \end{cases} \tag{2.16}$$

$$A_3 = \begin{cases} -\frac{c}{4}, \\ \frac{c}{4}, \end{cases} A_4 = \begin{cases} -\frac{d}{4} \text{ in } \overline{D^+}, \\ \frac{d}{4} \text{ in } \overline{D^-}, \end{cases}$$

if $H(y) = \sqrt{|y|^m h(y)}$, where m, $h(y)$ are as stated before, then

$$A_1 = \begin{cases} \frac{1}{4}\left[-\frac{a}{H} - ib + i\left(\frac{h_y}{2h} + \frac{m}{2y}\right)\right], \\ \frac{1}{4}\left[\frac{a}{H} - jb + j\left(\frac{h_y}{2h} + \frac{m}{2y}\right)\right], \end{cases} A_3 = \begin{cases} -\frac{c}{4}, \\ \frac{c}{4}, \end{cases}$$

$$A_2 = \begin{cases} \frac{1}{4}\left[-\frac{a}{H} + ib + i\left(\frac{h_y}{2h} + \frac{m}{2y}\right)\right], \\ \frac{1}{4}\left[\frac{a}{H} + jb + j\left(\frac{h_y}{2h} + \frac{m}{2y}\right)\right], \end{cases} A_4 = \begin{cases} -\frac{d}{4} \text{ in } \overline{D^+}, \\ \frac{d}{4} \text{ in } \overline{D^-}. \end{cases}$$

Hence the function

$$u(z) = \begin{cases} 2\text{Re} \int_0^z [\frac{U(z)}{H(y)} + iV(z)]dz + b_0 \text{ in } \overline{D^+} \\ 2\text{Re} \int_0^z [\frac{U(z)}{H(y)} - jV(z)]dz + b_0 \text{ in } \overline{D^-} \end{cases} \tag{2.17}$$

is a solution of equation (2.1), where b_0 is a real constant. Moreover we have

$$W_{\bar{\bar{z}}} = [H(U + iV)_x + i(U + iV)_y]/2$$

$$= [HU_x - V_y + i(HV_x + U_y)]/2 = H[U_x - V_y/H + i(V_x + U_y/H)]/2$$

$$= H\{(U + V)_x/2 - (U + V)_y/2H + (U - V)_x/2 + (U - V)_y/2H$$

$$+ i[(U + V)_x/2 + (U + V)_y/2H - (U - V)_x/2 + (U - V)_y/2H]\}/2$$

$$= H\{(U-V)_\mu+(U+V)_\nu+i[(U+V)_\mu-(U-V)_\nu]\}/2 = g(Z)$$

$$= iH[(U+V)-i(U-V)]_{\mu+i\nu} = iH\overline{[(U+V)+i(U-V)]_{\mu-i\nu}} \text{ in } \overline{D^+},$$

where

$$\mu = x + G(y) = x + \int_0^y H(y)dy,\ \nu = x-G(y),$$

$$\frac{\partial x}{\partial \mu} = \frac{\partial y}{\partial \mu} = \frac{1}{2},\ \frac{\partial y}{\partial \mu} = -\frac{\partial y}{\partial \mu} = \frac{1}{2H(y)} \text{ in } \overline{D^+}.$$

Especially, the complex equation

$$W_{\bar{\bar{z}}} = 0 \text{ in } \overline{D} \tag{2.18}$$

can be rewritten in the system

$$\begin{aligned} &[(U+V)+i(U-V)]_{\mu-i\nu} = 0 \text{ in } \overline{D^+}, \\ &(U+V)_\mu = 0,\ (U-V)_\nu = 0 \text{ in } \overline{D^-}. \end{aligned} \tag{2.19}$$

The boundary value problem for equation (2.16) with the boundary condition (2.8) ($W(z) = u_{\bar{z}} = U + iV$ in $\overline{D^+}$ and $W(z) = u_{\bar{\bar{z}}} = U + jV$ in $\overline{D^-}$) and the relation (2.17) will be called Problem A.

Now, we discuss the Riemann-Hilbert problem(Problem A) for the second system of (2.19) in $\overline{D^-}$ with the boundary conditions

$$\text{Re}[\overline{\lambda(z)}(U+jV)] = \begin{cases} H(y)\psi'(x)/2\sqrt{2} = R_1(z),\ z \in L_1, \\ R_0(x),\ x \in L_0 = (0,2), \end{cases} \tag{2.20}$$

$$u(0)=b_0,\ \text{Re}[\overline{\lambda(z)}(U+jV)]|_{z=z_0} = H(y_0)\psi'(x_0)/2\sqrt{2} = b_1,$$

in which $\lambda(z) = a(z) + jb(z)$ on L_1 and $\lambda(z) = 1 + j$ on L_0 and $R_0(x)$ on L_0 is an undetermined real function. It is easy to see that the solution of Problem A for (2.19) in $\overline{D^-}$ can be expressed as

$$\begin{aligned} &\xi = U(z) + V(z) = f(\nu),\ \ \eta = U(z) - V(z) = g(\mu), \\ &U(z) = [f(\nu) + g(\mu)]/2,\ V(z) = [f(\nu) - g(\mu)]/2, \text{ i.e.} \\ &W(z) = U(z) + jV(z) = [(1+j)f(\nu) + (1-j)g(\mu)]/2, \end{aligned}$$

where $f(t), g(t)$ are two arbitrary real continuous functions on $L_0 = (0,2)$. For convenience, denote by the functions $a(x), b(x), r(x)$ of x the functions

$a(z), b(z), r(z)$ of z in (2.20), thus (2.20) can be rewritten as

$$a(x)U(z)-b(x)V(z)=R_1(x) \text{ on } L_1, U(x)-V(x)=R_0(x) \text{ on } L_0, \text{ i.e.}$$

$$[a(x)-b(x)]f(x-G(y))+[a(x)+b(x)]g(x+G(y))$$

$$=2R_1(x) \text{ on } L_1,\ U(x)-V(x)=R_0(x) \text{ on } L_0, \text{ i.e.}$$

$$[a(x)-b(x)]f(2x)+[a(x)+b(x)]g(0)=2R_1(x), x\in[0,1],$$

$$U(x)-V(x)=R_0(x),\ x\in[0,2], \text{ i.e.}$$

$$[a(t/2)-b(t/2)]f(t)+[a(t/2)+b(t/2)]g(0)=2R_1(t/2), t\in[0,2],$$

$$U(t)-V(t)=R_0(t),\ t\in[0,2],$$

where

$$(a(1)+b(1))g(0)=(a(1)+b(1))(U(z_0)-V(z_0))$$
$$=R_1(1)-b_1=0,$$

in which from the boundary condition (2.8), we have $\lambda(z)=a(z)+jb(z)=(1-j)/\sqrt{2}$ on L_1, hence $[a(t/2)+b(t/2)]g(0)=0$, $t\in[0,2]$. Moreover we can derive

$$f(\nu)=f(x-G(y))=\frac{2R_1(\nu/2)-(a(\nu/2)+b(\nu/2))g(0)}{a(\nu/2)-b(\nu/2)},$$

$$g(\mu)=g(x+G(y))=R_0(\mu),$$

$$U(z)=\frac{1}{2}\{f(\nu)+R_0(\mu)\},\ V(z)=\frac{1}{2}\{f(\nu)-R_0(\mu)\},$$

if $a(x)-b(x)\neq 0$ on $[0,1]$. From the above formula, it follows that

$$\begin{aligned}&\mathrm{Re}[(1+j)W(x)]=U(x)+V(x)=\frac{2R_1(x/2)-(a(x/2)+b(x/2))g(0)}{a(x/2)-b(x/2)},\\&\mathrm{Re}[(1-j)W(x)]=U(x)-V(x)=R_0(x),\ x\in[0,2],\end{aligned} \tag{2.21}$$

if $a(x)-b(x)\neq 0$ on $[0,1]$. Thus we obtain

$$\begin{aligned}W(z)=&\frac{1}{2}\{(1+j)\frac{2R_1((x-G(y))/2)-M(x,y)}{a((x-G(y))/2)-b((x-G(y))/2)}+(1-j)\\&\times R_0(x+G(y))\},M(x,y)=[a((x-G(y))/2)+b((x-G(y))/2)]g(0)=0.\end{aligned} \tag{2.22}$$

In particular, we have

$$\begin{aligned}&\mathrm{Re}[\overline{(1+i)}(U(x)+iV(x))]=U(x)+V(x)\\&=\frac{2R_1(x/2)-[a(x/2)+b(x/2)]g(0)}{a(x/2)-b(x/2)}\ \text{on}\ L_0,\end{aligned}\tag{2.23}$$

if $R_1(z)=0$ on L_1, $\hat{R}_0(x)=0$ on L_0, then $W(z)=U(z)+jV(z)=0$ in $\overline{D^-}$. From (2.2), (2.7) and (2.22), the corresponding function $u(z)$ in (2.17) of the solution $W(z)$ and $f(\nu)$, $g(\mu)$ satisfy the estimates

$$\begin{aligned}&C_\delta[u(z),D^-]+C_\delta^1[u(z),D_\varepsilon^-]\le M_1,\\&C_\delta[f(x),L_0\cap D_\varepsilon^-]+C_\delta[g(x),L_0\cap D_\varepsilon^-]\le M_2,\end{aligned}\tag{2.24}$$

in which $L_0=(0,2)$, $D_\varepsilon^-=\overline{D^-}\cap\Pi_{l=1}^2\{|z-t_l|>\varepsilon\}(>0)\}$, δ, ε are sufficiently small positive constants, and $M_l=M_l(\delta,k_0,k_2,D_\varepsilon^-)\,(l=1,2)$ are positive constants. As stated in Subsection 2.1, we can assume that $R(z)=0$ on L_1, $b_0=b_1=0$ in (2.8), hence the estimates in (2.24) are true.

Next we find a solution of the Riemann-Hilbert boundary value problem for the first equation of (2.19) in D^+ with the boundary conditions (2.23) and

$$\mathrm{Re}[\overline{\lambda(z)}(U(z)+iV(z))]=\mathrm{Re}\lambda(z)U(z)+\mathrm{Im}\lambda(z)V(z)=R(z)\ \text{on}\ \Gamma.$$

Noting that the index of the above boundary condition is $K=-1/2$, by the method in [87]1), we know that the above Riemann-Hilbert problem has a unique solution $W(z)$ in D^+, and then

$$U(x)-V(x)=\mathrm{Re}[(1-j)(U(x)+jV(x)]=R_0(x)\ \text{on}\ L_0$$

is determined. We mention that if $\mathrm{Re}W(x)=U(x)=0$ on L_0, then $R_0(x)=\tilde{R}_0(x)$ on L_0. This shows that Problem A for equation (2.18) is uniquely solvable, namely

Theorem 2.1 *Problem A of equation* (2.18) *or system* (2.19) *in $\overline{D}$ has a unique solution as stated in* (2.22) *satisfying the estimate* (2.24).

The representation of solutions of Problem T for equation (2.1) is as follows.

Theorem 2.2 *Under Condition C, any solution $u(z)$ of Problem T for*

equation (2.1) *in* D *can be expressed as follows*

$$\begin{aligned}
&u(z)=u(x)-2\int_0^y V(z)dy\\
&=2\mathrm{Re}\int_0^z[\frac{\mathrm{Re}w}{H}+\binom{i}{-j}\mathrm{Im}w]dz+b_0 \text{ in } \binom{\overline{D^+}}{\overline{D^-}},\\
&w[z(Z)]=\Phi(Z)+\Psi(Z)=\hat{\Phi}(Z)+\hat{\Psi}(Z) \text{ in } D_Z^+,\\
&\Psi(Z)=-2\mathrm{Re}\frac{1}{\pi}\int\int_{D_t^+}\frac{g(t)/H}{t-Z}d\sigma_t,\\
&\hat{\Psi}(Z)=-2i\mathrm{Im}\frac{1}{\pi}\int\int_{D_t^+}\frac{g(t)/H}{t-Z}d\sigma_t \text{ in } \overline{D_Z^+},\\
&w(z)=\phi(z)+\psi(z)=\xi(z)e_1+\eta(z)e_2 \text{ in } \overline{D^-},\\
&\xi(z)=\int_0^\mu\frac{g_1(z)}{2H(y)}d\mu=\zeta(z)+\int_0^y g_1(z)dy\\
&=\int_{S_1}g_1(z)dy+\int_0^y g_1(z)dy=\int_{y_1}^{\hat{y}}\hat{g}_1(z)dy,\ z\in s_1,\\
&\eta(z)=\theta(z)+\int_0^y g_2(z)dy,\ z\in s_2,\\
&g_l(z)=\tilde{A}_l(U+V)+\tilde{B}_l(U-V)+2\tilde{C}_lU+\tilde{D}_lu+\tilde{E}_l, l=1,2,
\end{aligned}\tag{2.25}$$

where $U=Hu_x/2, V=-u_y/2$, $\Phi(Z), \hat{\Phi}(Z)$ *are analytic functions in* $D_Z^+=Z(D^+)$, *here* $Z(z)=x+iY=x+iY(y)$ *is a mapping from* $z(\in D^+)$ *to* Z, $\zeta(z)=\int_{S_1}g_1(z)dy$, $\zeta(z)e_1+\theta(z)e_2$ *is a solution of* (2.18) *in* D^-, *and* s_1, s_2 *are two families of characteristics in* D^-:

$$s_1:\ \frac{dx}{dy}=\sqrt{-K(y)}=H(y),\ s_2:\ \frac{dx}{dy}=-\sqrt{-K(y)}=-H(y)\tag{2.26}$$

passing through $z=x+jy\in D^-$, $\xi(x)+\eta(x)=[U(x)+V(x)]+[U(x)-V(x)]=2U(x)=H(0)u_x=\zeta(x)+\theta(x)=0$ on L_0, $\xi(z)=\int_0^\mu[g_1(z)/2H(y)]d\mu$ *is the integral along characteristic curve* s_1 *from a point* $z_1=x_1+jy_1$ *on* L_1 *to the point* $z=x+jy\in\overline{D}$, S_1 *is the characteristic curve from a point on* L_1 *to a point on* L_0, $\theta(z)=-\zeta(x+G(y))$, *and*

$$\begin{aligned}
&W(z)=U(z)+jV(z)=\frac{1}{2}[Hu_x-ju_y],\\
&\xi(z)=\mathrm{Re}\psi(z)+\mathrm{Im}\psi(z), \eta(z)=\mathrm{Re}\psi(z)-\mathrm{Im}\psi(z),\\
&\tilde{A}_1=\tilde{B}_2=\frac{1}{2}(\frac{h_y}{2h}-b),\ \tilde{A}_2=\tilde{B}_1=\frac{1}{2}(\frac{h_y}{2h}+b),
\end{aligned}$$

$$\tilde{C}_1=\frac{a}{2H}+\frac{m}{4y},\ \tilde{C}_2=-\frac{a}{2H}+\frac{m}{4y},$$
$$\tilde{D}_1=-\tilde{D}_2=\frac{c}{2},\ \tilde{E}_1=-\tilde{E}_2=\frac{d}{2}, \tag{2.27}$$

in which we choose $H(y)=[|y|^m h(y)]^{1/2}, h(y)$ *is as stated in* (2.14), *and*

$$d\mu=d[x+G(y)]=2H(y)dy \ \text{ on } s_1,$$
$$d\nu=d[x-G(y)]=-2H(y)dy \ \text{ on } s_2.$$

Proof From (2.14) it is easy to see that equation (2.1) in $\overline{D}$ can be reduced to the system of integral equations: (2.25). Moreover we can extend the equation (2.16) onto the the symmetrical domain $\hat{D}_Z$ of D_Z with respect to the real axis $\text{Im}Z=0$, namely introduce the function $\hat{W}(Z)$ as follows:

$$\hat{W}(Z)-\begin{cases} W[z(Z)], \\ -\overline{W[z(\overline{Z})]}, \end{cases} \hat{u}(z)-\begin{cases} u(Z) \text{ in } D_Z, \\ -u(\overline{Z}) \text{ in } \hat{D}_Z, \end{cases} \tag{2.28}$$

and then the equation (2.16) is extended as

$$\hat{W}_{\bar{\bar{z}}}=\hat{A}_1\hat{W}+\hat{A}_2\overline{\hat{W}}+\hat{A}_3\hat{u}+\hat{A}_4=\hat{g}(Z) \text{ in } \overline{D_Z}\cup\overline{\hat{D}_Z}, \tag{2.29}$$

where

$$\hat{A}_l(Z)=\begin{cases} A_l(Z), \\ \overline{\tilde{A}_l(\overline{Z})}, \end{cases} l=1,2,3,\ \hat{A}_4(Z)=\begin{cases} A_4(Z), \\ -\overline{A_4(\overline{Z})}, \end{cases}$$

$$\hat{g}_l(Z)=\begin{cases} g_l(z) \text{ in } \overline{D_{\bar{Z}}}, \\ -\overline{g_l(\overline{Z})} \text{ in } \overline{\hat{D}_Z}, \end{cases} l=1,2,$$

here $\tilde{A}_1(\overline{Z})=A_2(\overline{Z}), \tilde{A}_2(\overline{Z})=A_1(\overline{Z}), \tilde{A}_3(\overline{Z})=A_3(\overline{Z})$. It is easy to see that the system of integral equations (2.25) can be written in the form

$$\xi(z)=\zeta(z)+\int_0^y g_1(z)dy=\int_{y_1}^{\hat{y}}\hat{g}_1(z)dy,$$
$$\eta(z)=\theta(z)+\int_0^y g_2(z)dy=\int_{y_2}^{\hat{y}}\hat{g}_2(z)dy, \tag{2.30}$$
$$\hat{z}=x+j\hat{y}=x+j|y| \text{ in } \tilde{D}_Z,$$

where y_1, y_2 are the ordinates of intersection points of L_1 and characteristics lines of family s_1 in (2.26) emanating from $z = x_1, x_2(< x_1)$ and L_1, herein x_1, x_2 are the intersection points of two characteristic lines s_1, s_2 passing through $z = x + jy \in \overline{D^-}$ and $x-$axis respectively. The function $\theta(z)$ is determined by $\zeta(z)$, which can be defined by $\theta(z) = -\zeta(z) = -\zeta(x+G(y))$, for the extended integral, for convenience the above form $\hat{g}_2(z)$ is written, later on the numbers $\hat{y} - y_1, \hat{t} - y_1$ will be written by $\tilde{y}, \tilde{t}$ respectively, and for convenience $|y| + \delta$ is written as y, $0 \le |y| \le \delta$, herein δ is a small positive number.

2.3 Existence of solutions of Tricomi problem for degenerate equations of mixed type

In order to prove the existence of solutions of Problem T for equation (2.1) with some conditions, we try to discuss the problem by using the complex analytic method. As stated in Subsection 2.1, it suffices to discuss Problem $\tilde{T}$ for (2.1), it is clear that Problem $\tilde{T}$ is equivalent to Problem $\tilde{A}$ for the complex equation

$$W_{\bar{\bar{z}}} = A_1(z)W + A_2(z)\overline{W} + A_3(z)u + A_4(z) \text{ in } D, \tag{2.31}$$

with the relation

$$u(z) = \begin{cases} 2\text{Re}\displaystyle\int_0^z [\frac{\text{Re}W(z)}{H(y)} + i\text{Im}w(z)]dz + b_0 \text{ in } \overline{D^+}, \\ u(x) - \displaystyle\int_0^y \text{Im}W(z)dy \text{ in } \overline{D^-}, \end{cases} \tag{2.32}$$

where $H(y)$ is as stated in (2.14), and the coefficients in (2.31) are as before, and the boundary conditions

$$\text{Re}[\overline{\lambda(z)}W(z)]=R(z) \text{ on } \Gamma\cup L_1, u(0)=b_0, \text{Im}[\overline{\lambda(z_0)}W(z_0)]=b_1, \tag{2.33}$$

in which $\lambda(z), R(z), z_0, b_0, b_1$ are as stated in (2.10), and $R(z) = 0$ on $\Gamma \cup L_1$, $b_0 = b_1 = 0$. By Theorems 2.1 and 2.2, we see that Problem $\tilde{A}$ can be divided into two problems, i.e. Problem A_1 of equation (2.31), (2.32) in D^+ and Problem A_2 of equation (2.31), (2.32) in D^-, the boundary conditions of Problems A_1 and A_2 are as follows:

$$\text{Re}[\overline{\lambda(z)}W(z)] = 0 \text{ on } \Gamma \cup L_0,\ u(0) = b_0, \tag{2.34}$$

and

$$\text{Re}[\overline{\lambda(z)}W(z)]=R(z) \text{ on } L_1\cup L_0, \text{Im}[\overline{\lambda(z_0)}W(z_0)]=0, \tag{2.35}$$

where $\lambda(z) = i$, $R(x) = -\hat{R}_0(x)$ on L_0 in (2.10), (2.34), $R(z) = 0$ on L_1, $\lambda(z) = j$, $R(x) = \tilde{R}_0(x)$ on L_0, because $\text{Re}W = H(y)u_x/2 = 0$ on L_0, $1+j$ can be replaced by j. From the result in Section 2, Chapter II, we know that Problem A_1 for equation (2.31), (2.32), (2.34) in D^+ has a unique solution $W(z)$. Hence in the following we only prove the unique solvability of Problem A_2 for (2.31), (2.32), (2.35) in D^-, which is the Darboux type problem (see [12]3)). Actually after we find the solution of the above Problem A_2, the function u_y on L_0 is determined, then the solution of Problem A_1 can be found as stated before.

Theorem 2.3 *If equation* (2.31) *satisfies Condition C and* (2.36) *below, then there exists a unique solution* $[W(z), v(z)]$ *of Problem* A_2 *for* (2.31), (2.32), (2.35) *in* D^-.

Proof Denote $D_0 = \overline{D^-}\cap\{a_0 \le x \le b_0\}$, where $0 < a_0 = \delta_0 < b_0 < 2-\delta_0$, and δ_0 is a sufficiently small positive constant.

We consider $v(z) = u(z) - u_0(z)$ as stated in Subsection 2.1 and $K(y) = -|y|^m h(y)$ as stated in (2.14). In order to find a solution of the system of integral equations (2.25), we need to add the condition

$$a(x,y)|y|/H(y)=o(1), \text{ i.e. } |a(x,y)|=\varepsilon(y)H(y)/|y|,\ m\ge 2, \tag{2.36}$$

where $\varepsilon(y) \to 0$ as $y \to 0$ and $\max_{\overline{D^-}}\varepsilon(y) \le \varepsilon_0$, ε_0 is a positive number. It is clear that for two characteristics s_1, s_2 passing through a point $z = x+jy \in \overline{D^-}$ and x_1, x_2 are the intersection points of the characteristics and $x-$axis respectively, for any two points $\tilde{z}_1 = \tilde{x}_1 + j\tilde{y} \in s_1, \tilde{z}_2 = \tilde{x}_2 + j\tilde{y} \in s_2$, we have

$$\begin{aligned} &|\tilde{x}_1 - \tilde{x}_2| \le |x_1 - x_2| = 2\left|\int_0^y \sqrt{-K(t)}dt\right| \\ &\le \frac{2k_0}{m+2}|y|^{1+m/2} \le \frac{k_1}{6}|y|^{m/2+1} \le M|y|^{m/2+1}, \\ &|y|^{1+m/2} \le \frac{k_0(m+2)}{2}|x_1 - x_2|. \end{aligned} \tag{2.37}$$

From Condition C, we can assume that the coefficients in (2.25) are continuously differentiable with respect to $x \in L_0$ and satisfy the conditions

$$\begin{aligned} &|\tilde{A}_l|, |\tilde{A}_{lx}|, |\tilde{B}_l|, |\tilde{B}_{lx}|, |\tilde{D}_l|, |\tilde{D}_{lx}| \le k_0 \le \frac{k_1}{6}, |\tilde{E}_l|, |\tilde{E}_{lx}| \le \frac{k_1}{6}, \\ &2\sqrt{h},\ \frac{1}{\sqrt{h}},\ \left|\frac{h_y}{h}\right| \le k_0 \le \frac{k_1}{6} \text{ in } \bar{D},\ l = 1, 2, \end{aligned} \tag{2.38}$$

and we shall use the constants

$$
\begin{gathered}
M=4\max[M_1,M_2,M_3],\ \ M_1=\max[8(k_1d)^2,\frac{M_3}{k_1}],\\
M_2=\frac{(2+m)k_0d}{\delta^{2+m}}[4k_1+\frac{4\varepsilon_0+m}{\delta}],\ M_3=2k_1^2[d+\frac{1}{2H(y_1')}],\\
\gamma=\max[4k_1d\delta^\beta+\frac{4\varepsilon(y)+m}{2\beta'}]<1,\ \ 0\le|y|\le\delta,\\
\frac{2dM_0}{N+1}\le\gamma,\ 2M\frac{(M_0|\tilde t|)^n}{n!}\le M'\gamma^n,\ n=0,1...,N,N+1,...,
\end{gathered}\tag{2.39}
$$

and $M_l(l=1,2,3)$ are positive constants as stated in (2.43)–(2.47) below, d is the diameter of D, $\beta'=(1+m/2)(1-3\beta)$, $\varepsilon_0=\max_{\overline{D^-}}\varepsilon(z)$, $1/2H(y_1')\le k_0[(m+2)a_0/k_0]^{-m/(2+m)}$, δ, β are sufficiently small positive constants, and y_1' is as stated in (2.43), N, M' are sufficiently large positive integer and constant respectively. We choose $v_0=0,\xi_0=0,\eta_0=0$ and substitute them into the corresponding positions of v,ξ,η in the right-hand sides of (2.25), and obtain

$$
\begin{aligned}
&v_1(z)=v_1(x)-2\int_0^y V_0dy\ =v_1(x)+\int_0^y(\eta_0-\xi_0)\,dy,\\
&\xi_1(z)=\zeta_1(z)+\int_0^y g_{10}(z)dy=\zeta_1(z)+\int_0^y\tilde E_1dy=\int_{y_1}^{\hat y}\hat E_1dy,\\
&\eta_1(z)=\theta_1(z)+\int_0^y g_{20}(z)dy=\theta_1(z)+\int_0^y\hat E_2dy=\int_{y_1}^{\hat y}\hat E_2dy,\\
&g_{l0}=\tilde A_l\xi_0+\tilde B_l\eta_0+\tilde C_l(\xi_0+\eta_0)+\tilde D_lv_0+\tilde E_l=\tilde E_l,l=1,2,
\end{aligned}\tag{2.40}
$$

where $z_1=x_1+jy_1$ is a point on L_1, which is the intersection of L_1 and the characteristic curve s_1 passing through the point $z=x+jy\in\overline{D^-}$. By the successive approximation, we find the sequences of functions $\{v_k\},\{\xi_k\},\{\eta_k\}$, which satisfy the relations

$$
\begin{aligned}
&v_{k+1}(z)=v_{k+1}(x)-2\int_0^y V_k(z)dy=v_{k+1}(x)+\int_0^y(\eta_k-\xi_k)dy,\\
&\xi_{k+1}(z)=\zeta_{k+1}(z)+\int_0^y g_{1k}(z)dy=\int_{y_1}^{\hat y}\hat g_{1k}dy,\\
&\eta_{k+1}(z)=\theta_{k+1}(z)+\int_0^y g_{2k}(z)dy=\int_{y_1}^{\hat y}\hat g_{2k}(z)dy,\\
&g_{lk}(z)=\tilde A_l\xi_k+\tilde B_l\eta_k+\tilde C_l(\xi_k+\eta_k)+\tilde D_lv_k+\tilde E_l,\\
&l=1,2,\ k=0,1,2,....
\end{aligned}\tag{2.41}
$$

Setting that $\tilde{g}_{lk+1}(z)=g_{lk+1}(z)-g_{lk}(z)(l=1,2)$, and

$$\tilde{y}=\hat{y}-y_1,\ \tilde{t}=\hat{t}-y_1,\ \tilde{v}_{k+1}(z)=v_{k+1}(z)-v_k(z),$$
$$\tilde{\xi}_{k+1}(z)=\xi_{k+1}(z)-\xi_k(z),\ \tilde{\eta}_{k+1}(z)=\eta_{k+1}(z)-\eta_k(z),$$
$$\tilde{\zeta}_{k+1}(z)=\zeta_{k+1}(z)-\zeta_k(z),\ \tilde{\theta}_{k+1}(z)=\theta_{k+1}(z)-\theta_k(z),$$

we shall prove that $\{\tilde{v}_k\},\{\tilde{\xi}_k\},\{\tilde{\eta}_k\},\{\tilde{\zeta}_k\},\{\tilde{\theta}_k\}$ in D_0 satisfy the estimates

$$\begin{aligned}
&|\tilde{v}_k(z)-\tilde{v}_k(x)|,|\tilde{\xi}_k(z)-\tilde{\zeta}_k(z)|,|\tilde{\eta}_k(z)-\tilde{\theta}_k(z)|\le M'\gamma^{k-1}|y|^{1-\beta},0\le|y|\le\delta,\\
&|\tilde{\xi}_k(z)|,|\tilde{\eta}_k(z)|\le M(M_2|\tilde{y}|)^{k-1}/(k-1)!,y\le-\delta,\text{ or }M'\gamma^{k-1},0\le|y|\le\delta,\\
&|\tilde{\xi}_k(z_1)-\tilde{\xi}_k(z_2)-\tilde{\zeta}_k(z_1)-\tilde{\zeta}_k(z_2)|,\ |\tilde{\eta}_k(z_1)-\tilde{\eta}_k(z_2)-\tilde{\theta}_k(z_1)-\tilde{\theta}_k(z_2)|\\
&\le M'\gamma^{k-1}[|x_1-x_2|^{1-\beta}+|x_1-x_2|^{\beta}|y|^{\beta'}],\ 0\le|y|\le\delta,\ |\tilde{v}_k(z_1)-\tilde{v}_k(z_2)|,\\
&|\tilde{\xi}_k(z_1)-\tilde{\xi}_k(z_2)|,|\tilde{\eta}_k(z_1)-\tilde{\eta}_k(z_2)|\le M(M_2|\tilde{t}|)^{k-1}|x_1-x_2|^{1-\beta}\\
&/(k-1)!,\ y\le-\delta,\text{ or }M'\gamma^{k-1}[|x_1-x_2|^{1-\beta}+|x_1-x_2|^{\beta}|t|^{\beta'}],0\le|y|\le\delta,\\
&|\tilde{\xi}_k(z)+\tilde{\eta}_k(z)-\tilde{\zeta}_k(z)-\tilde{\theta}_k(z)|\le M'\gamma^{k-1}|x_1-x_2|^{\beta}|y|^{\beta'},\ |\tilde{\xi}_k(z)+\tilde{\eta}_k(z)|\\
&\le M(M_2|\tilde{y}|)^{k-1}|x_1-x_2|^{1-\beta}/(k-1)!\text{ or }M'\gamma^{k-1}|x_1-x_2|^{\beta}|y|^{\beta'},
\end{aligned}\tag{2.42}$$

where $z=x+jy$, $z=x+jt$ is the intersection point of two characteristics of family s_1 in (2.26) passing through two points z_1,z_2, $\beta'=(1+m/2)(1-3\beta)$, β, δ are appropriately small positive constants, such that $(2+m)\beta<1$ and $\gamma=\max_{-\delta\le y\le 0}[4k_1d\delta^\beta+(4\varepsilon(y)+m)/2\beta']<1$, d is the diameter of D, $|x_1-x_2|^{1-\beta}\le k_0\,\delta^\beta\,|x_1-x_2|^{\beta}|y|^{\beta'}\le|x_1-x_2|^{\beta}|y|^{\beta'}$, if $0\le|y|\le\delta$, and M' is a sufficiently large positive constant as stated in (2.39) and (2.44) below.

In fact, from (2.40), it follows that the first formula with $k=1$ holds, namely

$$\begin{aligned}
&|\xi_1(z)-\zeta_1(z)|\le\Big|\int_0^y\tilde{E}_1dt\Big|\le\Big|\int_0^y\frac{k_1}{2}dt\Big|\le\frac{k_1|y|}{2},\\
&|\zeta_1(z)|=\Big|\int_{S_1}\tilde{E}_1dt\Big|\le\Big|\int_{S_1}\frac{k_1}{2}dt\Big|\le\frac{k_1|y_1|}{2},|\xi_1(z)|\le\frac{k_1}{2}|\tilde{y}|\le M,\\
&|\eta_1(z)-\theta_1(z)|\le\Big|\int_0^y\tilde{E}_1dt\Big|\le\frac{k_1|y|}{2},|\theta_1(z)|=|-\xi_1(x+G(y))|\le\frac{k_1|y_1|}{2},\\
&|\eta_1(z)|\le|\theta_1(z)|+|\eta_1(z)-\theta_1(z)|\le\frac{k_1}{2}|\tilde{y}|\le M|\tilde{y}|.
\end{aligned}$$

From the first formula in (2.40) and the above formula, we can obtain

$$|v_1(z)| \le 2d \max_{\overline{D^-}} |\xi_0(z) - \eta_0(z)| \le 2k_1 d^2.$$

Moreover we can get

$$\begin{aligned}
&|\xi_1(z_1)-\xi_1(z_2)-\zeta_1(z_1)+\zeta_1(z_2)| \le \left|\int_0^t [\tilde{E}_1(x_1+jt)-\tilde{E}_1(x_2+jt)]dt\right| \\
&\le \left|\int_0^t |\tilde{E}_{1x}||x_1-x_2|dt\right| \le \frac{k_1}{2}\left|\int_0^t |x_1-x_2|dt\right| \le \frac{k_1}{2}|t||x_1-x_2| \\
&\le \frac{M_3}{4k_1}|x_1-x_2|, |\zeta_1(z_1)-\zeta_1(z_2)| \le \left|\int_{S_1}\tilde{E}_1(z)dt - \int_{S_1'}\tilde{E}_1(z)dt\right| \\
&\le \left|\int_{S_2}\tilde{E}_{1x}|x_1-x_2|dt\right| + \left|\int_{S_3}\tilde{E}_{1x}|x_1-x_2|dt\right| + \left|\int_{S_4}\frac{\tilde{E}_1}{2H}d\mu\right| \\
&\le \left|\int_{y_1'}^0 \frac{k_1}{2}(x_1-x_2)dt\right| + \frac{k_1}{2H(y_1')}|G(y_2')-G(y_1')| \le \frac{k_1}{2}|y_1'||x_1-x_2| \\
&+\frac{k_1}{4H(y_1')}|x_1-x_2| \le \frac{k_1|\tilde{y}|[d+1/2H(y_1')]}{2}|x_1-x_2| \le \frac{M_3|\tilde{y}|}{4k_1}|x_1-x_2| \\
&\le \frac{M|\tilde{y}|}{8k_1}|x_1-x_2|,\ |\xi_1(z_1)-\xi_1(z_2)|, |\eta_1(z_1)-\eta_1(z_2)| \le \frac{M_3|\tilde{y}|}{2k_1}|x_1-x_2|, \\
&|\xi_1(z)+\eta_1(z)-\zeta_1(z)-\theta_1(z)| \le \left|\int_0^y [\tilde{E}_1(z_1)-\tilde{E}_1(z_2)]dt\right| \\
&\le \left|\int_0^y \tilde{E}_{1x}[x_1-x_2]dt\right| \le \frac{k_1|y|}{2}|x_1-x_2|,\ |\zeta_1(z)+\theta_1(z)| \\
&= |\zeta_1(z)-\zeta_1(x+G(y))| = |\zeta_1(x-G(y))-\zeta_1(x+G(y))| \\
&\le \frac{M_3|y_1|}{2k_1}|x_1-x_2|,\ |\xi_1(z)+\eta_1(z)| = |\xi_1(z)+\eta_1(z) \\
&-\zeta_1(z)-\theta_1(z)+\zeta_1(z)+\theta_1(z)| \le \frac{M|\tilde{y}|}{4}|x_1-x_2|,\ y \le -\delta, \\
&|\xi_1(z)+\eta_1(z)-\zeta_1(z)-\theta_1(z)|,\ |\xi_1(z)+\eta_1(z)| \\
&\le \frac{M|\tilde{y}|}{4}|x_1-x_2|^\beta|y|^{\beta'},\ 0 \le |y| \le \delta,
\end{aligned} \tag{2.43}$$

where $z = x + jy$ is the same as in (2.37) and (2.45) below, the meaning of the integral $\int_0^\mu [\tilde{E}_1/2H]d\mu$ is as stated in (2.25), $M_3 = 2k_1^2[d + 1/2H(y_1')]$ is a positive constant, S_1, S_1' are two characteristics of family s_1 in (2.26) passing through two points $z_2' = x_2' + jy_2', z_1'' = x_1'' + jy_1''$ on L_1 to two points $z_0' = x_0', z_0'' = x_0''(< x_0')$ on L_0, $z_1 = x_1 + jt$, $z_2 = x_2 + jt$ are two points on S_1, S_1' respectively, denote by $z_1' = x_1' + jy_1' = x_1' + jy_1''$ the intersection of S_1 and the line $y = y_1' = y_1''$, it is clear that $S_1 = S_2 \cup S_3 \cup S_4$, $S_2 = S_1 \cap \{-\delta \le y \le 0\}$, $S_3 = S_1 \cap \{y_1' \le y \le -\delta\}$, $S_4 = S_1 \cap \{y < y_1'\}$, and $x_2' - G(y_2') = x_1' - G(y_1') = x_0', x_1'' - G(y_1'') = x_0'', x_1 - x_2 = x_0' - x_0'' = x_1' - x_1'' = 2(x_1' - x_2') = x_2' - x_1'' + G(y_1'') - G(y_2')$, i.e. $G(y_1'') - G(y_2') = x_1' - x_2' = (x_1 - x_2)/2 \le k_0|y_0'|^{(m+2)/2}/(m+2)$, herein $z_0' = (x_0' + x_0'')/2 + jy_0'$ is the intersection of S_1 and S_2 emanating from the point $z = x_0''$ on L_0, and

$$|\tilde{v}_1(z)| = |v_1(z) - v_1(x)| \le M_1|y|, |\tilde{\xi}_1(z) - \tilde{\zeta}_1(z)|, |\tilde{\eta}_1(z) - \tilde{\theta}_1(z)| \le M_1|y|,$$

$$|\tilde{\xi}_1(z) + \tilde{\eta}_1(z)| = |\xi_1(z) + \eta_1(z)| \le M|x_1 - x_2|/4, \text{ or } M|\tilde{y}||x_1 - x_2|^{\beta}|y|^{\beta'},$$

$$|\tilde{v}_1(z_1) - \tilde{v}_1(z_2)| = |v_1(z_1) - v_1(z_2)| \le M|\tilde{y}||x_1 - x_2|,$$

$$|\tilde{\xi}_1(z_1) - \tilde{\xi}_1(z_2)| = |\xi_1(z_1) - \xi_1(z_2)| \le M|\tilde{y}||x_1 - x_2|,$$

$$|\tilde{\eta}_1(z_1) - \tilde{\eta}_1(z_2)| = |\eta_1(z_1) - \eta_1(z_2)| \le M|\tilde{y}||x_1 - x_2|.$$

In the following we use the inductive method, namely suppose the estimates in (2.42) for $k = n$ are valid, then we can prove that they are true for $k = n+1$. In the following, we only give the estimates of $\tilde{\xi}_{n+1}(z_1) - \tilde{\xi}_{n+1}(z_2)$, $\tilde{\xi}_{n+1}(z) + \tilde{\eta}_{n+1}(z)$, the other estimates can be similarly given. Firstly we estimate the upper bound of $|\tilde{\xi}_{n+1}(z_1) - \tilde{\xi}_{n+1}(z_2)|$. Noting that

$$\begin{aligned}
&|\tilde{A}_1(z_1)\tilde{\xi}_n(z_1) - \tilde{A}_1(z_2)\tilde{\xi}_n(z_2)| \le |(\tilde{A}_1(z_1) - \tilde{A}_1(z_2))\tilde{\xi}_n(z_1)\\
&+\tilde{A}_1(z_2)(\tilde{\xi}_n(z_1) - \tilde{\xi}_n(z_2))| \le k_1 M(M_2|\tilde{t}|)^{n-1}[|x_1 - x_2|/6(n-1)!\\
&+|x_1 - x_2|^{1-\beta}/6(n-1)!] \le k_1 M(M_2|\tilde{t}|)^{n-1}|x_1 - x_2|^{1-\beta}/3(n-1)!,\\
&t \le -\delta, \text{ and } \le k_1 M'\gamma^{n-1}[|x_1 - x_2|/6 + |x_1 - x_2|^{\beta}|t|^{\beta'}/6], -\delta \le t \le 0,\\
&|\tilde{C}_1(z_1)(\tilde{\xi}_n(z_1) + \tilde{\eta}_n(z_1)) - \tilde{C}_1(z_2)(\tilde{\xi}_n(z_2) + \tilde{\eta}_n(z_2))|\\
&\le |\frac{1}{2H}|(a(z_1) - a(z_2))(\tilde{\xi}_n(z_1) + \tilde{\eta}_n(z_1))\\
&+a(z_2)(\tilde{\xi}_n(z_1) + \tilde{\eta}_n(z_1) - \tilde{\xi}_n(z_2) - \tilde{\eta}_n(z_2))|\\
&+|\frac{m}{4t}(\tilde{\xi}_n(z_1) + \tilde{\eta}_n(z_1) - \tilde{\xi}_n(z_2) - \tilde{\eta}_n(z_2))|\\
&\le \frac{M(M_2|\tilde{t}|)^{n-1}}{(n-1)!}\frac{4\varepsilon(t) + m}{2|t|}|x_1 - x_2|^{1-\beta}, \ t \le -\delta, \text{ and}\\
&\le M'\gamma^{n-1}|x_1 - x_2|^{\beta}\frac{4\varepsilon(t) + m}{2|t|}|t|^{\beta'}, \ 0 \le |t| \le \delta,
\end{aligned}$$

from (2.41), we can get

$$
\begin{aligned}
&|\tilde{\xi}_{n+1}(z_1)-\tilde{\xi}_{n+1}(z_2)-\tilde{\zeta}_{n+1}(z_1)+\tilde{\zeta}_{n+1}(z_2)|\\
&=|\int_0^t[\tilde{g}_{1n}(z_1)-\tilde{g}_{1n}(z_2)]dt|\le|\int_0^t M'\gamma^{n-1}[2k_1|x_1-x_2|^{1-\beta}+|x_1-x_2|^{\beta}\\
&\times(2k_1+\frac{4\varepsilon(t)+m}{2|t|})|t|^{\beta'}]dt|\le M'\gamma^{n-1}[2k_1|x_1-x_2|^{1-\beta}|t|+|x_1-x_2|^{\beta}\\
&\times(2k_1|t|+\frac{4\varepsilon(t)+m}{2\beta'})|t|^{\beta'}]\le M'\gamma^{n}|x_1-x_2|^{\beta}|t|^{\beta'},\ 0\le|t|\le\delta,\\
&|\tilde{\xi}_{n+1}(z_1)-\tilde{\xi}_{n+1}(z_2)-\int_{-\delta}^{t}[\tilde{g}_{1n}(z_1)-\tilde{g}_{1n}(z_2)]dt|=|\int_{y_1}^{-\delta}[\tilde{g}_{1n}(z_1)-\tilde{g}_{1n}(z_2)]dt|\\
&\le|\int_{y_1'}^{-\delta}\frac{M(M_2|t|)^{n-1}}{(n-1)!}[k_1+\frac{4\varepsilon(t)+m}{2|t|}]|x_1-x_2|^{1-\beta}dt|\\
&+|\int_{S_4}\frac{M(M_2|t|)^{n-1}}{(n-1)!}[k_1+\frac{4\varepsilon(t)+m}{2|t|}|x_1-x_2|^{1-\beta}]dt|\\
&\le M\frac{M_2^{n-1}|\tilde{t}|^n}{(n-1)!}[\frac{M_2}{2n}+\frac{2k_1+(4\varepsilon_0+m)/\delta}{2n}]|x_1-x_2|^{1-\beta}\\
&\le M[\frac{(M_2|\tilde{t}|)^n}{2n!}+\frac{(M_2|\tilde{t}|)^n}{2n!}]|x_1-x_2|^{1-\beta}\le\frac{M(M_2|t|)^n}{n!}|x_1-x_2|^{1-\beta}\\
&\le\frac{M'\gamma^n}{2}|x_1-x_2|^{1-\beta}\le\frac{M'\gamma^n}{2}|x_1-x_2|^{\beta}|y|^{\beta'},\ 0\le|t|\le\delta,\\
&|\int_{y_1}^{t}[\tilde{g}_{1n}(z_1)-\tilde{g}_{1n}(z_2)]dt|\le\frac{M(M_2|t|)^n}{n!}|x_1-x_2|^{1-\beta},\ \ t\le-\delta,
\end{aligned}
\tag{2.44}
$$

where $\tilde{t}=|t|-y_1$, M_2 is as stated in (2.39), δ, β are sufficiently small positive constants, x_1, x_2 are the intersection points of two characteristics s_1, s_2 passing through a point $z=x+jt\in\overline{D^-}$ and $x-$axis, and we use $|x_1-x_2|^{1-2\beta}\le k_0\,\delta^{\beta}\,|y|^{\beta'}\le|y|^{\beta'}$, if $0\le|y|\le\delta$. Similarly to the integral $\int_0^y[\tilde{g}_{1n}(z_1)-\tilde{g}_{2n}(z_2)]dy$, the estimates of $\int_{-\delta}^{|y|}[\tilde{g}_{1n}(z_1)-\tilde{g}_{2n}(z_2)]dy$ and $\int_{-\delta}^{0}[\tilde{g}_{1n}(z_1)-\tilde{g}_{2n}(z_2)]dy$ can be obtained. It is obvious that there exists a sufficiently large positive integer N such that $2dM_2/(N+1)\le\gamma$, it remains that we can discuss the problem only for $n\le N$. But the constant M can be replaced by a sufficiently large positive constant $M'=2M\max_{0\le n\le N}(2dM_2)^n/(n!\gamma^n)$, then we have $2M(M_2|\tilde{t}|)^n/n!\le M'\gamma^n\le M'$, $n=1,2,\ldots$. Secondly we consider

$$
I=I_1+I_2,\ I_2=\tilde{\zeta}_{n+1}(z)+\tilde{\theta}_{n+1}(z)=\tilde{\zeta}_{n+1}(x-G(y))-\tilde{\zeta}_{n+1}(x+G(y)),
$$

$$
I_1=\tilde{\xi}_{n+1}(z)+\tilde{\eta}_{n+1}(z)-\tilde{\zeta}_{n+1}(z)-\tilde{\theta}_{n+1}(z),
$$

noting that

$$
\begin{aligned}
&|\tilde A_1(z_1)\tilde\xi_n(z_1)+\tilde A_2(z_2)\tilde\xi_n(z_2)+\tilde B_1(z_1)\tilde\eta_n(z_1)+\tilde B_2(z_2)\tilde\eta_n(z_2)|\\
&\le|[\tilde A_1(z_1)-\tilde A_1(z_2)]\tilde\xi_n(z_1)+[\tilde A_1(z_2)+\tilde A_2(z_2)]\tilde\xi_n(z_1)\\
&+\tilde A_2(z_2)[\tilde\xi_n(z_2)-\tilde\xi_n(z_1)]+[\tilde B_1(z_1)-\tilde B_1(z_2)]\tilde\eta_n(z_1)\\
&+[\tilde B_1(z_2)+\tilde B_2(z_2)]\tilde\eta_n(z_1)+\tilde B_2(z_2)[\tilde\eta_n(z_2)-\tilde\eta_n(z_1)]|\\
&\le k_1M\gamma^{n-1}(|x_1-x_2|^{1-\beta}+|x_1-x_2|^{\beta}|t|^{\beta'})/2=I_3,\\
&|\tilde C_1(z_1)(\tilde\xi_n(z_1)+\tilde\eta_n(z_1))+\tilde C_2(z_2)(\tilde\xi_n(z_2)+\tilde\eta_n(z_2))|\\
&=|\tilde C_2(z_2)[\tilde\xi_n(z_2)+\tilde\eta_n(z_2)+\tilde\xi_n(z_1)+\tilde\eta_n(z_1)]+[\tilde C_1(z_1)-\tilde C_2(z_2)]\\
&\times[\tilde\xi_n(z_1)+\tilde\eta_n(z_1)]|\le M\gamma^{n-1}|x_1-x_2|^{\beta}\Big(\frac{2\varepsilon(t)}{|t|}+\frac{m}{2|t|}\Big)|t|^{\beta'}=I_4,
\end{aligned}
$$

if $0\le|y|\le\delta$, and then the inequalities

$$
\begin{aligned}
&|I_1|=\Big|\int_0^y[\tilde g_{1n}(z_1)+\tilde g_{2n}(z_2)]dt\Big|\\
&\le\Big|\int_0^y M'\gamma^{n-1}[2k_1|x_1-x_2|^{1-\beta}+|x_1-x_2|^{\beta}\Big(2k_1+\frac{4\varepsilon(t)+m}{2|t|}\Big)|t|^{\beta'}]dt\Big|\\
&\le M'\gamma^{n-1}[2k_1|x_1-x_2|^{1-\beta}|y|+|x_1-x_2|^{\beta}\Big(2k_1|y|+\frac{4\varepsilon(y)+m}{2\beta'}\Big)|y|^{\beta'}]\\
&\le M'\gamma^{n}|x_1-x_2|^{\beta}|y|^{\beta'},\ |I|=\Big|\int_0^y[\tilde g_{1n}(z)+\tilde g_{2n}(z)]dy+\tilde\zeta_{n+1}(z)+\tilde\theta_{n+1}(z)\Big|\\
&=\Big|\int_{y_1}^y[\tilde g_{1n}(z)+\tilde g_{2n}(z)]dy\Big|\le M'\gamma^{n}|x_1-x_2|^{\beta}|y|^{\beta'},\ 0\le|y|\le\delta,
\end{aligned}
\tag{2.45}
$$

can be derived, where $z=x+jy$, and the estimate of $|I_2|$ is used by the same way in (2.44). Moreover we can obtain the estimate

$$
\begin{aligned}
&|I|=\Big|\int_{y_1}^y[\tilde g_{1n}(z)+\tilde g_{2n}(z)]dy\Big|\le\Big|\int_{y_1}^y\frac{M(M_2|\tilde y|)^{n-1}}{(n-1)!}\frac{(2+m)k_0}{\delta^{2+m}}|x_1-x_2|^{1-\beta}\\
&\times\Big[6k_0d+\frac{4\varepsilon(t)+m}{2|t|}\Big]dt\Big|\le\Big|\int_{y_1}^y\frac{M(M_2|\tilde y|)^{n-1}}{(n-1)!}\frac{(2+m)k_0}{\delta^{2+m}}|x_1-x_2|^{1-\beta}\\
&\times\Big[2k_1d+\frac{4\varepsilon(t)+m}{2\delta}\Big]dt\Big|\le M\frac{(M_2|\tilde y|)^{n}}{n!}|x_1-x_2|^{1-\beta},\ y\le-\delta,
\end{aligned}
$$

similarly to (2.43), where x_1,x_2 are the intersection points of the characteristic curves s_1, s_2 passing through a point $z=x+jy\in\overline{D^-}$ and $x-$axis respectively, and we use $|x_1-x_2|^{\beta-1}\le(2+m)k_0\delta^{-2-m}$ if $y\le-\delta$. Finally we estimate

$$II=II_1+II_2, II_1=\tilde{\xi}_{n+1}(z)-\tilde{\zeta}_{n+1}(z)=\int_0^y\tilde{g}_{1n}(z)dy$$
$$=\int_0^y[\tilde{A}_l\tilde{\xi}_n+\tilde{B}_l\tilde{\eta}_n+\tilde{C}_l(\tilde{\xi}_n+\tilde{\eta}_n)+\tilde{D}_l\tilde{v}_n]dy,\ z\in s_1,$$

and can get

$$\begin{aligned}&|II_1|=|\int_0^y\tilde{g}_{1n}(z)dt|\le|\int_0^y M'\gamma^{n-1}d[k_1+\frac{2\varepsilon(t)+m}{4|t|}|t|^{\beta'}]dt|\\&\le M'\gamma^{n-1}d[k_1|y|+\frac{2\varepsilon(y)+m}{4\beta'}|y|^{\beta'}]\le M'\frac{\gamma^n}{2}|y|^{1-\beta},0\le|y|\le\delta,\\&|II_2|=|\tilde{\zeta}_{n+1}(x+G(y))|\le|\int_{-\delta}^0\tilde{g}_{1n}dt|+|\int_{y_1}^{-\delta}M\frac{(M_2|\tilde{t}|)^{n-1}}{(n-1)!}\\&\times[k_1d+\frac{2\varepsilon_0+m}{4\delta}d]dt|\le\frac{1}{2}[M'\gamma^n+M\frac{(M_2d)^n}{n!}],\\&|II|\le|II_1+II_2|\le M\frac{(M_2|\tilde{y}|)^n}{n!},\ y\le-\delta.\end{aligned}\tag{2.46}$$

In addition similarly to (2.46), we consider

$$III=III_1+III_2,\ III_2=\tilde{\theta}_{n+1}(z)=-\tilde{\zeta}_{n+1}(x+G(y)),$$
$$III_1=\tilde{\eta}_{n+1}(z)-\tilde{\theta}_{n+1}(z)=\int_0^y\tilde{g}_{2n}(z)dy,\ z\in s_2,$$

and can obtain

$$\begin{aligned}&|III_2|=|\tilde{\zeta}_{n+1}(x+G(y))|\le\frac{1}{2}[M'\gamma^n+M\frac{(M_2d)^n}{n!}],\\&|III_1|=|\int_0^y\tilde{g}_{2n}(z)dy|\le\frac{M'}{2}\gamma^n|y|^{1-\beta},\ 0\le|y|\le\delta,\\&|III|=|III_1+III_2|\le M\frac{(M_2|\tilde{y}|)^n}{n!},\ y\le-\delta.\end{aligned}\tag{2.47}$$

Here we mention that the functions $\tilde{\xi}_{n-1}(z)$, $\tilde{\xi}_{n-1}(z)+\tilde{\eta}_{n-1}(z)$, $\tilde{\eta}_{n-1}(z)$ can be successively chosen.

On the basis of the estimate (2.42), the convergence of two sequences $\{M(M_2|\tilde{y}|)^k/k!\}$, $\{M'\gamma^k|y|^{\beta'}\}$, and the comparison test, we can see that from $\{v_k(z)\}$, $\{\xi_k(z)\}$, $\{\eta_k(z)\}$, there exist the subsequences $\{v_k^l(z)\}$, $\{\xi_k^l(z)\}$, $\{\eta_k^l(z)\}$, which uniformly converge to $v_*(z)$, $\xi_*(z)$, $\eta_*(z)$ in

$$D_l = \overline{D^-} \cap \{|z| > 1/l\} \cap \{|z-2| > 1/l\},\ l = 1, 2, ...$$

satisfying

$$\begin{aligned}
&v_*(z) = v_*(x) - 2\int_0^y V_* dy = v_*(x) + \int_0^y (\eta_* - \xi_*)dt,\\
&\xi_*(z) = \zeta_*(z) + \int_0^y [\tilde{A}_1\xi_* + \tilde{B}_1\eta_* + \tilde{C}_1(\xi_* + \eta_*) + \tilde{D}_1 u_* + \tilde{E}_1]dt, z \in s_1,\\
&\eta_*(z) = \theta_*(z) + \int_0^y [\tilde{A}_2\xi_* + \tilde{B}_2\eta_* + \tilde{C}_2(\xi_* + \eta_*) + \tilde{D}_2 u_* + \tilde{E}_2]dt, z \in s_2,
\end{aligned}$$

and the function $[W_*(z), v_*(z)] = [(\xi_* + \eta_* + j\xi_* - j\eta_*)/2, v_*(z)]$ is a solution of Problem A_2 for equation (2.31) in D_l. Moreover from $\{v_k^l(z)\}$, $\{\xi_k^l(z)\}$, $\{\eta_k^l(z)\}$, we can choose the diagonal subsequence $\{v_l^l(z)\}$, $\{\xi_l^l(z)\}$, $\{\eta_l^l(z)\}$ in D_l, the limits of which are $v_*(z)$, $\xi_*(z)$, $\eta_*(z)$ in $\overline{D^-}\backslash\{0, 2\}$ respectively, thus $u(z) = v_*(z) + u_0(z)$ is a solution of Problem T for (2.1) in $\overline{D^-} \cap \{0, 2\}$. Hence the solution $u(z)$ of Problem T in $\overline{D^-}$ is obtained. In addition the boundary value $u_{0y}(x) = -2\text{Im}W$ of the solution $u_0(z)$ of Problem T on L_0 can be as a part of boundary value of Problem A_1, we can find the solution of Problem A_1 in $\overline{D^+}$. Thus the function $u(x)$ on L_0 about Problem A_2 is obtained. Hence the existence of solutions of Problem T for equation (2.1) is proved.

From the above discussion, we obtain the following theorem.

Theorem 2.4 *Let equation* (2.1) *satisfy Condition C and* (2.36). *Then the Tricomi problem* (*Problem T*) *for* (2.1) *is solvable.*

In the following we shall prove the uniqueness theorem of Problem T for equation (2.1).

Theorem 2.5 *Under the same conditions as in Theorem* 2.4, *Problem T for* (2.1) *has at most one solution in D.*

Proof Let $u_1(z), u_2(z)$ be any two solutions of Problem T for (2.1). By Theorem 2.2, it is easy to see that $u(z) = u_1(z) - u_2(z)$ and $w(z) = H(y)u_{\bar{z}}$ satisfy the homogeneous equation and boundary conditions

$$\begin{aligned}
&w_{\overline{Z}} = A_1 w + A_2\overline{w} + A_3 u \text{ in } D,\\
&\text{Re}[\overline{\lambda(z)}w(z)] = 0, z \in \Gamma \cup L_1,\ u(0) = 0.
\end{aligned} \tag{2.48}$$

Now we verify that the above solution $u(z) \equiv 0$ in D^+. If the maximum $M = \max_{\overline{D}} u(z) > 0$, it is clear that the maximum point $z^* \notin D \cup \Gamma$, and then $z^* = x_* \in L_0 = (0, 2)$. Denote by $U(x_*) = \{|z - x_*| < \varepsilon\ (> 0)\}$ a neighborhood of x_*, noting that u_x, u_y in $U(x_*) \cap \{y \geq 0\}$ are continuous and $U(x) = H(0)u_x/2 = 0$ on L_0 on the $x-$axis, we can symmetrically extend $W(z)$ from $U(x_*) \cap \{y > 0\}$ onto the symmetrical domain $U(x_*) \cap \{y < 0\}$ about $\mathrm{Im}z = y = 0$, according to Theorem 2.2 and the property of integral $2i\mathrm{Im}T[g(Z)/H(\mathrm{Im}Z)]$ over $U(x_*)$ (see Lemma 2.1, Chapter I), we know that $\mathrm{Re}W[z(Z)] = \mathrm{Re}\hat{\Phi}(Z) = H(y)u_x/2$ in $U(x_*) \cap \{y > 0\}$ is a harmonic function with the condition $\mathrm{Re}W(x) = H(y)u_x/2 = 0$ on $U(x_*) \cap \{y = 0\}$, hence

$$H(y)u_x = \sum_{k,l=0}^{\infty} c_{kl}(x - x_*)^k Y^l = YF,$$

where F is a continuous function in $U(x_*)$, and then we have $u_x = O(Y^{2/(m+2)}F)$ in $U(x_*) \cap \{y \geq 0\}$ and $u_x = 0$ on $U(x_*) \cap \{y = 0\}$, this shows that $u(x) = M$ on $U(x_*) \cap \{y = 0\}$, from this we can derive $u(x) = M$ on L_0, this contradicts $u(0) = 0$. Hence $\max_{\overline{D^+}} u(z) = 0$. Similarly we can verify $\min_{\overline{D^+}} u(z) = 0$. Thus $u(z) = 0$ in $\overline{D^+}$. Moreover similarly to the proof of Theorem 6.1, Chapter III, we have $u(z) = 0$ in $\overline{D^-}$. Therefore $u(z) = 0$, $u_1(z) = u_2(z)$ in $\overline{D}$.

From the above discussion, we obtain the following theorem.

Theorem 2.6 *Let equation* (2.1) *satisfy Condition C and the condition* (2.36). *Then the Tricomi problem* (*Problem T*) *for* (2.1) *has a unique solution.*

In [74] M. M. Smirnov mainly investigated the unique solvability of Tricomi problem for the Chaplygin equation, i.e. equation (2.1) with $a = b = c = d = 0$ in D. In [76]3), H. S. Sun discussed the uniqueness and existence of solutions of homogeneous Tricomi problem for general equation of mixed type with parabolic degenerate line, but he assumes that the coefficients of (2.1) satisfy some stronger conditions, for instance $b \geq 0$, $c \leq 0$, $K_y - a\max_{L_2}\sqrt{-K} > 0$, $c_y - \max_{L_2}\sqrt{-K}c_x \leq 0$ in D, and the inner angles of elliptic domain D^+ at $z = 0, 2$ are less than $\pi/2$.

Finally we mention that the coefficients $K(y)$ in equation (2.1) can be replaced by the function $K(x, y)$ with some conditions, for instance $K(x, y) = \mathrm{sgn}y|y|^m h(x, y)$, m is a positive number, and $h(x, y)$ is a continuously differentiable positive function. Besides if the boundary condition

(2.6) is replaced by the boundary conditions

$$u(z) = \phi(z) \text{ on } \Gamma, \ u(x) = \psi(x) \text{ on } L_2,$$

where the coefficients of the above boundary condition satisfy the conditions similar to those in (2.7), i.e.

$$C^2_\alpha[\phi(z),\Gamma] \le k_2, C^2_\alpha[\psi(x),L_2] \le k_2, \phi(2) = \psi(2),$$

in which $\alpha\,(0<\alpha<1), k_2$ are positive constants, then we can also discuss by the similar method.

3 The Discontinuous Oblique Derivative Problem for Second Order Degenerate Equations of Mixed Type

In [12]1),3), the author posed and discussed the discontinuous Tricomi problem of a special second order equation of mixed type without degenerate line. The present section deals with the discontinuous oblique derivative problem (general boundary value problem) for second order linear equations of mixed (elliptic-hyperbolic) type with parabolic degeneracy. Firstly, we give the formulation of the above boundary value problem, and then prove the existence of solutions for the above problem, in which the complex analytic method is used.

3.1 Formulation of discontinuous oblique derivative problem for equations of mixed type

Let D be a simply connected bounded domain in the complex plane $\mathbf{C}$ with the boundary $\partial D = \Gamma \cup L$, where $\Gamma(\subset \{y>0\}) \in C^2_\mu\ (0<\mu<1)$ is a curve with the end points $z = 0, 2$, and $L = L_1 \cup L_2$, where

$$\begin{aligned} L_1 &= \{x + \int_0^y \sqrt{|K(t)|}dt = 0, x \in [0,1]\}, \\ L_2 &= \{x - \int_0^y \sqrt{|K(t)|}dt = 2, x \in [1,2]\}, \end{aligned} \tag{3.1}$$

$K(y) = \operatorname{sgn} y|y|^m h(y)$ is as stated in Section 2. Denote $D^+ = D \cap \{y > 0\}$, $D^- = D\cap\{y<0\}$. Similarly to Section 2, there is no harm in assuming that the boundary Γ of the domain D^+ is a smooth curve with the form $x-\tilde{G}(y)=0$ and $x+\tilde{G}(y)=2$ near the points $z=0$ and 2 respectively as stated in Section 2. Denote $D^+=D\cap\{y>0\}$, $D^-=D\cap\{y<0\}$. Consider second order linear equation (2.1) of mixed type with parabolic degeneracy, i.e.

$$\begin{gathered}Lu=K(y)u_{xx}+u_{yy}+a(x,y)u_x+b(x,y)u_y\\ +c(x,y)u=-d(x,y) \text{ in } D.\end{gathered}\tag{3.2}$$

Denote $H(y)=\sqrt{|K(y)|}$, and suppose that the coefficients of (3.2) satisfy **Condition** C, namely

$$\begin{gathered}L_\infty[\eta,\overline{D^+}]\le k_0,\eta=a,b,c,L_\infty[d,\overline{D^+}]\le k_1,c\le 0 \text{ in } \overline{D^+},\\ \hat{C}[a,\overline{D^-}]=[a,\overline{D^-}]+C[a_x,\overline{D^-}]\le k_0,\\ \hat{C}[b,\overline{D^-}]\le k_0,\hat{C}[c,\overline{D^-}]\le k_0,\tilde{C}[d,\overline{D^-}]\le k_1,\end{gathered}\tag{3.3}$$

in which k_0, k_1 are positive constants. Let the functions $a(z)$, $b(z)$, $r(z)$ be continuous functions on Γ and $\lambda(z)=a(z)+ib(z)$ with the condition $|a(z)|+|b(z)|\ne 0$. Moreover there exist n points $E_1=a_1$, $E_2=a_2$, ..., $E_n=a_n$ on the segment $AB=[0,2]$ and $E_0=0, E_{n+1}=2$, where $a_0=0<a_1<a_2<\ldots<a_n<a_{n+1}=2$. Denote by $A=A_0=0, A_1=a_1/2-j|(-G)^{-1}(a_1/2)|$, $A_2=a_2/2-j|(-G)^{-1}(a_2/2)|$, ..., $A_n=a_n/2-j|(-G)^{-1}(a_n/2)|$, $A_{n+1}=C=1-j|(-G)^{-1}(1)|$ and $B_1=1-j|(-G)^{-1}(1)|+a_1/2+j|(-G)^{-1}(a_1/2)|$, $B_2=1-j|(-G)^{-1}(1)|+a_2/2+j|(-G)^{-1}(a_2/2)|$, ..., $B_n=1-j|(-G)^{-1}(1)|+a_n/2+j|(-G)^{-1}(a_n/2)|, B_{n+1}=B=2$ on the segments AC, CB respectively. Besides we denote $D_1^-=\overline{D^-}\cap\{\cup_{l=0}^{[n/2]}(a_{2l}\le x-G(y)\le a_{2l+1})\}$, $D_2^-=\overline{D^-}\cap\{\cup_{l=1}^{[(n+1)/2]}(a_{2l-1}\le x+G(y)\le a_{2l})\}$ and $\tilde{D}_{2l+1}^-=\overline{D^-}\cap\{a_{2l}\le x-G(y)\le a_{2l+1}\}$, $l=0,1,\ldots,[n/2]$, $\tilde{D}_{2l}^-=\overline{D^-}\cap\{a_{2l-1}\le x+G(y)\le a_{2l}\}$, $l=1,\ldots,[(n+1)/2]$, and $D_*^+=\overline{D^+}\backslash Z_0$, $D_*^-=\overline{D^-}\backslash Z_0$, $Z_0=\{a_0,a_1,\ldots,a_n,a_{n+1}\}$, and $D_*=D_*^+\cup D_*^-$.

The discontinuous oblique derivative boundary value problem for equation (3.2) may be formulated as follows:

Problem P Find a continuous solution $u(z)$ of (3.2) in $\bar{D}$, where u_x, u_y are continuous in $D_*=D_*^+\cup D_*^-$ and satisfy the boundary conditions

$$\begin{gathered}2\frac{\partial u}{\partial\nu}=\frac{1}{H(y)}\mathrm{Re}[\overline{\lambda(z)}u_{\bar z}]=\mathrm{Re}[\overline{\Lambda(z)}u_{\bar z}]=r(z),\ z\in\Gamma,\\ 2\frac{\partial u}{\partial\nu}=\frac{1}{H(y)}\mathrm{Re}[\overline{\lambda(z)}u_{\bar z}]=\mathrm{Re}[\overline{\Lambda(z)}u_z]=r(z),z\in L_3=\sum_{l=0}^{[n/2]}A_{2l}A_{2l+1},\end{gathered}$$

$$2\frac{\partial u}{\partial \nu}=\frac{1}{H(y)}\text{Re}[\overline{\lambda(z)}u_{\tilde{z}}]=r(z),\ z\in L_4=\sum_{l=1}^{[(n+1)/2]}B_{2l-1}B_{2l}, \tag{3.4}$$

$$\begin{aligned}&\frac{1}{H(y)}\text{Im}[\overline{\lambda(z)}u_{\tilde{z}}]|_{z=A_{2l+1}}=c_{2l+1}, l\in Z=\{0,1,...,[\frac{n}{2}]\}, u(0)=b_0,\\&\frac{1}{H(y)}\text{Im}[\overline{\lambda(z)}u_{\tilde{z}}]|_{z=B_{2l-1}}=c_{2l}, l\in Z'=\{1,...,[\frac{n+1}{2}]\}, u(2)=b_1,\end{aligned} \tag{3.5}$$

in which ν is a vector at every point on $\Gamma\cup L_3\cup L_4$, $b_l\,(l=0,1)$, $c_l\,(l=1,...,n+1)$ are real constants, $\Lambda(z)=a(z)+ib(z)=\cos(\nu,x)-i\cos(\nu,y)$, $z\in\Gamma$, $\Lambda(z)=a(z)+jb(z)=\cos(\nu,x)+j\cos(\nu,y)$, $z\in L_3\cup L_4$, the functions $\lambda(z)$, $r(z)$ and the constants $b_l\,(l=0,1)$, $c_l\,(l=1,...,n+1)$ satisfy the conditions

$$\begin{aligned}&C^1_\alpha[\lambda(z),\Gamma]\le k_0, C^1_\alpha[r(z),\Gamma]\le k_2, C^1_\alpha[\lambda(x),L_l]\le k_0,\\&C^1_\alpha[r(x),L_l]\le k_2, l=3,4,\ \cos(\nu,n)\ge 0\ \text{on}\ \Gamma\cup L_3\cup L_4,\\&|b_l|\le k_2,\ l=0,1,\ |c_l|\le k_2,\ l=1,...,n+1,\\&\max_{z\in L_3}\frac{1}{|a(x)-b(x)|}\le k_0,\ \max_{z\in L_4}\frac{1}{|a(x)+b(x)|}\le k_0,\end{aligned} \tag{3.6}$$

where n is the outward normal vector at every point on Γ, $\lambda(z)$, $r(z)$ are replaced by $\lambda(x)$, $r(x)$ on $L_3\cup L_4$, and $\alpha\,(0<\alpha<1)$, k_0,k_2 are positive constants. Moreover we give the some definition: Denote by $\lambda(t_l-0)$ and $\lambda(t_l+0)$ the left limit and right limit of $\lambda(z)$ as $z\to t_l=a_l\ (l=0,1,...,n+1)$ on $\Gamma\cup L_0\ (L_0=(0,2))$, and

$$\begin{aligned}&e^{i\phi_l}=\frac{\lambda(t_l-0)}{\lambda(t_l+0)},\ \gamma_l=\frac{1}{\pi i}\ln\frac{\lambda(t_l-0)}{\lambda(t_l+0)}=\frac{\phi_l}{\pi}-K_l,\\&K_l=\left[\frac{\phi_l}{\pi}\right]+J_l,\ J_l=0\ \text{or}\ 1,\ l=0,1,...,n+1,\end{aligned} \tag{3.7}$$

in which $[a]$ is the largest integer not exceeding the real number a, $\lambda(z)=\exp(i\pi/2)$ on $L_0\cap\overline{D_1^-}$,

$$\lambda(a_{2l}+0)=\lambda(a_{2l+1}-0)=\exp(i\pi/2),\ l=0,1,...,[n/2],$$

and $\lambda(z)=\exp(i\pi/2)$ on $L_0\cap\overline{D_2^-}$,

$$\lambda(a_{2l-1}+0)=\lambda(a_{2l}-0)=\exp(i\pi/2),\ l=1,...,[(n+1)/2],$$

where $0 \leq \gamma_l < 1$ when $J_l = 0$, and $-1 < \gamma_l < 0$ when $J_l = 1\,(l = 0, 1, ..., n+1)$, and

$$K = \frac{1}{2}(K_0 + K_1 + \cdots + K_{n+1}) = \sum_{l=0}^{n+1}\left(\frac{\phi_l}{2\pi} - \frac{\gamma_l}{2}\right) \tag{3.8}$$

is called the index of Problem P. Similarly to Section 3, Chapter II, we can choose $K = 0$ on Γ. Here we can assume that $-1/2 \leq \gamma_j < 1/2\,(j = 1, 2)$. Besides similarly to Section 2, Chapter II, we require that the solution $u(z)$ in D^+ satisfies the conditions

$$u_z = O(|Z - z_l|^{-\eta_l}),\ Z = x + iG(y),\ l = 0, 1, ..., n+1, \tag{3.9}$$

in the neighborhood D_l of $a_l\,(1 \leq l \leq n)$ in D^+, where $\eta_l = 2\delta$, if $\gamma_l \geq 0\,(l = 0$ or $n+1)$, $\eta_l = \max(0, -2\gamma_l)\,(l = 0$ or $n+1)$, $\eta_l = 1\,(l = 1, ..., n)$. and δ is a sufficiently small positive number. Now we choose $K_0 = \cdots = K_{n+1} = 0$, then $K = 0$. If $\cos(\nu, n) = 0$ on Γ, we can choose $K_0 = -1, K_1 = \cdots = K_{n+1} = 0$, and the index $K = -1/2$, in this case, the last point condition in (3.5) should be cancelled. If we consider $\mathrm{Re}[\overline{\lambda(z)}(U + jV)] = 0$ on L_0, where $\lambda(z) = 1$, then $\gamma_0 = -1/2$, $\gamma_2 = \cdots = \gamma_n = 0$, $\gamma_{n+1} = -1/2$, $K = 0$, and we need the last point condition in (3.5) such that the boundary value problem in D^+ is well-posed.

Furthermore, we need to introduce another oblique derivative boundary value problem.

Problem Q If $c = 0$ in equation (3.2), we find a continuously differentiable solution $u(z)$ of (3.2) in D_*, which is continuous in $\bar{D}$ and satisfies the boundary conditions (3.4), (3.5), but the last point conditions in (3.5) are replaced by

$$\frac{1}{H(y)}\mathrm{Im}[\overline{\lambda(z)}u_{\bar{z}}]|_{z=z_0'} = b_1, \tag{3.10}$$

where $z_0'(\notin Z_0) \in \Gamma$ is a point, b_1 is a real constant, in this case we do not assume $\cos(\nu, n) \geq 0$ on Γ.

Similarly to Subsection 2.1, there is no harm in assuming that the boundary condition in (3.4),(3.5) (or (3.10)) is the homogeneous boundary condition, because we can find two twice continuously differentiable functions $u_0^\pm(z)$ in $\overline{D^\pm}$, and then consider the functions $v(z) = v^\pm(z) = u(z) - u_0^\pm(z)$ in $\overline{D^\pm}$.

3.2 Representation of solutions of discontinuous oblique derivative problem for mixed equations

In this section, we first write the complex form of equation (3.2). Denote

$$Y = G(y) = \int_0^y H(y)dy = \pm\frac{2}{m+2}|y|^{(m+2)/2},$$

where $H(y) = |y|^{m/2}$, m is a positive number, and similarly to (2.12)-(2.16), we have

$$\begin{aligned}
&W(z)=U+iV=\frac{1}{2}[H(y)u_x-iu_y]=u_{\tilde{z}}=\frac{H(y)}{2}[u_x-iu_Y]=H(y)u_Z,\\
&H(y)W_{\overline{Z}}=\frac{H(y)}{2}[W_x+iW_Y]=\frac{1}{2}[H(y)W_x+iW_y]=W_{\bar{\tilde{z}}} \text{ in } \overline{D^+},\\
&W(z)=U+jV=\frac{1}{2}[H(y)u_x-ju_y]=u_{\tilde{z}}=\frac{H(y)}{2}[u_x-ju_Y]=H(y)u_Z,\\
&H(y)W_{\overline{Z}}=W_{\bar{\tilde{z}}}=He_1(U+V)_\mu+He_2(U-V)_\nu\\
&=\frac{e_1}{4}\{[\frac{a}{H}+\frac{H_y}{H}-b](U+V)+[\frac{a}{H}+\frac{H_y}{H}+b](U-V)+cu+d\}\\
&+\frac{e_2}{4}\{[\frac{a}{H}-\frac{H_y}{H}-b](U+V)+[\frac{a}{H}-\frac{H_y}{H}+b](U-V)+cu+d\},\\
&H(U+V)_\mu=\frac{1}{4}\{[\frac{a}{H}+\frac{H_y}{H}-b](U+V)\\
&+[\frac{a}{H}+\frac{H_y}{H}+b](U-V)+cu+d\},\\
&H(U-V)_\nu=\frac{1}{4}\{[\frac{a}{H}-\frac{H_y}{H}-b](U+V)\\
&+[\frac{a}{H}-\frac{H_y}{H}+b](U-V)+cu+d\} \text{ in } \overline{D^-},
\end{aligned} \tag{3.11}$$

in which $Z = x + iG(y)$ in $\overline{D^+}$, and $Z = x + jG(y)$ in $\overline{D^-}$, thus we can obtain

$$W_{\bar{\tilde{z}}} = A_1(z)W + A_2(z)\overline{W} + A_3(z)u + A_4(z) \text{ in } D,$$

$$A_1=\begin{cases}\frac{1}{4}[-\frac{a}{H}+\frac{iH_y}{H}-ib],\\ \frac{1}{4}[\frac{a}{H}+\frac{jH_y}{H}-jb],\end{cases} A_2=\begin{cases}\frac{1}{4}[-\frac{a}{H}+\frac{iH_y}{H}+ib],\\ \frac{1}{4}[\frac{a}{H}+\frac{jH_y}{H}+jb],\end{cases}$$

$$A_3 = \begin{cases} -\dfrac{c}{4}, \\ \dfrac{c}{4}, \end{cases} \quad A_4 = \begin{cases} -\dfrac{d}{4} \text{ in } \overline{D^+}, \\ \dfrac{d}{4} \text{ in } \overline{D^-}. \end{cases} \tag{3.12}$$

In particular, the complex equation

$$W_{\bar{z}} = 0, \text{ i.e. } W_{\overline{Z}} = 0 \text{ in } \overline{D} \tag{3.13}$$

can be rewritten in the system

$$\begin{gathered} W_{\overline{Z}} = 0 \text{ in } \overline{D^+_Z}, \\ (U+V)_\mu = 0, \ (U-V)_\nu = 0 \text{ in } \overline{D^-_\tau}. \end{gathered} \tag{3.14}$$

The boundary value problem for equations (3.12) with the boundary condition (3.4), (3.5) or (3.10) ($W(z) = u_{\tilde{z}}$) and the relation

$$u(z) = \begin{cases} 2\text{Re} \displaystyle\int_0^z [\frac{\text{Re}w(z)}{H(y)} + i\text{Im}w(z)]dz + b_0 \text{ in } \overline{D^+}, \\ 2\text{Re} \displaystyle\int_0^z [\frac{\text{Re}w(z)}{H(y)} - j\text{Im}w(z)]dz + b_0 \text{ in } \overline{D^-}, \end{cases} \tag{3.15}$$

will be called Problem A or B.

Now, we give the representation of solutions for the discontinuous oblique derivative problem (Problem Q) for equation (3.2) in $\overline{D}$. For this, we first discuss the Riemann-Hilbert problem (Problem B) for the second system of (3.14) in $\overline{D^-}$ with the boundary conditions

$$\text{Re}[\overline{\lambda(z)}(U+jV)] = \begin{cases} H(y)r(z) = R_1(z), \ z \in L_3 \cup L_4, \\ R_0(x), x \in L_0 = \{L_0 \cap D_1^-\} \cup \{L_0 \cap D_2^-\}, \end{cases}$$

$$\text{Im}[\overline{\lambda(z)}(U+jV)]|_{z=A_{2l+1}} = H(\text{Im}A_{2l+1})c_{2l+1} = c'_{2l+1}, l=0,1,...,[n/2],$$

$$\text{Im}[\overline{\lambda(z)}(U+jV)]|_{z=B_{2l-1}} = H(\text{Im}B_{2l-1})c_{2l} = c'_{2l}, l=1,...,[(n+1)/2], \tag{3.16}$$

in which $\lambda(z) = a(z) + jb(z)$ on L_1 and $\lambda(z) = 1 + j$ on $L'_0 = L_0 \cap D_1^-$, $\lambda(z) = 1 - j$ on $L''_0 = L_0 \cap D_2^-$, and $R_0(x)$ is an undetermined real function. It is clear that the solution of Problem B for (3.14) in $\overline{D^-}$ can be expressed as

$$\begin{gathered} \xi = U(z) + V(z) = f(\nu), \ \eta = U(z) - V(z) = g(\mu), \\ U(z) = [f(\nu) + g(\mu)]/2, \ V(z) = [f(\nu) - g(\mu)]/2, \text{ i.e.} \\ W(z) = U(z) + jV(z) = [(1+j)f(\nu) + (1-j)g(\mu)]/2, \end{gathered} \tag{3.17}$$

where $f(t), g(t)$ are two arbitrary real continuous functions on $[0,2]$. For convenience, sometimes we denote by the functions $a(x), b(x), r(x)$ of x the functions $a(z), b(z), r(z)$ of z in (3.16), thus (3.16) can be rewritten as

$$\begin{aligned}
&a(x)U(z) - b(x)V(z) = H(y)r(x) = R_1(x) \text{ on } L_3 \cup L_4,\\
&U(x) - V(x) = R_0(x) \text{ on } L_0', U(x) + V(x) = R_0(x) \text{ on } L_0'',\\
&[(a(z) - jb(z))(U(z) + jV(z))]|_{z=A_{2l+1}} = R_1(\mathrm{Re}A_{2l+1}) + jc_{2l+1}',\\
&\qquad l = 0, 1, ..., [n/2],\\
&[(a(z) - jb(z))(U(z) + jV(z))|_{z=B_{2l-1}} = R_1(\mathrm{Re}B_{2l-1}) + jc_{2l}',\\
&\qquad l - 1, ..., [(n+1)/2].
\end{aligned}$$

From the above formulas, we have

$$\begin{aligned}
&[a(x) - b(x)]f(2x) + [a(x) + b(x)]h_{2l} = 2R_1(x) \text{ on } L_3,\\
&[a(x) - b(x)]h_{2l-1} + [a(x) + b(x)]g(2x-2) = 2R_1(x) \text{ on } L_4,\\
&\mathrm{Im}[\overline{\lambda(z)}u_{\bar z}]|_{z=A_{2j+1}} = R_1(\mathrm{Re}A_{2l+1}) + jc_{2l+1}', l = 0, 1, .., [n/2],\\
&\mathrm{Im}[\overline{\lambda(z)}u_{\bar z}]|_{z=B_{2j-1}} = R_1(\mathrm{Re}B_{2l-1}) + jc_{2l}', l = 1, .., [(n+1)/2],\\
&[a(\mathrm{Re}A_{2l+1}) + b(\mathrm{Re}A_{2l+1})]h_{2l} = [a(\mathrm{Re}A_{2l+1}) + b(\mathrm{Re}A_{2l+1})][U(A_{2l+1})\\
&-V(A_{2l+1})] = R_1(\mathrm{Re}A_{2l+1}) - c_{2l+1}' \text{ or } 0, l = 0, 1, ..., [n/2],\\
&[a(\mathrm{Re}B_{2l-1}) - b(\mathrm{Re}B_{2l-1})]h_{2l-1} = [a(\mathrm{Re}B_{2l-1}) - b(\mathrm{Re}B_{2l-1})][U(B_{2l-1})\\
&+V(B_{2l-1})] = R_1(\mathrm{Re}B_{2l-1}) + c_{2l}' \text{ or } 0, l = 1, ..., [(n+1)/2].
\end{aligned} \tag{3.18}$$

The above formulas can be rewritten as

$$\begin{aligned}
f(x - G(y)) &= \frac{2R_1((x - G(y))/2)}{a((x - G(y))/2) - b((x - G(y))/2)}\\
&- \frac{[a((x - G(y))/2) + b((x - G(y))/2)]h_{2l}}{a((x - G(y))/2) - b((x - G(y))/2)} \text{ in } D_1^-,\\
g(x + G(y)) &= \frac{2R_1((x + G(y))/2 + 1)}{a((x + G(y))/2 + 1) + b((x + G(y))/2 + 1)}\\
&- \frac{[a((x+G(y))/2 + 1) - b((x+G(y))/2 + 1)]h_{2l-1}}{a((x+G(y))/2 + 1) + b((x+G(y))/2 + 1)} \text{ in } D_2^-.
\end{aligned}$$

In particular we have

$$\tilde{R}_0(x)=\begin{cases} f(x)=\dfrac{2R_1(x/2)-[a(x/2)+b(x/2)]h_{2l}}{a(x/2)-b(x/2)} \text{ on } L_0', \\ g(x)=\dfrac{2R_1(x/2+1)-[a(x/2+1)-b(x/2+1)]h_{2l-1}}{a(x/2+1)+b(x/2+1)} \text{ on } L_0''. \end{cases} \tag{3.19}$$

Thus the solution $w(z)$ of (3.13) can be expressed as

$$\begin{aligned} W(z) &= f(x-G(y))e_1 + g(x+G(y))e_2 \\ &= \frac{1}{2}\{f(x-G(y))+g(x+G(y))+j[f(x-G(y))-g(x+G(y))]\}, \end{aligned} \tag{3.20}$$

where $f(x-G(y))$, $g(x+G(y))$ are as stated in the above formula and h_{2l-1}, h_{2l} are as before.

Next we find a solution of the Riemann-Hilbert boundary value problem (Problem B) for equation (3.13) in D^+ with the boundary conditions

$$\begin{aligned} &\mathrm{Re}[\overline{\lambda(z)}(U(x)+iV(x))] = R_1(z) \text{ on } \Gamma,\ u(0)=b_0, \\ &\mathrm{Re}[\overline{\lambda(z)}(U(x)+iV(x))]=\tilde{R}_0(x) \text{ on } L_0'\cup L_0'', \\ &\mathrm{Im}[\overline{\lambda(z)}(U(x)+iV(x))]|_{z=z_0'}=H(\mathrm{Im}z_0')b_1 = b_1', \end{aligned} \tag{3.21}$$

where $\lambda(z)=1+i$ on L_0', $\lambda(z)=1-i$ on L'', $\tilde{R}_0(x)$ is as stated in (3.19). Noting that the index of the above boundary condition is $K=0$, by the result in Section 1, Chapter I, we know that the above Riemann-Hilbert problem has a unique solution $W(z)$ in D^+, and then the functions

$$U(x)-V(x)=\mathrm{Re}[(1-j)(U(x)+jV(x)]=R_0(x) \text{ on } L_0\cap D_1^-,$$

$$U(x)+V(x)=\mathrm{Re}[(1+j)(U(x)+jV(x)]=R_0(x) \text{ on } L_0\cap D_2^-,$$

are determined. This shows that Problem B for equation (3.13) is uniquely solvable, namely

Theorem 3.1 *Problem B of equation* (3.13) *or system* (3.14) *in $\overline{D}$ has a unique solution as stated in* (3.20), *and the solution satisfies the estimates*

$$\begin{aligned} &C_\delta[u(z),D^-]+C_\delta^1[u(z),D_\varepsilon^-]\le M_1 \\ &C_\delta[f(x),L_0\cap D_\varepsilon^-]+C_\delta[g(x),L_0\cap D_\varepsilon^-]\le M_2, \end{aligned} \tag{3.22}$$

in which $H(y)=|y|^{m/2}$, $\nu=x-G(y)$, $\mu=x+G(y)$, $u(z)$ is the corresponding function determined by (3.15), *$W(z)$ is as stated in* (3.20), *$D_\varepsilon^-=$*

$\overline{D^-} \cap \Pi_{l=0}^{n+1}\{|z - a_l| > \varepsilon\}(> 0)\}$, ε, $\delta(> 0)$ *are small enough, and* $M_l = M_l(\delta, k_0, k_2, D_\varepsilon^-)$ $(l = 1, 2)$ *are positive constants.*

The representation of solutions of Problem P for equation (3.2) is as follows.

Theorem 3.2 *Suppose that equation* (3.2) *satisfies Condition* C *and* $u(z)$ *is any solution of Problem* P *for equation* (3.2) *in* D. *Then the solution* $u(z)$ *can be expressed as follows*

$$
\begin{aligned}
&u(z) = -2\int_0^y V(z)dy + u(x)\\
&= 2\mathrm{Re}\int_0^z [\frac{\mathrm{Re}w}{H} + \begin{pmatrix} i \\ -j \end{pmatrix} \mathrm{Im}w]dz + b_0 \text{ in } \begin{pmatrix} \overline{D^+} \\ \overline{D^-} \end{pmatrix},\\
&w(z) = W(z) + \Phi(Z) + \Psi(Z),\ \Psi(Z) = -\mathrm{Re}\frac{2}{\pi}\iint_{D_t^+} \frac{f(t)}{t-Z} d\sigma_t \text{ in } \overline{D_Z^+},\\
&w(z) = \phi(z) + \psi(z) = \xi(z)e_1 + \eta(z)e_2 \text{ in } \overline{D^-},\\
&\xi(z) = \zeta(z) + \int_0^y g_1(z)dy = \zeta_0(z) + \int_{S_1} g_1(z)dy + \int_0^y g_1(z)dy,\\
&g_1(z) = \tilde{A}_1(U+V) + \tilde{B}_1(U-V) + 2\tilde{C}_1U + \tilde{D}_1u + \tilde{E}_1,\ z \in s_1,\\
&\eta(z) = \theta(z) + \int_0^y g_2(z)dy = \theta_0(z) + \int_{S_2} g_2(z)dy + \int_0^y g_2(z)dy,\\
&g_2(z) = \tilde{A}_2(U+V) + \tilde{B}_2(U-V) + 2\tilde{C}_2U + \tilde{D}_2u + \tilde{E}_2,\ z \in s_2,
\end{aligned}
\tag{3.23}
$$

in which $\zeta_0(z) = \mathrm{Re}W(z) + \mathrm{Im}W(z)$, $\zeta(z) = \zeta_0(z) + \int_{S_1} g_1(z)dy$ *in* D_1^-, $\theta(z) = \theta_0(z) + \int_{S_2} g_2(z)dy$ *in* D_2^-, $\theta(x) = -\zeta(x)$, $W(z)$ *is as stated in* (3.20), $f(Z) = g(Z)/H$, $U = Hu_x/2, V = -u_y/2$, *and* $\phi(z)$ *are solutions of* (3.13) *in* D^-, s_1, s_2 *are two families of characteristics in* D^-:

$$
s_1:\ \frac{dx}{dy} = \sqrt{-K(y)} = H(y),\ s_2:\ \frac{dx}{dy} = -\sqrt{-K(y)} = -H(y) \tag{3.24}
$$

passing through $z = x + jy \in D^-$, S_1, S_2 *are the characteristic curves from the points on* L_1, L_2 *to the points on* L_0 *respectively, and*

$$
\begin{aligned}
&w(z) = U(z) + jV(z) = \frac{1}{2}Hu_x - \frac{j}{2}u_y,\\
&\xi(z) = \mathrm{Re}\psi(z) + \mathrm{Im}\psi(z),\ \eta(z) = \mathrm{Re}\psi(z) - \mathrm{Im}\psi(z),
\end{aligned}
$$

$$\tilde{A}_1=\tilde{B}_2=\frac{1}{2}(\frac{h_y}{2h}-b),\,\tilde{A}_2=\tilde{B}_1=\frac{1}{2}(\frac{h_y}{2h}+b),\,\tilde{C}_1=\frac{a}{2H}+\frac{m}{4y},$$

$$\tilde{c}_2=-\frac{a}{2H}+\frac{m}{4y},\,\tilde{D}_1=-\tilde{D}_2=\frac{c}{2},\,\tilde{E}_1=-\tilde{E}_2=\frac{d}{2},$$

in which we choose $H(y)=[|y|^m h(y)]^{1/2}, h(y)$ *is a continuously differentiable positive function, and*

$$d\mu=d[x+G(y)]=2H(y)dy \text{ on } s_1,\,d\nu=d[x-G(y)]=-2H(y)dy \text{ on } s_2.$$

Proof From (3.11) we see that equation (3.2) in $\overline{D^-}$ can be reduced to the system of integral equations: (3.23). Noting (3.24), we have

$$ds_1=\sqrt{(dx)^2+(dy)^2}=-\sqrt{1+(dx/dy)^2}dy=-\sqrt{1-K}dy=-\frac{\sqrt{1-K}}{\sqrt{-K}}dx,$$
$$ds_2=\sqrt{(dx)^2+(dy)^2}=-\sqrt{1+(dx/dy)^2}dy=-\sqrt{1-K}dy=\frac{\sqrt{1-K}}{\sqrt{-K}}dx, \tag{3.25}$$

it is clear that the the formula (3.23) is true.

3.3 Solvability of discontinuous oblique derivative problem for degenerate equations of mixed type

In this section, we prove the existence of solutions of Problems P and Q for equation (3.12). Firstly we discuss the complex equation

$$w_{\bar{\bar{z}}}=A_1(z)w+A_2(z)\overline{w}+A_3(z)u+A_4(z) \text{ in } D, \tag{3.26}$$

with the relation

$$u(z)=\begin{cases}2\text{Re}\displaystyle\int_0^z[\frac{\text{Re}w(z)}{H(y)}+i\text{Im}w(z)]dz+b_0 \text{ in } \overline{D^+},\\ 2\text{Re}\displaystyle\int_0^z[\frac{\text{Re}w(z)}{H(y)}-j\text{Im}w(z)]dz+b_0 \text{ in } \overline{D^-},\end{cases} \tag{3.27}$$

where $H(y)$ is as stated before, and the coefficients in (3.26) are as stated in (3.12), and the boundary value problem (3.26), (3.27) with the boundary conditions

$$\begin{aligned}
&\mathrm{Re}[\overline{\lambda(z)}w(z)]=H(y)r(z)=R_1(z) \text{ on } \Gamma\cup L_3\cup L_4,\\
&\mathrm{Im}[\overline{\lambda(z)}w(z)]|_{z=z_0'} = H(\mathrm{Im}z_0')b_1 = b_1',\\
&\mathrm{Im}[\overline{\lambda(z)}w(z)]|_{z=A_{2l+1}} = c_{2l+1}',\ l=0,1,...,[n/2],\\
&\mathrm{Im}[\overline{\lambda(z)}w(z)]|_{z=B_{2l-1}} = c_{2l}', l=1,...,[(n+1)/2],
\end{aligned} \tag{3.28}$$

is called Problem A or B, where $\lambda(z), R_1(z), A_{2l+1}, B_{2l-1}, c_{2l-1}', c_{2l}'$ are as stated in (3.4), (3.5) and (3.16). Similarly to Section 2, we can give a transformation $v(z) = v^{\pm}(z) = u(z) - u_0^{\pm}(z)$ in $\overline{D^{\pm}}$, where $u_0^{\pm}(z)$ in $D^{\pm}$ are harmonic functions with the boundary condition (3.4) on $\Gamma\cup L_3\cup L_4$, and $u_0^{\pm}(z)$ in $\overline{D^{\pm}}$, then Problem Q for equation (2.1) is reduced to the boundary value problem (Problem $\tilde{Q}$) for equation

$$K(y)v_{xx} + v_{yy} + av_x + bv_y + cv + \tilde{d} = 0 \text{ in } D,$$

$$\tilde{d} = d + Lu_0^{\pm} \text{ in } D^{\pm},$$

with the boundary conditions

$$\begin{aligned}
&\mathrm{Re}[\overline{\lambda(z)}v_{\bar z}]=R_1(z)-\mathrm{Re}[\overline{\lambda(z)}u_{0\bar z}]=\tilde{R}_1(z) \text{ on } \Gamma\cup L_3\cup L_4,\\
&v(2)=b_1 \text{ or } \mathrm{Im}[\overline{\lambda(z)}v_{\bar z}]|_{z=z_0'}=b_1'-\mathrm{Im}[\overline{\lambda(z)}u_{0\bar z}]|_{z=z_0'}=b_1'',\\
&\mathrm{Im}[\overline{\lambda(z)}v_{\bar z}]|_{z=A_{2l+1}}=c_{2l+1}'', l=0,1,...,[n/2],\\
&\mathrm{Im}[\overline{\lambda(z)}v_{\bar z}]|_{z=B_{2l-1}}=c_{2l}'', l=1,...,[(n+1)/2],
\end{aligned}$$

where $R_1(z) = 0$ on $\Gamma\cup L_3\cup L_4$, $b_0 = b_1 = 0, c_l'' = 0, l = 1,...,n+1$ and the boundary value problem about $W(z) = v_{\bar z}$ is called Problem $\tilde{A}$ or $\tilde{B}$. It is not difficult to see that Problem $\tilde{B}$ can be divided into Problem B_1 for equation (3.26), (3.27) in D^+ with the boundary conditions

$$\mathrm{Re}[\overline{\lambda(z)}W(z)] = R_1(z) = 0 \text{ on } \Gamma,\ \mathrm{Re}[-iW(x)] = -\hat{R}_0(x) \text{ on } L_0,$$

and Problem B_2 for equation (3.26), (3.27) in D^- with the boundary conditions

$$\mathrm{Re}[\overline{\lambda(z)}W(z)] = 0 \text{ on } L_3\cup L_4,$$

$$\mathrm{Re}[(1-j)W(x)] = R(x) = \tilde{R}_0(x) \text{ on } L_0\cap D_1^-,$$

$$\mathrm{Re}[(1+j)W(x)]=R(x)=-\tilde{R}_0(x) \text{ on } L_0\cap D_2^-,$$

$$\mathrm{Im}[\overline{\lambda(z)}W(z)]|_{z=A_{2l+1}}=c''_{2l+1}=0, l=0,1,...,[n/2],$$

$$\mathrm{Im}[\overline{\lambda(z)}W(z)]|_{z=B_{2l}}=c''_{2l}=0,\ l=1,...,[(n+1)/2],$$

noting $\mathrm{Re}W(x)=H(y)v_x=0$ on L_0, the above number $1-j$, $1+j$ can be replaced by $-j$, j respectively. Similarly to Theorems 2.6 and 3.4, Chapter II, here $a_0,a_1,...,a_{n+1}$ are the discontinuous points of Problem B_1 on ∂D^+, but we can multiply the solution $W(z)$ by $X(Z)=\prod_{l=0}^{n+1}(Z-a_l)^{\eta_l}$, $\eta_l=1-2\gamma_l$ if $\gamma_l\ge 0\,(l=0$ or $n+1)$, $\eta_l=\max[0,-2\gamma_l]$ if $\gamma_l<0\,(l=0$ or $n+1)$, and $\eta_l=1\,(l=1,...,n)$, in this case the index of $X(Z)\lambda(z)$ on ∂D^+ is $\tilde{K}=n/2$, hence there are $n+1$ point conditions $u(a_l)=0\,(l=1,...,n+1)$ except $u(a_0)=b_0$, we can prove that Problem B_1 for (3.26), (3.27) has a unique solution $W(z)$. In the following we only prove the unique solvability of Problem B_2 for (3.26), (3.27) in D^-.

Theorem 3.3 *If equation* (3.2) *satisfies Condition C and* (3.29) *below, then there exists a solution* $[W(z),v(z)]$ *of Problem* B_2 *for* (3.26), (3.27).

Proof Denote $D_0=D^-\cap\Pi_{l=1}^{n+1}\{b_{l-1}\le x\le b_l\}$, in which $a_l+\delta_0=b_l<b_{l+1}=a_{l+1}-\delta_0\,(l=0,1,...,n)$ and δ_0 is a sufficiently small positive number.

We may only discuss the case of $K(y)=-|y|^m h(y)$. In order to find a solution of the system of integral equations (3.23), we need to add the condition

$$\frac{|y|a(x,y)}{H(y)}=o(1), \text{ i.e. } \frac{|a(x,y)|}{H(y)}=\frac{\varepsilon(y)}{|y|},\ m\ge 2. \tag{3.29}$$

It is clear that for two characteristics $s_1: x=x_1(y,z_0)$, $s_2: x=x_2(y,z_0)$ passing through $P_0=z_0=x_0+jy_0\in D^-$, we have

$$|x_1-x_2|\le 2|\int_0^{y_0}\sqrt{-K}dy|\le M|y_0|^{m/2+1} \text{ for } (x_0,y_0)\in\overline{D^-}, \tag{3.30}$$

for any $z_1=x_1+jy\in s_1, z_2=x_2+jy\in s_2$, in which $M\,(>\max[2\sqrt{h(y)},1])$ is a positive constant. From (3.3), we can assume that the coefficients of (3.23) possess continuously differentiable with respect to $x\in L_0$ and satisfy the conditions

$$\begin{gathered}|\tilde{A}_l|,|\tilde{A}_{lx}|,|\tilde{B}_l|,|\tilde{B}_{lx}|,|\tilde{D}_l|,|\tilde{D}_{lx}|,\\ |\tilde{E}_l|,|\tilde{E}_{lx}|,|1/\sqrt{h}|,|h_y/h|\le M, z\in\bar{D},\ l=1,2.\end{gathered} \tag{3.31}$$

According to the proof of Theorem 1.4, it is sufficient to find a solution of Problem B_2 for arbitrary segment $-\delta \le y \le 0$, where δ is a sufficiently small positive number, and we use the positive constant

$$M_0 \ge 2k_1 d + \frac{4\varepsilon_0 + m}{2\delta}, \tag{3.32}$$

where d is the diameter of D, $\varepsilon_0 = \max_{\overline{D^-}} \varepsilon(z)$, β is a sufficiently small positive constant. In the following we shall find a solution of Problem B_2 for (3.26), (3.27) on $-\delta < y < 0$. Firstly, similarly to Section 2, we choose v_0, ξ_0, η_0 and substitute them into the corresponding positions of v, ξ, η in the right-hand sides of (3.23), and obtain

$$\begin{aligned}
&v_1(z)=v_1(x)-2\int_0^y V_0 dy=v_1(x)+\int_0^y(\eta_0-\xi_0)dy,\\
&\xi_1(z)=\zeta_1(z)+\int_0^y[\tilde{A}_1\xi_0+\tilde{B}_1\eta_0+\tilde{C}_1(\xi_0+\eta_0)+\tilde{D}_1u_0+\tilde{E}_1]dy,\ z\in s_1,\\
&\eta_1(z)=\theta_1(z)+\int_0^y[\tilde{A}_2\xi_0+\tilde{B}_2\eta_0+\tilde{C}_2(\xi_0+\eta_0)+\tilde{D}_2u_0+\tilde{E}_2]dy,\ z\in s_2.
\end{aligned} \tag{3.33}$$

By the successive approximation, we find the sequences of functions $\{v_k\}$, $\{\xi_k\}$, $\{\eta_k\}$, which satisfy the relations

$$\begin{aligned}
&v_{k+1}(z)=v_{k+1}(x)-2\int_0^y V_k(z)dy=v_{k+1}(x)+\int_0^y(\eta_k-\xi_k)dy,\\
&\xi_{k+1}(z)=\zeta_{k+1}(z)+\int_0^y[\tilde{A}_1\xi_k+\tilde{B}_1\eta_k+\tilde{C}_1(\xi_k+\eta_k)+\tilde{D}_1u_k+\tilde{E}_1]dy,\ z\in s_1,\\
&\eta_{k+1}(z)=\theta_{k+1}(z)+\int_0^y[\tilde{A}_2\xi_k+\tilde{B}_2\eta_k+\tilde{C}_2(\xi_k+\eta_k)+\tilde{D}_2u_k+\tilde{E}_1]dy,\ z\in s_2,\\
&\qquad k=0,1,2,....
\end{aligned} \tag{3.34}$$

and similarly to (2.42) in Section 2, we can prove that $\{\tilde{v}_k\}$, $\{\tilde{\xi}_k\}$, $\{\tilde{\eta}_k\}$ in D_0 satisfy the estimates

$$\begin{aligned}
&|\tilde{\xi}_{k+1}(z)-\tilde{\zeta}_{k+1}(z)|\le M'\gamma^{k-1}|y|^{1-\beta},\\
&|\tilde{\eta}_{k+1}(z)-\tilde{\theta}_{k+1}(z)|,|\tilde{v}_{k+1}(z)-\tilde{v}_{k+1}(x)|\le M'\gamma^{k-1}|y|^{1-\beta},\\
&|\tilde{\xi}_{k+1}(z)+\tilde{\eta}_{k+1}(z)|\le M'\gamma^{k-1}|x_1-x_2|^{\beta}|y|^{\beta'},0\le|y|\le\delta,
\end{aligned} \tag{3.35}$$

where $z = x + jy$ is the intersection point of the characteristics s_1, s_2 in (3.24) emanating from two points x_1, x_2 on $x-$axis, $\beta' = (1+m/2)(1-3\beta)$,

δ, β are sufficiently small positive constants such that $(2+m)\beta < 1$, $\gamma\,(< 1)$ is a positive constant, and M' is a sufficiently large positive constant. On the basis of the above estimates, we can derive that $\{v_k\}$, $\{\xi_k\}$, $\{\eta_k\}$ in D_0 uniformly converge to v_*, ξ_*, η_* satisfying the system of integral equations

$$
\begin{aligned}
&v_*(z)=v_*(x)-2\int_0^y V_*dy=v_*(x)+\int_0^y(\eta_*-\xi_*)dy,\\
&\xi_*(z)=\zeta_*(z)+\int_0^y[\tilde{A}_1\xi_*+\tilde{B}_1\eta_*+\tilde{C}_1(\xi_*+\eta_*)+\tilde{D}_1u_*+\tilde{E}_1]dy,\, z\in s_1,\\
&\eta_*(z)=\theta_*(z)+\int_0^y[\tilde{A}_2\xi_*+\tilde{B}_2\eta_*+\tilde{C}_2(\xi_*+\eta_*)+\tilde{D}_2u_*+\tilde{E}_2]dy,\, z\in s_2,
\end{aligned}
$$

and the function $[v_*(z)]_{\bar{z}} = W^*(z)$ satisfies equation (3.26) and boundary condition (3.28), this shows that Problem B_2 in D_0 has a solution for equation (3.26). Due to $\delta_0(> 0)$ can be arbitrarily small and by the result in Section 1, hence $u(z) = v(z) + u_0(z)$ is a solution of Problem Q for (3.2) in D^-. Thus the existence of solutions of Problem Q for equation (3.2) with $c = 0$ in D is proved. From the above discussion, we can see that the solution of Problem Q for (3.2) with $c = 0$ in D is unique.

From the above result, we have the following theorem.

Theorem 3.4 *Let equation* (3.2) *satisfy Condition C and* (3.29). *Then the oblique derivative problem* (*Problem Q*) *for* (3.2) *with $c = 0$ has a unique solution.*

Finally, we prove the following theorem.

Theorem 3.5 *Under Condition C and* (3.29) *in D^-, the oblique derivative problem* (*Problem P*) *for* (3.2) *has a solution.*

Proof From Theorem 3.4, we see that Problem Q for (3.2) with $c = 0$ has a solution $u^*(z)$ in D, if $u^*(2) = b_1$, then the solution $u^*(z)$ is just a solution of Problem P for (3.2) with $c = 0$. Otherwise, $u^*(2) = b_1' \neq b_1$, we can find a solution $u_1(z)$ of Problem Q for the homogeneous equation

$$
K(y)u_{xx}+u_{yy}+a(x,y)u_x+b(x,y)u_y + c(x,y)u = 0 \text{ in } D \tag{3.36}
$$

with the boundary conditions

$$
\begin{aligned}
&\mathrm{Re}[\overline{\lambda(z)}u_{1\bar{z}}]=0,\, z\in\Gamma,\, \mathrm{Re}[\overline{\lambda(z)}u_{1\bar{z}}]=0,\, z\in L_3\cup L_4,\\
&u_1(0) = 0,\ \mathrm{Im}[\overline{\lambda(z)}u_{1\bar{\bar{z}}}]|_{z=z_0'} = 1,
\end{aligned}
$$

$$\begin{aligned}&\mathrm{Im}[\overline{\lambda(z)}u_{1\bar{z}}]|_{z=A_{2j+1}}=0,\ j=0,1,...,[n/2],\\&\mathrm{Im}[\overline{\lambda(z)}u_{1\bar{z}}]|_{z=B_{2j-1}}=0,\ j=1,...,[(n+1)/2].\end{aligned}\tag{3.37}$$

It is obvious that $u_1(z)\not\equiv 0$ in $\overline{D}$, thus there exists a real constant $d_1\neq 0$, such that

$$d_1u_1(2)=b_1'-b_1,$$

thus the function

$$u(z)=u^*(z)-d_1u_1(z)\ \text{in}\ \bar{D}$$

is just a solution of Problem P for the linear equation (3.2) with $c=0$. Moreover we can prove the solvability of Problem P for (3.2) by using the method of parameter extension. This completes the proof.

Remark 3.1 Actually the solution $W(z)=u_{\bar{z}}$ of Problem B_1 for equation (3.26) has $n+2$ discontinuous points in Z_0, hence in order to discuss the unique solvability of Problem B_1, we need $n+1$ point conditions in (3.4),(3.5) or (3.10) except $u(0)=b_0$. If the last point condition in (3.4),(3.5) or (3.10) is replaced by $n+1\,(n>0)$ point conditions as follows

$$u(0)=b_0,\ \ u(z_l)=b_l,\ \text{or}$$

$$\frac{1}{H(y)}\mathrm{Im}[\overline{\lambda(z)}u_{\bar{z}}]|_{z=z_l'}=b_l,\ l=1,...,n+1,$$

where $z_l\,(\in\Gamma\cup L_0\backslash\{0,2\})$, $z_l'\,(\in\Gamma\backslash Z_0,\ l=1,...,n+1)$ are $n+1$ distinct points respectively, $b_l\ (l=1,...,n+1)$ are all real constants satisfying the conditions $|b_l|\leq k_2\,(l=1,...,n+1)$, and choose the index $K=n/2$ of the boundary value problem in D^+, then the corresponding boundary value problems (Problems Q and P) are uniquely solvable.

4 The Exterior Tricomi-Rassias Problem for Second Order Degenerate Equations of Mixed Type

In [71]2), the author posed the exterior Tricomi-Rassias Problem for the mixed equation

$$K(y)u_{xx}+u_{yy}+r(x,y)u=f(x,y)$$

in a doubly connected domain and proved the uniqueness of solutions for the problem. In this section we discuss the exterior Tricomi-Rassias Problem for general second order equations of mixed type in a doubly connected domain. We first give the representation of solutions of the boundary value problem for the equations, and then prove the uniqueness and existence of solutions for the problem by a new method.

4.1 Formulation of exterior Tricomi-Rassias problem for degenerate equations of mixed type

Consider a function $K(y)$ as follows

$$K = K(y)\begin{cases} >0 \text{ for } \{y>0\}\cup\{y<-1\},\\ =0 \text{ for } \{y=0\}\cup\{y=-1\},\\ <0 \text{ for } \{-1<y<0\},\end{cases}$$

and $K(y)$ are continuous in $\overline{D}\cap\{y\geq -1/2\}$ and $\overline{D}\cap\{y\leq -1/2\}$, possesses the derivative $K'(y)$ in $y\neq 0$ and 1, the doubly connected domain $D = G_1\cup G_1'\cup G_2\cup G_2'\cup(A_1B_1)\cup(A_2B_2)$ possesses the exterior boundary $Ext(D)=\Gamma_0\cup\Gamma_0'\cup\Gamma_2\cup\Gamma_2'\cup\Delta_1\cup\Delta_1'$ and interior boundary $Int(D) = \Gamma_1\cup\Gamma_1'\cup\Delta_2\cup\Delta_2'$, where A_1B_1, A_2B_2 are two lines with end points $A_1 = (-1,0), B_1=(1,0), A_2=(-1,-1), B_2=(1,-1)$, and
Γ_0 is the elliptic arc for $y>0$ connecting points A_1, B_1,
Γ_0' is the elliptic arc for $y<-1$ connecting points A_2, B_2,
$\Gamma_1: x=-G_1(y)=-\int_0^y\sqrt{-K(t)}dt$ is a characteristic for $-1/2<y<0,\ 0<x<1$ emanating from the point $O_1=(0,0)$,
$\Gamma_1': x=G_2(y)=\int_{-1}^y\sqrt{-K(t)}dt$ is a characteristic for $-1<y<-1/2,\ 0<x<1$ emanating from the point $O_2=(0,-1)$,
$\Gamma_2: x=G_1(y)+1=\int_0^y\sqrt{-K(t)}dt+1$ is a characteristic for $-1/2<y<0,\ 0<x<1$ emanating from the point $B_1=(1,0)$,
$\Gamma_2': x=-G_2(y)+1=-\int_{-1}^y\sqrt{-K(t)}dt+1$ is a characteristic for $-1<y<-1/2,\ 0<x<1$ emanating from the point $B_2=(1,-1)$,
$\Delta_1: x=-G_1(y)-1=-\int_0^y\sqrt{-K(t)}dt-1$ is a characteristic for $-1/2<y<0,\ -1<x<0$ emanating from the point $A_1=(-1,0)$,
$\Delta_1': x=G_2(y)-1=\int_{-1}^y\sqrt{-K(t)}dt-1$ is a characteristic for $-1<y<-1/2,\ -1<x<0$ emanating from the point $A_2=(-1,-1)$,
$\Delta_2: x=G_1(y)=\int_0^y\sqrt{-K(t)}dt$ is a characteristic for $-1/2<y<0,\ -1<x<0$ emanating from the point $O_1=(0,0)$,
$\Delta_2': x=-G_2(y)=-\int_{-1}^y\sqrt{-K(t)}dt$ is a characteristic for $-1<y<-1/2,\ -1<x<0$ emanating from the point $O_2=(0,-1)$, and
$G_1=\{(x,y)\in D||x|<1, y>0\}$ is a upper elliptic domain, $G_1'=\{(x,y)\in D||x|<1, y<-1\}$ is a lower elliptic domain, $G_2=\{(x,y)\in D|0<x<$

$1, -1 < y < 0\}$ is a right hyperbolic domain, $G_2' = \{(x,y) \in D| -1 < x < 0, -1 < y < 0\}$ is a left hyperbolic domain with the boundaries $\partial G_1 = \Gamma_0 \cup (A_1B_1)$, $\partial G_1' = \Gamma_0' \cup (B_2A_2)$, $\partial G_2 = \Gamma_1 \cup \Gamma_1' \cup \Gamma_2 \cup \Gamma_2' \cup (B_1O_1) \cup (O_2B_2)$, $\partial G_2' = \Delta_1 \cup \Delta_1' \cup \Delta_2 \cup \Delta_2' \cup (O_1A_1) \cup (A_2O_2)$, respectively.
The above characteristic curves intersect at the points: $\Gamma_1 \cap \Gamma_1' = P_1 = (x_1, -1/2), \Gamma_2 \cap \Gamma_2' = P_2 = (x_2, -1/2), \Delta_1 \cap \Delta_1' = P_1' = (x_1', -1/2)$, $\Delta_2 \cap \Delta_2' = P_2' = (x_2', -1/2)$. We can assume that Γ_0, Γ_0' are two smooth curves vertical to the axis $\text{Im}z = 0$ near $z = \pm 1, \pm 1 - i$ respectively as similar to Section 2. Here note the difference between the domains $G_1, G_2, ...$ and the functions $G_1(y), G_2(y), ...$

Consider general second order equation of mixed type

$$Lu = K(y)u_{xx} + u_{yy} + a(x,y)u_x + b(x,y)u_y + c(x,y)u = \ d(x,y) \text{ in } D, \quad (4.1)$$

where the coefficients satisfy **Condition** C, namely

$$\begin{gathered} L_\infty[\eta, \overline{D^+}] \le k_0, \eta = a, b, c,\ L_\infty[d, \overline{D^+}] \le k_1, c \le 0 \text{ in } D^+, \\ \hat{C}[d, \overline{D^-}] = C[d, \overline{D^-}] + C[d_x, \overline{D^-}] \le k_1, \hat{C}[\eta, \overline{D^-}] \le k_0, \eta = a, b, c, \end{gathered} \quad (4.2)$$

where $D^+ = D \cap \{|y - 1/2| > 1/2\}, D^- = D \backslash \overline{D^+}$, k_0, $k_1 (\ge \max[1, 12k_0])$ are positive constants, and

$$G_1(y) = \int_0^y H(t)dt,\ G_2(y) = \int_{-1}^y H(t)dt,\ H(y) = \sqrt{|K(y)|}.$$

Now we can choose $\text{sgn}yK(y) = |y|^m h(y), y \ge -1/2, -\text{sgn}(1+y)K(y) = |y+1|^m h(y), y \le -1/2$, m is a positive number, $h(y)$ is a continuously differentiable positive function, if $h(y) = 1$, then $G_1(y) = 2\text{sgn}y|y|^{(m+2)/2}/(m+2), G_2(y) = -2\text{sgn}(1+y)|y+1|^{(m+2)/2}/(m+2)$. For simplicity, in the following denote by $G(y)$ the functions $G_1(y)$ in $D \cap \{y > -1/2\}$ and $G_2(y)$ in $D \cap \{y < -1/2\}$ respectively.

The exterior Tricomi-Rassias problem for equation (4.1) may be formulated as follows:

Problem TR Find a real continuous solution $u(z)$ of equation (4.1) in $\overline{D}$, where u_x, u_y are continuous in $D_* = \bar{D} \backslash T$ and $u(z)$ satisfies the boundary conditions

$$\begin{gathered} u(z) = \phi_1(s) \text{ on } \Gamma_0,\ u(z) = \phi_2(s) \text{ on } \Gamma_0', \\ u(z) = \psi_1(x) \text{ on } \Gamma_2,\ u(z) = \psi_2(x) \text{ on } \Gamma_2', \\ u(z) = \psi_3(x) \text{ on } \Delta_1,\ u(z) = \psi_4(x) \text{ on } \Delta_1', \end{gathered} \quad (4.3)$$

in which $T = \{\pm 1, \pm 1 - i, 0, -i, P_1P_2, P_1'P_2'\}$ and $\phi_l(s)$ $(l = 1, 2)$, $\psi_l(x)$ $(l=1,2,3,4)$ satisfy the conditions

$$\begin{aligned}
&C^2_\alpha[\phi_1(s), \Gamma_0] \le k_2,\ C^2_\alpha[\phi_2(s), \Gamma_0'] \le k_2,\\
&C^2_\alpha[\psi_1(x), \Gamma_2] \le k_2,\ C^2_\alpha[\psi_2(x), \Gamma_2'] \le k_2,\\
&C^2_\alpha[\psi_3(x), \Delta_1] \le k_2,\ C^2_\alpha[\psi_4(x), \Delta_1'] \le k_2,\\
&\phi_1(0) = \psi_3(-1), \phi_2(0) = \psi_2(1-i),\\
&\phi_1(S_1) = \psi_1(1), \phi_2(S_2) = \psi_4(-1-i),
\end{aligned} \tag{4.4}$$

in which S_1, S_2 are the lengths of Γ_0, Γ_0' respectively, and $\alpha\,(0 < \alpha < 1)$, k_2 are positive constants.

Now we mention that if Γ_0 is the curve with the form $x = \pm 1 \mp \tilde G_1(y)$ near $z = \pm 1$, and Γ_0' is the curve with the form $x = \pm 1 \pm \tilde G_2(y)$ including the line segments $\text{Re}z = \pm 1, \pm 1 - i$ near $z = \pm 1, \pm 1 - i$ respectively, then the boundary condition (4.3) can be rewritten in the form

$$\begin{aligned}
&\text{Re}[\overline{\lambda(z)}u_{\tilde z}] = r(z)/2,\ b_0 = \phi_1(0),\\
&r(z) = \phi_{1y} = \phi_1'(y)\ \text{ on } \Gamma_0 \text{ near } z = \pm 1,\\
&\text{Re}[\overline{\lambda(z)}u_{\tilde z}] = r(z)/2,\ b_{-1} = \phi_2(0),\\
&r(z) = \phi_{2y} = \phi_2'(y)\ \text{ on } \Gamma_0' \text{ near } z = \pm 1 - i,
\end{aligned} \tag{4.5}$$

where $u_{\tilde z} = [H(y)u_x - iu_y]/2$, $\lambda(z)$ is as stated in (4.7) below. In addition, we find the derivative for (4.3) with respect to the parameter $s = x$ on $\Delta_1, \Delta_1', \Gamma_2, \Gamma_2'$, and can obtain

$$\begin{aligned}
&u_s = u_x + u_y y_x = u_x + u_y/H(y) = \psi_1'(x) \text{ on } \Gamma_2,\\
&U(z) - V(z) = [H(y)u_x + u_y]/2\\
&= H(G_1^{-1}(x-1))\psi_1'(x)/2 = R_1(x) \text{ on } \Gamma_2,\\
&u_s = u_x + u_y y_x = u_x - u_y/H(y) = \psi_2'(x) \text{ on } \Gamma_2',\\
&U(z) + V(z) = [H(y)u_x - u_y]/2\\
&= H((-G_2)^{-1}(x-1))\psi_2'(x)/2 = R_2(x) \text{ on } \Gamma_2',\\
&u_s = u_x + u_y y_x = u_x - u_y/H(y) = \psi_3'(x) \text{ on } \Delta_1,\\
&U(z) + V(z) = [H(y)u_x - u_y]/2
\end{aligned}$$

$$
\begin{aligned}
&= H((-G_1)^{-1}(x+1))\psi_3'(x)/2 = R_3(x) \text{ on } \Delta_1,\\
&u_s = u_x + u_y y_x = u_x + u_y/H(y) = \psi_4'(x) \text{ on } \Delta_1',\\
&U(z) - V(z) = [H(y)u_x + u_y]/2\\
&= H(G_2^{-1}(x+1))\psi_4'(x)/2 = R_4(x) \text{ on } \Delta_1', \text{ i.e.}\\
&\mathrm{Re}[(1-j)(U+jV)] = U(z) - V(z) = R_1(x) \text{ on } \Gamma_2,\\
&\mathrm{Im}[(1-j)(U+jV)] = -[U-V]|_{z=z_1+0=P_2+0}\\
&= -H(-1/2)\psi_1'(x_2+0)/2 = b_1,\\
&\mathrm{Re}[(1+j)(U+jV)] = U(z) + V(z) = R_2(x) \text{ on } \Gamma_2',\\
&\mathrm{Im}[(1+j)(U+jV)] = [U+V]|_{z=z_3-0=P_2-0}\\
&= H(-1/2)\psi_2'(x_2+0)/2 = b_3,\\
&\mathrm{Re}[(1+j)(U+jV)] = U(z) + V(z) = R_3(x) \text{ on } \Delta_1,\\
&\mathrm{Im}[(1+j)(U+jV)] = [U+V]|_{z=z_2-0=P_1'-0}\\
&= H(-1/2)\psi_3'(x_1'-0)/2 = b_2,\\
&\mathrm{Re}[(1-j)(U+jV)] = U(z) - V(z) = R_4(x) \text{ on } \Delta_1',\\
&\mathrm{Im}[(1-j)(U+jV)] = -[U-V]|_{z=z_4+0=P_1'+0}\\
&= -H(-1/2)\psi_4'(x_1'-0)/2 = b_4,
\end{aligned} \tag{4.6}
$$

in which the function $H(y) = \sqrt{-K(y)}$, $(\pm G)_1^{-1}(x')$ is the inverse function of $x' = \pm G_1(y) = \pm \int_0^y H(t)dt$, $(\pm G)_2^{-1}(x')$ is the inverse function of $x' = \pm G_2(y) = \pm \int_{-1}^y H(t)dt$, and

$$
W(z) = \begin{cases} U + iV \text{ in } D^+,\ U = H(y)u_x/2 \\ U + jV \text{ in } \overline{D^-},\ V = -u_y/2 \end{cases} \text{ in } D,
$$

$$
\lambda(z) = \begin{cases} -i \text{ on } \Gamma_0 \cup \Gamma_0' \text{ near } z = -1 \text{ and } 1-i,\\ i \text{ on } \Gamma_0 \cup \Gamma_0' \text{ near } z = 1 \text{ and } -1-i,\\ 1-j,\ \mathrm{Re}\lambda = 1 \ne -\mathrm{Im}\lambda = -1 \text{ on } \Gamma_2' \cup \Delta_1,\\ 1+j,\ \mathrm{Re}\lambda = 1 \ne \mathrm{Im}\lambda = -1 \text{ on } \Gamma_2 \cup \Delta_1'. \end{cases} \tag{4.7}
$$

From the above formulas, we can write the complex forms of boundary conditions of $U+jV$:

$$\begin{aligned}
&\mathrm{Re}[\overline{\lambda(z)}W(z)]=R(z)\ \text{ on }\ \Gamma=\Gamma_0\cup\Gamma_0',\\
&\mathrm{Re}[\overline{\lambda(z)}(U+jV)]=\mathrm{Re}[\overline{\lambda(z)}W(z)]=H(y)r(z)\\
&=\tilde{R}(z)\ \text{ on }\ L=\Delta_1\cup\Delta_1'\cup\Gamma_2\cup\Gamma_2',\\
&\mathrm{Im}[\overline{\lambda(z)}(U+jV)]|_{z=z_l}=b_l,\ l=1,2,3,4,\\
&u(a_1)=\phi_1(-1)=b_0,\ u(a_2)=\phi_2(1-i)=b_{-1},\\
&\lambda(z)=\mp i,\ R(z)=\pm\phi_{1y}/2\ \text{ on }\ \Gamma_0\ \text{ near }\ \mp 1,\\
&\lambda(z)=\mp i,\ R(z)=\pm\phi_{2y}/2\ \text{ on }\ \Gamma_0'\ \text{ near }\ \pm 1-i,\\
&\lambda(z)=1+j,\tilde{R}(z)=R_1(x)=H(G_1^{-1}(x-1))\psi_1'(x)/2\ \text{on}\ \Gamma_2,\\
&\lambda(z)=1-j,\tilde{R}(z)=R_2(x)=H((-G_2)^{-1}(x-1))\psi_2'(x)/2\ \text{on}\ \Gamma_2',\\
&\lambda(z)=1-j,\tilde{R}(z)=R_3(x)=H((-G_1)^{-1}(x+1))\psi_3'(x)/2\ \text{on}\ \Delta_1,\\
&\lambda(z)=1+j,\tilde{R}(z)=R_4(x)=H((G_2)^{-1}(x+1))\psi_4'(x)/2\ \text{on}\ \Delta_1',\\
&b_1=-H(-1/2)\psi_1'(x_2+0)/2,\ b_3=H(-1/2)\psi_2'(x_2+0)/2,\\
&b_2=H(-1/2)\psi_3'(x_1'-0)/2,\ b_4=-H(-1/2)\psi_4'(x_1'-0)/2,
\end{aligned}\tag{4.8}$$

where $a_1=-1, a_2=1-i,\ b_0=u(a_1), b_{-1}=u(a_2)$, and

$$u(z)=2\mathrm{Re}\int_{-1}^{z}[\frac{U}{H(y)}+\begin{pmatrix} i\\ -j\end{pmatrix}V]dz+\phi_1(-1)\ \text{in}\ \begin{pmatrix}\overline{D^+}\\ \overline{D^-}\end{pmatrix}.$$

The number

$$K=\frac{1}{2}(K_1+K_2+K_3)$$

is called the index of Problem TR on ∂G_1, where

$$K_j=\left[\frac{\phi_j}{\pi}\right]+J_j,\ J_j=0\ \text{or}\ 1,$$

$$e^{i\phi_j}=\frac{\lambda(t_j-0)}{\lambda(t_j+0)},\ \gamma_j=\frac{\phi_j}{\pi}-K_j,\ j=1,2,3,$$

in which $t_1=-1, t_2=1, t_3=0$, and $\lambda(t_j-0), \lambda(t_j+0)$ are the left limit and right limit of $\lambda(t)$ at t_j on ∂G_1 respectively. Here $K=0$ on the boundary

∂G_1 of G_1 can be chosen. If $\cos(l,n) \equiv 0$ on Γ_0, then the value $u(1)$ can be derived, i.e.

$$u(1) = 2\mathrm{Re}\int_{-1}^{1} u_z dz + u(-1)$$

$$= 2\int_0^S \mathrm{Re}[z'(s)u_z]ds + \phi_1(-1) = 2\int_0^S r(z)ds + \phi_1(-1),$$

in which $\overline{\Lambda(z)} = z'(s)$ on Γ_0, $z(s)$ is a parameter expression of arc length s of Γ_0 with the condition $z(-1) = 0$, S is the length of the boundary Γ_0. We consider $\mathrm{Re}[\overline{\lambda(x)}W(z)] = 0$, $\lambda(x) = 1$ on $L_0' = \{-1 < x < 1, y = 0\}$, and $\lambda(t_1+0) = \lambda(t_2-0) = \lambda(t_3+0) = \lambda(t_3-0) = e^{0\pi i}$, thus we have

$$e^{i\phi_1} = \frac{\lambda(t_1-0)}{\lambda(t_1+0)} = e^{-\pi i/2 - 0\pi i} = e^{-\pi i/2}, \gamma_1 = \frac{-\pi/2}{\pi} - K_1 = -\frac{1}{2},$$

$$e^{i\phi_2} = \frac{\lambda(t_2-0)}{\lambda(t_2+0)} = e^{0\pi i - \pi i/2} = e^{-\pi i/2}, \gamma_2 = \frac{-\pi/2}{\pi} - K_2 = -\frac{1}{2},$$

$$e^{i\phi_3} = \frac{\lambda(t_3-0)}{\lambda(t_3+0)} = e^{0\pi i - 0\pi i} = e^{0\pi i}, \gamma_3 = 0\pi - K_3 = 0,$$

hence $K_1 = K_2 = K_3 = 0$, and then the index of $\lambda(z)$ on G_1 is $K = (K_1+K_2+K_3)/2 = 0$, in this case we can add a point condition $u(0) = 0$ in the boundary condition (4.8), but we shall omit to write it later on. Moreover we can get the index $K = 0$ of $\lambda(z)$ on $\partial G_1'$. In the following we mainly discuss the case of $\tilde{D} = D \cap \{y \geq -1/2\}$, $\tilde{D}_1^- = D \cap \{x < 0, -1/2 \leq y \leq 0\}$, $\tilde{D}_2^- = D \cap \{x > 0, -1/2 \leq y \leq 0\}$, and the case of $\hat{D} = D \cap \{y \leq -1/2\}$ can be similarly discussed.

For the above Problem *TR*, we can first discuss the problem in $\overline{D} \cap \{y \geq -1/2\}$, and then discuss the problem in $\overline{D} \cap \{y \leq -1/2\}$. Noting that $\phi_1(s) \in C_\alpha^2(\Gamma_0), \phi_2(s) \in C_\alpha^2(\Gamma_0')$, $\psi_1(x) \in C_\alpha^2(\Gamma_2)$, $\psi_2(x) \in C_\alpha^2(\Gamma_2')$, $\psi_3(x) \in C_\alpha^2(\Delta_1)$, $\psi_4(x) \in C_\alpha^2(\Delta_1')$ $(0 < \alpha < 1)$, we can find two twice continuously differentiable functions $u_0^\pm(z)$ in $\overline{D^\pm} \backslash \{P_1', P_2\}$, for instance, which are the solutions of the Dirichlet problem with the boundary condition on $\Gamma_0 \cup \Gamma_0' \cup \Gamma_2 \cup \Gamma_2' \cup \Delta_1 \cup \Delta_1'$ in (4.3) for harmonic functions in $D^\pm$, thus the functions $v(z) = v^\pm(z) = u(z) - u_0^\pm(z)$ in $D^\pm$ is the solution of Problem $\tilde{T}$, i.e. the equation

$$K(y)v_{xx} + v_{yy} + av_x + bv_y + cv + \tilde{d} = 0 \text{ in } D \tag{4.9}$$

and the corresponding boundary conditions

$$v(z) = 0 \text{ on } \Gamma_0 \cup \Gamma_0' \cup \Gamma_2 \cup \Gamma_2' \cup \Delta_1 \cup \Delta_1', \text{ i.e.}$$

$$\mathrm{Re}[\overline{\lambda(z)}W(z)] = R(z) = 0 \text{ on } \Gamma_0 \cup \Gamma_0' \cup \Gamma_2 \cup \Gamma_2' \cup \Delta_1 \cup \Delta_1',$$

$$u(a_l) = b_l = 0,\, l = 1,2,3,4,\ \mathrm{Im}[\overline{\lambda(z_l)}W(z_l)]|_{z=z_l} = b_l' = 0, l = 1,2, \tag{4.10}$$

where $a_1 = -1, a_2 = 1-i, a_3 = 0, a_4 = -i$, $\tilde{d} = d + Lu_0^{\pm}$ in $\overline{D}$, $W(z) = U + iV = v_{\bar{z}}^{+}$ in D^+ and $W(z) = U + jV = v_{\bar{z}}^{-}$ in D^-, hence we only discuss the homogeneous boundary condition (4.10) and the case of index $K = 0$ on ∂G_1. From $v(z) = v^{\pm}(z) = u(z) - u_0^{\pm}(z)$ in $\overline{D^{\pm}}$, we have $u(z) = v^{-}(z) + u_0^{-}(z)$ in $\overline{D^-}$, $u(z) = v^{+}(z) + u_0^{+}(z)$ in $\overline{D^+}$, $v^{+}(z) = v^{-}(z) - u_0^{+}(z) + u_0^{-}(z)$ on $L_0 = L_0' \cup L_0''$, and $v_y^{+} = v_y^{-} - u_{0y}^{+} + u_{0y}^{-} = 2\hat{R}_0(x)$ on $L_0 = L_0' \cup L_0''$, $L_0' = \{-1 < x < 0, y = 0\}$, $L_0'' = \{0 < x < 1, y = 0\}$, and $v_y^{-} = 2\tilde{R}_0(x)$ on L_0.

4.2 Representation of solutions of exterior Tricomi-Rassias Problem

First of all, for the latter requirement we first discuss the following problem: because the intersection points of characteristic boundary $P_1 = \Gamma_1 \cap \Gamma_1'$, $P_2 = \Gamma_2 \cap \Gamma_2'$, and $P_1' = \Delta_1 \cap \Delta_1'$, $P_2' = \Delta_2 \cap \Delta_2'$ are not equal, we need to give a transformation, such that the bounded domain $\hat{G}$ will be reduced to another bounded domain $\tilde{G}$, where $\hat{G}$ is bounded by the boundary $O_1B_1 \cup \Gamma_1 \cup \Gamma_2 \cup P_1P_2$, and $\tilde{G}$ is bounded by the boundary $O_1B_1 \cup \tilde{\Gamma}_1 \cup \tilde{\Gamma}_2$, herein

$$\tilde{\Gamma}_1 = \{x = -G_1(y), 0 \le x \le 1/2\}, \tilde{\Gamma}_2 = \{x = G_1(y) + 1, 1/2 \le x \le 1\}. \tag{4.11}$$

Setting $P_1 = (x_1, -1/2), P_2 = (x_2, -1/2)$, it is clear that $0 < x_1 < 1/2 < x_2 < 1$, and denote

$$L_1' = \{y = -\gamma(x),\, 0 \le x \le x_2\} = \{x = -G_1(y),\, 0 \le x \le x_1\}$$
$$\cup \{y = -\gamma(x) = -1/2, x_1 \le x \le x_2\}, L_2' = \{x = G_1(y) + 1, x_2 \le x \le 1\}, \tag{4.12}$$

where $G_1(y)$ is as stated in Subsection 4.1. Let the domain $\hat{G}$ be a simply connected domain with the boundary $L_1' \cup L_2' \cup O_1B_1$. Obviously the curve L_1' can be expressed by $x = \sigma(\nu) = (\mu + \nu)/2$, herein $\mu = x + G_1(y) = x + Y, \nu = x - G_1(y) = x - Y$, i.e. $\mu = 2\sigma(\nu) - \nu$, $0 \le \nu \le x_2 + \gamma(x_2)$. We make a transformation

$$\tilde{\mu} = [\mu - 2\sigma(\nu) + \nu]/[1 - 2\sigma(\nu) + \nu], \tilde{\nu} = \nu, 2\sigma(\nu) - \nu \le \mu \le 1, 0 \le \nu \le 1, \tag{4.13}$$

where μ, ν are real variables, its inverse transformation is

$$\mu = [1 - 2\sigma(\nu) + \nu]\tilde{\mu} + 2\sigma(\nu) - \nu,\ \nu = \tilde{\nu},\ 0 \le \tilde{\mu} \le 1,\ 0 \le \tilde{\nu} \le 1. \quad (4.14)$$

It is not difficult to see that the transformation in (4.13) maps the domain $\hat{G}$ onto $\tilde{G}$. The transformation (4.13) and its inverse transformation (4.14) can be rewritten as

$$\begin{cases} \tilde{x} = \dfrac{1}{2}(\tilde{\mu}+\tilde{\nu}) = \dfrac{2x-(1+x-Y)[2\sigma(x+\gamma(x))-x-\gamma(x)]}{2-4\sigma(x+\gamma(x))+2x+2\gamma(x)}, \\ \tilde{Y} = \dfrac{1}{2}(\tilde{\mu}-\tilde{\nu}) = \dfrac{2Y-(1-x+Y)[2\sigma(x+\gamma(x))-x-\gamma(x)]}{2-4\sigma(x+\gamma(x))+2x+2\gamma(x)}, \end{cases} \quad (4.15)$$

and

$$\begin{cases} x = \dfrac{1}{2}(\mu+\nu) = \dfrac{[1-2\sigma(x+\gamma(x))+x+\gamma(x)](\tilde{x}+\tilde{Y})}{2} \\ \quad +\sigma(x+\gamma(x)) - \dfrac{x+\gamma(x)-\tilde{x}+\tilde{Y}}{2}, \\ Y = \dfrac{1}{2}(\mu-\nu) = \dfrac{[1-2\sigma(x+\gamma(x))+x+\gamma(x)](\tilde{x}+\tilde{Y})}{2} \\ \quad +\sigma(x+\gamma(x)) - \dfrac{x+\gamma(x)+\tilde{x}-\tilde{Y}}{2}. \end{cases} \quad (4.16)$$

Denote by $\tilde{Z} = \tilde{x} + j\tilde{Y} = f(Z)$, $Z = x + jY = f^{-1}(\tilde{Z})$ the transformation (4.15) and the inverse transformation (4.16) respectively. In this case, the corresponding equation (4.21) below in $\hat{G}$ can be rewritten in the form

$$\xi_\mu = \tilde{A}_1\xi + \tilde{B}_2\eta + \tilde{C}_1 u + \tilde{D}_1,\ \eta_\nu = \tilde{A}_2\xi + \tilde{B}_2\eta + \tilde{C}_1 u + D_2,\ z \in \hat{G}. \quad (4.17)$$

Through the transformation (4.13), we obtain $\xi_{\tilde{\mu}} = [1-2\sigma(\nu)+\nu]\xi_\mu$, $\eta_{\tilde{\nu}} = \eta_\nu$ in $\hat{G}$, where $\xi = u + v, \eta = u - v$, and then

$$\left.\begin{aligned} &\xi_{\tilde{\mu}} = [1-2\sigma(\nu)+\nu][\tilde{A}_1\xi + \tilde{B}_2\eta + \tilde{C}_1 u + \tilde{D}_1], \\ &\eta_{\tilde{\nu}} = \tilde{A}_2\xi + \tilde{B}_2\eta + \tilde{C}_1 u + \tilde{D}_2, \end{aligned}\right\} \text{ in } \tilde{G}. \quad (4.18)$$

Moreover the boundary condition on Γ_2 in (4.3) can be reduced to the form

$$\begin{aligned} &\mathrm{Re}[(1-j)(U+jV)] = U(z) - V(z) = \tilde{R}(z), \text{ i.e.} \\ &\mathrm{Re}[\overline{\lambda(z)}w(z)] = \tilde{R}(z) = R_1(x) \text{ on } \Gamma_2, \end{aligned} \quad (4.19)$$

which $R_1(x)$ is as stated in (4.8). Through the transformation (4.15), the above boundary condition (4.19) is transformed into

$$\begin{gathered}\mathrm{Re}[\overline{\lambda(f^{-1}(\tilde{Z}))}w(f^{-1}(\tilde{Z}))]=R_1(f^{-1}(\tilde{Z})),\ \tilde{Z}=\tilde{x}+j\tilde{Y}\in f(\Gamma_2),\\ \mathrm{Im}[\overline{\lambda(f^{-1}(\tilde{Z}_1))}w(f^{-1}(\tilde{Z}_1))]=b_1,\end{gathered}\tag{4.20}$$

in which $\tilde{Z}=f(Z)$, $\tilde{Z}_1=f(Z_1')$, $Z_1'=x_2+jG[-\gamma(x_2)]$. Therefore the boundary value problem (4.17), (4.19) is reduced to the boundary value problem (4.18), (4.20), which is called Problem A^- in $\tilde{G}$. On the basis of Theorem 4.6 below, we see that the boundary value problem (4.18) (in $\tilde{G}$), (4.20) has a unique solution $w(\tilde{Z})$, and then

$$w=w[\tilde{Z}(z)]\ \text{ in }\hat{G}$$

is just a solution of the corresponding Problem A^- for (4.17) in $\hat{G}$ with the boundary condition (4.19). Similarly we can discuss the domain bounded by $A_1O_1\cup\Delta_1\cup\Delta_2\cup P_1'P_2'$. As for other characteristic curves Γ_1',Γ_2' and Δ_1',Δ_2', we can similarly handle, but it needs to choose the above known value $u(z)=u_0(z)$ on the boundary $\{x_1'\le x\le x_2',x_1\le x\le x_2,y=-1/2\}$ as the boundary value on $\{x_1'\le x\le x_2',x_1\le x\le x_2,y=-1/2\}$ of $D^-\cap\{y<-1/2\}$, because we first have found the solution $u_0(z)$ of Problem T in $D^-\cap\{y\ge -1/2\}$.

In this section, we first give the representation of solutions for the exterior Tricomi-Rassias problem (Problem TR) for equation (4.1) in D. Noting that $W(z)=U+iV=[H(y)u_x-iu_y]/2$ in D^+ and $W(z)=U+jV=[H(y)u_x-ju_y]/2$ in D^-, we have

$$\begin{aligned}&W_{\overline{\tilde{z}}}=[H(y)W_x+iW_y]/2=\frac{1}{4}[H^2u_{xx}+u_{yy}+iH_yu_x]=\frac{1}{4}[(\frac{iH_y}{H}-\frac{a}{H})Hu_x\\ &-bu_y-cu-d]=\frac{1}{4}[(\frac{iH_y}{H}-\frac{a}{H}-ib)W+(\frac{iH_y}{H}-\frac{a}{H}+ib)\overline{W}-cu-d]\\ &=W_{\overline{Z}}=A_1(z)W+A_2\overline{W}+A_3(z)u+A_4(z)=g(Z)\ \text{ in } D^+,\\ &W_{\overline{\tilde{z}}}=W_{\overline{Z}}=\frac{1}{2}[H(y)W_x+jW_y]=\frac{1}{2}[H[(U+jV)_x+j(U+jV)_y]\\ &=H[e_1(U+V)_\mu+e_2(U-V)_\nu]=\frac{e_1}{4}[(\frac{H_y}{H}+\frac{a}{H}-b)(U+V)\\ &+(\frac{H_y}{H}+\frac{a}{H}+b)(U-V)+cu+d]+\frac{e_2}{4}[(-\frac{H_y}{H}+\frac{a}{H}-b)(U+V)\\ &+(-\frac{H_y}{H}+\frac{a}{H}+b)(U-V)+cu+d]\ \text{ in } D^-,\end{aligned}\tag{4.21}$$

where $Z = Z(z)$, $\tau = \mu + i\nu = x + G(y) + j[x - G(y)] = \tau(z)$ are the mappings from D onto the domains D_Z and D_τ respectively. Especially, the complex equation

$$W_{\bar{\bar{z}}} = 0 \text{ in } \overline{D} \tag{4.22}$$

can be rewritten in the system

$$\begin{aligned} &[(U+V)+i(U-V)]_{\mu-i\nu} = 0 \text{ in } \overline{D^+}, \\ &(U+V)_\mu = 0, \ (U-V)_\nu = 0 \text{ in } \overline{D^-}. \end{aligned} \tag{4.23}$$

The boundary value problem for equation (4.21) with the boundary condition (4.8) $(W(z) = u_{\bar{z}})$ and the relation

$$u(z) = \begin{cases} 2\text{Re}\displaystyle\int_{-1}^{z}[\frac{\text{Re}W(z)}{H(y)} + i\text{Im}W(z)]dz + b_0 \text{ in } \overline{D^+}, \\ 2\text{Re}\displaystyle\int_{-1}^{z}[\frac{\text{Re}W(z)}{H(y)} - j\text{Im}W(z)]dz + b_0 \text{ in } \overline{D^-}, \end{cases} \tag{4.24}$$

will be called Problem A. By the above discussion at the beginning of this subsection, we can consider that characteristics Γ_1, Γ_2 and Δ_1, Δ_2 possess the intersection points $z_0 = (x_0, jy_0) = (1/2, jy_0), z_0' = (x_0', jy_0) = (-1/2, jy_0)$ respectively.

Now, we give the representation of solutions for the exterior Tricomi-Rassias problem (Problem *TR*) for system (4.23) in $\overline{D}$. It is obvious that Problem *TR* for equation (4.1) is equivalent to the following boundary value problem (Problem A) for (4.21) with the boundary conditions

$$\begin{aligned} &\text{Re}[\overline{\lambda(z)}W(z)] = \tilde{R}(z) \text{ on } L = \Delta_1 \cup \Gamma_2, \\ &\text{Re}[\overline{\lambda(z)}W(z)] = R_0(z) \text{ on } L_0 = L_0' \cup L_0'', \\ &\text{Im}[\overline{\lambda(z)}W(z)]|_{z=z_l} = b_l, u(a_l) = b_{1-l}, l = 1, 2, \end{aligned} \tag{4.25}$$

where $\lambda(x) = 1+j = a+jb$ on $L_0' = \{-1<x<0, y=0\}$, $\lambda(x) = 1-j = a+jb$ on $L_0'' = \{0 < x < 1, y = 0\}$, and $R_0(z)$ on L_0 is an undetermined real function. It is clear that the solution of problem A for (4.23) in $\overline{D^-}$ can be expressed as

$$\begin{aligned} &\xi(z) = U(z)+V(z) = f(\nu), \ \eta(z) = U(z)-V(z) = g(\mu), \\ &U(z) = [f(\nu) + g(\mu)]/2, \ V(z) = [f(\nu) - g(\mu)]/2, \ \text{ i.e.} \\ &W(z) = U(z) + jV(z) = [(1+j)f(\nu) + (1-j)g(\mu)]/2, \end{aligned} \tag{4.26}$$

in which $\mu = x + G(y), \nu = x - G(y)$, $G(y) = \int_0^y H(t)dt$, $f(t), g(t)$ are two arbitrary real continuous functions of $t \in (-1, 0) \cup (0, 1)$. For convenience, denote by the functions $a(x), b(x), r(x)$ of x the functions $a(z), b(z), r(z)$ of z in (4.25), and from (4.8), (4.25), we have

$$\begin{aligned}
&a(z)u(z) - b(z)v(z) = \tilde{R}(z) \text{ on } L,\\
&\overline{\lambda(x_l)}W(z_l) = \tilde{R}(x_l) + jb_l,\ l = 1, 2, \text{ i.e.}\\
&[a(x)-b(x)]f(2x+1)+[a(x)+b(x)]g(-1)=2\tilde{R}(x) \text{ on } [-1, -1/2],\\
&[a(x) - b(x)]f(1) + [a(x) + b(x)]g(2x - 1) = 2\tilde{R}(x) \text{ on } [1/2, 1],\\
&[a(x_0) - b(x_0)]f(1) = [a(x_0) - b(x_0)][U(z_0) + V(z_0)] = 0,\\
&[a(x_0')+b(x_0')]g(-1) = [a(x_0') + b(x_0')][U(z_0') - V(z_0')] = 0,
\end{aligned} \tag{4.27}$$

if $f(x) = U(x) + V(x) = V(x) = -[U(x) - V(x)] = -g(x)$ on L_0, then we can replace $f(\nu)$, $g(\mu)$ by $-g(\nu)$, $-f(\mu)$ respectively. The above formula can be rewritten in the form

$$\begin{aligned}
&[a((t-1)/2)-b((t-1)/2)]f(t)+[a((t-1)/2)+b((t-1)/2)]g(-1)\\
&\qquad = 2\tilde{R}((t-1)/2),\ t \in [-1, 0],\\
&[a((t+1)/2)-b((t+1)/2)]f(1)+[a((t+1)/2)+b((t+1)/2)]g(t)\\
&\qquad = 2\tilde{R}((t+1)/2),\ t \in [0, 1], \text{ i.e.}
\end{aligned}$$

$$\begin{aligned}
f(x - G(y)) &= \frac{2\tilde{R}((x - G(y) - 1))/2)}{a((x - G(y) - 1)/2) - b((x - G(y) - 1)/2)}\\
&- \frac{[a((x - G(y) - 1)/2) + b((x - G(y) - 1)/2)]g(-1)}{a((x - G(y) + 1)/2) - b((x - G(y) + 1)/2)},\\
&\qquad -1 \le x - G(y) \le 0,\\
g(x + G(y)) &= \frac{2\tilde{R}((x + G(y) + 1)/2)}{a((x + G(y) + 1)/2) + b((x + G(y) + 1)/2)}\\
&- \frac{[a((x + G(y) + 1)/2) - b((x + G(y) + 1)/2)]f(1)}{a((x + G(y) + 1)/2) + b((x + G(y) + 1)/2)},\\
&\qquad 0 \le x + G(y) \le 1.
\end{aligned} \tag{4.28}$$

Thus the solution $w(z)$ of (4.22) can be expressed as (4.26), where $f(x - G(y))$, $g(x + G(y))$ are as stated in (4.28) and $f(1), g(-1)$ are as stated before. Finally we find a solution $W(z)$ of the Riemann-Hilbert boundary value problem for equation (4.22) in D^+ with the boundary conditions

$$\begin{gathered}\mathrm{Re}[\overline{\lambda(z)}W(z)] = R(z) \ \text{ on } \Gamma = \Gamma_0,\\ \mathrm{Re}[\overline{\lambda(z)}W(z)]=-\hat{R}_0(x) \ \text{ on } L_0',\ \mathrm{Re}[\overline{\lambda(z)}W(z)]=\hat{R}_0(x) \ \text{ on } L_0'',\end{gathered} \tag{4.29}$$

in which $\lambda(z) = 1 + i$ on $L_0' = \{-1 < x < 0, y = 0\}$, and $\lambda(z) = 1 - i$ on $L_0'' = \{0 < x < 1, y = 0\}$. Noting that the index of the above boundary condition is $K = 0$, on the basis of the result in [87]1), we know that the above Riemann-Hilbert problem has a unique solution $w(z)$ in D^+, and then

$$\mathrm{Re}[\overline{\lambda(z)}W(z)]=\tilde{R}_0(x) \ \text{ on } L_0',\ \mathrm{Re}[\overline{\lambda(z)}W(z)]=-\tilde{R}_0(x) \ \text{ on } L_0'', \tag{4.30}$$

is determined, where $\lambda(z) = 1 + j$ on L_0, and $\lambda(x) = 1 - j$ on L_0''. Hence Problem A for equation (4.22) has a unique solution $w(z)$ in D. The above results can be written as a theorem.

Theorem 4.1 *Problem TR of equation* (4.22) *or system* (4.23) *in* $\overline{D}$ *has a unique solution* $u(z)$ *as stated in* (4.26), (4.28).

The representation of solutions of Problem TR for equation (4.1) can be written the following theorem.

Theorem 4.2 *Under Condition C, any solution* $u(z)$ *of Problem TR for equation* (4.1) *in* D *can be expressed as follows*

$$\begin{aligned}
&u(z)=u(x)-2\int_{d_l}^{y}V(z)dy=2\mathrm{Re}\int_{c_l}^{z}[\frac{\mathrm{Re}w}{H}+\begin{pmatrix} i\\ -j\end{pmatrix}\mathrm{Im}w]dz+b_0 \ \text{ in } \begin{pmatrix}\overline{D^+}\\ \overline{D^-}\end{pmatrix},\\
&w[z(Z)] = \Phi(Z) + \Psi(Z), \Psi(Z)=-2\mathrm{Re}\frac{1}{\pi}\iint_{D_t^+}\frac{f(t)}{t-Z}d\sigma_t \ \text{ in } \overline{D_Z^+},\\
&w(z) = \phi(z) + \psi(z) = \xi(z)e_1 + \eta(z)e_2 \ \text{ in } \overline{D^-},\\
&\xi(z)=\int_0^{\mu}\frac{g_1(z)}{2H(y)}d\mu=\zeta(z)+\int_{d_l}^{y}g_1(z)dy=\int_{S_1}g_1(z)dy+\int_{d_l}^{y}g_1(z)dy\\
&=\int_{y_1}^{|y|}\hat{g}_1(z)dy, z\in s_1,\ \eta(z) = \theta(z) + \int_{d_l}^{y}g_2(z)dy,\ z\in s_2,\\
&g_l(z)=\tilde{A}_l(U+V)+\tilde{B}_l(U-V)+2\tilde{C}_lU+\tilde{D}_lu+\tilde{E}_l, l=1,2,
\end{aligned} \tag{4.31}$$

where $c_1 = -1, c_2 = 1 - j, d_1 = 0, d_2 = -1, U = Hu_x/2, V = -u_y/2$, $\Phi(Z)$ *is an analytic function in* $D_Z^+ = Z(D^+)$, *in which* $Z(z) = x+iY = x+iG(y)$

is a mapping from $z(\in D^+)$ *to* $Z(\in D_Z^+)$, $Z(z) = x + jY = x + jG(y)$ *in* D^-, $D_Z^- = Z(D^-)$, *and* s_1, s_2 *are two families of characteristics in* D^-*:*

$$s_1: \frac{dx}{dy} = \sqrt{-K(y)} = H(y),\ s_2: \frac{dx}{dy} = -\sqrt{-K(y)} = -H(y) \tag{4.32}$$

passing through $z = x + jy \in D^-$, S_1, S_2 *are the characteristic curves from the points on* Δ_1, Γ_2' *and* Γ_2, Δ_1' *to the points on* L_0, $\phi(z) = \zeta(z)e_1 + \theta(z)e_2$ *is a solution of* (4.23) *in* D^-, $\zeta(z) = \int_{S_1} g_1(z)dy$, $\theta(z) = -\zeta(x + G(y))$ *in* $\tilde{D}_1^- = D^- \cap \{x < 0, y > -1/2\}$, *and* $\eta(z) = \int_{S_2} g_2(z)dy$, $\zeta(z) = -\theta(x - G(y))$ *in* $\tilde{D}_2^- = D^- \cap \{x > 0, y > -1/2\}$, *and*

$$w(z) = U(z) + jV(z) = \frac{1}{2}[Hu_x - ju_y],$$

$$\xi(z) = \mathrm{Re}\psi(z) + \mathrm{Im}\psi(z), \eta(z) = \mathrm{Re}\psi(z) - \mathrm{Im}\psi(z),$$

$$\tilde{A}_1 = \tilde{B}_2 = \frac{1}{2}(\frac{h_y}{2h} - b),\ \tilde{A}_2 = \tilde{B}_1 = \frac{1}{2}(\frac{h_y}{2h} + b),$$

$$\tilde{C}_1 = \frac{a}{2H} + \frac{m}{4y},\ \tilde{C}_2 = -\frac{a}{2H} + \frac{m}{4y},$$

$$\tilde{D}_1 = -\tilde{D}_2 = \frac{c}{2},\ \tilde{E}_1 = -\tilde{E}_2 = \frac{d}{2},$$

and

$$d\mu = d[x + G(y)] = 2H(y)dy \text{ on } s_1,$$

$$d\nu = d[x - G(y)] = -2H(y)dy \text{ on } s_2.$$

Proof From (4.21) it is easy to see that equation (4.1) in $\overline{D^-}$ can be reduced to the system of integral equations: (4.31). Moreover we can extend the equation (4.21) onto the the symmetrical domain $\hat{D}_Z$ of D_Z^- with respect to the real axis $\mathrm{Im}Z = 0$, namely introduce the function $\hat{W}(Z)$ as follows:

$$\hat{W}(Z) = \begin{cases} W[z(Z)], \\ -\overline{W[z(\overline{Z})]}, \end{cases} \quad \hat{u}(z) = \begin{cases} u(Z) \text{ in } D_Z^-, \\ -u(\overline{Z}) \text{ in } \hat{D}_Z, \end{cases} \tag{4.33}$$

and then the equation (4.31) is extended as

$$\hat{W}_{\bar{z}} = \hat{A}_1\hat{W} + \hat{A}_2\overline{\hat{W}} + \hat{A}_3\hat{u} + \hat{A}_4 = \hat{g}(Z) \text{ in } \overline{D_Z^-} \cup \overline{\hat{D}_Z}, \tag{4.34}$$

where

$$\hat{A}_l(Z)=\begin{cases}A_l(Z),\\ \overline{\tilde{A}_l(\overline{Z})},\end{cases} l=1,2,3,\ \hat{A}_4(Z)=\begin{cases}A_4(Z),\\ \overline{-A_4(\overline{Z})},\end{cases}$$

$$\hat{g}_l(Z)=\begin{cases}g_l(z) \text{ in } \overline{D_Z^-},\\ \overline{-g_l(\overline{Z})} \text{ in } \overline{\hat{D}_Z},\end{cases} l=1,2,$$

in which $\tilde{A}_1(\overline{Z})=A_2(\overline{Z}), \tilde{A}_2(\overline{Z})=A_1(\overline{Z}), \tilde{A}_3(\overline{Z})=A_3(\overline{Z})$, the system of integral equations (4.31) can be written in the form

$$\begin{aligned}\xi(z)&=\zeta(z)+\int_0^y g_1(z)dy=\int_{y_1}^{\hat{y}}\hat{g}_1(z)dy,\\ \eta(z)&=\theta(z)+\int_0^y g_2(z)dy=\int_{y_1}^{\hat{y}}\hat{g}_2(z)dy,\\ \hat{z}&=x+j\hat{y}=x+j|y| \text{ in } \overline{D_Z^-}\cup\hat{D}_Z,\end{aligned}\tag{4.35}$$

where x_1+jy_1 is the intersection point of Δ_1 or Γ_2 and the characteristic curve s_1 or s_2 passing through $z=x+jy$, the function $\theta(z)$ is determined by $\zeta(z)$, namely the function $\theta(z)$ will be defined by $\theta(z)=-\zeta(x+G(y))$ in $\tilde{D}_1^-=D^-\cap\{x<0,y>-1/2\}$, and the function $\zeta(z)$ will be defined by $\zeta(z)=-\theta(x-G(y))$ in $\tilde{D}_2^-=D^-\cap\{x>0,y>-1/2\}$, for the extended integral, for convenience later on the above form $\hat{g}_2(z)$ is written, and the numbers $\hat{y}-y_1,\hat{t}-y_1$ will be written by $\tilde{y},\tilde{t}$ respectively.

4.3 Unique solvability of solutions of exterior Tricomi-Rassias problem

In this subsection, we prove the uniqueness and existence of solutions of Problem *TR* for equation (4.1).

Theorem 4.3 *Suppose that equation* (4.1) *satisfies the above conditions. Then Problem TR for* (4.1) *has at most one solution in D.*

Proof Let $u_1(z),u_2(z)$ be any two solutions of Problem *TR* for (4.1). By Theorem 4.2, it is easy to see that $u(z)=u_1(z)-u_2(z)$ and $w(z)=u_{\tilde{z}}$ satisfy the homogeneous equation and boundary conditions

$$\begin{aligned}&K(y)u_{xx}+u_{yy}+au_x+bu_y+cu=0, \text{ i.e.}\\ &w_{\overline{\tilde{z}}}=A_1w+A_2\overline{w}+A_3u \text{ in } D,\end{aligned}\tag{4.36}$$

$$\frac{1}{2}\frac{\partial u}{\partial \nu}=\frac{1}{H(y)}\mathrm{Re}[\overline{\lambda(z)}u_{\bar z}]=\mathrm{Re}[\overline{\lambda(z)}w(z)]=0 \text{ on } \Gamma\cup L,$$
$$\mathrm{Im}[\overline{\lambda(z)}u_{\bar z}]|_{z=z_l}=0,\ l=1,2,\ u(a_l)=0, l=1,2, \tag{4.37}$$

where the function $w(z)=U(z)+jV(z)=[Hu_x-ju_y]/2$ in the hyperbolic domain D^- can be expressed in the form

$$w(z)=\phi(x)+\psi(z)=\xi(z)e_1+\eta(z)e_2,$$
$$\xi(z)=\zeta(z)+\int_0^y[\tilde A_1(U+V)+\tilde B_1(U-V)+2\tilde C_1U+2\tilde D_1u]dy,\ z\in s_1, \tag{4.38}$$
$$\eta(z)=\theta(z)+\int_0^y[\tilde A_2(U+V)+\tilde B_2(U-V)+2\tilde C_2U+2\tilde D_2u]dy,\ z\in s_2,$$

where $\phi(z)$ is a solution of (4.23). On the basis of Theorem 4.4 below, we can derive that $\phi(z)+\psi(z)=0$, $w(z)=0$, $z\in\overline{D^-}$. Thus the solution

$$u(z)=2\,\mathrm{Re}\int_{-1}^z[\frac{\mathrm{Re}w(z)}{H}+\binom{i}{-j}\mathrm{Im}w]dz \text{ in } \binom{\overline{D^+}}{\overline{D^-}}, \tag{4.39}$$

is the solution of the homogeneous equation of (4.1) with homogeneous boundary conditions of (4.3):

We first verify that the above solution $u(z)\equiv 0$ in D^+. If the maximum $M=\max_{\overline{D^+}}u(z)>0$, it is clear that the maximum point $z^*\notin G_1\cup\{0\}$. If $u(z)$ attains its maximum at a point $z^*=x^*\in L_0=L_0'\cup L_0''$. Now we verify the maximum point $z^*\notin L_0$. In fact, $u(z)$ can be expressed as in (4.31), it is easy to see that $\mathrm{Re}W(Z)=\mathrm{Re}\Phi(Z)=U(Z)=H(0)u_x/2=0$, $x\in L_0$, we can extend the analytic function $\Phi(Z)$ from $G_1\cap\{Y>0\}$ onto the symmetrical domain $\tilde D^+$ about the real axis $\mathrm{Im}Z=0$, the extended function is denoted by $\Phi(Z)$ again, obviously $\mathrm{Re}\Phi(Z)$ is a harmonic function in $D'=G_1\cup\tilde D^+\cup L_0$, thus $\mathrm{Re}\Phi(Z)=\mathrm{Re}W(Z)=H(y)u_x/2=YF$ in $D'_\varepsilon=D'\cap\{|Z-x^*|<\varepsilon$, ε is a sufficiently small positive constant and F is Hölder continuous in D'_ε, and then $u_x=O(Y^{2/(m+2)})$, it follows that $u(x)=M$ in $D'_\varepsilon\cap\{Y=0\}$, which can be derived $u(x)=M$ on L_0', this contracts $u(1)=u(-1)=0$. Hence $\max_{\overline{D^+}}u(z)=0$. By the similar method, $\min_{\overline{D^+}}u(z)=0$ can be derived. From Theorem 4.4 below, we can get $u(z)=0$ in $\overline{D^-}$. Therefore $u(z)=0$, $u_1(z)=u_2(z)$ in $\overline{D^+}$.

Theorem 4.4 *Let D^- be given as above and equation* (4.1) *satisfy Condition C and* (4.50) *below. Then the exterior Tricomi-Rassias problem* (*Problem TR*) *for* (4.1) *in D^- at most has a solution.*

Proof We assume that m is a positive number, denote by $u_1(z), u_2(z)$ two solutions of Problem TR for (4.1), by Theorem 4.2, we see that the function $u_{\bar z}(z)=u_{1\bar z}(z)-u_{2\bar z}(z)=U(z)+jV(z)$ in $\overline{D^-}$ is a solution of the homogeneous system of integral equations

$$\begin{aligned}
&u(z)=u(x)-2\int_0^y V(z)dy=\int_1^z[\frac{\text{Re}w}{H(y)}-j\text{Im}w]dz \text{ in } D^-,\\
&w(z)=\phi(z)+\psi(z)=\xi(z)e_1+\eta(z)e_2,\\
&\xi(z)=\zeta(z)+\int_0^y[\tilde A_1(U+V)+\tilde B_1(U-V)+2\tilde C_1U+\tilde D_1u]dy,\ z\in s_1,\\
&\eta(z)=\theta(z)+\int_0^y[\tilde A_2(U+V)+\tilde B_2(U-V)+2\tilde C_2U+\tilde D_2u]dy,\ z\in s_2.
\end{aligned}\tag{4.40}$$

Noting that u_x, u_y are continuous in $\overline{D}\backslash(\{\pm1,\pm1-j,0,-j,P_1P_2,P_1'P_2'\}$ and $\phi(z)$ is a solution of (4.22) in D^-, we can prove $w(z)=\phi(z)+\psi(z)=0$ in D^-.

In fact, choose any closed set

$$D_0=\overline{D^-}\cap\{-1\le a_0=-1+\delta_0\le x<b_0=-\delta_0\le 0,-\delta\le y\le 0\} \text{ or}$$
$$\overline{D^-}\cap\{0\le a_0=\delta_0<x<b_0=1-\delta_0\le 1,-\delta\le y\le 0\},$$

where δ, δ_0 are sufficiently small positive constants, a_0, b_0 are real numbers, noting that the continuity of u_x, u_y in D_0 and the estimate in (4.56) below, there exist positive numbers $N\,(>1), \gamma(<1), \beta=\min(1,m/2)-\varepsilon_0$ dependent on $m, u(z), D_0, \varepsilon_0$, such that

$$\begin{aligned}
&|u(z)-u(x)|\le N\gamma|y|^\beta,\ |u(z)|\le N\gamma,\ |\xi(z)|\le N\gamma,\\
&|\eta(z)|\le N\gamma,\ |\xi(z)+\eta(z)|\le N\gamma|y|^{1+m/2-\varepsilon_0},
\end{aligned}\tag{4.41}$$

in which ε_0 is a sufficiently small positive constant. From (40), we can obtain

$$\begin{aligned}
&|\xi(z)-\zeta(z)|=|\int_0^y[\tilde A_1\xi+\tilde B_1\eta+\tilde C_1(\xi+\eta)+\tilde D_1u]dy|\\
&\le|\int_0^y N\gamma[|\tilde A_1|+|\tilde B_1|+|\tilde D_1|+|\tilde C_1||y|^{1+m/2-\varepsilon_0}]dy|\\
&\le|\int_0^y N\gamma[2k_3+(\frac{|\varepsilon(y)|}{2}+\frac{m}{4})|y|^{m/2-\varepsilon_0}]dy|\\
&\le N\gamma[2k_3|y|+\frac{|\varepsilon(y)|+m/2}{2+m-2\varepsilon_0}|y|^{1+m/2-\varepsilon_0}]\le N\gamma^2|y|^\beta \text{ on } s_1,
\end{aligned}$$

where $\varepsilon(y)$ is as stated in (4.50) below, and k_3 is a constant such that $|h_y/h| \le k_3(\ge \max[1, k_0, k_1, k_2])$. Similarly we have

$$|\eta(z) - \theta(z)| = |\int_0^y [\tilde{A}_2\xi + \tilde{B}_2\eta + \tilde{C}_2(\xi+\eta) + \tilde{D}_2 u]dy| \le N\gamma^2|y|^\beta \text{ on } s_2.$$

Applying the repeated insertion, as stated in the proof of Theorem 4.5 below, the inequalities

$$\begin{gathered}|u(z)-u(x)|\le N\gamma^k|y|^\beta,\ |\xi(z)-\zeta(z)|\le N\gamma^k|y|^\beta,\\ |\eta(z)-\theta(z)|\le N\gamma^k|y|^\beta,\ k=2,3,...\end{gathered} \tag{4.42}$$

are obtained. This shows that $u(z) = 0, \xi(z) = 0, \eta(z) = 0$ in D_0. Taking into account the arbitrariness of a_0, b_0, by using the method in Section 1, we can derive $u(z) = 0, \xi(z) = 0, \eta(z) = 0$ in D.

In order to prove the existence of solutions of Problem TR for equation (4.1), it suffices to discuss Problem $\tilde{T}$ for (4.1), it is clear that Problem $\tilde{T}$ is equivalent to Problem A for the complex equation

$$w_{\bar{\bar{z}}} = A_1(z)w + A_2(z)\overline{w} + A_3(z)u + A_4(z) \text{ in } D, \tag{4.43}$$

with the relation

$$u(z) = \begin{cases} 2\mathrm{Re}\displaystyle\int_{c_l}^{z}[\frac{\mathrm{Re}w(z)}{H(y)} + i\mathrm{Im}w(z)]dz + b_0 \text{ in } \overline{D^+}, \\ u(x) - \displaystyle\int_0^y \mathrm{Im}w(z)dy \text{ in } \overline{D^-}, \end{cases} \quad l = 1, 2, \tag{4.44}$$

and the boundary conditions

$$\begin{gathered}\mathrm{Re}[\overline{\lambda(z)}w(z)] = R(z) \text{ on } \Gamma\cup L,\\ \mathrm{Im}[\overline{\lambda(z)}u_{\bar{\bar{z}}}]|_{z=z_l} = b_l, l=1,2, u(a_l) = b_{1-l}, l=1,2,\end{gathered} \tag{4.45}$$

where $c_1 = -1, c_2 = 1 - i$ or $1 - j$, $H(y) = \sqrt{|K(y)|}$, the coefficients in (4.43) are as stated in (4.21), $\lambda(z), R(z), z_l, b_l$ are as stated before, but

$$R(z) = 0 \text{ on } \Gamma\cup L,\ b_l = 0,\ l = -1, 0, 1, 2. \tag{4.46}$$

Similarly to Section 2, we can divided Problem A into two boundary value problems, i.e. Problem A^+: (4.43), (4.44) in D^+ with the boundary conditions

$$\begin{gathered}\mathrm{Re}[\overline{\lambda(z)}w(z)] = R(z) \text{ on } \Gamma_0,\\ \mathrm{Re}[(1+i)w(z)] = \hat{R}_0(x) \text{ on } L_0',\end{gathered} \tag{4.47}$$

and Problem A^-: (4.43), (4.44) in D^- with the boundary conditions

$$\begin{gathered}\text{Re}[\overline{\lambda(z)}w(z)]=R(z) \text{ on } L,\ \text{Im}[\overline{\lambda(z_l)}w(z_l)]=b_l, l=1,2,\\ \text{Re}[(1-j)w(z)] = \tilde{R}_0(x) \text{ on } L_0'\end{gathered} \tag{4.48}$$

where $L_0 = L_0' \cup L_0''$, we mention that the corresponding functions $f(x) = -g(x)$ on L_0 in (4.48), and

$$R(z) = 0 \text{ on } \Gamma \cup L,\ b_0 = b_{-1} = b_1 = b_2 = 0. \tag{4.49}$$

On the basis of the result in Section 2, Chapter II (here $z = \pm 1, \pm 1-i, 0, -i$ are the discontinuous points of Problem A^+), we can prove the existence of solutions of Problem A^+. In the following we shall find a solution of the above Problem A^- in D^- as follows.

Theorem 4.5 *If equation* (4.1) *satisfies Condition C and* (4.50) *below, then there exists a solution $[w(z), u(z)]$ of Problem A^- for* (4.43), (4.48), (4.49).

Proof We first discuss the problem in $D_0 = D^- \cap (\{a_0 \le x \le b_0\} \cup \{a_1 \le x \le b_1\})$, where $-1 < a_0 = -1 + \delta_0 < b_0 = -\delta_0 < 0$ or $0 < a_1 = \delta_0 < b_1 = 1 - \delta_0 < 1$), and δ_0 is a sufficiently small positive constant.

We choose $v(x) = u(x) - u_0(x)$ on L_0 and discuss the case of $K(y) = -|y|^m h(y)$, m, $h(y)$ is as stated in (4.2) and $u_0(z)$ is as stated in Subsection 4.1. In order to find a solution of the system of integral equations (4.31), we need to add a condition, namely

$$a(x,y)|y|/H(y) = o(1), \text{ i.e. } |a(x,y)|/H(y) = \varepsilon(y)/|y|,\ m \ge 2, \tag{4.50}$$

where $\varepsilon(y) \to 0$ as $y \to 0$. It is clear that for two characteristics s_1, s_2 passing through a point $z = x + jy \in \overline{D^-}$ and x_1, x_2 are the intersection points with the axis $y = 0$ respectively, for any two points $\tilde{z}_1 = \tilde{x}_1 + j\tilde{y} \in s_1, \tilde{z}_2 = \tilde{x}_2 + j\tilde{y} \in s_2$, we have

$$\begin{gathered}|\tilde{x}_1-\tilde{x}_2| \le |x_1-x_2| = 2\left|\int_0^y \sqrt{-K(t)}dt\right|\\ \le \frac{2k_0}{m+2}|y|^{1+m/2} \le \frac{k_1}{12}|y|^{m/2+1} \le M|y|^{m/2+1}.\end{gathered} \tag{4.51}$$

From Condition C, we can assume that the coefficients in (4.31) are continuously differentiable with respect to $x \in L_0$ and satisfy the conditions

$$\begin{gathered}|\tilde{A}_l|, |\tilde{A}_{lx}|, |\tilde{B}_l|, |\tilde{B}_{lx}|, |\tilde{D}_l|, |\tilde{D}_{lx}| \le k_0 \le k_1/12, |\tilde{E}_l|, |\tilde{E}_{lx}| \le k_1/12,\\ 2\sqrt{h}, 1/\sqrt{h}, |h_y/h| \le k_0 \le k_1/12 \text{ in } \bar{D},\ l = 1, 2,\end{gathered} \tag{4.52}$$

and later on we shall use the constants

$$M = 4\max[M_1, M_2, M_3],\ M_1 = \max[8(k_1d)^2, \frac{M_3}{k_1}],$$
$$M_2 = \frac{(2+m)k_0d}{\delta^{2+m}}[4k_1 + \frac{4\varepsilon_0+m}{\delta}],\ M_3 = 2k_1^2[d + \frac{1}{2H(y_1')}], \tag{4.53}$$
$$\gamma = \max[4k_1d\delta^\beta + \frac{4\varepsilon(y)+m}{2\beta'}] < 1,\ 0 \le |y| \le \delta,$$

in which $\beta' = (1+m/2)(1-3\beta)$, $\varepsilon_0 = \max_{\overline{D^-}}\varepsilon(z)$, δ, β are sufficiently small positive constants, d is the diameter of $\overline{D^-}$, and

$$1/2H(y_1') \le k_0[(m+2)a_0/k_0]^{-m/(2+m)},$$

herein y_1' is similar to that as in (2.43). We choose $v_0 = 0, \xi_0 = 0, \eta_0 = 0$ and substitute them into the corresponding positions of v, ξ, η in the right-hand sides of (4.31), and obtain

$$v_1(z) = v_1(x) - 2\int_0^y V_0dy = v_1(x) + \int_0^y(\eta_0 - \xi_0)dy,$$
$$\xi_1(z) = \zeta_1(z) + \int_0^y g_{10}(z)dy = \zeta_1(z) + \int_0^y \tilde{E}_1dy = \int_{y_1}^{\hat{y}} \hat{E}_1dy,$$
$$\eta_1(z) = \theta_1(z) + \int_0^y g_{20}(z)dy = \theta_1(z) + \int_0^y \hat{E}_2dy = \int_{y_1}^{\hat{y}} \hat{E}_2dy, \tag{4.54}$$
$$g_{l0} = \tilde{A}_l\xi_0 + \tilde{B}_l\eta_0 + \tilde{C}_l(\xi_0+\eta_0) + \tilde{D}_lv + \tilde{E}_l = \tilde{E}_l, l=1,2,$$

where $z_1 = x_1 + jy_1$ is a point on Δ_1, which is the intersection of Δ_1 and the characteristic curve s_1 passing through the point $z = x + jy \in \overline{D^-}$. By the successive approximation, we find the sequences of functions $\{v_k\}, \{\xi_k\}, \{\eta_k\}$, which satisfy the relations

$$v_{k+1}(z) = v_{k+1}(x) - 2\int_0^y V_k(z)dy = v_{k+1}(x) + \int_0^y(\eta_k - \xi_k)dy,$$
$$\xi_{k+1}(z) = \zeta_{k+1}(z) + \int_0^y g_{1k}(z)dy = \int_{y_1}^{\hat{y}} \hat{g}_{1k}dy,$$
$$\eta_{k+1}(z) = \theta_{k+1}(z) + \int_0^y g_{2k}(z)dy = \int_{y_1}^{|y|} \hat{g}_{2k}(z)dy, \tag{4.55}$$
$$g_{lk}(z) = \tilde{A}_l\xi_k + \tilde{B}_l\eta_k + \tilde{C}_l(\xi_k + \eta_k) + \tilde{D}_lv_k + \tilde{E}_l,$$
$$l = 1,2,\ k = 0,1,2,....$$

Setting that $\tilde{g}_{lk+1}(z) = g_{lk+1}(z) - g_{lk}(z)$ and

$$\tilde{y} = \hat{y} - y_1,\ \tilde{t} = \hat{y} - y_1,\ \tilde{v}_{k+1}(z) = v_{k+1}(z) - v_k(z),$$

$$\tilde{\xi}_{k+1}(z) = \xi_{k+1}(z) - \xi_k(z), \tilde{\eta}_{k+1}(z) = \eta_{k+1}(z) - \eta_k(z),$$

$$\tilde{\zeta}_{k+1}(z) = \zeta_{k+1}(z) - \zeta_k(z), \tilde{\theta}_{k+1}(z) = \theta_{k+1}(z) - \theta_k(z),$$

we can prove that $\{\tilde{v}_k\}$, $\{\tilde{\xi}_k\}$, $\{\tilde{\eta}_k\}$, $\{\tilde{\zeta}_k\}$, $\{\tilde{\theta}_k\}$ in D_0 satisfy the estimates

$$\begin{aligned}
&|\tilde{v}_k(z) - \tilde{v}_k(x)|, |\tilde{\xi}_k(z) - \tilde{\zeta}_k(z)|, |\tilde{\eta}_k(z) - \tilde{\theta}_k(z)| \le M'\gamma^{k-1}|y|^{1-\beta}, 0 \le |y| \le \delta,\\
&|\tilde{\xi}_k(z)|, |\tilde{\eta}_k(z)| \le M(M_2|\tilde{y}|)^{k-1}/(k-1)!, y \le -\delta, \text{ or } M'\gamma^{k-1}, 0 \le |y| \le \delta,\\
&|\tilde{\xi}_k(z_1) - \tilde{\xi}_k(z_2) - \tilde{\zeta}_k(z_1) - \tilde{\zeta}_k(z_2)|,\ |\tilde{\eta}_k(z_1) - \tilde{\eta}_k(z_2) - \tilde{\theta}_k(z_1) - \tilde{\theta}_k(z_2)|\\
&\le M'\gamma^{k-1}[|x_1 - x_2|^{1-\beta} + |x_1 - x_2|^{\beta}|y|^{\beta'}],\ 0 \le |y| \le \delta,\ |\tilde{v}_k(z_1) - \tilde{v}_k(z_2)|,\\
&|\tilde{\xi}_k(z_1) - \tilde{\xi}_k(z_2)|, |\tilde{\eta}_k(z_1) - \tilde{\eta}_k(z_2)| \le M(M_2|\tilde{t}|)^{k-1}|x_1 - x_2|^{1-\beta}\\
&/(k-1)!,\ y \le -\delta, \text{ or } M'\gamma^{k-1}[|x_1 - x_2|^{1-\beta} + |x_1 - x_2|^{\beta}|t|^{\beta'}], 0 \le |y| \le \delta,\\
&|\tilde{\xi}_k(z) + \tilde{\eta}_k(z) - \tilde{\zeta}_k(z) - \tilde{\theta}_k(z)| \le M'\gamma^{k-1}|x_1 - x_2|^{\beta}|y|^{\beta'}, |\tilde{\xi}_k(z) + \tilde{\eta}_k(z)|\\
&\le M(M_2|\tilde{y}|)^{k-1}|x_1 - x_2|^{1-\beta}/(k-1)! \text{ or } M'\gamma^{k-1}|x_1 - x_2|^{\beta}|y|^{\beta'},
\end{aligned} \tag{4.56}$$

in which $z = x + jy$, $z = x + jt$ is a intersection point of s_1, s_2 in (4.32) passing through z_1, z_2; β', β are as stated before, δ is a sufficiently small positive constant, moreover $\gamma = 4k_1 d\delta^{\beta} + (4\max_{-\delta \le y \le 0}|\varepsilon(z)| + m)/2\beta' < 1$, d is the diameter of D, and $M_2 = (2+m)k_0 d\delta^{-2-m}[4k_1\delta + 4\varepsilon_0 + m]/\delta$, and M' is a sufficiently large positive constants a stated in (2.53).

From the estimate (4.56), we see that for any two points $z_1 = x_1 + jG(y_1)$, $z_2 = x_2 + jG(y_2) \in D_0$, where $x_1 < x_2$, $y_1 = y < y_2(< 0)$, setting that $z_3 = x_3 + jG(y_1)$ is the intersection point of the characteristic line $s_1 : x - G(y) = x_2 - G(y_2)$ and the straight line $G(y) = G(y_1)$, then

$$|\xi_k(z_1) - \xi_k(z_3)| \le 2M\gamma^{k-1}|x_1 - x_3|^{\beta}|t|^{\beta'},$$

$$|\xi_k(z_2) - \xi_k(z_3)| \le M\gamma^{k-1}|y|^{1-\beta}|y_1 - y_2|^{\beta},$$

and then

$$\begin{aligned}
&|\xi_k(z_1) - \xi_k(z_2)| \le |\xi_k(z_1) - \xi_k(z_3)| + |\xi_k(z_2) - \xi_k(z_3)|\\
&\le M\gamma^{k-1}[2|x_1 - x_3|^{\beta}|y|^{\beta'} + |y|^{1-\beta}|y_1 - y_2|^{\beta}]\\
&\le M\gamma^{k-1}(2|y|^{\beta'} + |y|^{1-\beta})|z_1 - z_2|^{\beta} \le M''M\gamma^{k-1}|z_1 - z_2|^{\beta},
\end{aligned} \tag{4.57}$$

where $M''=\max_{\overline{D_-}}[2|y|^{\beta'}+|y|^{1-\beta}]$. For other case, we can similarly get the estimates:

$$|v_k(z_1)-v_k(z_2)|,|\eta_k(z_1)-\eta_k(z_2)|\le M''M\gamma^{k-1}|z_1-z_2|^{\beta}. \tag{4.58}$$

The formula (4.57) shows that these sequences of functions $\{v_k(z)\}$, $\{\xi_k(z)\}$, $\{\eta_k(z)\}$ in $D_l=\overline{D^-}\cap\{|z\pm 1|\ge 1/l\}\cap\{|z|\ge 1/l\}$ $(l>2)$ are uniformly bounded and equicontinuous. In particular for $\delta_0=1/l$, $l\,(>2)$ is a positive integer, from these sequences, we can choose the subsequences $\{v_k^l(z)\}$, $\{\xi_k^l(z)\}$, $\{\eta_k^l(z)\}$, which uniformly converge to $v_*(z),\xi_*(z),\eta_*(z)$ in D_l respectively, and $v_*(z),\xi_*(z),\eta_*(z)$ satisfy the system of integral equations

$$\begin{aligned}
&v_*(z)=v_*(x)-2\int_0^y V_*dy=u_*(x)+\int_0^y(\eta_*-\xi_*)dy,\\
&\xi_*(z)=\zeta_*(z)+\int_0^y[\tilde A_1\xi_*+\tilde B_1\eta_*+\tilde C_1(\xi_*+\eta_*)+\tilde D_1u_*+\tilde E_1]dy,\, z\in s_1,\\
&\eta_*(z)=\theta_*(z)+\int_0^y[\tilde A_2\xi_*+\tilde B_2\eta_*+\tilde C_2(\xi_*+\eta_*)+\tilde D_2u_*+\tilde E_2]dy,\, z\in s_2,
\end{aligned}$$

and the function $v(z)=v_*(z)$ is a solution of Problem A_0 for equation (4.43) in D_l. Moreover from $\{v_k^l(z)\}$, $\{\xi_k^l(z)\}$, $\{\eta_k^l(z)\}$ $(l=1,2...)$, we can select the diagonal sequence $\{v_l^l(z)\}$, $\{\xi_l^l(z)\}$, $\{\eta_l^l(z)\}$, which uniformly converge to $v_*(z),\xi_*(z),\eta_*(z)$ in D_0 respectively. Hence the function $u(z)=v(z)+u_0(z)$ is a solution of Problem *TR* for (4.1) in D^-. Thus the existence of solutions of Problem *TR* for equation (4.1) in D is proved.

From the above discussion, we obtain the following theorem.

Theorem 4.6 *Let equation* (4.1) *satisfy Condition C and* (4.50). *Then the Problem TR for* (4.1) *has a solution.*

Remark 4.1 If the coefficients $K(y)$ in equation (4.1) can be replaced by function $K(x,y)=\mathrm{sgn}y|y|^m h(x,y)$ in $\overline{D}\cap\{y\ge -1/2\}$ and $K(x,y)=-\mathrm{sgn}(1+y)|1+y|^m h(x,y)$ in $\overline{D}\cap\{y\le -1/2\}$, where m is a positive number and $h(x,y)$ is a continuously differentiable positive function in $\overline{D}$. Moreover, the above characteristics $\Gamma_l,\Gamma_l',\Delta_l,\Delta_l'\,(l=1,2)$ can be replaced by more general curves, which can discussed by the similar method as stated in Section 3, Chapter IV and Section 2, Chapter VI below. Besides we can discuss the unique solvability of some oblique derivative problem for equation (4.1) in an $N(2<N<\infty)$-connected domain D by using the method of parameter extension and the repeated insertion similar to Section 2.

5 The Frankl Boundary Value Problem for Second Order Degenerate Equations of Mixed Type

This section deals with the Frankl boundary value problem for second order linear equations of mixed (elliptic-hyperbolic) type with parabolic degeneracy. We first give the representation formula and prove the uniqueness of solutions for the above boundary value problem, moreover by the method of parameter extension, the existence of solutions is proved. In the books [12]1),3), the Frankl problem was discussed for the special mixed equations of second order: $u_{xx} + \text{sgn}y\, u_{yy} = 0$. In the book [74], the Frankl problem was discussed for the mixed equation with parabolic degeneracy: $\text{sgn}y|y|^m u_{xx} + u_{yy} = 0$, which is a mathematical model of problem of gas dynamics, where the existence of solutions of Frankl problem was proved by using the method of integral equations. In this section, we shall not use this method. We first prove the uniqueness of solutions of the Frankl problem for general mixed equations with parabolic degeneracy, give a priori estimate of its solutions, and then prove the solvability of the problem for the equations, which is generalized the corresponding result from [12]1),3).

5.1 Formulation of Frankl problem for second order equations of mixed type

Let D be a simply connected bounded domain in the complex plane $\mathbf{C}$ with the boundary $\partial D = \Gamma \cup A'A \cup A'C \cup CB$, where $\Gamma(\subset \{x>0, y>0\}) \in C^2_\mu(0<\mu<1)$ with the end points $A=i$ and $B=b_1, A'A=\{x=0, -1\le y\le 1\}$, $A'C=\{x-G(y)=c_1, x>0, y<0\}$ is the characteristic line and $CB=\{c_1\le x\le b_1, y=0\}$, and denote $D^+=D\cap\{y>0\}, D^-=D\cap\{y<0\}$. Let $K(y)$ be an odd continuous function, possess the derivative $K'(y)$ and $yK(y)>0$ on $y\ne 0$, and $K(0)=0$. Without loss of generality, we may assume that Γ possesses the form $x=b_1-\tilde{G}(y)$, $y=\tilde{G}(x)$, $x>0, y>0$ near $z=b_1, i$ and are orthogonal with the real axis and imaginary axis at $z=i$ respectively as stated in Section 3, otherwise through a conformal mapping from D^+ onto the above domain, such that three boundary points $i, 0, c_1$ are not changed, then the above requirement can be realized.

Frankl Problem Find a continuous solution $u(z)$ of equation (2.1) in $\overline{D}$, where u_x, u_y are continuous in $D^* = \overline{D}\backslash\{0, c_1, b_1, i, -i\}$ and satisfy the boundary conditions

$$u = \psi_1(s) \text{ on } \Gamma, \tag{5.1}$$

$$u = \psi_2(x) \text{ on } CB, \tag{5.2}$$

$$\frac{\partial u}{\partial x} = 0 \text{ on } A'A, \tag{5.3}$$

$$u(iy) - u(-iy) = \phi(y), \ -1 \le y \le 1. \tag{5.4}$$

Here $\psi_1(s)$, $\psi_2(x)$, $\phi(y)$ are given real-valued functions satisfying the conditions

$$\begin{gathered} C_\alpha^2[\psi_1(s), \Gamma] \le k_2, \ C_\alpha^2[\psi_2(x), CB] \le k_2, \\ C_\alpha^2[\phi(y), A'A] \le k_2, \ \psi_1(0) = \psi_2(b_1) = b_0, \end{gathered} \tag{5.5}$$

in which s is the arc length parameter on Γ normalized such that $s = 0$ at the point B, S_0 is the length of Γ, and $\alpha\,(0 < \alpha < 1)$, k_2 are positive constants. The above boundary value problem is called Problem F and the corresponding homogeneous problem is called Problem F_0.

Let

$$U = \frac{1}{2}H(y)u_x, V = -\frac{1}{2}u_y, \text{Re}W = U, \text{Im}W = V \text{ in } D, \tag{5.6}$$

then equation (2.1) can be written as the complex equation

$$\begin{gathered} W_{\bar{z}} = A_1 W + A_2 \overline{W} + A_3 u + A_4 \text{ in } \overline{D}, \\ u(z) = 2\text{Re}\int_{b_1}^{z} [\frac{\text{Re}W(z)}{H(y)} + \begin{pmatrix} i \\ -j \end{pmatrix} \text{Im}W(z)]dz + \psi_1(0) \text{ in } \begin{pmatrix} D^+ \\ D^- \end{pmatrix}. \end{gathered} \tag{5.7}$$

The boundary conditions of the Frankl problem can be rewritten as

$$\begin{gathered} \frac{1}{2}\frac{\partial u}{\partial l} = \frac{1}{H(y)}\text{Re}[\overline{\lambda(z)}W(z)] = \text{Re}[\overline{\Lambda(z)}W(z)] = R(z), z \in \Gamma \cup CB, \\ U(0,y) = \frac{1}{2}\frac{\partial u}{\partial x} = R(z) = \frac{1}{H(y)}\text{Re}[\overline{\lambda(iy)}W(iy)] = \\ = \text{Re}[\overline{\Lambda(iy)}W(iy)] = 0, -1 \le y \le 1, \ u(b_1) = b_0 = \psi_1(0), \end{gathered} \tag{5.8}$$

$$\begin{gathered} \frac{1}{H(y)}\text{Re}[\overline{\lambda(x)}W(x)] = \text{Re}[\overline{\Lambda(x)}W(x)] = R(x) \\ = \frac{1}{\sqrt{2}}[F(-x) - \frac{1}{2}\phi'(-x)], \ x \in L_0 = (0, c_1), \end{gathered} \tag{5.9}$$

in which l is the tangent vector on the boundary Γ, $F(-x)$ is an undetermined function, and

$$\lambda(z)=\begin{cases} x_s/H-iy_s, \\ 1, \\ 1 \\ (1+i)/\sqrt{2}, \end{cases} \quad R(z)=\begin{cases} \psi_1'(s)/2 \text{ on } \Gamma, \\ 0 \text{ on } A'A, \\ H(0)\psi_2'(x)/2=0 \text{ on } CB, \\ [F(-x)-\phi'(-x)/2]/\sqrt{2} \text{ on } OC, \end{cases}$$

the above last boundary condition is derived as follows. From (5.4) and (5.3), for system (5.25) below we have

$$\begin{aligned} -V(0,y) &= \frac{1}{2}[u(0,y)]_y = \frac{1}{2}[u(0,-y)]_y + \frac{1}{2}\phi'(y) \\ &= V(0,-y) + \frac{1}{2}\phi'(y) = -F(y) + \frac{1}{2}\phi'(y), -1 \le y \le 1, \\ U(0,y) &= \frac{1}{2}H(y)u_x = \frac{1}{2}[f(-G(y)) + g(G(y))] = 0,\ V(0,y) = -\frac{1}{2}u_y \\ &= \frac{1}{2}[f(-G(y)) - g(G(y))] = F(y) - \frac{1}{2}\phi'(y),\ -1 \le y \le 1, \\ U(0,y)+V(0,y) &= f(-G(y)) = F(y) - \frac{1}{2}\phi'(y),\ -1 \le y \le 0, \\ U(0,y)-V(0,y) &= g(G(y)) = g(-G(-y)) = -F(y) + \frac{1}{2}\phi'(y), 0 \le y \le 1, \\ U(x,0)+V(x,0) &= f(x) = F[(-G)^{-1}(x)] - \frac{1}{2}\phi'[(-G)^{-1}(x)], 0 \le x \le c_1, \\ U(x,0)-V(x,0) &= g(x) = -F[-(-G)^{-1}(x)] + \frac{1}{2}\phi'[-(-G)^{-1}(x)], \\ U(x,0) &= 0,\ 0 \le x \le c_1, \end{aligned} \tag{5.10}$$

in which $F(y) = -V(0,-y)$, $y = (-G)^{-1}(x)$ is the inverse function of $x = -G(y) = -\int_0^y H(t)dt$, $-1 \le y \le 0$, $0 \le x = -G(y) \le c_1$, and from $K(y) = -K(-y)$, it follows that $G(y) = -G(-y)$. We shall prove the solvability of the Frankl problem for equation (2.1) by using the methods of parameter extension and symmetry extension.

We can choose the index $K = 0$ of $\lambda(z)$ on the boundary ∂D^+ of D^+.

In fact, due to the boundary condition

$$\mathrm{Re}[\overline{\lambda(z)}W(z)]=\frac{1}{2}\mathrm{Re}[\overline{\lambda(z)}(H(y)u_x-iu_y)]=R(z) \text{ on } \partial D^+, \tag{5.11}$$

where $\partial D^+ = AO\cup OB\cup\Gamma$, and $\lambda(z)=1$ on $AO\cup CB$, $\lambda(z)=\exp(\pi i/4)$ on OC, $\lambda(z)=\mathrm{Re}\lambda(z)+i\mathrm{Im}\lambda(z)$ on Γ, $R(z)$ is as stated in (5.9). Denote $t_1=0, t_2=c_1, t_3=b_1, t_4=i$, it is clear that $\lambda(t_1+0)=\lambda(t_2-0)=\exp(\pi i/4), \lambda(t_2+0)=\lambda(t_3-0)=\lambda(t_4-0)=1=\exp(0\pi i)$, hence we have

$$\begin{aligned}
&K_j=\left[\frac{\phi_j}{\pi}\right]+J_j, J_j=0 \text{ or } 1, e^{i\phi_j}=\frac{\lambda(t_j-0)}{\lambda(t_j+0)}, \gamma_j=\frac{\phi_j}{\pi}-K_j, j=1,...,4,\\
&e^{i\phi_1}=\frac{\lambda(t_1-0)}{\lambda(t_1+0)}=\frac{e^{i0\pi}}{e^{i\pi/4}}=e^{-i\pi/4}, -1<\gamma_1=\frac{\phi_1}{\pi}-K_1=-\frac{1}{4}-K_1=-\frac{1}{4},\\
&e^{i\phi_2}=\frac{\lambda(t_2-0)}{\lambda(t_2+0)}=\frac{e^{i\pi/4}}{e^{0i\pi}}=e^{i\pi/4}, 0<\gamma_2=\frac{\phi_2}{\pi}-K_2=\frac{1}{4}-K_2=\frac{1}{4}<1,\\
&e^{i\phi_3}=\frac{\lambda(t_3-0)}{\lambda(t_3+0)}=\frac{e^{0i\pi}}{e^{i\pi/2}}, -1<\gamma_3=\frac{\phi_3}{\pi}-K_3=-\frac{1}{2}-K_3=-\frac{1}{2}<0,\\
&e^{i\phi_4}=\frac{\lambda(t_4-0)}{\lambda(t_4+0)}=\frac{e^{i0}}{e^{i0}}=e^{i0}=1, 0\le\gamma_4=\frac{\phi_4}{\pi}-K_4=0-K_4=0<1,
\end{aligned} \tag{5.12}$$

here $[b]$ is the largest integer not exceeding the real number b, we choose $K_1=K_2=K_3=K_4=0$. Under these conditions, the index K of $\lambda(z)$ on the boundary ∂D^+ of D^+ is just as follows:

$$K=\frac{1}{2}(K_1+K_2+K_3+K_4)=0. \tag{5.13}$$

If the last boundary condition in (5.9) is replaced by that in (5.10), then $\lambda(x)=1$, $R(x)=0$ on OC, in this case we can choose $\gamma_1=\gamma_2=0, \gamma_3=-1/2, \gamma_4=0, K_1=K_2=K_3=K_4=0$, hence the index of $\lambda(z)$ on ∂D^+ is also $K=(K_1+K_2+K_3+K_4)/2=0$. In the following we choose this case and $K=0$, and can add one point condition $u(0)=0$ in the boundary conditions (5.8)-(5.10), but we omit to write it later on.

Noting that $U(0,y)=0$ on $A'A$, we can extend $W(z)$ onto the reflected domain $\tilde{D}$ of D about the segment $A'A$. In fact, we introduce the function

$$\tilde{W}(z)=\begin{cases} W(z) \text{ in } D,\\ -\overline{W(-\bar{z})} \text{ in } \tilde{D}, \end{cases} \tag{5.14}$$

this function $\tilde{W}(z)$ is a solution of the equation

$$\begin{gathered}\tilde{W}_{\bar{\tilde{z}}}=\tilde{A}_1\tilde{W}+\tilde{A}_2\overline{\tilde{W}}+\tilde{A}_3\tilde{u}+\tilde{A}_4 \text{ in } D,\\ \tilde{u}(z)=2\text{Re}\int_{b_1}^{z}[\frac{\text{Re}\tilde{W}(z)}{H}+\begin{pmatrix} i\\ -j\end{pmatrix}\text{Im}W]dz+\psi_1(0) \text{ in } \begin{pmatrix}\overline{D^+}\\ \overline{D^-}\end{pmatrix},\end{gathered} \tag{5.15}$$

with the boundary conditions

$$\begin{gathered}\text{Re}[\overline{\tilde{\lambda}(z)}\tilde{W}(z)]=\tilde{R}_1(z), z\in\tilde{\Gamma}, u(b_1)=b_0=\psi_1(0),\\ \text{Re}[\overline{\tilde{\lambda}(x)}\tilde{W}(x)]=\tilde{R}_0(x),\ x\in\tilde{L},\end{gathered} \tag{5.16}$$

in which $\tilde{\Gamma}=\Gamma\cup CB\cup\tilde{\Gamma}\cup\widetilde{BC}$, $\tilde{L}=(0,c_1)\cup(-c_1,0)$, and

$$\tilde{A}_l=\begin{cases}A_l(z),\\ -\overline{A_l(-\bar{z})},\end{cases} l=1,2,\ \tilde{A}_l=\begin{cases}A_l(z),\\ A_l(-\bar{z}),\end{cases} l=3,4, \text{ in } \begin{pmatrix}D\\ \tilde{D}\end{pmatrix}, \tag{5.17}$$

and

$$\begin{gathered}\tilde{\lambda}(z)=\begin{cases}\lambda(z),\\ \overline{\lambda(-\bar{z})},\end{cases}\quad \tilde{R}(z)=\begin{cases}R(z) \text{ on } \Gamma\cup CB,\\ -R(-\bar{z}) \text{ on } \tilde{\Gamma}\cup\tilde{BC},\end{cases}\\ \tilde{\lambda}(z)=\begin{cases}\dfrac{1+i}{\sqrt{2}},\\ \dfrac{1-i}{\sqrt{2}},\end{cases}\quad \tilde{R}_0(x)=\begin{cases}R(x) \text{ on } OC=(0,c_1),\\ -R(-x) \text{ on } \tilde{CO}=(-c_1,0),\end{cases}\end{gathered} \tag{5.18}$$

in which $\tilde{\Gamma},\tilde{BC}=(-b_1,-c_1),\tilde{CO}$ and $\tilde{AB}$ are the reflected curves of Γ,CB,OC and BA about the imaginary axis respectively. We choose the index of the function $\tilde{\lambda}(z)$ on the boundary $\partial(\hat{D}^+)$ of the elliptic domain $\hat{D}^+=D^+\cup\tilde{D}^+\cup AO$ as $K=0$, where $\tilde{D}^+$ is the symmetric domain of D^+ with respect to $\text{Re}z=0$. In fact, noting that $\tilde{\lambda}(z)=1$ on $CB\cup\tilde{BC}$, $\tilde{\lambda}(z)=\exp(i\pi/4)$ on OC, $\tilde{\lambda}(z)=\exp(-i\pi/4)$ on $\tilde{CO}$, denote $t_1=0, t_2=c_1, t_3=b_1, t_4=i, t_5=-b_1, t_6=-c_1$, we have $\tilde{\lambda}(b_1+0)=\exp(i\pi/2)$, $\tilde{\lambda}(i-0)=\tilde{\lambda}(i+0)=\exp(0i)$, $\tilde{\lambda}(-b_1-0)=\exp(-i\pi/2)$, hence we have

$$
\begin{aligned}
&K_j=\left[\frac{\phi_j}{\pi}\right]+J_j,\,J_j=0\text{ or }1,\,e^{j\phi_j}=\frac{\tilde\lambda(t_j-0)}{\tilde\lambda(t_j+0)},\,\gamma_j=\frac{\phi_j}{\pi}-K_j,\,j=1,...,6,\\
&e^{i\phi_1}=\frac{\tilde\lambda(t_1-0)}{\tilde\lambda(t_1+0)}=\frac{e^{-i\pi/4}}{e^{i\pi/4}}=e^{-i\pi/2},\,-1<\gamma_1=\frac{\phi_1}{\pi}-K_1=-\frac{1}{2}-K_1=-\frac{1}{2},\\
&e^{i\phi_2}=\frac{\tilde\lambda(t_2-0)}{\tilde\lambda(t_2+0)}=\frac{e^{i\pi/4}}{e^{0i\pi}}=e^{i\pi/4},\,-1<\gamma_2=\frac{\phi_2}{\pi}-K_2=\frac{1}{4}-K_2=\frac{1}{4},\\
&e^{i\phi_3}=\frac{\tilde\lambda(t_3-0)}{\tilde\lambda(t_3+0)}=\frac{e^{0i\pi}}{e^{i\pi/2}}=e^{-i\pi/2},\,-1<\gamma_3=\frac{\phi_3}{\pi}-K_3=-\frac{1}{2}-K_3=-\frac{1}{2},\\
&e^{i\phi_4}=\frac{\tilde\lambda(t_4-0)}{\tilde\lambda(t_4+0)}=\frac{e^{0i}}{e^{0i}}=e^{i0},\,-1<\gamma_4=\frac{\phi_4}{\pi}-K_4=0-K_4=0,\\
&e^{i\phi_5}=\frac{\tilde\lambda(t_5-0)}{\tilde\lambda(t_5+0)}=\frac{e^{-i\pi/2}}{e^{0i\pi}}=e^{-i\pi/2},\,-1<\gamma_5=\frac{\phi_5}{\pi}-K_5=-\frac{1}{2}-K_5=-\frac{1}{2},\\
&e^{i\phi_6}=\frac{\tilde\lambda(t_6-0)}{\tilde\lambda(t_6+0)}=\frac{e^{0i\pi}}{e^{-i\pi/4}}=e^{i\pi/4},\,0<\gamma_6=\frac{\phi_6}{\pi}-K_6=\frac{1}{4}-K_6=\frac{1}{4}.
\end{aligned}
\tag{5.19}
$$

If we choose $K_1=K_2=K_3=K_4=K_5=K_6=0$, the index K of $\tilde\lambda(z)$ is just

$$K=\frac{1}{2}(K_1+K_2+\cdots+K_6)=0. \tag{5.20}$$

We can discuss the solvability of the corresponding boundary value problem (5.15), (5.16), and then derive the existence of solutions of the Frankl problem for equation (2.1).

5.2 Representation and uniqueness of solutions of Frankl problem for degenerate mixed equations

In this section, we first write the complex form of equation (2.1). Denote

$$Y=G(y)=\int_0^y H(y)dy=\pm\frac{2}{m+2}|y|^{(m+2)/2}, \tag{5.21}$$

where $H(y)=|y|^{m/2}$, m is a positive number, and

$$W(z)=U+iV=\frac{1}{2}[H(y)u_x-iu_y]=u_{\tilde{z}}=\frac{H(y)}{2}[u_x-iu_Y]=H(y)u_Z,$$

$$H(y)W_{\overline{Z}}=\frac{H(y)}{2}[W_x+iW_Y]=\frac{1}{2}[H(y)W_x+iW_y]=W_{\overline{\tilde{z}}} \text{ in } \overline{D^+},$$

$$W(z)=U+jV=\frac{1}{2}[H(y)u_x-ju_y]=u_{\tilde{z}}=\frac{H(y)}{2}[u_x-ju_Y]=H(y)u_Z,$$

$$H(y)W_{\overline{Z}}=\frac{H(y)}{2}[W_x+jW_Y]=\frac{1}{2}[H(y)W_x+jW_y]=W_{\overline{\tilde{z}}} \text{ in } \overline{D^-}, \tag{5.22}$$

in which $Z=x+iG(y)$ in $\overline{D^+}$. Similarly to (2.12), (2.15), we have

$$H(y)W_{\overline{Z}}=A_1(z)W+A_2(z)\overline{W}+A_3(z)u+A_4(z)=g(Z) \text{ in } D_Z. \tag{5.23}$$

In particular the complex equation

$$W_{\overline{\tilde{z}}}=0, \text{ i.e. } W_{\overline{Z}}=0 \text{ in } \overline{D} \tag{5.24}$$

can be rewritten in the system

$$\begin{aligned} &[(U+V)+i(U-V)]_{\mu-i\nu}=0 \text{ in } \overline{D^+},\\ &(U+V)_\mu=0,\ (U-V)_\nu=0 \text{ in } \overline{D^-}. \end{aligned} \tag{5.25}$$

The boundary value problem for equations (5.23) with the boundary condition (5.1)-(5.4)($W(z)=u_{\tilde{z}}$) and the relation

$$u(z)=\begin{cases} 2\mathrm{Re}\displaystyle\int_{b_1}^{z}[\frac{\mathrm{Re}W(z)}{H_1(y)}+i\mathrm{Im}W(z)]dz+b_0 \text{ in } \overline{D^+},\\ 2\mathrm{Re}\displaystyle\int_{b_1}^{z}[\frac{\mathrm{Re}W(z)}{H_1(y)}-j\mathrm{Im}W(z)]dz+b_0 \text{ in } \overline{D^-}, \end{cases} \tag{5.26}$$

will be called Problem A.

Now, we give the representation of solutions for the Frankl problem (Problem F) for system (5.25) in $\overline{D}$. For this, we first discuss the Riemann-Hilbert problem (Problem A) for the second system of (5.25) in $\overline{D^-}$ with the boundary conditions

$$\begin{aligned} &\mathrm{Re}[\overline{\tilde{\lambda}(z)}\tilde{W}(z)]=\tilde{R}_1(z),\ z\in\hat{\Gamma},\ u(b_1)=b_0,\\ &\mathrm{Re}[\overline{\tilde{\lambda}(x)}\tilde{W}(x)]=\tilde{R}_0(x), x\in\hat{L}=(0,c_1)\cup(-c_1,0), \end{aligned} \tag{5.27}$$

in which $\tilde{\lambda}(x) = e^{i\pi/4}$ on $(0, c_1)$ and $\tilde{\lambda}(x) = e^{-i\pi/4}$ on $(-c_1, 0)$, $\hat{\Gamma} = \Gamma \cup CB \cup \tilde{\Gamma} \cup \widetilde{BC}$, $\tilde{\lambda}(z)$, $\tilde{R}_1(z)$, $\tilde{R}_0(x)$, b_0 are as stated in (5.16). On the basis of the result in [86]33), the solution $\tilde{W}(z)$ of Problem A for the second system of (5.25) and the function $F(-x)$ in (5.9) can be found. Moreover we can find a solution $\tilde{W}(z)$ of Problem A for (5.24) in $\overline{D^+} \cup \tilde{D}^+$ by using Theorem 1.3, Chapter I, thus the function $\tilde{R}_0(x)$ in the boundary condition

$$\begin{gathered}\mathrm{Re}[\tilde{W}(z)]=0 \text{ on } OA', \mathrm{Re}[(1-j)\tilde{W}(x)]/\sqrt{2}=\tilde{R}_0(x),\ x\in(0,c_1),\\ \mathrm{Re}[(1+j)\tilde{W}(x)]/\sqrt{2} = \tilde{R}_0(x),\ x \in (-c_1, 0)\end{gathered} \tag{5.28}$$

is found. The result can be written the following theorem.

Theorem 5.1 *Problem A of equation* (5.24) *or system* (5.25) *in $\overline{D}$ has a unique solution $\tilde{W}(z)$, which satisfies the estimates*

$$|\mathrm{Re}\tilde{W}(z)| \le M_1,\ |\mathrm{Im}\tilde{W}(z)| \le M_1 \text{ in } D_\varepsilon^-,$$

in which $D_\varepsilon^- = \overline{D^-} \cap \{|z| > \varepsilon\} \cap \{|z - c_1| < \varepsilon\} \cap \{|z - b_1| < \varepsilon(> 0)\}$, $M_1 = M_1(\alpha, k_0, k_2, D_\varepsilon^-)$ is a positive constant.

Now we state and can verify the representation of solutions of Problem F for equation (2.1).

Theorem 5.2 *Under Condition C, any solution $u(z)$ of Problem F for equation* (2.1) *in D^- can be expressed as follows*

$$\begin{aligned}
&u(z)=u(x)-2\int_0^y V(z)dy=2\mathrm{Re}\int_{b_1}^z \Big[\frac{\mathrm{Re}w}{H}+\begin{pmatrix} i\\ -j\end{pmatrix}\mathrm{Im}w\Big]dz+b_0 \text{ in } \begin{pmatrix}\overline{D^+}\\ \overline{D^-}\end{pmatrix},\\
&w(z)=W(z)+\Phi(Z)+\Psi(Z), \Psi(Z)=-2\mathrm{Re}\frac{1}{\pi}\iint_{D_t^+}\frac{f(t)}{t-Z}d\sigma_t \text{ in } \overline{D_Z^+},\\
&w(z) = \phi(z) + \psi(z) = \xi(z)e_1 + \eta(z)e_2,\\
&\xi(z)=\zeta(z)+\int_0^y g_1(z)dy = \zeta_0(z)+\int_{S_1} g_1(z)dy+\int_0^y g_1(z)dy,\\
&g_1(z)=\tilde{A}_1(U+V)+\tilde{B}_1(U-V)+2\tilde{C}_1U+\tilde{D}_1u+\tilde{E}_1,\ z \in s_1,\\
&\eta(z)=\theta(z)+\int_0^y [\tilde{A}_2(U+V)+\tilde{B}_2(U-V)+2\tilde{C}_2U+2\tilde{D}_2u+\tilde{E}_2]dy, z \in s_2,
\end{aligned} \tag{5.29}$$

in which $U = Hu_x/2, V = -u_y/2$, $\xi_0(z) = \mathrm{Re}W(z) + \mathrm{Im}W(z)$, $W(z)$ is as stated in Theorem 5.1, *$\phi(z)$ is a solution of* (5.25) *in D^-, $u(x) = \psi_2(x)$ on*

CB, s_1, s_2 are two families of characteristics in D^- as follows:

$$s_1: \frac{dx}{dy}=\sqrt{-K(y)}=H(y),\ s_2: \frac{dx}{dy}=-\sqrt{-K(y)}=-H(y) \tag{5.30}$$

passing through $z = x + jy \in D^-$, $\zeta(z) = \zeta_0(z)+\int_{S_1} g_1(z)dy$, S_1 is the characteristic curve from a point on $x + G(y) = 0$ to a point on OC, $\theta(z) = -\zeta(x + G(y))$, and

$$\begin{aligned}
&w(z) = U(z) + jV(z) = \frac{1}{2}[Hu_x - ju_y],\\
&\xi(z) = \mathrm{Re}\psi(z) + \mathrm{Im}\psi(z),\ \eta(z) = \mathrm{Re}\psi(z) - \mathrm{Im}\psi(z),\\
&\tilde{A}_1=\tilde{B}_2=\frac{1}{2}(\frac{h_y}{2h}-b),\ \tilde{A}_2=\tilde{B}_1=\frac{1}{2}(\frac{h_y}{2h}+b),\ \tilde{C}_1=\frac{a}{2H}+\frac{m}{4y},\\
&\tilde{C}_2 = -\frac{a}{2H}+\frac{m}{4y},\ \tilde{D}_1 = -\tilde{D}_2 = \frac{c}{2},\ \tilde{E}_1 = -\tilde{E}_2 = \frac{d}{2},
\end{aligned} \tag{5.31}$$

in which we choose $H(y) = [|y|^m h(y)]^{1/2}, h(y)$ is a continuously differentiable positive function and

$$\begin{aligned}
&d\mu = d[x + G(y)] = 2H(y)dy \ \text{ on } s_1,\\
&d\nu = d[x - G(y)] = -2H(y)dy \ \text{ on } s_2.
\end{aligned} \tag{5.32}$$

Theorem 5.3 *The solution of Problem F of equation (2.1) in $\overline{D}$ is unique, and the solution $u(z)$ and $W(z) = u_{\tilde{z}}$ satisfy the estimate*

$$\begin{aligned}
\hat{C}_\delta[W(z),\overline{D^+}]=C_\delta[X(Z)(\mathrm{Re}W/H+i\mathrm{Im}W),\overline{D^+}]+C_\delta[u(z),\overline{D^+}]\le M_2,\\
\hat{C}_\delta[W(z),\overline{D^+}] \le M_3(k_1 + k_2),
\end{aligned} \tag{5.33}$$

in which $X(Z) = \Pi_{l=1}^4 (Z - t_l)^{\eta_l}$, if $\eta_1 = \eta_2 = 1$, $\eta_3 = \max(-2\gamma_3, 0)$ if $\gamma_3 < 0$, $\eta_4 = 0$, γ_l, t_l $(l = 1, ...4)$ are as stated in (5.12), and $M_2 = M_2(\delta, k, H, D)$, $M_3 = M_3(\delta, k_0, H, D)$ are non-negative constants, and $k = (k_0, k_1, k_2)$.

Proof We first prove the uniqueness of solutions of Problem F for equation (2.1) in D. Let $u_1(z), u_2(z)$ be any two solutions of Problem F for (2.1). It is easy to see that $u(z) = u_1(z) - u_2(z)$ and $W(z) = u_{\tilde{z}} = U + iV$ in D^+ satisfy the homogeneous equation and boundary conditions

$$\begin{aligned}
&W_{\overline{\tilde{z}}} = A_1W + A_2\overline{W} + A_3u \ \text{ in } D^+,\\
&u(z) = 0 \text{ on } \Gamma\cup CB,\ u_x=0 \text{ on } AO.
\end{aligned} \tag{5.34}$$

If the maximum $M = \max_{\overline{D}} u(z) > 0$, it is clear that the maximum point $z^* \notin D^+ \cup BA \cup CB$. From the method in the proof of Theorem 3.2, Chapter II, we can derive that the maximum M cannot be attained at a point $z^* = iy^* \in AO$. If the maximum M attains at a point $z^* = x^* \in OC$, from the same proof of Theorem 5.2, Chapter II, we can derive that $u_x = 0$ on OC, and then $u(x) = M$ on OC, this contradicts $u(x) = 0$ on CB. Hence $\max_{\overline{D^+}} u(z) = 0$. Similarly we can derive $\min_{\overline{D^+}} u(z) = 0$. Moreover on the basis of the uniqueness of Cauchy problem of the homogeneous equation of (2.1) in D^- with the initial conditions $u(x) = 0$, $u_y = 0$ on OC, we can prove $\min_{\overline{D^-}} u(z) = 0$(see [86]33)). Therefore $u(z) = 0$, $u_1(z) = u_2(z)$ in $\overline{D}$. In addition by using the method in the proofs of Theorems 2.4 and 3.3, Chapter II, we can obtain the estimate (5.33), here we mention that noting that $\mathrm{Re}W = U = H(y)u_x/2 = 0$ on $OB = [0, b_1]$, the function $W(Z)$ in D_Z^+ can be symmetrically extended in the symmetrical domain $\tilde{D}_Z$ about $\mathrm{Im}Z = 0$.

5.3 Existence of solutions of Frankl problem for degenerate equations of mixed type

In this section, we prove the existence of solutions of Problem F for equation (2.1). Firstly we discuss the boundary value problem (Problem A), i.e. the complex equation (5.23), the relation (5.26) and the boundary value problem (5.27), (5.28).

Making a transformation $v(z) = u(z) - u_0(z)$, where $u_0(z)$ is a function as stated in (5.26), in which $u_{0\bar{z}} = W_0(z)$ is a solution of Problem A for equation (5.24), then Problem F for equation (2.1) is reduced to the boundary value problem (Problem F_0) for equation

$$\begin{gathered} K(y)v_{xx} + v_{yy} + av_x + bv_y + cu + \tilde{d} = 0, \\ \tilde{d} = [a - \begin{pmatrix} i \\ -j \end{pmatrix} H'(y)]u_{0x} + bu_{0y} + cu_0 + d \text{ in } \begin{pmatrix} \overline{D^+} \\ \overline{D^-} \end{pmatrix}. \end{gathered} \tag{5.35}$$

with the homogeneous boundary conditions

$$\begin{gathered} \mathrm{Re}[\overline{\tilde{\lambda}(z)}\tilde{W}(z)] = 0, z \in \Gamma \cup \tilde{\Gamma} \cup CB \cup \tilde{B}C, u(b_1) = 0, \\ \mathrm{Re}[\overline{\tilde{\lambda}(x)}\tilde{W}(x)] = R(x),\ x \in L = (0, c_1) \cup (-c_1, 0), \end{gathered} \tag{5.36}$$

where $\tilde{W} = v_{\bar{z}}$, $R(x) = F(-x)/\sqrt{2}$ is as stated in (5.9). According to the property of solutions of the discontinuous Riemann-Hilbert boundary value problem for analytic functions, we can verify that the orders of poles at $z = 0, c_1, b_1, i$ are less than 1, hence we can only discuss Problem F_0 for equation (5.23) with the boundary conditions (5.27), (5.28). Problem F_0 can be divided into two problems, i.e. Problem A_1 of equation (5.23), (5.26) in D^+ and Problem A_2 of equation (5.23), (5.26) in D^-, the boundary conditions of Problems A_1 and A_2 are as follows:

$$\begin{gathered}\operatorname{Re}[\overline{\tilde{\lambda}(z)}\tilde{W}(z)] = 0 \text{ on } \Gamma\cup\tilde{\Gamma}\cup CB\cup\tilde{B}C,\ u(b_1) = 0,\\ \operatorname{Re}[\overline{\tilde{\lambda}(x)}\tilde{W}(x)] = \hat{R}_0(x) \text{ or } \tilde{R}_0(x) \text{ on } L = (0, c_1)\cup(-c_1, 0),\end{gathered} \tag{5.37}$$

and

$$u(b_1) = 0,\ \operatorname{Re}[\overline{\lambda(x)}W(x)] = R(x) \text{ on } L,\ \operatorname{Re}[\overline{\tilde{\lambda}(z)}\tilde{W}(z)] = 0 \text{ on } OA', \tag{5.38}$$

in which $\tilde{\lambda}(x)$, $R(x) = \hat{R}_0(x)$ or $\tilde{R}_0(x)$ on $L = (0, c_1)\cup(-c_1, 0)$ are as stated before. As stated in Section 2, $\hat{R}_0(x)$ and $\tilde{R}_0(x)$ is as stated in (2.10). The solvability of Problem A_1 can be obtained by using the estimate (5.33) and the method of parameter extension as stated in Theorem 5.5 below, and the result of Problem A_2 will be proved as follows.

Theorem 5.4 *Let equation* (2.1) *satisfy Condition C and* (5.40) *below. Then there exists a unique solution* $[w(z), u(z)]$ *of Problem* A_2 *for* (5.23), (5.26) *in* D^-.

Proof Denote $D_0 = \overline{D^-}\cap(\{a_0 \le x \le a_1\}$, where $a_0 = \delta_0 < a_1 = c_1 - \delta_0$, and the positive constant δ_0 is small enough. We choose $v_0 = 0, \xi_0 = 0, \eta_0 = 0$ and substitute them into the corresponding positions of v, ξ, η in the right-hand sides of (5.29), and by the successive approximation, we find the sequences of functions $\{v_k\}$, $\{\xi_k\}$, $\{\eta_k\}$, which satisfy the relations

$$\begin{gathered}v_{k+1}(z) = v_{k+1}(x) - 2\int_0^y V_k(z)dy = v_{k+1}(x) + \int_0^y (\eta_k - \xi_k)dy,\\ \xi_{k+1}(z) = \zeta_{k+1}(z) + \int_0^y [\tilde{A}_1\xi_k + \tilde{B}_1\eta_k + \tilde{C}_1(\xi_k + \eta_k) + \tilde{D}_1 v_k + \tilde{E}_1]dy,\ z\in s_1,\\ \eta_{k+1}(z) = \theta_{k+1}(z) + \int_0^y [\tilde{A}_2\xi_k + \tilde{B}_2\eta_k + \tilde{C}_2(\xi_k + \eta_k) + \tilde{D}_2 v_k + \tilde{E}_2]dy,\ z\in s_2,\\ k = 0, 1, 2, \ldots.\end{gathered} \tag{5.39}$$

We may only discuss the case of $K(y) = -|y|^m h(y)$, because otherwise we can similarly discuss. In order to find a solution of the system of integral equations (5.29), we can assume that

$$a(x,y)|y|/H(y)=o(1),\ \text{i.e. } |a(x,y)|/H(y)=\varepsilon(y)/|y|,\ m\geq 2. \quad (5.40)$$

It is clear that for two characteristics $s_1: x=x_1(y,z_0)$, $s_2: x=x_2(y,z_0)$ passing through $P_0=z_0=x_0+jy_0\in D$, we have

$$|x_1-x_2|\leq 2\left|\int_0^{y_0}\sqrt{-K}dy\right|\leq M|y_0|^{m/2+1} \text{ for } y_1<y<0, \quad (5.41)$$

for any $z_1=x_1+jy\in s_1$, $z_2=x_2+jy\in s_2$, in which $M\ (>\max[2\sqrt{h(y)},1])$ is a positive constant. From (2.2), we can assume that the coefficients of (5.39) satisfy the conditions

$$\begin{gathered}|\tilde{A}_l|,|\tilde{A}_{lx}|,|\tilde{B}_l|,|\tilde{B}_{lx}|,|\tilde{D}_l|,|\tilde{D}_{lx}|,\\ |\tilde{E}_l|,|\tilde{E}_{lx}|,|1/\sqrt{h}|,|h_y/h|\leq M, z\in\bar{D}, l=1,2.\end{gathered} \quad (5.42)$$

According to the proof of Theorem 2.3, we can prove that $\{v_k\}$, $\{\xi_k\}$, $\{\eta_k\}$ in D_0 uniformly converge to v_*,ξ_*,η_*, thus Problem A_2 in D_0. Noting the arbitrariness of δ_0, we see that Problem A_2 in D^- has a unique solution.

Now we give the sketch of the proof of solvability of Problem A_1 for (2.1) in D^+.

Theorem 5.5 *Suppose that equation* (2.1) *satisfies Condition C. Then there exists a solution of* (*Problem* A_1) *for* (5.23), (5.26) *in* D^+.

Proof We first consider the boundary value problem (Problem A_t) with the parameter $t\in[0,1]$ for the linear case of equation (5.23), i.e.

$$w_{\bar{z}}-tF(z,u,w)=E(z) \text{ in } D^+, \quad (5.43)$$

$$\mathrm{Re}[\overline{\lambda(z)}w(z)]=tR(z)+S(z) \text{ on } \partial D^+=\Gamma\cup AO\cup OB, \quad (5.44)$$

where $F(z,u,w)$ in D^+_Z, $R(z)$ on $\partial D^+=\Gamma\cup AO\cup OB$ are as stated in (5.23) and (5.8) – (5.10), and $y^\tau X(Z)E(z)\in\hat{C}(\overline{D^+})$, $X(Z)S(z)\in C_\alpha(\partial D^+)$, $\tau=\max(0,1-m/2)$. When $t=0$, the unique solution of Problem A_0 for (5.43), (5.44)) can be found. Suppose that when $t=t_0\ (0\leq t_0<1)$, Problem A_{t_0} is solvable, i.e. Problem A_{t_0} for (5.43), (5.44)) has a solution $[W_0(z),u_0(z)]$ ($W_0(z)\in\hat{C}(\overline{D})$), we can find a neighborhood $T_\varepsilon=\{|t-t_0|\leq\varepsilon, 0\leq t\leq 1\}(0<\varepsilon<1)$ of t_0 such that for every $t\in T_\varepsilon$, Problem A_t is solvable. In fact, Problem A_t can be written in the form

$$w_{\bar{z}}-t_0F(z,u,w)=(t-t_0)F(z,u,w)+E(z) \text{ in } D^+, \quad (5.45)$$

$$\operatorname{Re}[\overline{\lambda(z)}w(z)] - t_0R(z) = (t-t_0)R(z) + S(z) \text{ in } \partial D^+, u(b_1) = 0, \quad (5.46)$$

Replace $[u(z), w(z)]$ in the right-hand side of (5.45), (5.46) by a system $[u_0(z), w_0(z)]$ ($w_0(z) \in \hat{C}(\overline{D^+})$) and the corresponding function $u_0(z)$ in the first formula of (5.29), especially, we select $w_0(z) = 0, u_0(z) = 0$ and substitute them into the right-hand sides of (5.45), (5.46), it is obvious that the boundary value problem for such equation then has a solution $[w_1(z), u_1(z)]$ ($w_1(z) \in \hat{C}(\overline{D})$). Using successive approximation, we obtain a sequence of solutions $[w_n(z), u_n(z)]$ ($w_n(z) \in \hat{C}(\overline{D^+})$, $n = 1, 2, ...$), which satisfy the equations

$$\begin{aligned} &w_{n+1\bar{\bar{z}}} - t_0F(z, u_{n+1}, w_{n+1}) = (t-t_0)F(z, u_n, w_n) + E(z) \text{ in } \overline{D^+}, \\ &\operatorname{Re}[\overline{\lambda(z)}w_{n+1}(z)] - t_0R(z) = (t-t_0)R(z) + S(z) \text{ on } \partial D^+, u_{n+1}(b_1) = 0. \end{aligned} \quad (5.47)$$

From the above formulas, it follows that

$$\begin{aligned} &[w_{n+1} - w_n]_{\bar{\bar{z}}} - t_0[F(z, u_{n+1}, w_{n+1}) - F(z, u_n, w_n)] \\ &= (t-t_0)[F(z, u_n, w_n) - F(z, u_{n-1}, w_{n-1})] \text{ in } D, \\ &\operatorname{Re}[\overline{\lambda(z)}(w_{n+1}(z) - w_n(z))] = 0 \text{ on } \partial D^+, \ u_{n+1}(b_1) - u_n(b_1) = 0. \end{aligned} \quad (5.48)$$

Noting that Condition C and for the above systems of functions $[u_n(z), w_n]$, $[u_{n-1}, w_{n-1}]$, we can obtain the estimate

$$\begin{aligned} L_\infty[(t-t_0)[|y|^\tau X(Z)(F(z, u_n, w_n) - F(z, u_{n-1}, w_{n-1})), \overline{D^+}] \\ \le 2|t-t_0|M_4\hat{C}[w_n - w_{n-1}, \overline{D^+}], \end{aligned}$$

and by Theorem 5.3 we can derive

$$\hat{C}[w_{n+1} - w_n, \overline{D_Z}] \le 2|t-t_0|M_3M_4\hat{C}[w_n - w_{n-1}, \overline{D_Z}],$$

in which the constant $M_3 = M_3(\delta, k_0, H, D)$ is as stated in (5.33), and $M_4 = M_4(\delta, k, H, D)$ is a positive constant. Choosing the constant ε so small that $2\varepsilon M_3M_4 \le 1/2$ and $|t-t_0| \le \varepsilon$, it follows that

$$\hat{C}[w_{n+1} - w_n), \overline{D_Z}] \le 2\varepsilon M_3M_4\hat{C}[w_n - w_{n-1}, \overline{D}] \le \frac{1}{2}\hat{C}, \overline{D}],$$

and when $n, m \ge N_0 + 1$, N_0 is a positive integer, the inequality

$$\hat{C}[w_{n+1} - w_n, \overline{D}] \le 2^{-N_0}\sum_{j=0}^{\infty} 2^{-j}\hat{C}[w_1 - w_0, \overline{D}] \le 2^{-N_0+1}\hat{C}[w_1 - w_0, \overline{D}],$$

holds. Hence $\{w_n(z)\}, \{u_n(z)\}$ are Cauchy sequences, according to the completeness of the Banach space $\hat{C}(\overline{D})$, there exists a function $w_*(z) \in \hat{C}(\overline{D^+})$, and $w_*(z) = u_{*\bar{z}}(z)$, so that $\hat{C}[w_n(z) - w_*(z), \overline{D^+}]$ to 0 as $n \to \infty$. we can see that $[w_*(z), u_*(z)]$ is a solution of Problem A_t for every $t \in T_\varepsilon = \{|t - t_0| \le \varepsilon\}$. Because the constant ε is independent of $t_0\,(0 \le t_0 < 1)$, therefore from the solvability of Problem A_{t_0} when $t_0 = 0$, we can derive the solvability of Problem A_t when $t = \varepsilon, 2\varepsilon, ..., [1/\varepsilon]\,\varepsilon, 1$. In particular, when $t = 1$ and $E(z) = 0, S(z) = 0$, Problem A_1 for the equation (5.23) has a solution $[W(z), u(z)]$. This completes the proof.

Combining the above results, we have the following theorem.

Theorem 5.6 *Suppose that equation* (2.1) *satisfies Condition C and* (5.40). *Then the Frankl problem (Problem F) for* (2.1) *has a unique solution.*

CHAPTER VI

SECOND ORDER QUASILINEAR EQUATIONS OF MIXED TYPE

This chapter mainly deals with the Tricomi problem and oblique derivative boundary value problem for second order quasilinear equations of mixed (elliptic-hyperbolic) type with parabolic degeneracy. We shall discuss the boundary value problems for general second order quasilinear equations of mixed (elliptic-hyperbolic) type and the problems in general domains including multiply connected domains. Moreover we shall consider the oblique derivative problem for second order quasilinear equations of mixed type with nonsmooth degenerate line and degenerate rank 0. In the meantime, we shall give a priori estimates of solutions for above problem.

1 The Oblique Derivative Problem for Second Order Quasilinear Equations of Mixed Type

In this section, we first give the representation of solutions for some boundary value problem of second order quasilinear equations of mixed type without parabolic degeneracy, and then prove the uniqueness and existence of solutions of the problem and give a priori estimates of solutions of the above problem.

1.1 Formulation of oblique derivative problem for second order equations of mixed type

Let D be a simply connected bounded domain D in the complex plane $\mathbf{C}$ as stated in Chapter V. We consider the second order quasilinear equation of mixed type

$$u_{xx} + \mathrm{sgn}y\, u_{yy} = au_x + bu_y + cu + d \text{ in } D, \tag{1.1}$$

where a, b, c, d are real functions of $z(\in D), u, u_x, u_y\,(\in \mathbf{R})$, its complex form is the following complex equation of second order

$$Lu_z = u_{z\bar{z}} = F(z, u, u_z), F = \mathrm{Re}[A_1 u_z] + A_2 u + A_3 \text{ in } D, \tag{1.2}$$

where $A_j = A_j(z, u, u_z)$, $j = 1, 2, 3$, and

$$u_{z\bar{z}} = \begin{cases} [(u_z)_x + i(u_z)_y]/2 = [u_{xx} + u_{yy}]/4 \text{ in } D^+, \\ [(u_z)_x + j(u_z)_y]/2 = [u_{xx} - u_{yy}]/4 \text{ in } D^-, \end{cases}$$

$$A_1 = \begin{cases} [a + ib]/2 \text{ in } D^+, \\ [a - jb]/2 \text{ in } D^-, \end{cases} \quad A_2 = \frac{c}{4}, A_3 = \frac{d}{4} \text{ in } D.$$

Suppose that the equation (1.2) satisfies the following conditions, namely

Condition C.

1) $A_l(z, u, u_z)\,(l = 1, 2, 3)$ are measurable in $z \in D^+$ and continuous in $\overline{D^-}$ for all continuously differentiable functions $u(z)$ in $D^* = \bar{D}\backslash\{0, 2\}$, and satisfy

$$\begin{gathered} L_p[A_l, \overline{D^+}] \le k_0,\ l = 1, 2,\ L_p[A_3, \overline{D^+}] \le k_1,\ A_2 \ge 0 \text{ in } D^+, \\ \hat{C}[A_l, \overline{D^-}] = C[A_l, \overline{D^-}] + C[A_{lx}, \overline{D^-}] \le k_0, l = 1, 2, \hat{C}[A_3, \overline{D^-}] \le k_1. \end{gathered} \tag{1.3}$$

2) For any continuously differentiable functions $u_1(z), u_2(z)$ in D^*, the equality

$$F(z, u_1, u_{1z}) - F(z, u_2, u_{2z}) = \mathrm{Re}[\tilde{A}_1(u_1 - u_2)_z] + \tilde{A}_2(u_1 - u_2) \text{ in } D \tag{1.4}$$

holds, where $\tilde{A}_l = \tilde{A}_l(z, u_1, u_2)\,(j = 1, 2)$ satisfy the conditions

$$L_p[\tilde{A}_l, \overline{D^+}] \le k_0,\ \hat{C}[\tilde{A}_l, \overline{D^-}] \le k_0,\ l = 1, 2, \tag{1.5}$$

in (1.3),(1.5), $p\,(> 2)$, k_0, k_1 are positive constants. In particular, when (1.2) is a linear equation, the condition (1.4) obviously holds.

Problem P The oblique derivative boundary value problem for equation (1.2) is to find a real continuously differentiable solution $u(z)$ of (1.2) in $D^* = \overline{D}\backslash\{0, 2\}$, which is continuous in $\overline{D}$ and satisfies the boundary conditions

$$\begin{gathered} \frac{1}{2}\frac{\partial u}{\partial \nu} = \mathrm{Re}[\overline{\lambda(z)} u_z] = r(z), z \in \Gamma, \frac{1}{2}\frac{\partial u}{\partial \nu} = \mathrm{Re}[\overline{\lambda(z)} u_z] = r(z), z \in L_1, \\ \mathrm{Im}[\overline{\lambda(z)} u_z]_{z=z_0} = b_1,\ u(0) = b_0,\ u(2) = b_2, \end{gathered} \tag{1.6}$$

where ν is a given vector at every point on $\Gamma \cup L_1, \lambda(z) = a(x) + ib(x) = \cos(\nu, x) - i\cos(\nu, y)$, $z \in \Gamma$, and $\lambda(z) = a(x) + ib(x) = \cos(\nu, x) + j\cos(\nu, y)$,

z_0 is the intersection point of L_1 and L_2, $z \in L_1$, b_0, b_1, b_2 are real constants, and $\lambda(z)$, $r(z)$, b_0, b_1, b_2 satisfy the conditions

$$\begin{gathered} C^1_\alpha[\lambda(z),\Gamma\cup L_1]\le k_0, C^1_\alpha[r(z),\Gamma\cup L_1]\le k_2, \cos(\nu,n)\ge 0 \text{ on } \Gamma, \\ |b_0|, |b_1|, |b_2| \le k_2, \ \max_{z\in L_1}[1/|a(z)-b(z)|] \le k_0, \end{gathered} \tag{1.7}$$

in which n is the outward normal vector at every point on Γ, $\alpha\,(0<\alpha<1), k_0, k_2$ are positive constants. For convenience, we may assume that $u_z|_{z=z_0} = w(z_0) = 0$.

The boundary value problem for (1.2) with $A_3(z,u,u_z)=0$, $z\in D, u\in \mathbf{R}$, $u_z \in \mathbf{C}$, $r(z)=0$, $z\in\partial D$ and $b_0=b_1=b_2=0$ will be called Problem P_0. The number

$$K = \frac{1}{2}(K_1+K_2) \tag{1.8}$$

is called the index of Problem P and Problem P_0, where

$$K_l=\left[\frac{\phi_l}{\pi}\right]+J_l, J_l=0 \text{ or } 1, e^{i\phi_l}=\frac{\lambda(t_l-0)}{\lambda(t_l+0)}, \gamma_l=\frac{\phi_l}{\pi}-K_l, l=1,2, \tag{1.9}$$

in which $[a]$ is the largest integer not exceeding the real number a, $t_1=0$, $t_2=2$, $\lambda(t)=\exp(i\pi/4)$ on L_0, and $\lambda(t_1+0)=\lambda(t_2-0)=\exp(i\pi/4)$, here we only discuss the case of $K=0$ on ∂D^+ if $\cos(\nu,n)\not\equiv 0$ on Γ, or $K=-1/2$ if $\cos(\nu,n)\equiv 0$ on Γ, because in this case the last point condition in (1.6) can be eliminated, and the solution of Problem P is unique.

Besides, if $A_2=0$ in D, the last condition in (1.6) is replaced by

$$\operatorname{Im}[\overline{\lambda(z)u_z}]|_{z=z_1} = b_2, \tag{1.10}$$

where $z_1(\ne 0,2)\in\Gamma$, and b_2 is a real constant, and here the condition $\cos(l,n)\ge 0$ can be cancelled, then the boundary value problem for (1.2) will be called Problem Q.

1.2 Existence and uniqueness of solutions for oblique derivative problem

Similarly to Section 1, Chapter V, we can prove the following results.

Lemma 1.1 *Let equation* (1.2) *satisfy Condition C. Then any solution of Problem P for* (1.2) *can be expressed as*

$$u(z) = 2\operatorname{Re}\int_0^z \hat{w}(z)dz + b_0, \ w(z) = w_0(z) + W(z) \text{ in } D, \tag{1.11}$$

here and later on we denote $\hat{w} = w$ *in* $\overline{D^+}$ *and* $\hat{w} = \overline{w}$ *in* $\overline{D^-}$, $w_0(z)$ *is a solution of Problem A for the equation*

$$w_{\bar{z}} = 0 \text{ in } D \tag{1.12}$$

with the boundary condition (1.6) $(w_0(z) = u_{0z})$, *and* $W(z)$ *possesses the form*

$$\begin{aligned}
&W(z) = w(z) - w_0(z) \text{ in } D,\ w(z) = \tilde{\Phi}(z)e^{\tilde{\phi}(z)} + \tilde{\psi}(z) \text{ in } D^+,\\
&\tilde{\phi}(z) = \tilde{\phi}_0(z) + Tg = \tilde{\phi}_0(z) - \frac{1}{\pi}\iint_{D^+}\frac{g(\zeta)}{\zeta - z}d\sigma_\zeta,\ \tilde{\psi}(z) = Tf \text{ in } D^+,\\
&W(z) = \Phi(z) + \Psi(z),\ \Psi(z) = \int_0^\mu g_1(z)d\mu e_1 + \int_2^\nu g_2(z)d\nu e_2 \text{ in } D^-,
\end{aligned} \tag{1.13}$$

where $\mathrm{Im}\tilde{\phi}(z) = 0$ *on* $L_0 = \{0 \le x < 2, y = 0\}$, $e_1 = (1+j)/2$, $e_2 = (1-j)/2$, $\mu = x + y, \nu = x - y$, $\tilde{\phi}_0(z)$ *is an analytic function in* D^+ *and continuous in* $\overline{D^+}$, *such that* $\mathrm{Im}\tilde{\phi}(x) = 0$ *on* L_0, *and*

$$\begin{aligned}
&g(z) = \begin{cases} \dfrac{A_1}{2} + \dfrac{\overline{A_1\tilde{W}}}{2\tilde{W}}, \tilde{W}(z) \ne 0, \\ 0,\ \tilde{W}(z) = 0,\ z \in D^+, \end{cases} \quad f(z) = \mathrm{Re}[A_1\tilde{\phi}_z] + A_2 u + A_3 \text{ in } D^+,\\
&g_1(z) = g_2(z) = A\xi + B\eta + Cu + D, \xi = \mathrm{Re}w + \mathrm{Im}w, \eta = \mathrm{Re}w - \mathrm{Im}w,\\
&A = \frac{\mathrm{Re}A_1 + \mathrm{Im}A_1}{2}, B = \frac{\mathrm{Re}A_1 - \mathrm{Im}A_1}{2},\ C = A_2,\ D = A_3 \text{ in } D^-,
\end{aligned} \tag{1.14}$$

where $\tilde{W}(z) = w(z) - \tilde{\psi}(z)$, $\tilde{\Phi}(z)$ *is an analytic function in* D^+ *and* $\Phi(z)$ *is a solution of equation* (1.12) *in* D^- *satisfying the boundary conditions*

$$\begin{aligned}
&\mathrm{Re}[\overline{\lambda(z)}(\tilde{\Phi}(z)e^{\tilde{\phi}(z)} + \tilde{\psi}(z))] = r(z),\ z \in \Gamma,\\
&\mathrm{Re}[\overline{\lambda(x)}(\tilde{\Phi}(x)e^{\tilde{\phi}(x)} + \tilde{\psi}(x))] = s(x),\ x \in L_0,\\
&\mathrm{Re}[\overline{\lambda(x)}\Phi(x)] = \mathrm{Re}[\overline{\lambda(x)}(W(x) - \Psi(x))],\ z \in L_0,\\
&\mathrm{Re}[\overline{\lambda(z)}\Phi(z)] = -\mathrm{Re}[\overline{\lambda(z)}\Psi(z)],\ z \in L_1,\\
&\mathrm{Im}[\overline{\lambda(z_0)}\Phi(z_0)] = -\mathrm{Im}[\overline{\lambda(z_0)}\Psi(z_0)],
\end{aligned} \tag{1.15}$$

in which $\lambda(x) = 1+i$ *on* L_0 *in the second formula of* (1.15) *and* $\lambda(x) = 1+j$ *on* L_0 *in the third formula of* (1.15), *and* $s(x)$ *is as stated in* (1.33), *Chapter*

V. *Moreover by Theorem* 1.2, *Chapter* V, *the solution* $w_0(z)$ *of Problem A for* (1.12) *and* $u_0(z)$ *satisfy the estimate in the form*

$$C_\delta[u_0(z),\overline{D}]+C_\delta[X(z)w_0(z),\overline{D^+}]$$
$$+C_\delta[Y^\pm(\mu,\nu)w_0^\mp(\mu,\nu),\overline{D^-}]\le M_1(k_1+k_2), \tag{1.16}$$

where

$$X(z)=\prod_{l=1}^{2}|z-t_l|^{\eta_l},\ Y^\pm(z)=\prod_{l=1}^{2}Y^\pm(\mu,\nu)=\prod_{l=1}^{2}[|\nu-t_l||\mu-t_l|]^{\eta_l},$$
$$\eta_l=\max[-2\gamma_l,0]+8\delta,\ \ l=1,2, \tag{1.17}$$

herein $w_0^\pm(\mu,\nu)=\mathrm{Re}w_0(z)\pm\mathrm{Im}w_0(z)$, $w_0(z)=w_0(\mu,\nu)$, $\mu=x+y$, $\nu=x-y$, *and* γ_1,γ_2 *are the real constants in* (1.9), $\delta\,(\le\alpha)$ *is a sufficiently small positive constant, and*

$$u_0(z)=2\mathrm{Re}\int_0^z \hat{w}_0(z)dz+b_0\ \text{ in }\bar{D} \tag{1.18}$$

where $p_0\ (2<p_0\le p)$, $M_1=M_1(p_0,\delta,k_0,D)$ *are positive constants.*

Theorem 1.2 *Suppose that equation* (1.2) *satisfies Condition C. Then Problem P for* (1.2) *has a unique solution* $u(z)$ *in D.*

Theorem 1.3 *Suppose that the equation* (1.2) *satisfies Condition C. Then any solution* $u(z)$ *of Problem P for* (1.2) *satisfies the estimates*

$$\tilde{C}^1_\delta[u,\overline{D^+}]=C_\delta[u(z),\overline{D^+}]+C_\delta[X(z)u_z,\overline{D^+}]\le M_2,$$
$$\tilde{C}^1[u,\overline{D^-}]=C_\delta[u(z),\overline{D^-}]+C[Y^\pm(\mu,\nu)u_z^\pm(\mu,\nu),\overline{D^-}]\le M_3, \tag{1.19}$$
$$\tilde{C}^1_\delta[u,\overline{D^+}]\le M_4(k_1+k_2),\ \tilde{C}^1[u,\overline{D^-}]\le M_4(k_1+k_2),$$

where $X(z),Y^\pm(\mu,\nu)$ *are stated in* (1.17), *and* $M_l=M_l\,(p_0,\delta,k_0,D)$ $(l=2,3,4)$ *are positive constants* (see [86]17)).

1.3 $C^1_\alpha(\bar{D})$-estimate of solutions of oblique derivative problem for second order mixed equations

Now, we give the $C^1_\alpha(\bar{D})$-estimate of solutions $u(z)$ for Problem P for (1.2), but it needs to assume the following conditions: For any real numbers u_1,u_2

and complex numbers w_1, w_2, the inequalities

$$|A_l(z_0,u_1,w_1)-A_l(z_2,u_2,w_2)|\le k_0[|z_0-z_2|^\alpha+|u_1-u_2|^\alpha$$
$$+|w_1-w_2|],\ l=1,2,\ |A_3(z_0,u_1,w_1)-A_3(z_2,u_2,w_2)| \tag{1.20}$$
$$\le k_1[|z_0-z_2|^\alpha+|u_1-u_2|^\alpha+|w_1-w_2|],\ z_0,z_2\in\overline{D^-},$$

hold,where $\alpha\,(0<\alpha<1)$, k_0, k_1 are positive constants, which is called **Condition C'**.

Theorem 1.4 *If Condition C' holds, then any solution $u(z)$ of Problem P for equation (1.2) in $\overline{D^-}$ satisfies the estimates*

$$\tilde C^1_\delta[u,\overline{D^-}]=C_\delta[u,\overline{D^-}]+C_\delta[Y^\pm(\mu,\nu)u^\pm_z(\mu,\nu),\overline{D^-}]\le M_5,$$
$$\tilde C^1_\delta[u,\overline{D^-}]\le M_6(k_1+k_2), \tag{1.21}$$

where $u^\pm_z(\mu,\nu)=\operatorname{Re}u_z\pm\operatorname{Im}u_z$, $\delta\,(0<\delta\le\alpha)$, $M_5=M_5(p_0,\delta,k,D)$, $M_6=M_6(p_0,\delta,k_0,D)$ are positive constants, $k-(k_0,k_1.k_2)$.

Proof Similarly to Theorem 1.3, it suffices to prove the first estimate in (1.21). Due to the solution $u(z)$ of Problem P for (1.2) is found by the successive approximation through the integral expressions (1.11), (1.13) and (1.14), we first choose the solution of Problem A of (1.12), i.e.

$$w_0(z)=\xi_0(z)e_1+\eta_0(z)e_2,\ \text{and}\ u_0(z)=2\operatorname{Re}\int_0^z\hat w_0(z)dz+b_0\ \text{in}\ D. \tag{1.22}$$

Substitute them into the positions of u_0, w_0 in the right-hand side of (1.14), we can obtain $\Psi_1(z), w_1(z), u_1(z)$ as stated in (1.11)-(1.14). Denote

$$u_1(z)=2\operatorname{Re}\int_0^z\hat w_1(z)dz+b_0,\ w_1(z)=w_0(z)+\Phi_1(z)+\Psi_1(z),$$
$$\Psi^1_1(z)=\int_0^\mu G_1(z)d\mu,\ G_1(z)=A\xi_0+B\eta_0+Cu_0+D, \tag{1.23}$$
$$\Psi^2_1(z)=\int_2^\nu G_2(z)d\nu,\ G_2(z)=A\xi_0+B\eta_0+Cu_0+D,$$

from the last two equalities in (1.23), it is not difficult to see that $G_1(z)=G_1(\mu,\nu)$, $\Psi^1_1(z)=\Psi^1_1(\mu,\nu)$ and $G_2(z)=G_2(\mu,\nu)$, $\Psi^2_1(z)=\Psi^2_1(\mu,\nu)$ satisfy the Hölder continuous estimates about μ,ν respectively, namely

$$C_\delta[G_1(\mu,\cdot),\overline{D^-}]\le M_7,\ C_\delta[\Psi^1_1(\mu,\cdot),\overline{D^-}]\le M_7R,$$
$$C_\delta[G_2(\cdot,\nu),\overline{D^-}]\le M_8,\ C_\delta[\Psi^2_1(\cdot,\nu),\overline{D^-}]\le M_8R, \tag{1.24}$$

in which $M_l = M_l(p_0, \delta, k, D)$ $(l = 7, 8)$ and $R = 2$. Moreover, from (1.20) we can derive that $\Psi_1^1(\mu,\nu)$, $\Psi_1^2(\mu,\nu)$ about ν, μ satisfy the Hölder conditions respectively, namely

$$C_\delta[\Psi_1^1(\cdot,\nu),\overline{D^-}]\le M_9R,\ C_\delta[\Psi_1^2(\mu,\cdot),\overline{D^-}]\le M_9R, \tag{1.25}$$

where $M_9 = M_9(p_0, \delta, k, D)$. Besides we can obtain the estimate of $\Phi_1(z)$, i.e.

$$C_\delta[\Phi_1(z),\overline{D^-}]\le M_{10}R = M_{10}(p_0,\delta,k,D^-)R, \tag{1.26}$$

in which $\Phi_1(z)$ satisfies equation (1.12) and the boundary conditions in (1.15), but in which the function $\Psi(z)$ is replaced by $\Psi_1(z)$. Setting $w_1(z) = w_0(z) + \Phi_1(z) + \Psi_1(z)$ and by the first formula in (1.23), the function $u_1(z)$ from $w_1(z)$ can be found. Furthermore from (1.25), (1.26), we can derive that the functions $\tilde{w}_1^\pm(z) = \tilde{w}_1^\pm(\mu,\nu) = \text{Re}\tilde{w}_1(z)\pm\text{Im}\tilde{w}_1(z)$ $(\tilde{w}_1(z) = w_1(z) - w_0(z))$ and $\tilde{u}_1(z) = u_1(z) - u_0(z)$ satisfy the estimates

$$C_\delta[\tilde{w}_1^\pm(\mu,\nu)Y^\pm(\mu,\nu),\overline{D^-}]\le M_{11}R,\ C_\delta[\tilde{u}_1(z),\overline{D^-}]\le M_{11}R, \tag{1.27}$$

where $M_{11} = M_{11}(p_0, \delta, k, D)$. Thus, according to the successive approximation, we can obtain the estimates of functions $\tilde{w}_n^\pm(z) = \tilde{w}_n^\pm(\mu,\nu) = \text{Re}\tilde{w}_n(z)\pm\text{Im}\tilde{w}_n(z)$ $(\tilde{w}_n(z) = w_n(z)-w_{n-1}(z))$ and the corresponding function $\tilde{u}_n(z) = u_n(z) - u_{n-1}(z)$ satisfy the estimates

$$\begin{aligned}&C_\delta[Y^\pm(\mu,\nu)\tilde{w}_n^\pm(\mu,\nu),\overline{D^-}]\le (M_{11}R)^n/n!,\\ &C_\delta[\tilde{u}_n(z),\overline{D^-}]\le (M_{11}R)^n/n!.\end{aligned} \tag{1.28}$$

Therefore the sequences of functions

$$w_n(z)=\sum_{m=1}^{n}\tilde{w}_m(z)+w_0(z),\ u_n(z)=\sum_{m=1}^{n}\tilde{u}_m(z)+u_0(z),\ n=1,2,\ldots \tag{1.29}$$

uniformly converges to $w(z)$, $u(z)$ in any close subset of D^* respectively, and $w(z)$, $u(z)$ satisfy the estimates

$$C_\delta[Y^\pm(\mu,\nu)w^\pm(\mu,\nu),\overline{D^-}]\le e^{M_{11}R},\ \tilde{C}_\delta^1[u(z),\overline{D^-}]\le M_5, \tag{1.30}$$

this is just the first estimate in (1.21).

From the estimates (1.19) and (1.21), we can see the regularity of solutions of Problem P for (1.2).

As for Problem Q for (1.2), we can similarly discuss its unique solvability (see [86]33)).

By using a similar method as before, we can discuss the solvability of Problem P' for equation (1.2) with the boundary conditions

$$\mathrm{Re}[\overline{\lambda(z)}u_z] = r(z),\ z \in \Gamma,\ u(0) = b_0,\ u(2) = b_2,$$

$$\mathrm{Re}[\overline{\lambda(z)}u_z] = r(z), z\in L_2,\ \mathrm{Im}[\overline{\lambda(z)}u_z]|_{z=z_0} = b_1,$$

and the corresponding Problem Q', in which the coefficients $\lambda(z)$, $r(z)$, b_0, b_1, b_2 satisfy the condition (1.7), but where the conditions $C^1_\alpha[\lambda(z), L_1] \le k_0$, $C^1_\alpha[r(z), L_1] \le k_2$, $\max_{z\in L_1}[1/|a(x) - b(x)|] \le k_0$ are replaced by $C^1_\alpha[\lambda(z), L_2] \le k_0$, $C^1_\alpha[r(z), L_2] \le k_2$, $\max_{z\in L_2}[1/|a(x) + b(x)|] \le k_0$ and in (1.9) the condition $\lambda(t) = e^{i\pi/4}$ on $L_0 = (0, 2)$ and $\lambda(t_1 - 0) = \lambda(t_2 + 0) = \exp(i\pi/4)$ is replaced by $\lambda(t) = e^{-i\pi/4}$ on $L_0 = (0, 2)$ and $\lambda(t_1 - 0) = \lambda(t_2 + 0) = e^{-i\pi/4}$.

2 The Oblique Derivative Problem for Degenerate Equations of Mixed Type in General Domains

This section deals with the oblique derivative problem for second order quasilinear equations of mixed (elliptic-hyperbolic) type with parabolic degeneracy. Firstly, we give a representation theorem and prove the uniqueness of solutions for the above boundary value problem, and then by using the methods of successive approximation and parameter extension, the existence of solutions for the above problem is proved. Moreover we discuss the above boundary value problem in general domains, in particular, the Tricomi problem for Chaplygin equation in general domain is called Frankl problem (see [71]2)).

2.1 Formulation of oblique derivative problem for mixed equations with parabolic degeneracy

Let D be a simply connected bounded domain in the complex plane $\mathbf{C}$ with the boundary $\partial D = \Gamma \cup L$, where $\Gamma(\subset \{y > 0\}) \in C^2_\mu\,(0 < \mu < 1)$ is a smooth curve with the form $x - \tilde{G}(y) = 0$ and $x + \tilde{G}(y) = 2$ near the points $z = 0$ and 2 respectively, here $\tilde{G}(y)$ is the same as that in Section 2, Chapter V, and $L = L_1 \cup L_2 \cup L_3 \cup L_4$, where

$$L_1 = \{x + \int_0^y \sqrt{-K(t)}dt = 0,\ x \in [0, a/2]\},$$

$$L_2=\{x-\int_0^y\sqrt{-K(t)}dt=a,\ x\in[a/2,a]\},$$

$$L_3=\{x+\int_0^y\sqrt{-K(t)}dt=a,\ x\in[a,1+a/2]\},$$

$$L_4=\{x-\int_0^y\sqrt{-K(t)}dt=2,\ x\in[1+a/2,2]\},$$

in which $a\,(0<a<2)$ is a constant, $K(y)=\mathrm{sgn}y|y|^m h(y)$, m is a positive number, $h(y)$ is a continuously differentiable positive function, and $z_1=a/2+jy_1, z_2=1+a/2+jy_2$ are the intersection points of L_1,L_2 and L_3,L_4 respectively. Denote $D^+=D\cap\{y>0\}$, $D^-=D\cap\{y<0\}, D_1^-=D_-\cap\{x<a\}$, $D_2^-=D_-\cap\{x>a\}$. Consider second order quasilinear equation of mixed type with parabolic degeneracy

$$K(y)u_{xx}+u_{yy}+au_x+bu_y+cu+d=0 \text{ in } D. \tag{2.1}$$

where a,b,c,d are real functions of $z(\in\overline{D}), u, u_x, u_y(\in\mathbf{R})$, and suppose that the equation (2.1) satisfies the following conditions, namely **Condition** C.

1) a,b,c,d are measurable in D^+ and continuous in $\overline{D^-}$ for any continuously differentiable function $u(z)$ in $D^*=\overline{D}\backslash\{0,a,2\}$, and satisfy

$$\begin{gathered}L_\infty[\eta,\overline{D^+}]\le k_0,\ \eta=a,b,c,\ L_\infty[d,\overline{D^+}]\le k_1,\ c\le 0 \text{ in } D^+,\\ \tilde{C}[d,\overline{D^-}]=C[d,\overline{D^-}]+C[d_x,\overline{D^-}]\le k_1,\tilde{C}[\eta,\overline{D^-}]\le k_0,\eta=a,b,c.\end{gathered}\tag{2.2}$$

2) For any two continuously differentiable functions $u_1(z),u_2(z)$ in D^*, $F(z,u,u_z)=au_x+bu_y+cu+d$ satisfies the following condition

$$F(z,u_1,u_{1z})-F(z,u_2,u_{2z})=\tilde{a}(u_1-u_2)_x+\tilde{b}(u_1-u_2)_y+\tilde{c}(u_1-u_2) \text{ in } \overline{D},$$

in which $\tilde{a},\tilde{b},\tilde{c}$ satisfy the conditions

$$\begin{gathered}L_\infty[\tilde{a},\overline{D^+}],L_\infty[\tilde{b},\overline{D^+}],L_\infty[\tilde{c},\overline{D^+}]\le k_0,\tilde{c}\le 0 \text{ in } D^+,\\ \tilde{C}[\tilde{a},\overline{D^-}]\le k_0,\tilde{C}[\tilde{b},\overline{D^-}]\le k_0,\tilde{C}[\tilde{c},\overline{D^-}]\le k_0,\end{gathered}\tag{2.3}$$

where $H(y)=\sqrt{|K(y)|}$, k_0, $k_1(\ge 6\max[1,k_0])$ are positive constants.

In order to write the complex form of the above equation, we denote

$W(z)$ and $W_{\overline{z}}$ as follows

$$W(z)=U+iV=\frac{1}{2}[H(y)u_x-iu_y]=u_{\tilde{z}}=\frac{H(y)}{2}[u_x-iu_Y]=H(y)u_Z,$$
$$H(y)W_{\overline{Z}}=\frac{H(y)}{2}[W_x+iW_Y]=\frac{1}{2}[H(y)W_x+iW_y]=W_{\overline{\tilde{z}}} \text{ in } \overline{D^+},$$

in which $H(y)=\sqrt{|K(y)|}$, $Z=x+iG(y)$ in $\overline{D^+}$, $G(y)=\int_0^z H(y)dy$, we have

$$\begin{aligned}
&K(y)u_{xx}+u_{yy}=H(y)[H(y)u_x-iu_y]_x+i[H(y)u_x-iu_y]_y\\
&-iH_yu_x=2\{H[U+iV]_x+i[U+iV]_y\}-i[H_y/H]Hu_x\\
&=4H(y)W_{\overline{Z}}-i[H_y/H]Hu_x=-[au_x+bu_y+cu+d], \text{ i.e.}\\
&H(y)W_{\overline{Z}}=H[W_x+iW_Y]/2=H[(U+iV)_x+i(U+iV)_Y]/2\\
&=H\{U_x-V_Y+i[V_x+U_Y]\}/2\\
&=\{i[H_y/H]H(y)u_x-[au_x+bu_y+cu+d]\}/4\\
&=\{[iH_y/H-a/H]H(y)u_x-bu_y-cu-d\}/4\\
&=\{i[H_y/H-a/H](W+\overline{W})-ib(W-\overline{W})+cu+d]\}/4\\
&=A_1W+A_2\overline{W}+A_3u+A_4=g(Z) \text{ in } D_Z^+,
\end{aligned} \tag{2.4}$$

in which D_Z^+ is the image domains of D^+ with respect to the mapping $Z=Z(z)$ respectively. Moreover denote

$$W(z)=U+jV=\frac{1}{2}[H(y)u_x-ju_y]=\frac{H(y)}{2}[u_x-ju_Y]=H(y)u_Z,$$
$$H(y)W_{\overline{Z}}=\frac{H(y)}{2}[W_x+jW_Y]=\frac{1}{2}[H(y)W_x+jW_y]=W_{\overline{\tilde{z}}} \text{ in } \overline{D^-},$$

in which $H(y)=\sqrt{|K(y)|}$, we get

$$\begin{aligned}
&-K(y)u_{xx}-u_{yy}=H(y)[H(y)u_x-ju_y]_x+j[H(y)u_x-ju_y]_y\\
&-jH_yu_x=2\{H[U+jV]_x+j[U+jV]_y\}-j[H_y/H]Hu_x\\
&=4H(y)W_{\overline{Z}}-j[H_y/H]Hu_x=au_x+bu_y+cu+d, \text{ i.e.}\\
&H(y)W_{\overline{Z}}=H[W_x+jW_Y]/2=H[(U+jV)_x+j(U+jV)_Y]/2
\end{aligned}$$

$$
\begin{aligned}
&=H\{e_1[U_x+V_Y+V_x+U_Y]/2+e_2[U_x+V_Y-V_x-U_Y]/2\}\\
&=H\{e_1[(U+V)_x+(U+V)_Y]/2+e_2[(U-V)_x-(U-V)_Y]/2\}\\
&=H[e_1(U+V)_\mu+e_2(U-V)_\nu]=\{j[H_y/H]Hu_x+au_x+bu_y+cu\\
&+d\}/4=\{[iH_y/H+a/H](W+\overline{W})-jb(W-\overline{W})+cu+d\}/4\\
&=(e_1-e_2)[H_y/H]Hu_x/4+(e_1+e_2)[au_x+bu_y+cu+d]/4\\
&=e_1\{[H_y/H]Hu_x+au_x+bu_y+cu+d\}/4\\
&+e_2\{-[H_y/H]Hu_x+au_x+bu_y+cu+d\}/4\\
&=e_1\{[H_y/H+a/H]Hu_x+bu_y+cu+d\}/4\\
&+e_2\{-[H_y/H-a/H]Hu_x+bu_y+cu+d\}/4\\
&=e_1\{[H_y/H+a/H]U-bV+cu/2+d/2\}/2\\
&+e_2\{-[H_y/H-a/H]U-bV+cu/2+d/2\}/2, \text{ i.e.}\\
&(U+V)_\mu=\frac{1}{4H}\{2[\frac{H_y}{H}+\frac{a}{H}]U-2bV+cu+d\},\\
&(U-V)_\nu=\frac{1}{4H}\{-2[\frac{H_y}{H}-\frac{a}{H}]U-2bV+cu+d\} \text{ in } D_\tau^-,
\end{aligned}
\tag{2.5}
$$

in which $e_1=(1+j)/2, e_2=(1-j)/2$. Hence the complex form of (2.1) can be written as

$$
\begin{aligned}
&W_{\overline{z}}=A_1(z,u,W)W+A_2(z,u,W)\overline{W}+A_3(z,u,W)u+A_4(z,u,W),\\
&u(z)=\begin{cases} 2\mathrm{Re}\int_0^z[\frac{U(z)}{H(y)}+iV(z)]dz+b_0 \text{ in } \overline{D^+}\\ 2\mathrm{Re}\int_0^z[\frac{U(z)}{H(y)}-jV(z)]dz+b_0 \text{ in } \overline{D^-},\end{cases}
\end{aligned}
\tag{2.6}
$$

where $b_0=u(0)$, and the coefficients $A_l(z,u,W)(l=1,2,3,4)$ are similar to (2.16), Chapter V, namely

$$
A_1=\begin{cases}\frac{1}{4}[-\frac{a}{H}+\frac{iH_y}{H}-ib],\\ \frac{1}{4}[\frac{a}{H}+\frac{jH_y}{H}-jb],\end{cases}\quad A_2=\begin{cases}\frac{1}{4}[-\frac{a}{H}+\frac{iH_y}{H}+ib],\\ \frac{1}{4}[\frac{a}{H}+\frac{jH_y}{H}+jb],\end{cases}
$$

$$A_3=\begin{cases}-\dfrac{c}{4},\\ \dfrac{c}{4},\end{cases}\quad A_4=\begin{cases}-\dfrac{d}{4}\ \text{ in } \overline{D^+},\\ \dfrac{d}{4}\ \text{ in } \overline{D^-}.\end{cases}\tag{2.7}$$

The oblique derivative boundary value problem for equation (2.1) may be formulated as follows:

Problem P Find a continuous solution $u(z)$ of (2.1) in $\bar{D}$, where u_x, u_y are continuous in $D^*=\overline{D}\backslash\{0,a,2\}$, and satisfy the boundary conditions

$$\begin{gathered}\frac{1}{2}\frac{\partial u}{\partial\nu}=\frac{1}{H(y)}\text{Re}[\overline{\lambda(z)}u_{\bar z}]=\text{Re}[\overline{\Lambda(z)}u_z]=r(z)\ \text{ on } \Gamma\cup L_1\cup L_4,\\ \frac{1}{H(y)}\text{Im}[\overline{\lambda(z)}u_{\bar z}]|_{z=z_l}=\text{Im}[\overline{\Lambda(z)}u_z]|_{z=z_l}=b_l, l=1,2, u(0)=b_0, u(2)=b_3,\end{gathered}\tag{2.8}$$

in which ν is a given vector at every point $z\in\Gamma\cup L_1\cup L_4$, $u_{\bar z}=[H(y)u_x-iu_y]/2$, $\Lambda(z)=\cos(\nu,x)-i\cos(\nu,y)$, $\lambda(z)=\text{Re}\lambda(z)+i\text{Im}\lambda(z)$, if $z\in\Gamma$, and $u_{\bar z}=[H(y)u_x-ju_y]/2$, $\lambda(z)=\text{Re}\lambda(z)+j\text{Im}\lambda(z)$, if $z\in L_1$, $b_l\ (l=0,1,2,3)$ are real constants, and $r(z), b_l\ (l=0,1,2,3)$ satisfy the conditions

$$\begin{gathered}C^1_\alpha[\lambda(z),\Gamma]\le k_0, C^1_\alpha[\lambda(x),L_1]\le k_0, C^1_\alpha[r(z),\Gamma]\le k_2, C^1_\alpha[r(x),L_1]\le k_2,\\ \cos(\nu,n)\ge 0\ \text{ on } \Gamma\cup L_1,\ |b_l|\le k_2,\ l=0,1,2,3,\\ \max_{z\in L_1}\frac{1}{|\text{Re}\lambda(z)-\text{Im}\lambda(z)|},\ \max_{z\in L_4}\frac{1}{|\text{Re}\lambda(z)+\text{Im}\lambda(z)|}\le k_0,\end{gathered}\tag{2.9}$$

in which n is the outward normal vector at every point on $\Gamma, \alpha(0<\alpha<1)$, k_0, k_2 are positive constants. We consider the boundary condition (2.8) and

$$\text{Re}[\overline{\lambda(z)}(U+jV)]=0,\ \lambda(z)=e^{i\pi/2}\ \text{ on } L_0,$$

the number

$$K=\frac{1}{2}(K_1+K_2+K_3)$$

is called the index of Problem P on ∂D^+, in which

$$K_l=\left[\frac{\phi_l}{\pi}\right]+J_l, J_l=0 \text{ or } 1, e^{i\phi_l}=\frac{\lambda(t_l-0)}{\lambda(t_l+0)}, \gamma_l=\frac{\phi_l}{\pi}-K_l, l=1,2,3,$$

in which $t_1=0, t_2=2, t_3=a, \lambda(t)=e^{i\pi/2}$ on $L_0'=(0,a)$ and $L_0''=(a,2)$ on the $x-$axis, and $\lambda(t_1+0)=\lambda(t_2-0)=\lambda(t_3-0)=\lambda(t_3+0)=$

$e^{i\pi/2}$. Similarly to Section 3, Chapter II, $K = 0$ on the boundary ∂D^+ of D^+ can be chosen, for instance we can choose $K_1 = K_2 = K_3 = 0$ and $K = 0$, and require that $-1/2 \le \gamma_l < 1/2$, $(l = 1, 2)$. For the latter sections, we have the similar requirement. When $\cos(\nu, n) \equiv 0$ on Γ and $\text{Re}[\overline{\lambda(x)}(U + jV)] = 0$, $\lambda(x) = i = e^{i\pi/2}$ on L_0, we can also select $\gamma_1 = \gamma_2 = \gamma_3 = 0$, $K_1 = -1, K_2 = K_3 = 0$, $K = -1/2$. We mention that from the boundary condition (2.8), we can determine the value $u(2)$ by the value $u(0)$, namely

$$u(2)=2\text{Re}\int_0^2 u_z dz+u(0)=2\int_0^S \text{Re}[z'(s)u_z]ds+b_0=2\int_0^S r(z)ds+b_0,$$

in which $\overline{\Lambda(z)} = z'(s)$ on Γ, $z(s)$ is a parameter expression of arc length s of Γ with the condition $z(0) = 0$, and S is the length of the boundary Γ. If we consider $\cos(l, n) \equiv 0$ on Γ, and

$$\text{Re}[\overline{\lambda(z)}(U + jV)] = 0,\ \lambda(x) = 1 = e^{0\pi i} \text{ on } L_0,$$

then $\gamma_1 = \gamma_2 = -1/2$, $\gamma_3 = 0$, $K_1 = K_2 = K_3 = 0$, and $K = 0$. In this section, we choose the case $K = 0$.

When $c = 0$ in equation (2.1), the last point condition in (2.8) can be replaced by

$$\text{Im}[\overline{\lambda(z)}u_{\bar{z}}]|_{z=z_3} = H(\text{Im}z_3)b_3 = b_3', \tag{2.10}$$

where z_3 is a point on $\Gamma\backslash\{0, 2\}$, and b_3 is a constant satisfying the conditions $|b_3| \le k_2$, in this case the condition $\cos(\nu, n) \ge 0$ on Γ can be cancelled. The boundary value problem is called Problem Q.

2.2 Representation of solutions of oblique derivative problem for degenerate mixed equations

It is clear that the complex equation

$$W_{\bar{\bar{z}}} = 0 \text{ in } \overline{D} \tag{2.11}$$

can be rewritten in the system

$$\begin{aligned}&[(U + V) + i(U - V)]_{\mu - i\nu} = 0 \text{ in } \overline{D^+},\\&(U + V)_\mu = 0,\ (U - V)_\nu = 0 \text{ in } \overline{D^-}.\end{aligned} \tag{2.12}$$

The corresponding boundary value problems of Problems P and Q for equation (2.1) with the boundary conditions (2.8), (2.10) ($W = u_{\bar z}$) will be called Problems A and B respectively.

Now, we give the representation of solutions for the oblique derivative problem (Problem P) for system (2.12) in $\overline{D}$. For this, we first discuss the Riemann-Hilbert problem (Problem A) for system: the second system of (2.12) in $\overline{D^-}$ with the boundary conditions

$$\mathrm{Re}[\overline{\lambda(z)}(U+jV)]=\begin{cases} H(y)r(z)=R_1(z),\, z\in L_1\cup L_4, \\ R_0(x),\ x\in L_0=L_0'\cup L_0'', \end{cases} \tag{2.13}$$

$$\mathrm{Im}[\overline{\lambda(z)}(U+jV)]|_{z=z_l}=b_l,\ l=1,2,\ u(0)=b_0,\ u(2)=b_3,$$

in which $\lambda(z)=a(z)+jb(z)$ on $L_1\cup L_4$, and $\lambda(z)=1+j$ on $L_0'=(0,a)$, $\lambda(x)=1-j$ on $L_0''=(a,2)$ and $R_0(x)$ is an undetermined real function. It is clear that the solution of Problem A for (2.12) in $\overline{D^-}$ can be expressed as

$$\begin{aligned} &\xi=U(z)+V(z)=f(\nu),\ \ \eta=U(z)-V(z)=g(\mu),\\ &U(z)=[f(\nu)+g(\mu)]/2,\ V(z)=[f(\nu)-g(\mu)]/2,\ \text{i.e.} \\ &W(z)=U(z)+jV(z)=[(1+j)f(\nu)+(1-j)g(\mu)]/2, \end{aligned} \tag{2.14}$$

where $f(t), g(t)$ are two arbitrary real continuous functions on $[0,2]$. For convenience, denote by the functions $a(x), b(x), r(x)$ of x the functions $a(z), b(z), r(z)$ of z in (2.13), thus (2.13) can be rewritten as

$$\begin{aligned} &a(x)U(x,y)-b(x)V(x,y)=R_1(z)\ \text{ on } L_1\cup L_4,\\ &U(x)-V(x)=R_0(x)\ \text{ on } L_0'=(0,a),\\ &U(x)+V(x)=R_0(x)\ \text{ on } L_0''=(a,2),\ \text{i.e.}\\ &[a(x)-b(x)]f(x-G(y))+[a(x)+b(x)]g(x+G(y))=2R_1(z)\text{ on } L_1\cup L_4,\\ &U(x)-V(x)=R_0(x)\ \text{ on } L_0'=(0,a),\\ &U(x)+V(x)=R_0(x)\text{ on } L_0''=(a,2),\ \text{i.e.}\\ &[a(\frac{t}{2})-b(\frac{t}{2})]f(t)+[a(\frac{t}{2})+b(\frac{t}{2})]g(0)=2R_1(\frac{t}{2}),\ t\in[0,a],\\ &[a(\frac{t}{2}+1)-b(\frac{t}{2}+1)]f(2)+[a(\frac{t}{2}+1)+b(\frac{t}{2}+1)]g(t)=2R_1(\frac{t}{2}+1), t\in[a,2], \end{aligned}$$

$$U(t) - V(t) = R_0(t) \text{ on } L_0' = (0, a),$$

$$U(t) + V(t) = R_0(t) \text{ on } L_0'' = (a, 2),$$

where

$$[a(a/2) + b(a/2)]g(0) = r(a/2) - b_1' \text{ or } 0,$$

$$a(a/2+1) - b(a/2+1)]f(2) = r(a/2+1) + b_2' \text{ or } 0.$$

Moreover we can derive

$$f(\nu) = f(x - G(y)) = \frac{2R_1(\nu/2) - (a(\nu/2) + b(\nu/2))g(0)}{a(\nu/2) - b(\nu/2)},$$

$$g(\mu) = g(x + G(y)) = R_0(\mu),$$

$$U(z) = \frac{1}{2}\{f(\nu) + R_0(\mu)\}, V(z) = \frac{1}{2}\{f(\nu) - R_0(\mu)\}, \ x \le a,$$

$$f(\nu) = f(x - G(y)) = R_0(\nu),$$

$$g(\mu) = \frac{2R_1(\mu/2+1) - (a(\mu/2+1) - b(\mu/2+1))f(2)}{a(\mu/2+1) + b(\mu/2+1)},$$

$$U(z) = \frac{1}{2}\{R_0(\nu) + g(\mu)\}, V(z) = \frac{1}{2}\{R_0(\nu) - g(\mu)\}, \ x > a,$$

if $a(x) - b(x) \neq 0$ on $[0, a]$, and $a(x) + b(x) \neq 0$ on $(a, 2]$. From the above formula, it follows that

$$\begin{aligned} &\text{Re}[(1 + j)W(x)] = U(x) + V(x) \\ &= \frac{2R_1(x/2) - (a(x/2) + b(x/2))g(0)}{a(x/2) - b(x/2)}, \ x \in [0, a], \\ &\text{Re}[(1 - j)W(x)] = U(x) - V(x) = R_0(x), \ x \in [0, a], \\ &\text{Re}[(1 + j)W(x)] = U(x) + V(x) = R_0(x), x \in (a, 2], \\ &\text{Re}[(1 - j)W(x)] = U(x) - V(x) \\ &= \frac{2R_1(x/2+1) - (a(x/2+1) - b(x/2+1))f(2)}{a(x/2+1) + b(x/2+1)}, x \in (a, 2], \end{aligned}$$

if $a(x) - b(x) \neq 0$ on $[0, a]$ and $a(x) + b(x) \neq 0$ on $(a, 2]$. Thus we can obtain

$$W(z)=\begin{cases}\frac{1}{2}\{(1+j)\frac{2R_1((x-G(y))/2)-M(x,y)}{a((x-G(y))/2)-b((x-G(y))/2)}\\ +(1-j)R_0(x+G(y))\},\\ M(x,y)=[a((x-G(y))/2)+b((x-G(y))/2)]g(0), x\le a,\\ \frac{1}{2}\{(1+j)R_0(x-G(y))\\ +(1-j)\frac{2R_1((x+G(y))/2+1)-N(x,y)}{a((x+G(y))/2+1)+b((x+G(y))/2+1)}\},\\ N(x,y)=[a((x+G(y))/2+1)-b((x+G(y))/2)+1]\\ \times f(2),\ x>a,\end{cases} \tag{2.15}$$

in particular, we have

$$\begin{aligned}&\text{Re}[\overline{(1+i)}(U(x)+iV(x))]=U(x)+V(x)=-\hat{R}_0(x)\\ &=\frac{2R_1(x/2)-[a(x/2)+b(x/2)]g(0)}{a(x/2)-b(x/2)}\ \text{ on } L_0',\\ &\text{Re}[\overline{(1-i)}(U(x)+iV(x))]=U(x)-V(x)=\hat{R}_0(x)\\ &=\frac{2R_1(x/2{+}1)-[a(x/2{+}1)+b(x/2{+}1)]f(2)}{a(x/2+1)+b(x/2+1)}\ \text{ on } L_0''.\end{aligned} \tag{2.16}$$

If $R_1(z)=0$ on $L_1\cup L_4$, $\hat{R}_0(x)=0$ on L_0, then $W(z)=U(z)+jV(z)=0$ in $\overline{D^-}$. Next we find a solution of the Riemann-Hilbert boundary value problem for equation (2.11) in D^+ with the above boundary conditions and

$$\text{Re}[\overline{\lambda(z)}(U(z)+iV(z))]=H(y)r(z)=R_1(z)\ \text{ on } \Gamma.$$

Noting that the index of the above boundary condition is $K=0$, by the result in [87]1), we know that the above Riemann-Hilbert problem has a unique solution $W(z)$ in D^+, and then

$$\begin{aligned}U(x)-V(x)&=\text{Re}[(1-j)(U(x)+jV(x)]=R_0(x)\ \text{ on } L_0',\\ U(x)+V(x)&=\text{Re}[(1+j)(U(x)+jV(x)]=R_0(x)\ \text{ on } L_0''\end{aligned} \tag{2.17}$$

are determined. This shows that Problem A for equation (2.11) is uniquely solvable, namely

Theorem 2.1 *Problem A of equation* (2.11) *or system* (2.12) *in* $\overline{D}$ *has a unique solution* $w(z)$, *which can be expressed as in* (2.14), (2.15) *in* D^- *satisfying the following estimates*

$$C_\delta[u(z),\overline{D^-}]+C^1_\delta[u(z),D^-_\varepsilon]\le M_1,\ C_\delta[f(x),L_\varepsilon]+C_\delta[g(x),L_\varepsilon]\le M_2,$$

in which $\nu=x-G(y)$, $\mu=x+G(y)$, $u(z)$ *in* D^- *is the corresponding function determined by the first formula in* (2.18) *below, where* $W(z)$ *is as stated in* (2.15), $D^-_\varepsilon=\overline{D^-}\cap\{|z|>\varepsilon\}\cap\{|z-a|>\varepsilon\}\cap\{|z-2|>\varepsilon(>0)\}$, $L_\varepsilon=\{0\le x\le a-\varepsilon, y=0\}]\cup\{a+\varepsilon\le x\le 2, y=0\}$, ε, δ *are sufficiently small positive constants, and* $M_1=M_1(\delta,k_0,k_1,D^-_\varepsilon)$, $M_2=M_2(k_0,k_1,L_\varepsilon)$ *are non-negative constants.*

The representation of solutions of Problem P for equation (2.1) is as follows.

Theorem 2.2 *Under Condition C, any solution* $u(z)$ *of Problem P for equation* (2.1) *in* D^- *can be expressed as follows*

$$\begin{aligned}
&u(z)=-2\int_0^y V(z)dy+u(x)=2\text{Re}\int_0^z[\frac{\text{Re}w}{H}+\binom{i}{-j}\text{Im}w]dz+b_0 \text{ in } \binom{\overline{D^+}}{\overline{D^-}},\\
&w(z)=W(z)+\Phi(Z)+\Psi(Z), \Psi(Z)=-\text{Re}\frac{2}{\pi}\iint_{D_t^+}\frac{f(t)}{t-Z}d\sigma_t \text{ in } \overline{D_Z^+},\\
&w(z)=\phi(z)+\psi(z)=\xi(z)e_1+\eta(z)e_2 \text{ in } \overline{D^-},\\
&\xi(z)=\zeta(z)+\int_0^y g_1(z)dy=\zeta_0(z)+\int_{y_1}^{|y|}\hat g_1(z)dy, z\in s_1,\\
&\eta(z)=\theta(z)+\int_0^y g_2(z)dy=\theta_0(z)+\int_{y_1}^{|y|}\hat g_2(z)dy, z\in s_2,\\
&g_l(z)=\tilde A_l(U+V)+\tilde B_l(U-V)+2\tilde C_lU+\tilde D_lu+\tilde E_l, l=1,2,
\end{aligned}\tag{2.18}$$

in which $Z=x+jG(y)$, $f(Z)=g(Z)/H$, $U=Hu_x/2, V=-u_y/2$, $\phi(z)=\zeta(z)e_1+\theta(z)e_2$ *is a solution of* (2.11) *in* D^-, $\zeta_0(z)=\text{Re}W(z)+\text{Im}W(z)$, $W(z)$ *is as stated in* (2.14), (2.15), $\hat g_1(z),\hat g_2(z)$ *are as stated in* (2.20) *below,* $\zeta(z)=\zeta_0(z)+\int_{S_1}g_1(z)dy$, $\theta(z)=-\zeta(x+G(y))$ *in* $D_1^-=\overline{D^-}\cap\{x\le a\}$, $\theta_0(z)=\text{Re}W(z)-\text{Im}W(z)$, $\theta(z)=\theta_0(z)+\int_{S_2}g_2(z)dy$, $\zeta(z)=-\theta(x-G(y))$ *in* $D_2^-=\overline{D^-}\cap\{x\ge a\}$, *and* s_1,s_2 *are two families of characteristics in* D^-:

$$s_1:\ \frac{dx}{dy}=\sqrt{-K(y)}=H(y),\ s_2:\ \frac{dx}{dy}=-\sqrt{-K(y)}=-H(y)$$

passing through $z = x + jy \in D^-$, S_1, S_2 *are the characteristic curves from the point* $z_1 = x_1 + jy_1$ *on* L_1 *and* L_4 *to the point* $z = x + jy \in \overline{D_Z^-}$ *respectively, and*

$$\begin{aligned} &w(z) = U(z) + jV(z) = \frac{1}{2}Hu_x - \frac{j}{2}u_y, \\ &\xi(z) = \mathrm{Re}\psi(z) + \mathrm{Im}\psi(z), \eta(z) = \mathrm{Re}\psi(z) - \mathrm{Im}\psi(z), \\ &\tilde{A}_1 = \tilde{B}_2 = \frac{1}{2}(\frac{h_y}{2h} - b),\ \tilde{A}_2 = \tilde{B}_1 = \frac{1}{2}(\frac{h_y}{2h} + b), \\ &\tilde{C}_1 = \frac{a}{2H} + \frac{m}{4y},\ \tilde{C}_2 = -\frac{a}{2H} + \frac{m}{4y}, \\ &\tilde{D}_1 = -\tilde{D}_2 = \frac{c}{2},\ \tilde{E}_1 = -\tilde{E}_2 = \frac{d}{2}, \end{aligned} \tag{2.19}$$

in which we choose $H(y) = [|y|^m h(y)]^{1/2}, h(y)$ *is a continuously differentiable positive function and*

$$\begin{aligned} d\mu &= d[x + G(y)] = 2H(y)dy \text{ on } s_1, \\ d\nu &= d[x - G(y)] = -2H(y)dy \text{ on } s_2. \end{aligned}$$

Proof From (2.5), we see that equation (2.1) in $\overline{D^-}$ can be reduced to the system of integral equations: (2.18), where $\zeta_0(z) = \mathrm{Re}W(z) + \mathrm{Im}W(z), \theta_0(z) = \mathrm{Re}W(z) - \mathrm{Im}W(z)$, $W(z)$ is a solution of Problem A for equation (2.11) as stated in (2.15), and the coefficients of (2.18) are as stated in (2.19). Moreover

$$\begin{aligned} ds_1 &= \sqrt{(dx)^2 + (dy)^2} = -\sqrt{1 + (dx/dy)^2}dy = -\sqrt{1-K}dy = -\frac{\sqrt{1-K}}{\sqrt{-K}}dx, \\ ds_2 &= \sqrt{(dx)^2 + (dy)^2} = -\sqrt{1 + (dx/dy)^2}dy = -\sqrt{1-K}dy = \frac{\sqrt{1-K}}{\sqrt{-K}}dx. \end{aligned}$$

It is clear that $\zeta(z) = \zeta_0(z) + \int_{S_1} g_1(z)dy$, $\theta(z) = -\zeta(x + G(y))$ in D_1^-, and $\theta(z) = \theta_0(z) + \int_{S_2} g_2(z)dy$, $\zeta(z) = -\theta(x - G(y))$ in D_2^-, especially $\zeta(x) + \theta(x) = 0$ on L_0.

Now we can extend the equation (2.6) onto the the symmetrical domain $\tilde{D}_Z$ of D_Z^- with respect to the real axis $\mathrm{Im}Z = 0$, namely introduce the function $\hat{W}(Z)$ as follows:

$$\hat{W}(Z) = \begin{cases} W[z(Z)], \\ -\overline{W[z(\overline{Z})]}, \end{cases} \quad \hat{u}(z) = \begin{cases} u(Z) \text{ in } D_Z^-, \\ -u(\overline{Z}) \text{ in } \tilde{D}_Z, \end{cases}$$

and then the equation (2.6) is extended as

$$\hat{W}_{\bar{\bar{z}}} = \hat{A}_1\hat{W} + \hat{A}_2\overline{\hat{W}} + \hat{A}_3\hat{u} + \hat{A}_4 = \hat{g}(Z) \text{ in } \hat{D}_Z = \overline{D_Z^-} \cup \overline{\hat{D}_Z},$$

where

$$\hat{A}_l(Z) = \begin{cases} A_l(Z), \\ \overline{\tilde{A}_l(\overline{Z})}, \end{cases} l = 1,2,3, \ \hat{A}_4(Z) = \begin{cases} A_4(Z), \\ \overline{-A_4(\overline{Z})}, \end{cases}$$
$$\hat{g}_l(z) = \begin{cases} g_l(z) \text{ in } \overline{D_Z^-}, \\ \overline{-g_l(\overline{z})} \text{ in } \overline{\tilde{D}_Z}, \end{cases} l = 1,2, \tag{2.20}$$

here $\tilde{A}_1(\overline{Z}) = A_2(\overline{Z}), \tilde{A}_2(\overline{Z}) = A_1(\overline{Z}), \tilde{A}_3(\overline{Z}) = A_3(\overline{Z})$. It is easy to see that the system of integral equations (2.18) can be written in the form

$$\xi(z) = \zeta(z) + \int_0^{\tilde{y}} g_1(z)dy = \int_{y_1}^{\hat{y}} \hat{g}_1(z)dy \text{ in } D_1^-,$$
$$\eta(z) = \theta(z) + \int_0^{y} g_2(z)dy = \int_{y_1}^{\hat{y}} \hat{g}_2(z)dy \text{ in } D_2^-,$$

where $\hat{z} = \hat{x} + j\hat{y}$, $x_1 + jy_1$ is the intersection point of L_1 and the characteristic curve s_1 passing through $z = x + jy \in \overline{D_1^-}$, for the extended integral, which can be appropriately defined in $\overline{D_2^-}$. For convenience the above form $\hat{g}_1(z)$, $\hat{g}_2(z)$ are written, and the numbers $\hat{y} - y_1, \hat{t} - y_1$ will be written by $\tilde{y}, \tilde{t}$ respectively.

2.3 Existence of solutions of oblique derivative problem for degenerate mixed equations

In this subsection, we prove the existence of solutions of Problems P and Q for equation (2.1). As stated in Subsection 2.1, we can discuss the equivalent Riemann-Hilbert boundary value problems (Problems A) for equation (2.6), i.e. the equation

$$w_{\bar{\bar{z}}} = A_1(z,u,w)w + A_2(z,u,w)\overline{w} + A_3(z,u,w)u + A_4(z,u,w) \text{ in } D, \tag{2.21}$$

the relation

$$u(z)=\begin{cases}2\mathrm{Re}\displaystyle\int_0^z[\frac{\mathrm{Re}w(z)}{H(y)}+i\mathrm{Im}w(z)]dz+b_0 \text{ in } \overline{D^+},\\ 2\mathrm{Re}\displaystyle\int_0^z[\frac{\mathrm{Re}w(z)}{H(y)}-j\mathrm{Im}w(z)]dz+b_0 \text{ in } \overline{D_1^-},\\ 2\mathrm{Re}\displaystyle\int_2^z[\frac{\mathrm{Re}w(z)}{H(y)}-j\mathrm{Im}w(z)]dz+b_3 \text{ in } \overline{D_2^-},\end{cases} \tag{2.22}$$

and the boundary conditions

$$\begin{aligned}&\mathrm{Re}[\overline{\lambda(z)}w(z)]=R(z) \text{ on } \Gamma\cup L_1\cup L_4,\\ &\mathrm{Im}[\overline{\lambda(z_l)}w(z_l)]=b_l'=H(\mathrm{Im}z_l)b_l, l=1,2, u(0)=b_0, u(2)=b_3,\end{aligned} \tag{2.23}$$

where $H(y)$ is as stated in (2.4), (2.5), the coefficients of (2.21) are the same as in (2.7), $\lambda(z)$, $r(z), z_l$, b_l $(l=0,1,2,3)$ are as stated in (2.8), (2.9), and $R(z)=R_1(z)=0$ on $\Gamma\cup L_1\cup L_4$, $b_0=b_1=b_2=b_3=0$. Moreover (2.21), (2.22) in D^+ with the boundary conditions (2.23) and

$$\mathrm{Re}[(-i)w(x)]=R(x)=-\hat{R}_0(x) \text{ on } L_0=(0,a)\cup(a,2)$$

is called Problem A_1, and (2.21), (2.22) in D^- with the boundary conditions (2.23) and

$$\mathrm{Re}[(1-j)w(x)]=R(x)=\tilde{R}_0(x) \text{ on } L_0=(0,a)\cup(a,2)$$

is called Problem A_2, because $\mathrm{Re}w(x)=0$ on $L_0=(0,2)$, then the above numbers $1-j$ can be replaced by $-j$ on L_0. The solvability of Problem A_1 can be obtained by the result in Sections 2 and 3, Chapter II, and Problem A_2 will be proved as follows. If we use the method as stated in Section 3, Chapter V to reduce the nonhomogeneous boundary condition (2.8) to the homogeneous boundary condition, then we can handle the problem more simply.

Theorem 2.3 *If equation* (2.1) *satisfy Condition C and the condition* (2.24) *below, then there exists a solution* $[w(z),u(z)]$ *of Problem* A_2 *for* (2.21), (2.22).

Proof We first prove the solvability of Problem A_2 for (2.21), (2.22). Denote $D_*=D^-\cap\{(\delta_0\le x\le a)\cup(a\le x\le 2-\delta_0)\}\cap\{-\delta\le y\le 0\}$, and s_1,s_2 are the characteristics of families in Theorem 2.2 emanating from any two points $(a_0,0),(a_1,0)(\delta_0=a_0<a_1=a)$ or $(a\le a_0<a_1=2-\delta_0)$ respectively, which intersect at a point $z=x+jy\in\overline{D^-}$, where δ_0,δ are sufficiently small positive numbers. In order to find a solution of the system

of integral equations (2.18), we need to add the condition for the coefficient a in equation (2.1) as follows

$$\frac{ay}{H(y)}=o(1),\ \text{ i.e. }\ \frac{|a|}{H(y)}=\frac{\varepsilon(y)}{|y|},\ m\geq 2, \tag{2.24}$$

where $\varepsilon(y)\to 0$ as $y\to 0$ and $\max_{\overline{D}}\varepsilon(y)\leq\varepsilon_0$, ε_0 is a positive number. It is clear that for two characteristics s_1, s_2 passing through a point $z=x+jy\in\overline{D^-}$ and x_1,x_2 are the intersection points with the axis $y=0$ respectively, for any two points $\tilde{z}_1=\tilde{x}_1+j\tilde{y}\in s_1,\tilde{z}_2=\tilde{x}_2+j\tilde{y}\in s_2$, $-\delta\leq y\leq 0$, we have

$$\begin{aligned}&|\tilde{x}_1-\tilde{x}_2|\leq|x_1-x_2|=2\Big|\int_0^y\sqrt{-K(t)}dt\Big|\leq\frac{2k_0}{m+2}|y|^{1+m/2}\\&\leq\frac{k_1}{6}|y|^{m/2+1}\leq M|y|^{m/2+1},\ |y|^{1+m/2}\leq\frac{k_0(m+2)}{2}|x_1-x_2|.\end{aligned}\tag{2.25}$$

From Condition C, we can assume that the coefficients in (2.21) are continuously differentiable with respect to $x\in L_0$ and satisfy the following conditions

$$\begin{aligned}&|\tilde{A}_l|,|\tilde{A}_{lx}|,|\tilde{B}_l|,|\tilde{B}_{lx}|,|\tilde{D}_l|,|\tilde{D}_{lx}|\leq k_0\leq k_1/6,|\tilde{E}_l|,|\tilde{E}_{lx}|\leq k_1/2,\\&2\sqrt{h},1/\sqrt{h},|h_y/h|\leq k_0\leq k_1/6\ \text{ in }\bar{D},\ l=1,2,\end{aligned}\tag{2.26}$$

and similarly to Section 2, Chapter V, we shall use several constants as follows

$$\begin{aligned}&M=4\max[M_3,M_4,M_5],\ M_3=\max[8(k_1d)^2,\frac{M_3}{k_1}],\\&M_4=\frac{(2+m)k_0d}{\delta^{2+m}}[4k_1+\frac{4\varepsilon_0+m}{\delta}],\ M_5=2k_1^2[d+\frac{1}{2H(y_1')}],\\&\gamma=\max[4k_1d\delta^\beta+\frac{4\varepsilon(y)+m}{2\beta'}]<1,\ 0\leq|y|\leq\delta,\\&\frac{2dM_4}{N+1}\leq\gamma,\ 2M\frac{(M_4|\tilde{t}|)^n}{n!}\leq M'\gamma^n,\ n=0,1...,N,N+1,...,\end{aligned}\tag{2.27}$$

where $\beta'=(1+m/2)(1-3\beta)$, $\varepsilon_0=\max_{\overline{D^-}}\varepsilon(z)$, d is the diameter of D, the positive numbers δ, β are small enough, and N, M' are sufficiently large positive integer and constant respectively. We choose $v_0=0,\xi_0=0,\eta_0=0$ and substitute them into the corresponding positions of v,ξ,η in the right-hand sides of (2.18), and obtain

$$\begin{aligned}&v_1(z)=v_1(x)-2\int_0^y V_0dy=v_1(x)+\int_0^y(\eta_0-\xi_0)dy,\\&\xi_1(z)=\zeta_1(z)+\int_0^y g_{10}(z)dy=\zeta_1(z)+\int_0^y\tilde{E}_1dy=\int_{y_1}^{\hat{y}}\hat{E}_1dy,\\&\eta_1(z)=\theta_1(z)+\int_0^y g_{20}(z)dy=\theta_1(z)+\int_0^y\hat{E}_2dy=\int_{y_1}^{\hat{y}}\hat{E}_2dy,\\&g_{l0}=\tilde{A}_l\xi_0+\tilde{B}_l\eta_0+\tilde{C}_l(\xi_0+\eta_0)+\tilde{D}_lv+\tilde{E}_l=\tilde{E}_l,l=1,2,\end{aligned}\tag{2.28}$$

where $v(z)=u(z)-u_0^-(z)$ in D^- is as stated before, $z_1=x_1+jy_1$ is a point on L_1, which is the intersection of L_1 and the characteristic curve s_1 passing through the point $z=x+jy\in\overline{D^-}$. For convenience we can only discuss the case later on. By the successive approximation, we find the sequences of functions $\{v_k\},\{\xi_k\},\{\eta_k\}$, which satisfy the relations

$$\begin{aligned}&v_{k+1}(z)=v_{k+1}(x)-2\int_0^y V_k(z)dy=v_{k+1}(x)+\int_0^y(\eta_k-\xi_k)dy,\\&\xi_{k+1}(z)=\zeta_{k+1}(z)+\int_0^y g_{1k}(z)dy=\int_{y_1}^{\hat{y}}\hat{g}_{1k}dy,\\&\eta_{k+1}(z)=\theta_{k+1}(z)+\int_0^y g_{2k}(z)dy=\int_{y_1}^{\hat{y}}\hat{g}_{2k}(z)dy,\\&g_{lk}(z)=\tilde{A}_l\xi_k+\tilde{B}_l\eta_k+\tilde{C}_l(\xi_k+\eta_k)+\tilde{D}_lv_k+\tilde{E}_l,\\&l=1,2,\ k=0,1,2,....\end{aligned}\tag{2.29}$$

Setting that $\tilde{g}_{lk+1}(z)=g_{lk+1}(z)-g_{lk}(z)\,(l=1,2)$ and

$$\begin{aligned}&\tilde{y}=\hat{y}-y_1,\,\tilde{t}=\hat{t}-y_1,\,\tilde{v}_{k+1}(z)=v_{k+1}(z)-v_k(z),\\&\tilde{\xi}_{k+1}(z)=\xi_{k+1}(z)-\xi_k(z),\tilde{\eta}_{k+1}(z)=\eta_{k+1}(z)-\eta_k(z),\\&\tilde{\zeta}_{k+1}(z)=\zeta_{k+1}(z)-\zeta_k(z),\tilde{\theta}_{k+1}(z)=\theta_{k+1}(z)-\theta_k(z),\end{aligned}$$

we shall prove that the sequences $\{\tilde{v}_k\},\{\tilde{\xi}_k\},\{\tilde{\eta}_k\},\{\tilde{\zeta}_k\},\{\tilde{\theta}_k\}$ in D_0 satisfy the estimates

$$|\tilde{v}_k(z)-\tilde{v}_k(x)|,|\tilde{\xi}_k(z)-\tilde{\zeta}_k(z)|,|\tilde{\eta}_k(z)-\tilde{\theta}_k(z)|\le M'\gamma^{k-1}|y|^{1-\beta},0\le|y|\le\delta,$$

$$
\begin{aligned}
&|\tilde{\xi}_k(z)|, |\tilde{\eta}_k(z)| \le M(M_4|\tilde{y}|)^{k-1}/(k-1)!, y \le -\delta, \text{ or } M'\gamma^{k-1}, 0 \le |y| \le \delta,\\
&|\tilde{\xi}_k(z_1) - \tilde{\xi}_k(z_2) - \tilde{\zeta}_k(z_1) - \tilde{\zeta}_k(z_2)|, |\tilde{\eta}_k(z_1) - \tilde{\eta}_k(z_2) - \tilde{\theta}_k(z_1) - \tilde{\theta}_k(z_2)|\\
&\le M'\gamma^{k-1}|x_1 - x_2|^{\beta}|y|^{\beta'}, 0 \le |y| \le \delta, |\tilde{v}_k(z_1) - \tilde{v}_k(z_2)|,\\
&|\tilde{\xi}_k(z_1) - \tilde{\xi}_k(z_2)|, |\tilde{\eta}_k(z_1) - \tilde{\eta}_k(z_2)| \le M(M_4|\tilde{t}|)^{k-1}|x_1 - x_2|^{1-\beta}\\
&/(k-1)!, y \le -\delta, \text{ or } M'\gamma^{k-1}|x_1 - x_2|^{\beta}|t|^{\beta'}, 0 \le |y| \le \delta,\\
&|\tilde{\xi}_k(z) + \tilde{\eta}_k(z) - \tilde{\zeta}_k(z) - \tilde{\theta}_k(z)| \le M'\gamma^{k-1}|x_1 - x_2|^{\beta}|y|^{\beta'}, |\tilde{\xi}_k(z) + \tilde{\eta}_k(z)|\\
&\le M(M_4|\tilde{y}|)^{k-1}|x_1 - x_2|^{1-\beta}/(k-1)! \text{ or } M'\gamma^{k-1}|x_1 - x_2|^{\beta}|y|^{\beta'},
\end{aligned}
\tag{2.30}
$$

where $z = x + jy, z = x + jt$ is the intersection point of two characteristics of families in Theorem 2.2 passing through z_1, z_2, $\beta' = (1 + m/2)(1 - 3\beta)$, δ, β are sufficiently small positive constants, such that $(2 + m)\beta < 1$, moreover $\gamma = \max_{-\delta \le y \le 0}[4k_1 d\delta^{\beta} + (4\varepsilon(y) + m)/2\beta'] < 1$, here d is the diameter of D, and $M_4 = (2 + m)k_0 d\delta^{-2-m}(4k_1\delta + 4\varepsilon_0 + m)/\delta$, and M' is a sufficiently large positive constant as stated in (2.44), Chapter V. As for the estimate in $\overline{D^-} \cap (\{0 \le x \le a_0 = \delta_0, -\delta \le y \le 0\} \cup \{a_1 = 2 - \delta_0 \le x \le 2, -\delta \le y \le 0\})$, which can be obtained by the simpler successive approximation, because it is only discussed the case: $-\delta \le y \le 0$.

By means of the estimates (2.30), we see that $\{u_k\}, \{\xi_k\}, \{\eta_k\}$ in D_* uniformly converge to u_*, ξ_*, η_* satisfying the system of integral equations

$$
\begin{aligned}
&u_*(z) = u_*(x) - 2\int_0^y V_* dy = u_*(x) + \int_0^y (\eta_* - \xi_*) dy,\\
&\xi_*(z) = \zeta_*(z) + \int_0^y [\tilde{A}_1\xi_* + \tilde{B}_1\eta_* + \tilde{C}_1(\xi_* + \eta_*) + \tilde{D}_1 u_* + \tilde{E}_1] dy, z \in s_1,\\
&\eta_*(z) = \theta_*(z) + \int_0^y [\tilde{A}_2\xi_* + \tilde{B}_2\eta_* + \tilde{C}_2(\xi_* + \eta_*) + \tilde{D}_2 u_* + \tilde{E}_2] dy, z \in s_2,
\end{aligned}
$$

and the function $W_*(z) = e_1\xi_*(x) + e_2\eta_*(z)$ satisfies equation (2.21) and boundary condition (2.23), similarly to Theorem 2.3, Chapter V, the solvability of Problem A_2 for (2.21) in D^- can be proved. Thus the existence of solutions of Problem P for equation (2.1) is proved. From the above discussion and by using the method in the proof of Theorem 3.2, Chapter II, we can prove the uniqueness of solutions of Problem P for (2.1) in D. Similarly we can discuss Problem Q for (2.1).

From the above result, we have the following theorem.

Theorem 2.4 *Let equation* (2.1) *satisfy Condition C and* (2.24). *Then the oblique derivative problems* (*Problems P and Q*) *for* (2.1) *have a unique solution.*

Finally we mention that the coefficients $K(y)$ in equation (2.1) can be replaced by functions $K(x,y)$ with some conditions. Besides if the boundary condition (2.8) on $L_1 \cup L_4$ is replaced by $L_2 \cup L_3$, then we can also discuss by the similar method.

2.4 Oblique derivative problem for degenerate mixed equations in general domains

Now we consider some general domains with non-characteristic boundary and prove the unique solvability of Problem P for equation (2.1).

1. Let D be a simply connected bounded domain D in the complex plane $\mathbf{C}$ with the boundary $\partial D = \Gamma \cup L$, where Γ, L are as stated before, and Γ can be replaced by another smooth curve Γ' ($\in C^2_\mu$ $(0 < \mu < 1)$ similar to Section 2, Chapter II. Moreover, we consider the domain D' with the boundary $\Gamma \cup L_1' \cup L_2' \cup L_3' \cup L_4'$, where the parameter equations of the curves L_1', L_2', L_3', L_4' are as follows:

$$\begin{aligned}&L_1'=\{\gamma_1(s)+y=0, 0\le s\le s_1'\}, L_2'=\{x-G(y)=a, l_1\le x\le a\},\\&L_3'=\{\gamma_2(s)+y=0, 0\le s\le s_2'\}, L_4'=\{x-G(y)=2, l_2\le x\le 2\},\end{aligned} \tag{2.31}$$

in which $Y = G(y) = \int_0^y \sqrt{|K(y)|}dy$ in $\overline{D}$, s is the parameter of arc length of L_1' or L_3', and $\gamma_k(s)$ on $\{0 \le s \le s_k'\}$ $(k = 1,2)$ are continuously differentiable, and $\gamma_k(0) = 0$ $(k = 1,2)$, the slopes of the curves L_1', L_3' at z_1^*, z_2^* are not equal to those of the characteristic curves of $s_1 :\ dy/dx = 1/H(y)$ at the points, where z_k^* $(k = 1,2)$ are L_1', L_3' and the characteristic curves of s_1, and $z_k' = l_k - j\gamma_k(s_k')(k = 1,2)$ are the intersection points of L_1', L_2' and L_3', L_4' respectively, hence $\gamma_k(s)$ $(k = 1,2)$ can be expressed by $\gamma_k[s(\nu)]$ $(k = 1,2)$. Actually we can permit that the intersection point of the curve L_1' or L_3' and any characteristic curve of $s_1 :\ dy/dx = 1/H(y)$ is not greater than one. Here we mention that in [12]1),3), the author considers the case $G(y) = y$ and assumes that the derivative of $\gamma(x)$ satisfies $0 < \gamma'(x) \le 1$.

We consider the oblique derivative boundary value problem (Problem

P') for equation (2.1) in D' with the boundary conditions

$$\begin{aligned}&\frac{1}{2}H_1(y)\frac{\partial u}{\partial \nu}=\mathrm{Re}[\overline{\lambda(z)}u_{\tilde z}]=H_1(y)r(z),\ z\in\Gamma\cup L_1'\cup L_4',\\&\mathrm{Im}[\overline{\lambda(z)}u_{\tilde z}]|_{z=z_k}=b_k,\ k=1,2,\ u(0)=b_0,\ u(2)=b_3,\end{aligned}\tag{2.32}$$

where $\lambda(z)=a(x)+ib(x)$ on $\Gamma\cup L_1'\cup L_4'$, and $\lambda(z),r(z),b_k(k=0,1,2,3)$ satisfy the conditions

$$\begin{aligned}&C^1_\alpha[\lambda(z),\Gamma]\le k_0,C^1_\alpha[r(z),\Gamma]\le k_2,C^1_\alpha[\lambda(z),L_1'\cup L_4']\le k_0,\\&C^1_\alpha[r(z),L_1'\cup L_4']\le k_2,\ |b_l|\le k_2,\ l=0,1,2,3,\\&\max_{z\in L_1'}\frac{1}{|a(x)-b(x)|},\ \max_{z\in L_4'}\frac{1}{|a(x)+b(x)|}\le k_0,\end{aligned}\tag{2.33}$$

in which $\alpha\,(0<\alpha<1),k_0,k_2$ are non-negative constants.

Setting $Y=G(y)=\int_0^y\sqrt{|K(t)|}dt$. By the conditions in (2.31), the inverse function $x=\sigma(\nu)=(\mu+\nu)/2$ of $\nu=x-G(y)$ can be found, i.e. $\mu=2\sigma(\nu)-\nu,0\le\nu\le2$, and the curve L_1',L_3' can be expressed by $\mu=2\sigma(\nu)-\nu=2\sigma(x+\gamma_1(s))-x-\gamma_1(s)$ on $L_1'\cup L_3'$. We make a transformation

$$\begin{aligned}&\tilde\mu=(t_k-t_{k-1})[\mu-2\sigma(\nu)+\nu]/[t_k-2\sigma(\nu)+\nu]+t_{k-1},\tilde\nu=\nu,\\&2\sigma(\nu)-\nu\le\mu\le t_k,\ 0\le\nu\le2,k=1,2,\end{aligned}\tag{2.34}$$

where $t_0=0,t_1=a,t_2=2$, μ, ν are real variables, its inverse transformation is

$$\begin{aligned}&\mu=[t_k-2\sigma(\nu)+\nu](\tilde\mu-t_{k-1})/(t_k-t_{k-1})+2\sigma(\nu)-\nu,\ \nu=\tilde\nu,\\&0\le\tilde\mu\le2,\ ,0\le\tilde\nu\le2,k=1,2.\end{aligned}\tag{2.35}$$

It is not difficult to see that the transformation in (2.34) maps the domain D' onto D. The transformation (2.34) and its inverse transformation (2.35) can be rewritten as

$$\begin{cases}\tilde x=\dfrac{(t_k-t_{k-1})(x+Y-2\sigma(x+\gamma_k(s))+x+\gamma_k(s)]}{2t_k-4\sigma(x+\gamma_k(s))+2x+2\gamma_k(s)}+\dfrac{t_{k-1}+x-Y}{2},\\\tilde Y=\dfrac{(t_k-t_{k-1})(x+Y-2\sigma(x+\gamma_k(s))+x+\gamma_k(s)]}{2t_k-4\sigma(x+\gamma_k(s))+2x+2\gamma_k(s)}+\dfrac{t_{k-1}-x+Y}{2},\end{cases}\tag{2.36}$$

$k=1,2$, and

$$\begin{cases} x=\dfrac{1}{2}(\mu+\nu)=\dfrac{[t_k-2\sigma(x+\gamma_k(s))+x+\gamma_k(s)](\tilde{x}+\tilde{Y}-t_{k-1})}{2(t_k-t_{k-1})} \\ +\sigma(x+\gamma_k(s))-\dfrac{x+\gamma_k(s)-\tilde{x}+\tilde{Y}}{2}, \\ Y=\dfrac{1}{2}(\mu-\nu)=\dfrac{[t_k-2\sigma(x+\gamma_k(s))+x+\gamma_k(s)](\tilde{x}+\tilde{Y}-t_{k-1})}{2(t_k-t_{k-1})} \\ +\sigma(x+\gamma_k(s))-\dfrac{x+\gamma_k(s)+\tilde{x}-\tilde{Y}}{2}, k=1,2. \end{cases} \tag{2.37}$$

Denote by $\tilde{Z}=\tilde{x}+j\tilde{Y}=f(Z)$, $Z=x+jY=f^{-1}(\tilde{Z})$ the transformation (2.36) and the inverse transformation (2.37) respectively. In this case, the system of equations (2.5) can be rewritten as

$$\begin{matrix} \xi_\mu=A_1\xi+B_1\eta+C_1(\xi+\eta)+Du+E, \\ \eta_\nu=A_2\xi+B_2\eta+C_2(\xi+\eta)+Du+E, \end{matrix} \quad z\in D'. \tag{2.38}$$

Suppose that (2.1) in D' satisfies Condition C, through the transformation (2.34), we obtain $\xi_{\tilde{\mu}}=[t_k-2\sigma(\nu)+\nu]\xi_\mu/(t_k-t_{k-1})$, $\eta_{\tilde{\nu}}=\eta_\nu, k=1,2$, in D'^-, where $\xi=U+V,\eta=U-V$, and then

$$\begin{aligned} &\xi_{\tilde{\mu}}=[t_k-2\sigma(\nu)+\nu][A_1\xi+B_1\eta+C_1(\xi+\eta)+Du+E]/(t_k-t_{k-1}), \\ &\eta_{\tilde{\nu}}=A_2\xi+B_2\eta+C_2(\xi+\eta)+Du+E \text{ in } D,\ k=1,2, \end{aligned} \tag{2.39}$$

and through the transformation (2.36), the boundary condition (2.32) is reduced to

$$\begin{aligned} &\text{Re}[\overline{\lambda(f^{-1}(\tilde{Z}))}W(f^{-1}(\tilde{Z}))]=H[y(Y)]r(f^{-1}(\tilde{Z})), \tilde{Z}=\tilde{x}+j\tilde{Y}\in L_1\cup L_4, \\ &\text{Im}[\overline{\lambda(f^{-1}(\tilde{Z}_k))}W(f^{-1}(\tilde{Z}_k)]=b_k,\ k=1,2,\ u(0)=b_0,\ u(2)=b_3, \end{aligned} \tag{2.40}$$

in which $Z=f^{-1}(\tilde{Z}),\tilde{Z}_k=f(Z'_k),Z'_k=l_k+jG[-\gamma_1(s'_k)],k=1,2$. Therefore the boundary value problem (2.1), (2.32) (Problem A') is transformed into the boundary value problem (2.39), (2.40), i.e. the corresponding Problem A in D. On the basis of Theorem 2.4, we see that the boundary value problem (2.39),(2.40) has a unique solution $w(\tilde{Z})$, and

$$u(z)=2R\int_0^z[\frac{\text{Re}W}{H(y)}+\binom{i}{-j}\text{Im}W]dz+b_0 \text{ in } \binom{D^+}{\overline{D^-}}$$

is just a solution of Problem P for (2.1) in D' with the boundary conditions (2.32), where $W = W(\tilde{Z}(z)]$.

Theorem 2.5 *If equation* (2.1) *in* D' *satisfies Condition* C *and* (2.24) *in the domain* D' *with the boundary* $\Gamma \cup L_1' \cup L_2' \cup L_3' \cup L_4'$, *where* L_1', L_2', L_3', L_4' *are as stated in* (2.31), *then Problem* P' *for* (2.1) *with the boundary conditions* (2.32) *has a unique solution* $u(z)$.

2. Next let the domain D'' be a simply connected domain with the boundary $\Gamma \cup L_1'' \cup L_2'' \cup L_3'' \cup L_4''$, where Γ is as stated before, and the parameter equations of arc length of curves $L_1'', L_2'', L_3'', L_4''$ are as follows:

$$\begin{aligned}&L_1''=\{\gamma_1(s)+y=0, 0\le s\le s_1'\}, L_2''=\{\gamma_3(s)+y=0, 0\le s\le s_3'\},\\&L_3''=\{\gamma_2(s)+y=0, 0\le s\le s_2'\}, L_4'=\{\gamma_4(s)+y=0, 0\le s\le s_4'\},\end{aligned}\tag{2.41}$$

in which $\gamma_k(0) = 0$, $\gamma_k(s) > 0$ on $\{0 \le s \le s_k'\}$ are continuously differentiable, and the slopes of the curve L_2'', L_4'' at the intersection points z_3^*, z_4^* of L_2'', L_4'' and the characteristic curves of $s_2 : dy/dx = -1/H(y)$ should be not equal to those of the characteristic curves at the points. Moreover denote $z_k'' = l_k - i\gamma_k(s_k) = l_{k+2} - i\gamma_{k+2}(s_{k+2}')$ $(k = 1, 2)$ are the intersection points of L_1'', L_2'' and L_3'', L_4'' respectively. We consider the Riemann-Hilbert problem (Problem A'') for equation (1) in D'' with the boundary conditions

$$\begin{aligned}&\mathrm{Re}[\overline{\lambda(z)}W(z)] = R_1(z),\ z\in\Gamma\cup L_1''\cup L_4'',\\&\mathrm{Im}[\overline{\lambda(z_k'')}W(z_k'')]=b_k, k=1,2, u(0)=b_0, u(2)=b_3,\end{aligned}\tag{2.42}$$

where $\lambda(z)$, $r(z)$ satisfy the corresponding conditions

$$\begin{aligned}&C_\alpha^1[\lambda(z),\Gamma]\le k_0, C_\alpha^1[r(z),\Gamma]\le k_2, C_\alpha^1[\lambda(z), L_1''\cup L_4'']\le k_0,\\&C_\alpha^1[r(z), L_1''\cup L_4'']\le k_2,\ |b_l|\le k_2,\ l=0,1,2,3,\\&\max_{z\in L_1''}\frac{1}{|a(x)-b(x)|},\ \max_{z\in L_4''}\frac{1}{|a(x)+b(x)|}\le k_0,\end{aligned}\tag{2.43}$$

in which $\alpha\,(0 < \alpha < 1), k_0, k_2$ are non-negative constants. By the conditions in (2.41), the inverse function $x = \tau(\mu) = (\mu + \nu)/2$ of $\mu = x + G(y)$ can be found, i.e. $\nu = 2\tau(\mu) - \mu$, $0 \le \mu \le 2$, and the curves L_2'', L_4'' can be expressed by

$$\nu=2\tau(\mu)-\mu=2\tau(x-\gamma_k(s))-x+\gamma_k(s)\ \text{ on } \{0\le s\le s_k'\},\ k=3,4.\tag{2.44}$$

We make a transformation

$$\begin{aligned}&\tilde{\mu}=\mu,\ \tilde{\nu}=\frac{a\nu}{2\tau(\mu)-\mu},\ 0\le\mu\le 2,\ 0\le\nu\le 2\tau(\mu)-\mu,\\&\tilde{\nu}=\frac{2(a-\nu)+a(\nu-2\tau(\mu)+\mu)}{a-2\tau(\mu)+\mu},\ a\le\nu\le 2\tau(\mu)-\mu,\end{aligned}\tag{2.45}$$

where the first transformation in (2.45) can be discussed similar to that as in (3.37), Chapter IV, but the number 2 is replaced by a. Now we only discuss the second transformation, it is clear that its inverse transformation is

$$\begin{aligned}&\nu=\frac{a(2-2\tau(\mu)+\mu)-\tilde{\nu}(a-2\tau(\mu)+\mu)}{2-a},\\&\mu=\tilde{\mu}=\tilde{x}+\tilde{Y},\ a\le\tilde{\mu}\le 2,\ a\le\tilde{\nu}\le 2.\end{aligned}\tag{2.46}$$

Hence $\tilde{x}=(\tilde{\mu}+\tilde{\nu})/2$, $\tilde{Y}=(\tilde{\mu}-\tilde{\nu})/2$, $x=(\mu+\nu)/2$, $Y=(\mu-\nu)/2$ possess the forms

$$\begin{aligned}&\tilde{x}=\frac{2ax+2(a-x+Y)-(a+x+Y)[2\tau(x-\gamma_4(s))-x+\gamma_4(s)]}{2[a-2\tau(x-\gamma_4(s))+x-\gamma_4(s)]},\\&\tilde{Y}=\frac{2aY-2(a-x+Y)+(a-x-Y)[2\tau(x-\gamma_4(s))-x+\gamma_4(s)]}{2[a-2\tau(x-\gamma_4(s))-x+\gamma_4(s)]},\\&x=\frac{2(\tilde{x}+\tilde{Y})-2a(\tilde{x}-1)+(\tilde{x}-\tilde{Y}-a)(2\tau(x-\gamma_4(s))-x+\gamma_4(s))}{2(2-a)},\\&Y=\frac{2(\tilde{x}+\tilde{Y})-2a(\tilde{Y}+1)-(\tilde{x}-\tilde{Y}-a)(2\tau(x-\gamma_4(s))-x+\gamma_4(s))}{2(2-a)}.\end{aligned}\tag{2.47}$$

Denote by $\tilde{Z}=\tilde{x}+j\tilde{Y}=g(z)$, $Z=x+jY=g^{-1}(\tilde{Z})$ the transformation and its inverse transformation in (2.47) respectively. Through the transformation (2.45), we obtain

$$\left.\begin{aligned}&(u+v)_{\tilde{\mu}}=(u+v)_{\mu},\\&(u-v)_{\tilde{\nu}}=[2\tau(\mu)-\mu-a](u-v)_{\nu}/(2-a)\end{aligned}\right\}\ \text{in } D''.\tag{2.48}$$

System (2.38) in D'' is reduced to

$$\left.\begin{aligned}&\xi_{\tilde{\mu}}=A_1\xi+B_1\eta+C_1(\xi+\eta)+Du+E\\&\eta_{\tilde{\nu}}=[2\tau(\mu)-\mu-a][A_2\xi+B_2\eta+C_2(\xi+\eta)+Du+E]/(2-a)\end{aligned}\right\}\ \text{in } D'.\tag{2.49}$$

Moreover, through the transformation (2.47), the boundary condition (2.42) on L_1'', L_4'' is reduced to

$$\begin{gathered}\mathrm{Re}[\overline{\lambda(g^{-1}(\tilde{Z}))}W(g^{-1}(\tilde{Z}))]=H_1[y(Y)]r[g^{-1}(\tilde{Z})],\ z=x+jy\in L_1'\cup L_4',\\ \mathrm{Im}[\overline{\lambda(g^{-1}(Z_k'))}W(g^{-1}(Z_k')]=b_k,\ k=1,2,\ u(0)=b_0,\ u(2)=b_3,\end{gathered}\tag{2.50}$$

in which $Z=g^{-1}(\tilde{Z})$, $\tilde{Z}_k'=g(Z_k'')$, $Z_k''=l_{k+2}+jG[-\gamma_{k+2}(s_{k+2}')], k=1,2$. Therefore the boundary value problem (2.38), (2.42) in D'' is transformed into the boundary value problem (2.49), (2.50). According to the method in the proof of Theorem 2.5, we can see that the boundary value problem (2.49), (2.50) has a unique solution $u(\tilde{Z})$, and then the corresponding $u=u(z)$ is a solution of Problem P'' of equation (2.1).

Theorem 2.6 *If the mixed equation* (2.1) *satisfies Condition C and* (2.24) *in the domain D'' with the boundary $L_0\cup L_1''\cup L_2''\cup L_3''\cup L_4''$, where $L_1'', L_2'', L_3'', L_4''$ are as stated in* (2.41), *then Problem P'' for* (2.1) *in D'' has a unique solution $u(z)$.*

3 Boundary Value Problems for Degenerate Equations of Mixed Type in Multiply Connected Domains

In this section we mainly discuss the Tricomi boundary value problem for second order degenerate equations of mixed type in multiply connected domains, this problem was posed by L. Bers in [9]2). We first give the representation and estimates of solutions of the boundary value problem for the equations, and then prove the uniqueness and existence of solutions for the problem by a new method. In [86]33), the author discussed the existence and uniqueness of solutions for oblique derivative problem for mixed equations of second order, which is uniformly mixed (elliptic-hyperbolic) type.

3.1 Formulation of oblique derivative problem in multiply connected domains

Denote

$$x=G(y)=\int_0^y\sqrt{|K(t)|}dt,\ x=H(y)=G'(y)=\sqrt{|K(y)|}\ \text{in}\ \overline{D^{\pm}},$$

and their inverse functions are $y = \pm|G^{-1}(x)|$ and $y = \pm|H^{-1}(x)|$ respectively, where D is a bounded domain, $D^+ = D\cap\{y > 0\}$, $D^- = D\cap\{y < 0\}$, we consider $K(y) = y^m h(y)$ in D^+ and $-K(y) = |y|^m h(y)$ in D^-, here m is a positive number, $h(y)$ in $\overline{D}$ is a continuously differentiable positive function, and $k_0(\geq \max[1, 2\sqrt{h(y)}])$ is a positive constant, then

$$|x|=|G(y)|\leq\frac{k_0}{m+2}|y|^{(m+2)/2},|x|=H(y)=G'(y)=\sqrt{|K(y)|}\leq\frac{k_0}{2}|y|^{m/2},$$

$$|y|=|G^{-1}(x)|\leq\frac{k_0(m+2)}{2}|x|^{2/(m+2)},\ |y|=|H^{-1}(x)|\leq[k_0|x|^{2/m}.$$

In this section we consider that D is an N-connected bounded domain in the complex plane $\mathbf{C}$ with the boundary $\partial D = \Gamma \cup L$, where $\Gamma = \sum_{l=1}^{N} \Gamma_l \in C^2_\mu$ $(0 < \mu < 1)$ in $\{y > 0\}$ and $\Gamma_l\,(l = 1, ..., N)$ are curves with the end points $z = a_1 = 0, b_1, a_2, b_2, ..., a_N, b_N = 2$ respectively, and $L = \cup_{l=1}^{2N} L_j$, $L_1 = \{x = -G(y), 0 \leq x \leq 1\}$, $L_2 = \{x = -G(y)+b_1, b_1 \leq x \leq b_1 + (a_2 - b_1)/2\}$, $L_3 = \{x = G(y)+a_2, b_1+(a_2-b_1)/2 \leq x \leq a_2\}$, $L_4 = \{x = -G(y)+b_2, b_2 \leq x \leq b_2+(a_3-b_2)/2\}, ..., L_{2N-1} = \{x = G(y)+a_N, b_{N-1}+(a_N-b_{N-1})/2 \leq x \leq a_N\}$, $L_{2N} = \{x = G(y) + 2, 1 \leq x \leq 2\}$, in which $a_1 = 0 < b_1 < a_2 < b_2 < ... < a_N < b_N = 2$, and denote $D^+ = D\cap\{y > 0\}$, $D^- = D\cap\{y < 0\}$, $D_1^- = D^- \cap\{x + G(y) < b_1\}$, $D_2^- = D^- \cap \{b_1 < x + G(y) < a_2\}$, $D_3^- = D^- \cap \{a_2 < x + G(y) < b_2\}, ..., D_{2N-2}^- = D^- \cap \{b_{N-1} < x + G(y) < a_N\}$, $D_{2N-1}^- = D^- \cap \{a_N < x + G(y)\}$, and $z_1 = 1 + iy_1$ is the intersection point of L_1, L_{2N}, $z_2 = (a_2 + b_1)/2 + jy_2, ..., z_N = (a_N + b_{N-1})/2 + jy_N$, herein $y_l = -|G^{-1}[(a_l - b_{l-1})/2]|, l = 2, ..., N$. For convenience, we can assume that the boundary $\Gamma_l\,(l = 1, ..., N)$ of the domain D^+ are smooth curves with the form $x - \tilde{G}(y) = a_1$, $x + \tilde{G}(y) = b_N$ near the points $z = a_1, b_N$, and $x - \tilde{G}(y) = b_l$ and $x + \tilde{G}(y) = a_l$ near the points $z = b_l\,(l = 1, ..., N-1), a_l\,(l = 2, ..., N)$ respectively, where $\tilde{G}(y)$ is similar to that as that in (5.35), Chapter II, but if we discuss the estimates of solutions near the degenerate line for the boundary value problem, then the method of conformal mappings can be used, i.e. the domain D^+ can be divided into N subdomains $D_l^+\,(l = 1, ..., N)$ with partial boundaries $a_l b_l\,(l = 1, ..., N)$, and then the conformal mapping can be applied such that new subdomains $\tilde{D}_l^+\,(1 \leq l \leq N)$ with the partial boundaries in $\{y > 0\}$ include the line segments $\mathrm{Re} z = a_l, b_l\,(1 \leq l \leq N)$ near the points $a_l, b_l\,(1 \leq l \leq N)$.

Consider generalized Chaplygin equation

$$K(y)u_{xx} + u_{yy} + au_x + bu_y + cu + d = 0 \text{ in } D, \tag{3.1}$$

where a, b, c, d are real functions of $z(\in\overline{D})$, $u, u_x, u_y(\in\mathbf{R})$, and suppose that the equation (3.1) satisfies **Condition** C:

1) The coefficients a, b, c, d are measurable in D^+ and continuous in $\overline{D^-}$ for any continuously differentiable function $u(z)$ in $D^* = \overline{D}\backslash T$, $T = \{a_l, b_l, l = 1, ..., N\}$, and satisfy

$$\begin{gathered}L_\infty[\eta, \overline{D^+}] \leq k_0,\ \eta = a, b, c,\ L_\infty[d, \overline{D^+}] \leq k_1,\ c \leq 0 \text{ in } D^+,\\ \hat{C}[d, \overline{D^-}] = C[d, \overline{D^-}] + C[d_x, \overline{D^-}] \leq k_1, \hat{C}[\eta, \overline{D^-}] \leq k_0, \eta = a, b, c,\end{gathered} \tag{3.2}$$

in which k_0, k_1 are positive constants.

2) For any two continuously differentiable functions $u_1(z), u_2(z)$ in D^*, $F(z, u, u_z) = au_x + bu_y + cu + d$ satisfies the following condition

$$F(z, u_1, u_{1z}) - F(z, u_2, u_{2z}) = \tilde{a}(u_1 - u_2)_x + \tilde{b}(u_1 - u_2)_y + \tilde{c}(u_1 - u_2) \text{ in } \overline{D},$$

where $\tilde{a}, \tilde{b}, \tilde{c}$ satisfy the conditions as those of a, b, c. It is clear that (3.1) with $a = b = c = d = 0$ is the famous Chaplygin equation.

The oblique derivative boundary value problem or general Tricomi-Bers problem for equation (3.1) may be formulated as follows:

Problem P or *GTB* Find a continuous solution $u(z)$ of equation (3.1) in $\overline{D}$, which is continuously differentiable in $D_* = \bar{D}\backslash T$ and satisfies the boundary conditions

$$\begin{gathered}\frac{1}{2}\frac{\partial u}{\partial \nu} = \frac{1}{H(y)} \mathrm{Re}\,[\overline{\lambda(z)} u_{\tilde{z}}] = \mathrm{Re}\,[\overline{\Lambda(z)} u_z] = r(z),\ z \in \Gamma \cup L',\\ \frac{1}{H(y)} \mathrm{Im}[\overline{\lambda(z)} u_{\tilde{z}}]|_{z=z_l} = c_l, u(a_l) = d_l, u(b_l) = d_{N+l}, l = 1, ..., N,\end{gathered} \tag{3.3}$$

where ν is the vector of $\Gamma \cup L'$, $L' = \cup_{l=1}^N L_{2l-1}$, ν is a vector at every point on Γ, $\lambda(z) = \cos(\nu, x) - i\cos(\nu, y), z \in \Gamma$, $\lambda(z) = \cos(\nu, x) + j\cos(\nu, y), z \in L'$, $c_l, d_l (l = 1, ..., N)$ are real constants, and $\lambda(z)$, $r(z)$, c_l, d_l $(l = 1, ..., N)$ satisfy the conditions

$$\begin{gathered}C^1_\alpha[\lambda(z), \Gamma] \leq k_0, C^1_\alpha[r(z), \Gamma] \leq k_2, C^1_\alpha[\lambda(x), L'] \leq k_0, C^1_\alpha[r(x), L'] \leq k_2,\\ \cos(\nu, n) \geq 0 \text{ on } \Gamma \cup L', |c_l|, |d_l|, |d_{N+l}| \leq k_2, l = 1, ..., N,\\ \max_{z \in L_1} \frac{1}{|a(x) - b(x)|} \leq k_0,\ \max_{z \in L''} \frac{1}{|a(x) + b(x)|} \leq k_0,\end{gathered} \tag{3.4}$$

where n is the outward normal vector on Γ, $L'' = \cup_{l=2}^N L_{2l-1}$, $\alpha\,(0 < \alpha < 1), k_0, k_2$ are positive constants. Here we mention that if $c = 0$ of equation

(3.1), then we can cancel the condition $\cos(\nu,n)\ge 0$ on Γ, and the last $2N-1$ conditions in (3.3) are replaced by

$$\mathrm{Im}[\overline{\lambda(z)}u_{\bar z}]|_{z=z_l'}=H(\mathrm{Im}z_l')d_l=d_l',l=2,3,...,2N, \tag{3.5}$$

where $z_l'\,(l=2,...,2N)$ are distinct points, such that $z_l'\in\Gamma\backslash T\,(l=2,...,2N)$ and $d_l\,(l=2,...,2N)$ are real constants satisfying the conditions $|d_l|\le k_2, l=2,...,2N$. The above boundary value problem is called Problem Q or Problem GTB'.

The number $K=(K_1+K_2+\cdots+K_{2N})/2$ is called the index of Problem P and Problem P_0 on the boundary ∂D^+ of D^+, where

$$K_l=\left[\frac{\phi_l}{\pi}\right]+J_l, J_l=0 \text{ or } 1, e^{j\phi_l}=\frac{\lambda(t_l-0)}{\lambda(t_l+0)},\gamma_l=\frac{\phi_l}{\pi}-K_l,l=1,...,2N, \tag{3.6}$$

in which $[a]$ is the largest integer not exceeding the real number a, and $t_1=a_1=0$, $t_2=b_1$, $t_3=a_2$, $t_4=b_2,...,t_{2N-1}=a_N,t_{2N}=b_N=2$, and

$$\lambda(t)=e^{i\pi/2} \text{ on } \tilde L_l, \lambda(t_{2l-1}+0)=\lambda(t_{2l}-0)=e^{i\pi/2}, l=1,...,N, \tag{3.7}$$

where $\tilde L_l=\{a_l<x<b_l,y=0\},l=1,...,N$. If $\cos(\nu,n)\not\equiv 0$ on each of $\Gamma_l\,(l=1,...,N)$, then we need to select the index $K=N-1$ on ∂D^+. More simply if we rewrite the boundary condition on $\tilde L=\cup_{l=1}^N\tilde L_l$ in an appropriate form, for instance setting that

$$\theta(x)=\theta_l(x)=\arg\lambda(a_l-0)-\pi+$$
$$+[\arg\lambda(b_l+0)+2\pi-\arg\lambda(a_l-0)]\frac{x-a_l}{b_l-a_l},\ x\in\tilde L_l, l=1,...,N,$$

but $\arg\lambda(a_l-0)-\pi$ is replaced by $\arg\lambda(a_l-0)$ if $l=1$, and $\arg\lambda(b_l+0)+2\pi$ is replaced by $\arg\lambda(b_l+0)+\pi$ if $l=N$, for the function $\lambda(x)=e^{i\theta(x)}$ and $\mathrm{Re}[\overline{\lambda(x)}W(x)]=-\sin\theta(x)u_y(x)/2$ on $\tilde L$, we have $\gamma_l=0,l=1,...,2N,K_1=K_N=0,K_2=\cdots=K_{2N-1}=1$ and $K=N-1$. If $\cos(\nu,n)\equiv 0$ on $\Gamma_l\,(l=1,...,N)$, Problem P includes Tricomi problem (Problem T) or Tricomi-Bers problem (Problem TB) as a special case and the last N point conditions in (3.3), (3.5) should be cancelled.

Now we explain that the above problems include the Tricomi problem (Problem T or TB) as a special case. In fact, the boundary condition of the Tricomi problem is

$$u(z)=\phi(z) \text{ on } \Gamma,\ u(z)=\psi(x) \text{ on } L', \tag{3.8}$$

where $C^2_\alpha[\phi(z),\Gamma]\le k_2$, $C^2_\alpha[\psi(x),L']\le k_2$, $\alpha(<1)$, k_2 are non-negative constants. In 1958, L. Bers posed to investigate the Tricomi problem for

Chaplygin equation in the multiply connected domain (see [12]2), P.89). In this section, we shall prove the unique solvability of the problem. It is easy to see that the boundary condition (3.8) can be rewritten as

$$\begin{gathered}\mathrm{Re}[\overline{\lambda(z)}w(z)]=r(z)\ \text{ on } \Gamma, \mathrm{Im}[\overline{\lambda(z_l)}w(z_l)]=H(\mathrm{Im}z_l)c_l=c_l', l=1,...,N,\\ \mathrm{Re}[\overline{\lambda(z)}w(z)]=R(z)\ \text{ on }\ L',\ u(z)=2\mathrm{Re}\int_0^z \hat{w}(z)dz+d_1\ \text{ in }\ D,\end{gathered}\tag{3.9}$$

in which $\hat{w}(z)$ is the same as (3.11), $d_1=\phi(0), \phi(z)=\phi(y)$ on Γ near $z=a_l,b_l,l=1,...,N$, $\lambda(z)=a+ib$ on Γ, $\lambda(z)=a+jb$ on L', and

$$\lambda(z)=\begin{cases}-i \text{ on } \Gamma_l \text{ if } x_y=\tilde{H}(y)=\tilde{G}'(y) \text{ at } a_1,b_l,l=1,...,N-1,\\ i\ \text{ on } \Gamma_l \text{ if } x_y=-\tilde{H}(y)=\tilde{G}'(y) \text{ at } a_l\ (l=2,...,N), b_N,\\ i\ \text{ on }\ \tilde{L}=U_{l=1}^N\tilde{L}_j,\ \tilde{L}_j=(a_l,b_l),\ l=1,...,N,\\ 1-j\ \text{ on }\ L_1,\ \text{if } x_y=-H(y),\\ 1+j\ \text{ on }\ L''=\cup_{l=2}^N L_{2l-1},\ \text{if } x_y=H(y),\end{cases}$$

and

$$\begin{gathered}R(z)=R_1(z)=\begin{cases}\phi_y/2\ \text{ on } \Gamma_l \text{ at } a_1,b_l\ (l=1,...,N-1),\\ -\phi_y/2\ \text{ on } \Gamma_l \text{ at } a_l\ (l=2,...,N), b_N,\\ -H(y)\psi_x/2\ \text{ on }\ L_1,\\ H(y)\psi_x/2\ \text{ on }\ L''=\cup_{l=2}^N L_{2l-1},\end{cases}\\ c_1=\mathrm{Im}[(1+j)u_{\bar{z}}(z_1-0)]=H(y)\psi_x/2|_{z=z_1-0},\\ c_l=\mathrm{Im}[(1-j)u_{\bar{z}}(z_l+0)]=-H(y)\psi_x/2|_{z=z_l+0}, l=2,...,N,\end{gathered}$$

in which $a=1\neq b=-1$ on L' and $a=1\neq -b=-1$ on L_N.

If the index of Problem T on ∂D^+ is $K=N/2-1$, we can argue as follows: The boundary conditions of Problem T in D^+ possess the form

$$\mathrm{Re}[\overline{\lambda(z)}w(z)]=R(z), z\in\Gamma,\ \mathrm{Re}[\overline{\lambda(x)}w(x)]=\hat{R}_0(x),\ x\in\tilde{L}=\cup_{l=1}^N\tilde{L}_l,$$

where $\lambda(x)=e^{i\pi/2}$, $\hat{R}_0(x)$ is an undetermined real function. It is clear that the possible points of discontinuity of $\lambda(z)$ on ∂D^+ are $t_1=a_1=0$, $t_2=$

b_1, $t_3 = a_2$, $t_4 = b_2, ..., t_{2N-1} = a_N$, $t_{2N} = b_N = 2$, and

$$\lambda(a_1 - 0) = \lambda(b_l + 0) = e^{-i\pi/2},\ l = 1, 2, ..., N - 1,$$

$$\lambda(b_N + 0) = \lambda(a_l + 0) = e^{i\pi/2},\ l = 2, ..., N,$$

$$\lambda(a_l + 0) = \lambda(b_l - 0) = e^{i\pi/2},\ l = 1, ..., N,$$

$$\frac{\lambda(t_1 - 0)}{\lambda(t_1 + 0)} = e^{-i\pi} = e^{i\phi_1}, 0 \le \gamma_1 = \frac{\phi_1}{\pi} - K_1 = -1 + K_1 = 0,$$

$$\frac{\lambda(t_l - 0)}{\lambda(t_l + 0)} = e^{i\pi} = e^{i\phi_l}, \gamma_l = \frac{\phi_l}{\pi} - K_l = 1 - 1 = 0, l = 2, 4, ..., 2N - 2,$$

$$\frac{\lambda(t_l - 0)}{\lambda(t_l + 0)} = e^{0\pi i} = e^{i\phi_l}, \gamma_l = \frac{\phi_l}{\pi} - K_l = 0 - K_l = 0, l = 3, 5, ..., 2N - 1,$$

$$\frac{\lambda(t_{2N} - 0)}{\lambda(t_{2N} + 0)} = e^{0\pi i} = e^{i\phi_{2N}}, -1 < \gamma_{2N} = \frac{\phi_{2N}}{\pi} - K_{2N} = 0.$$

Thus

$$K_1 = -1, K_3 = \cdots = K_{2N-1} = K_{2N} = 0, K_2 = K_4 = \cdots = K_{2N-2} = 1,$$

hence the index of Problem T on ∂D^+ is

$$K = \frac{1}{2}(K_1 + K_2 + \cdots + K_{2N}) = \frac{N}{2} - 1.$$

For Problem T, from (3.3) and $u(a_l) = d_l (l = 2, ..., N)$, the values of $u(z)$ at $b_{l-1} (l = 2, ..., N)$ are determined. In this section we mainly discuss the case $K = N/2 - 1$. In Remark 3.1 below, we shall explain the case of the general oblique derivative problem (Problem P).

Noting that $\phi(z) \in C^2_\alpha(\Gamma), \psi(x) \in C^2(L')\ (0 < \alpha < 1)$, we can find two twice continuously differentiable functions $u_0^\pm(z)$ in $\overline{D}^\pm$, for instance, which are the solutions of the Dirichlet problem with the boundary condition on $\Gamma \cup L'$ in (3.8) for harmonic equations in $D^\pm$, thus the functions $v(z) = v^\pm(z) = u(z) - u_0^\pm(z)$ in $\overline{D}^\pm$ is the solution of Problem $\tilde{T}$, which satisfies the equation in the form

$$K_1(y)v_{xx} + v_{yy} + \tilde{a}v_x + \tilde{b}v_y + \tilde{c}v + \tilde{d} = 0 \text{ in } D$$

and the boundary conditions

$$\begin{gathered} v(z) = 0 \text{ on } \Gamma \cup L', \text{ i.e. } \mathrm{Re}[\overline{\lambda(z)}W(z)] = R(z) = 0 \text{ on } \Gamma \cup L', \\ \mathrm{Im}[\overline{\lambda(z)}u_{\bar{z}}]|_{z=z_l} = c_l' = 0,\ u(a_l) = d_l = 0,\ l = 1, ..., N, \end{gathered} \tag{3.10}$$

where the coefficients of the above equation satisfy the conditions similar to Condition C, $W(z) = U+iV = v_{\bar z}^+$ in D^+ and $W(z) = U+jV = v_{\bar z}^-$ in $\overline{D^-}$, hence later on we only discuss the above homogeneous boundary condition and the case of index $K = N/2-1$, the other case can be similarly discussed. From $v(z) = v^{\pm}(z) = u(z) - u_0^{\pm}(z)$ in $\overline{D}^{\pm}$, we have $u(z) = v^-(z) + u_0^-(z)$ in $\overline{D}^-$, $u(z) = v^+(z) + u_0^+(z)$ in $\overline{D}^+$, and $u_y = v_y^{\pm} + u_{0y}^{\pm}$ in $D^{\pm}$, and $v^+(z) = v^-(z) - u_0^+(z) + u_0^-(z)$, $v_y^+ = v_y^- - u_{0y}^+ + u_{0y}^- = 2\hat R_0(x)$, $v_y^- = 2\tilde R_0(x)$ on $\tilde L = D\cap\{y=0\}$.

3.2 Representation of solutions of Tricomi problem for degenerate equations of mixed type

In this subsection, we shall give the representation of solutions for the Tricomi problem (Problem T) for equation (3.1) in D. Noting that

$$W(z)=U+iV=\frac{1}{2}[H(y)u_x-iu_y]=\frac{H(y)}{2}[u_x-iu_Y]=H(y)u_Z=u_{\bar z},$$

$$H(y)W_{\overline{Z}}=\frac{H(y)}{2}[W_x+iW_Y]=\frac{1}{2}[H(y)W_x+iW_y]=W_{\bar z} \text{ in } \overline{D^+},$$

$$W(z)=U+jV=\frac{1}{2}[H(y)u_x-ju_y]=u_{\bar z}=\frac{H(y)}{2}[u_x-ju_Y]=H(y)u_Z,$$

$$H(y)W_{\overline{Z}}=\frac{H(y)}{2}[W_x+jW_Y]=\frac{1}{2}[H(y)W_x+jW_y]=W_{\bar z} \text{ in } \overline{D^-},$$

we have the complex equation (2.4), namely

$$\begin{aligned}&H(y)W_{\overline{Z}}=A_1W+A_2\overline{W}+A_3u+A_4=g(Z) \text{ in } D,\\ &\left.\begin{aligned}&(U+V)_\mu=\frac{1}{4H}\{2[H_y/H+a/H]U-2bV+cu+d\},\\ &(U-V)_\nu=\frac{1}{4H}\{-2[H_y/H-a/H]U-2bV+cu+d\},\end{aligned}\right\} \text{ in } D_\tau^-,\end{aligned}\tag{3.11}$$

in which $Z=x+iG(y)$ in $\overline{D^+}$, $Z=x+jG(y)$, $\tau=e_1\mu+e_1\nu=e_1(x+G(y))+e_2(x-G(y))=\tau(z)$ in $\overline{D^-}$, and $e_1=(1+j)/2, e_2=(1-j)/2$. Especially, the complex equation

$$W_{\bar z}=0 \text{ in } \overline{D}\tag{3.12}$$

can be rewritten in the system

$$W_{\overline{Z}}=0 \text{ in } \overline{D_Z^+}, \text{ i.e. } (U+V)_\mu=0, (U-V)_\nu=0 \text{ in } \overline{D_\tau^-}.\tag{3.13}$$

The boundary value problem for equation (3.11) with the boundary condition (3.3) ($W(z) = u_{\bar z}$) and the relation

$$u(z) = \begin{cases} 2\mathrm{Re}\displaystyle\int_{a_1}^{z}[\frac{\mathrm{Re}W(z)}{H(y)} + i\mathrm{Im}W(z)]dz + d_1 \ \text{ in } \overline{D^+}, \\ 2\mathrm{Re}\displaystyle\int_{a_1}^{z}[\frac{\mathrm{Re}W(z)}{H(y)} - j\mathrm{Im}W(z)]dz + d_1 \ \text{ in } \overline{D^-} \end{cases} \tag{3.14}$$

will be called Problem A, the above second formula can be replaced by the formula in D^- in (3.22) below.

Now, we give the representation of solutions for the Tricomi problem (Problem T) for equation (3.1) in D. It is obvious that Problem T for equation (3.1) is equivalent to the following boundary value problem (Problem B) for equation (3.11) and the relation (3.14) with the boundary conditions

$$\begin{aligned} &\mathrm{Re}[\overline{\lambda(z)}W(z)] = H(y)r(z) = R_1(z) \ \text{ on } \Gamma\cup L', \\ &\mathrm{Im}[\overline{\lambda(z)}W(z)]|_{z=z_l} = H(\mathrm{Im}z_l)c_l = c_l',\ l=1,...,N,\ u(a_1) = d_1, \\ &u(a_l) = d_l \ \text{ or } \ \mathrm{Im}[\overline{\lambda(z)}W(z)]|_{z=z_l'} = H(\mathrm{Im}z_l')d_l = d_l',\ l=2,...,N. \end{aligned} \tag{3.15}$$

Now we first give the representation of solutions of Problem B for equation (3.12). For convenience, denote the functions $a(z), b(z), r(z)$ in (3.15) by the functions $a(x), b(x), r(x)$ of x, from

$$\mathrm{Re}[\overline{\lambda(z)}W(z)] = R_1(z) \ \text{ on } L',\ \mathrm{Im}[\overline{\lambda(z)}W(z)]|_{z=z_l} = c_l',\ l=1,...,N, \tag{3.16}$$

by the way as that in the proof of Theorem 2.1, Chapter V, we can find

$$\begin{aligned} &\mathrm{Re}[\overline{\lambda(x)}W(x)] = U(x) + V(x) = -\tilde R_0(x) \ \text{ on } \tilde L = \cup_{l=1}^{N}\tilde L_l,\ \tilde L_l = (a_l, b_l), \\ &\tilde R_0(x) = -\frac{2R_1(x/2) - [a(x/2) + b((x/2)]k_1}{a(x/2) - b(x/2)} \ \text{ on } \tilde L, \end{aligned} \tag{3.17}$$

where $[a(1) + b(1)]k_1 = R(1) - c_1' = 0$ and $[a(x/2) + b(x/2)]k_1 = 0$. On the basic of the result in Chapter I, we can find a unique solution $w(z)$ of (3.13) in D^+ with the boundary conditions (3.17) and

$$\begin{aligned} &\mathrm{Re}[\overline{\lambda(z)}W(z)] = H(y)r(z) = R_1(z) \ \text{ on } \Gamma, \\ &\mathrm{Im}[\overline{\lambda(z)}W(z)]|_{z=z_l'} = d_l',\ l = 2,...,N. \end{aligned} \tag{3.18}$$

Thus the boundary condition

$$\begin{aligned} U(x)-V(x) &= \mathrm{Re}[(1-j)(U(x)+jV(x))] \\ &= \mathrm{Re}[(1-j)W(x)] = R_0(x) \ \text{ on } \tilde{L} \end{aligned} \tag{3.19}$$

is determined. Furthermore similarly to the proof of Theorem 2.1, we can find a solution of Problem B for (3.13) in $\overline{D_1^-}$, and the solution $w(z)$ in $\overline{D_1^-}$ possesses the form

$$\begin{aligned} &W(z) = \frac{1}{2}[(1+j)f_1(x-G(y)) + (1-j)g_1(x+G(y))], \\ &f_1(x-G(y)) = \frac{2R_1[(x-G(y))/2] - K[(x-G(y))/2]}{a((x-G(y))/2] - b((x-G(y))/2]}, \\ &K[(x-G(y))/2] = [a((x-G(y))/2 + b((x-G(y))/2)]k_1 = 0, \\ &g_1(x+G(y)) = \mathrm{Re}[(1-j)W(x+G(y))] \ \text{ in } \overline{D_1^-}\backslash\{a_1, b_1\}. \end{aligned} \tag{3.20}$$

Similarly we can write the solution of Problem B in $\overline{D_l^-}$ $(l = 2, 3, ..., 2N-1)$ as

$$\begin{aligned} &W(z) = \frac{1}{2}[(1+j)f_l(x-G(y)) + (1-j)g_l(x+G(y))] \\ &\qquad\qquad \text{in } D_l^-, \ l = 2, ..., 2N-1, \\ &f_l(x-G(y)) = \frac{2R_1((x-G(y))/2) - K((x-G(y))/2)}{a((x-G(y))/2) - b((x-G(y))/2)}, \\ &K((x-G(y))/2) = [a((x-G(y))/2) + b((x-G(y))/2)]k_l = 0, \\ &[a(x_1)+b(x_1)]k_l = R_1(x_1) - c_1' = 0 \ \text{ in } D_l^-, l = 2, ..., 2N-1, \\ &g_{2l}(x+G(y)) = \frac{2R_1((x+G(y)+a_{l+1})/2)}{a((x+G(y)+a_{l+1})/2) + b(x+G(y)+a_{l+1})/2)} \\ &\quad - \frac{[a((x+G(y)+a_{l+1})/2) - b((x+G(y)+a_{l+1})/2)]h_{2l}}{a((x+G(y)+a_{l+1})/2) + b((x+G(y)+a_{l+1})/2)}, \\ &[a(x_{l+1}) - b(x_{l+1})]h_{2l} = R_1(x_{l+1}) + c_{l+1}' = 0 \ \text{ in } D_{2l}^-, l = 1, ..., N-1, \\ &g_{2l-1}(x+G(y)) = \mathrm{Re}[(1-j)W(x+G(y))] \ \text{ in } D_{2l-1}^-, \ l = 2, ..., N, \end{aligned} \tag{3.21}$$

in which $\overline{D_l^-}$ $(l = 2, 3, ..., 2N-1)$ are as stated in Subsection 3.1. This shows that Problem B for equation (3.12) is uniquely solvable, namely

Theorem 3.1 *Problem B of equation (3.12) or system (3.13) in $\overline{D}$ has a unique solution as in (3.21), and the solution of Problem T or TB for (3.12) can be represented as in (3.14), where* $\text{Re}W(z)+iV=u_{\bar z}$ *in* $\overline{D^+}$ *and* $\text{Im}W(z)+jV=u_{\bar z}$ *in* $\overline{D^-}$.

The representation of solutions of Problem T (or Problem TB) for equation (3.1) is as follows, which can be proved by the same method as in Theorem 2.2.

Theorem 3.2 *Under Condition C, any solution $u(z)$ of Problem T or TB for equation* (3.1) *in D^- can be expressed as follows*

$$
\begin{aligned}
&u(z)=u(x)-2\int_0^y V(z)dy=2\text{Re}\int_0^z[\frac{\text{Re}w}{H}+\begin{pmatrix} i\\ -j\end{pmatrix}\text{Im}w]dz+d_1 \text{ in}\begin{pmatrix}\overline{D^+}\\ \overline{D^-}\end{pmatrix},\\
&w(z)=\Phi[Z(z)]+\Psi[Z(z)],\ \Psi(Z)=-\text{Re}\frac{2}{\pi}\iint_{D_t^+}\frac{f(t)}{t-Z}d\sigma_t \text{ in } \overline{D_Z^+},\\
&w(z)=\phi(z)+\psi(z)=\xi(z)e_1+\eta(z)e_2 \text{ in } \overline{D^-},\\
&\xi(z)=\zeta(z)+\int_0^y g_1(z)dy=\int_{S_1}g_1(z)dy+\int_0^y g_1(z)dy,\\
&\eta(z)=\theta(z)+\int_0^y g_2(z)dy,\ z\in s_2,\\
&g_l(z)=\tilde A_l(U+V)+\tilde B_l(U-V)+2\tilde C_lU+\tilde D_lu+\tilde E_l,\ l=1,2,
\end{aligned}
\tag{3.22}
$$

where $f(Z)=g(Z)/H$, $U=Hu_x/2$, $V=-u_y/2$, $\phi(z)=\zeta(z)e_1+\theta(z)e_2$ *is a solution of* (3.13) *in* D^-, $\zeta(z)=\int_{S_1}g_1(z)dy$, $\theta(z)=-\zeta(x+G(y))$ *in* D^-, s_1,s_2 *are two families of characteristics in* D^-:

$$
s_1:\frac{dx}{dy}=\sqrt{-K(y)}=H(y),\ s_2:\frac{dx}{dy}=-\sqrt{-K(y)}=-H(y) \tag{3.23}
$$

passing through $z=x+jy\in\overline{D^-}$, S_1 *is the characteristic curve from a point on* L_1 *to a point on* L_0, *and*

$$
\begin{aligned}
&w(z)=U(z)+jV(z)=\frac{1}{2}Hu_x-\frac{j}{2}u_y,\\
&\xi(z)=\text{Re}\psi(z)+\text{Im}\psi(z),\ \eta(z)=\text{Re}\psi(z)-\text{Im}\psi(z),\\
&\tilde A_1=\tilde B_2=\frac{1}{2}(\frac{h_y}{2h}-b),\ \tilde A_2=\tilde B_1=\frac{1}{2}(\frac{h_y}{2h}+b),\\
&\tilde C_1=\frac{a}{2H}+\frac{m}{4y},\ \tilde C_2=-\frac{a}{2H}+\frac{m}{4y},
\end{aligned}
$$

$$\tilde{D}_1 = -\tilde{D}_2 = \frac{c}{2}, \ \ \tilde{E}_1 = -\tilde{E}_2 = \frac{d}{2}, \tag{3.24}$$

in which we choose $H(y) = [|y|^m h(y)]^{1/2}$, m, $h(y)$ *are as stated before, and*

$$\begin{aligned} d\mu &= d[x + G(y)] = 2H(y)dy \ \text{ on } s_1, \\ d\nu &= d[x - G(y)] = -2H(y)dy \ \text{ on } s_2. \end{aligned} \tag{3.25}$$

For the homogeneous boundary condition of (3.8), from Theorem 3.1 we can derive $\zeta(z)=\int_{S_1} g_1(z)dy$, $\theta(z)=\int_{S_2} g_2(z)dy$, and $\theta(x)=-\zeta(x)$ on $\tilde{L}$.

3.3 Uniqueness of solutions of Tricomi problem for degenerate equation of mixed type

In this subsection, we prove the uniqueness of solutions of Problem T or TB for equation (3.1).

Theorem 3.3 *Suppose that equation* (3.1) *satisfies the above conditions. Then Problem* T *or* TB *for* (3.1) *in* D^+ *has at most one solution.*

Proof Let $u_1(z), u_2(z)$ be any two solutions of Problem T for (3.1). It is easy to see that $u(z) = u_1(z)-u_2(z)$ and $w(z) = u_{\tilde{z}}$ satisfy the homogeneous equation and boundary conditions

$$w_{\bar{\tilde{z}}} = A_1 w + A_2 \overline{w} + A_3 u \ \text{ in } D, \tag{3.26}$$

$$\begin{aligned} &\mathrm{Re}[\overline{\lambda(z)}w(z)] = 0, \ \ z \in \Gamma \cup L', \\ &\mathrm{Im}[\overline{\lambda(z)}w(z)]|_{z=z_l}=0, u(a_l)=0, l=1,...,N, \end{aligned} \tag{3.27}$$

where the function $w(z) = U(z)+jV(z) = [Hu_x - ju_y]/2$ in the hyperbolic domain D^- can be expressed in the form

$$\begin{aligned} &w(z) = \phi(x) + \psi(z) = \phi(x) + \xi(z)e_1 + \eta(z)e_2, \\ &\xi(z)=\zeta(z) +\int_0^y [\tilde{A}_1(U+V)+\tilde{B}_1(U-V)+2\tilde{C}_1U+\tilde{D}_1u]dy, z\in s_1, \\ &\eta(z)=\theta(z)+\int_0^y [\tilde{A}_2(U+V)+\tilde{B}_2(U-V)+2\tilde{C}_2U+\tilde{D}_2u]dy, z\in s_2, \end{aligned} \tag{3.28}$$

where $\phi(z) = \zeta(z)e_1 + \theta(z)e_2$ is a solution of (3.12). Thus the solution

$$u(z) = 2\,\mathrm{Re}\int_0^z [\frac{\mathrm{Re}w(z)}{H(y)} + \begin{pmatrix} i \\ -j \end{pmatrix} \mathrm{Im}w]dz \ \text{ in } \begin{pmatrix} \overline{D^+} \\ \overline{D^-} \end{pmatrix},$$

is the solution of the homogeneous equation of (3.1) with homogeneous boundary conditions of (3.3):

$$\begin{gathered}\frac{\partial u}{\partial l}=2\mathrm{Re}[\overline{\lambda(z)}u_{\bar z}]=0 \ \text{ on } \Gamma,\\ 2\mathrm{Re}[\overline{-i}u_{\bar z}(x)]=u_y=\tilde R(x) \text{ on } \tilde L, u(a_l)=0, l=1,...,N.\end{gathered}\tag{3.29}$$

Now we verify that the above solution $u(z)\equiv 0$ in D^+. If the maximum $M=\max_{\overline{D^+}}u(z)>0$, it is clear that the maximum point $z^*\notin D^+\cup\Gamma$. Thus $u(z)$ attains its maximum at a point $z^*=x^*\in\tilde L$. By using the method in the proof of Theorem 2.5, Chapter V, we can derive that $u(x)=0$ on $\tilde L$. Hence $\max_{\overline{D^+}}u(z)=0$. By the similar method, we can prove $\min_{\overline{D^+}}u(z)=0$. Moreover by using Theorem 3.4 below, we have $u(z)=0$ in $\overline{D^-}$. Therefore $u(z)=0$, $u_1(z)=u_2(z)$ in $\overline{D^+}$. This completes the proof.

Theorem 3.4 *Let D^- be given as above and equation* (3.1) *satisfy Condition C and* (2.24). *Then the Tricomi problem (Problem T) or Tricomi-Bers problem (Problem TB) for* (3.1) *in D^- at most has a solution.*

Proof We assume that m is a positive number, denote by $u_1(z),u_2(z)$ two solutions of Problem T or TB for (3.1), by Theorem 3.2, we see that the function $u_{\bar z}(z)=u_{1\bar z}(z)-u_{2\bar z}(z)=U(z)+jV(z)$ in $\overline{D^-}$ is a solution of the homogeneous system of integral equations

$$\begin{aligned}&u(z)=u(x)-2\int_0^y V(z)dy=\int_0^z[\frac{\mathrm{Re}w}{H(y)}-j\mathrm{Im}w]dz \text{ in } D^-,\\ &w(z)=\phi(z)+\psi(z)=\xi(z)e_1+\eta(z)e_2,\\ &\xi(z)=\zeta(z)+\int_0^y[\tilde A_1(U+V)+\tilde B_1(U-V)+2\tilde C_1U+\tilde D_1u]dy,\ z\in s_1,\\ &\eta(z)=\theta(z)+\int_0^y[\tilde A_2(U+V)+\tilde B_2(U-V)+2\tilde C_2U+\tilde D_2u]dy,\ z\in s_2.\end{aligned}\tag{3.30}$$

Noting that the functions u_x,u_y are continuous in $\overline{D}\backslash T$ and $\phi(z)$ is a solution of (3.12) in D^-, by Theorem 3.1, we can prove $w(z)=\phi(z)+\psi(z)=0$ in D^-.

In fact, choose any closed set $D_0=D\cap\{a_l<a_l'\le x\le b_l'<b_l, 1\le l\le N\}$, where $0<a_l'<b_l'<2$, $l=1,...,N)$. Noting that the continuity of u_x,u_y in D_0 and (3.53) below, hence there exists a sufficiently large positive number M dependent on $u(z),\xi(z),\eta(z),D_0$, and $M_0=d[4k_1\delta+4\varepsilon_0+m_1]/\delta$, $\max_{\overline{D^-}}\varepsilon(y)=\varepsilon_0<\infty$, β is a sufficiently small positive constant, such that

$$|u(z)| \le M,\ |\xi(z)| \le M,\ |\eta(z)| \le M,\ |\xi(z)+\eta(z)| \le M|x_1 - x_2|^{1-\beta}, \tag{3.31}$$

where ε_0 is a positive constant. From (3.30) and (3.31), we can obtain

$$\begin{aligned}|\xi(z)| &= |\int_{y_1}^{y}[\tilde{A}_1\xi + \tilde{B}_1\eta + \tilde{C}_1(\xi+\eta) + \tilde{D}_1 u]dy|\\ &\le |\int_{y_1}^{y} M(|\tilde{A}_1|+|\tilde{B}_1|)+|\tilde{C}_1||x_1 - x_2|^{1-\beta}+|\tilde{D}_1|]dy|\\ &\le |\int_{y_1}^{y} M[k_1+(\frac{|\varepsilon(y)|}{2}+\frac{m}{4})d]dy| \le M(M_0|\tilde{y}|)^k/k! \text{ on } s_1, k=1.\end{aligned}$$

Similarly we have

$$|\eta(z)| = |\int_{y_2}^{y}[\tilde{A}_2\xi+\tilde{B}_2\eta+\tilde{C}_2(\xi+\eta)+\tilde{D}_2 u]dy| \le M(M_0|\tilde{y}|)^k/k! \text{ on } s_2, k=1.$$

Here $y \le -\delta$, $\max_{D_0}[|\tilde{A}_l|, |\tilde{B}_l|, |\tilde{D}_l|] \le k_1/3, l = 1,2$, $\tilde{y} = y - y_1$ or $y - y_2$, y_1, y_2 are the ordinates of intersection points of L_1 and characteristics lines of family s_1 in (3.23) emanating from $z = x_1, x_2 (< x_1)$ and L_1, herein x_1, x_2 are the intersection points of two characteristics lines s_1, s_2 passing through $z = x + jy \in \overline{D^-}$ and $x-$axis respectively. Applying the repeated insertion, the inequalities

$$|u(z)| \le M(M_0|\tilde{y}|)^k/k!, |\xi(z)| \le M(M_0|\tilde{y}|)^k/k!,$$
$$|\eta(z)| \le M(M_0|\tilde{y}|)^k/k!,\ k = 2, 3, \ldots$$

are obtained. As for the case $-\delta < y \le 0$, we can discuss still. This shows that $u(z) = 0, \xi(z) = 0, \eta(z) = 0$ in D_0. Taking into account the arbitrariness of δ_0, we can derive $u(z) = 0, \xi(z) = 0, \eta(z) = 0$ in D.

In addition, we can also prove that the solution of Problem Q for equation (3.1) is unique.

3.4 Solvability of Tricomi problem for degenerate equation of mixed type

In this subsection, we prove the existence of solutions of Problem T or TB for equation (3.1). As stated in Subsection 3.1, we can discuss Problem $\tilde{T}$ of the equation in the form

$$K(y)v_{xx}+v_{yy}+\tilde{a}v_x+\tilde{b}v_y+\tilde{c}v+\tilde{d} = 0, \text{ in } D, \tag{3.32}$$

satisfying the homogeneous boundary condition of (3.3). Denote $w(z) = v_{\tilde{z}}$, then Problem $\tilde{T}$ is transformed into the boundary value problem (Problem $\tilde{B}$), i.e. the complex equation

$$W_{\tilde{\bar{z}}} = A_1W + A_2\overline{W} + A_3u + \tilde{A}_4 \ \text{ in } D, \tag{3.33}$$

with the relation

$$u(z)=\begin{cases} 2\mathrm{Re}\displaystyle\int_{a_1}^{z}[\frac{\mathrm{Re}W(z)}{H(y)}+i\mathrm{Im}W(z)]dz \ \text{ in } \overline{D^+}, \\ 2\mathrm{Re}\displaystyle\int_{a_1}^{z}[\frac{\mathrm{Re}W(z)}{H(y)}-j\mathrm{Im}W(z)]dz \ \text{ in } \overline{D^-}, \end{cases} \tag{3.34}$$

and the boundary conditions

$$\begin{gathered} \mathrm{Re}[\overline{\lambda(z)}W(z)]=H(y)r(z)=R(z)=0 \ \text{ on } \Gamma\cup L', \\ \mathrm{Im}[\overline{\lambda(z_l)}W(z_l)] = c_l' = 0, \ l = 1, ..., N, \\ u(a_1)=0, \mathrm{Im}[\overline{\lambda(z_l')}W(z_l')]=d_l'=0, \ l=2, ..., N, \end{gathered} \tag{3.35}$$

where the coefficients in (3.33) are as stated in (2.7), and $\lambda(z), z_l\,(l = 1, ..., N)$ are as stated (3.3)-(3.5), and $R(z) = 0$ on Γ. It is not difficult to see that Problem $\tilde{B}$ can be divided into two problems, i.e. Problem B_1 of equation (3.33), (3.34) in D^+ and Problem B_2 of equation (3.33), (3.34) in D^-, the boundary conditions of Problems B_1 and B_2 as follows:

$$\begin{gathered} \mathrm{Re}[\overline{\lambda(z)}W(z)]=0 \text{ on } \Gamma, \ \mathrm{Re}[\overline{\lambda(z)}W(z)]=\hat{R}_0(x) \text{ on } \tilde{L}, \\ u(a_1) = 0, \ \mathrm{Im}[\overline{\lambda(z_l')}W(z_l')] = 0, \ l = 2, ..., N, \end{gathered} \tag{3.36}$$

where $\lambda(z) = -i$ on $\tilde{L}$, and

$$\begin{gathered} \mathrm{Re}[\overline{\lambda(z)}W(z)]=0 \text{ on } L', \mathrm{Re}[\overline{\lambda(z)}W(z)]=\tilde{R}_0(x) \text{ on } \tilde{L}, \\ \mathrm{Im}[\overline{\lambda(z)}W(z)]|_{z=z_l} = c_l' = 0, \ l = 1, ..., N, \end{gathered} \tag{3.37}$$

in which $\lambda(z) = a(z) + jb(z$ on L' and $\lambda(z) = j$ on $\tilde{L}$. The solvability of Problem B_1 can be proved by using the method in Sections 2 and 3, Chapter II, and the result of Problem B_2 will be proved by using the way as stated in the proof of Theorem 2.3. In the following, we first discuss Problem B_1.

Introduce a function

$$X(Z) = \prod_{l=1}^{2N}(Z - t_l)^{\eta_l}, \tag{3.38}$$

where $\eta_l = 1\,(l = 1, ..., 2N)$. Obviously that $X(Z)W[z(Z)]$ satisfies the complex equation

$$\begin{aligned}[X(Z)W]_{\overline{Z}} &= X(Z)[A_1W + A_2\overline{W} + \\ &+A_3u + A_4]/H = X(Z)g(Z)/H \text{ in } D_Z^+,\end{aligned} \tag{3.39}$$

and the boundary conditions

$$\begin{gathered}\operatorname{Re}[\overline{\hat{\lambda}(z)}X(Z)W(z)] = R(z) \text{ on } \Gamma, \operatorname{Im}[\overline{\lambda(z_l')}W(z_l')] = d_l' = 0,\ l = 2, ..., N, \\ \operatorname{Re}[\overline{\hat{\lambda}(x)}X(x)W(x)] = |X(x)|\hat{R}_0(x) \text{ on } \tilde{L},\ u(a_1) = d_1 = 0,\end{gathered} \tag{3.40}$$

where $R(z) = 0$ on Γ, and $\hat{\lambda}(z) = \lambda(z)e^{i \arg X(Z)}$. Noting that

$$e^{i\tilde{\phi}_l} = \frac{\hat{\lambda}(t_l - 0)}{\hat{\lambda}(t_l + 0)} = \frac{\lambda(t_l - 0)}{\lambda(t_l + 0)} \frac{e^{i \arg X(t_l - 0)}}{e^{i \arg X(t_l + 0)}} = e^{i(\phi_l + \eta_l \pi/2)},$$

$$\tau_l = \frac{\tilde{\phi}_l}{\pi} - K_l = 0,\ l = 1, 2, ..., 2N,$$

where the numbers $\tau_l\,(1 \le l \le 2N)$ about $\hat{\lambda}[z(Z)] = \lambda[z(Z)]e^{i \arg X(Z)}$ are equal to $\tau_l = 1/2\,(l = 1 \le l \le 2N)$ when we consider $\operatorname{Im}[X(Z)W(x)] = |X(Z)|\hat{R}_0(x)$ on $\tilde{L}$, which are corresponding to the numbers $\gamma_l\,(1 \le l \le 2N)$ in (3.6), and the index of $\hat{\lambda}(z)$ is $\tilde{K} = N/2 - 1$, in this case we need $N - 1$ point conditions in (3.10) such that Problem T (or TB) is well posed. Now we prove a lemma.

Theorem 3.5 Let equation (3.1) satisfy Condition C. Then any solution of Problem B_1 for (3.1) satisfies the estimate

$$\begin{gathered}\hat{C}_\delta[W(z),\overline{D}] = C_\delta[X(Z)(\operatorname{Re}W/H + i\operatorname{Im}W),\overline{D}] + C_\delta[u(z),\overline{D}] \le M_1, \\ \hat{C}_\delta[W(z),\overline{D}] \le M_2(k_1 + k_2),\end{gathered} \tag{3.41}$$

where $X(Z)$ is as stated in (3.38), $\delta\,(\le \min[2, m]/(m + 2))$ is a sufficiently small positive constant, $M_1 = M_1(\delta, k, H, D)$, $M_2 = M_2(\delta, k_0, H, D)$ are non-negative constants, and $k = (k_0, k_1, k_2)$.

Proof We first assume that any solution $[W(z),\ u(z)]$ of Problem B_1 satisfies the estimate

$$\hat{C}[W(z),\overline{D}] = C[X(Z)(\operatorname{Re}W/H + i\operatorname{Im}W),\overline{D_Z}] + C[u(z),\overline{D}] \le M_3, \tag{3.42}$$

in which M_3 is a non-negative constant. Now, Substituting the solution $[W(z), u(z)]$ into equation (3.39) and noting $\operatorname{Re}W(Z) = R(x) = 0$ on

$\tilde{L}$, $b_l=0$, $j=0,1,...,N-1$, we can extend the function $X(Z)W[z(Z)]$ onto the symmetrical domain $\tilde{D}_Z$ of D_Z with respect to the real axis $\text{Im}Z=0$, namely set

$$\tilde{W}(Z)=\begin{cases} X(Z)W[z(Z)] \text{ in } D_Z,\\ -\overline{X(\overline{Z})W[z(\overline{Z})]} \text{ in } \tilde{D}_Z,\end{cases}$$

which satisfies the boundary conditions

$$\text{Re}[\overline{\tilde{\lambda}(Z)}\tilde{W}(Z)]=0 \text{ on } \Gamma\cup\tilde{\Gamma},$$

$$\tilde{\lambda}(Z)=\begin{cases}\lambda[z(Z)],\\ \overline{\lambda[z(\overline{Z})]},\\ 1,\end{cases}\quad \tilde{R}(Z)=\begin{cases}0 \text{ on } \Gamma,\\ 0 \text{ on } \tilde{\Gamma},\\ 0 \text{ on } \tilde{L},\end{cases}$$

where $\tilde{\Gamma}$ is the symmetrical curve of Γ about $\text{Im}Z=0$. It is clear that the corresponding function $u(z)$ in (3.34) can be extended to the function $\tilde{u}(Z)$, where $\tilde{u}(Z)=u[z(Z)]$ in D_Z and $\tilde{u}(Z)=-u[z(\overline{Z})]$ in $\tilde{D}_Z$. Noting Condition C and the condition (3.42), we see that the function $\tilde{f}(Z)=X(Z)g(Z)$ in D_Z and $\tilde{f}(Z)=-\overline{X(\overline{Z})g(\overline{Z})}$ in $\tilde{D}_Z$ satisfies the condition $L_\infty[y^\tau\tilde{f}(Z),D'_Z]\le M_4$, in which $D'_Z=D_Z\cup\tilde{D}_Z\cup\tilde{L}$, $\tau=\max(1-m/2,0)$, $M_4=M_4(\delta,k,H,D,M_3)$ is a positive constant. On the basis of Lemma 2.1, Chapter I, we can verify that the function $\tilde{\Psi}(Z)=2\text{Re}Tf=-\text{Re}\{(2/\pi)\iint_{D_t}[\tilde{f}(t)/(t-Z)]d\sigma_t\}$ satisfies the estimates

$$\begin{gathered}C_\beta[\tilde{\Psi}(Z),\overline{D_Z}]\le M_5,\\ \tilde{\Psi}(Z)-\tilde{\Psi}(t_l)=O(|Z-t_l|^{\beta_l}), 1\le j\le 2N,\end{gathered}\tag{3.43}$$

in which $\tilde{f}(Z)=f(Z)/H(y)$, $\beta=\min(2,m)/(m+2)-\delta=\beta_l\ (1\le l\le 2N)$, δ is a constant as stated in (3.41), and $M_5=M_5(\delta,k,H,D,M_3)$ is a positive constant. On the basis of Theorem 3.2, the solution $X(Z)W(z)$ can be expressed as $X(Z)W(Z)=\tilde{\Phi}(Z)+\tilde{\Psi}(Z)$, where $\tilde{\Phi}(Z)$ is an analytic function in D_Z satisfying the boundary conditions

$$\text{Re}[\overline{\tilde{\lambda}(Z)}\tilde{\Phi}(Z)]=\tilde{R}(Z)-\text{Re}[\overline{\tilde{\lambda}(Z)}\tilde{\Psi}(Z)]=\hat{R}(Z) \text{ on } \Gamma\cup\tilde{L},$$

$$u(a_1)=0,\ \text{Im}[\overline{\lambda(z'_l)}W(z'_l)]=0,\ l=2,...,N,$$

taking the index $\tilde{K}=N/2-1$ into account, the above formula includes $N-1$ point conditions expect the first one. It is easy to see that $\text{Im}\tilde{\Psi}(Z)=0$, $\text{Im}X(Z)W(Z)=\text{Im}\tilde{\Phi}(Z)$ in D_Z, $\text{Im}X(x)W(x)=\text{Im}\tilde{\Phi}(x)=$

$-X(x)\tilde{u}_y(x)/2$ on $\tilde{L}$, there is no harm in assuming that $\tilde{\Psi}(t_l)=0$, otherwise it suffices to replace $\tilde{\Psi}(Z)$ by $\tilde{\Psi}(Z)-\tilde{\Psi}(t_l)\,(1\le l\le 2N)$. For giving the estimates of $\tilde{\Phi}(Z)$ in $D_Z\cap\{\text{dist}(Z,\Gamma)\ge\varepsilon(>0)\}$, from the integral expression of solutions of the discontinuous Riemann-Hilbert problem for analytic functions, we can write the representation of the solution $\tilde{\Phi}(Z)$ of Problem A for analytic functions, namely

$$\tilde{\Phi}[Z(\zeta)]=\frac{X_0(\zeta)}{2\pi i}\left[\int_{\partial D_t}\frac{(t+\zeta)\tilde{\lambda}[Z(t)]\hat{R}[Z(t)]dt}{(t-\zeta)tX_0(t)}+Q(z)\right],$$

$$Q(z)=i\sum_{k=0}^{[\tilde{K}]}(c_k\zeta^k+\overline{c_k}\zeta^{-k})+\begin{cases}0, \text{ when } 2\tilde{K}=N-2 \text{ is even},\\ ic_*\dfrac{\zeta_1+\zeta}{\zeta_1-\zeta}, c_*=i\displaystyle\int_{\partial D_t}\frac{\tilde{\lambda}_n[Z(t)]\hat{R}_n[Z(t)]dt}{X_0(t)t},\\ \text{when } 2\tilde{K}=N-2 \text{ is odd},\end{cases}$$

(see [86]33),[87]1)), where $X_0(\zeta)=\Pi_{j=1}^{2}(\zeta-t_l)^{\tau_l}$, $\tau_l\,(l=1,2,...,2N)$ are as before, $Z=Z(\zeta)$ is the conformal mapping from the unit disk $D_\zeta=\{|\zeta|<1\}$ onto the domain D_Z such that the three points $\zeta=-1,i,1$ are mapped onto $Z=0,Z'(\in\Gamma),2$ respectively, and the $2\tilde{K}+1=N-1$ real constants $c_k(k=0,...,[\tilde{K}])$ are determined by the last $N-1$ points conditions in (3.40). Taking into account

$$|X_0(\zeta)|=O(|\zeta-t_l|^{\tau_l}),|\tilde{\lambda}[Z(\zeta)]\hat{R}[Z(\zeta)]/X_0(\zeta)|=O(|\zeta-t_l|^{\eta_l\pi/2-\tau_l}),$$

and according to the results in [87]1), we see that the function $\tilde{\Phi}(Z)$ determined by the above integral in $D_Z\cap\{\text{dist}(Z,\Gamma)\ge\varepsilon(>0)\}$ is Hölder continuous and $\tilde{\Phi}(t_l)=0$.

For giving the estimates of $X(Z)u_x$, $X(Z)u_x$ in $\tilde{D}_l=D_l\cap D_Z\,(D_l=\{|Z-t_l|<\varepsilon(>0)\},\,1\le l\le 2N)$ separately, when $t_l\in\tilde{L}_k=(a_k,b_k)\,(1\le k\le N)$, denote $X(Z)=\tilde{X}+i\tilde{Y}$ as in (3.38), we first conformally map the domain $D'_Z=D_Z\cup\tilde{D}_Z\cup\tilde{L}$ onto a domain D_ζ, such that $\tilde{L}_k$ is mapped onto himself where D_ζ is a domain with the partial boundary $\Gamma\cup\tilde{\Gamma}$, and $\Gamma\cup\tilde{\Gamma}$ is a smooth curve including the segment line $\text{Re}\,\zeta=t_l$ near $\zeta=t_l\,(0\le l\le 2N)$, through the above mapping, the index $\tilde{K}=N/2-1$ is not changed, and the function $\tilde{\Psi}[Z(\zeta)]$ in the neighborhood $\zeta(D_l)$ of $t_l\,(0\le l\le 2N)$ is Hölder continuous. For convenience denote by D_Z, D_l, $\tilde{W}(Z)$ the domains and function D_ζ, $\zeta(D_l)$, $\tilde{W}[Z(\zeta)]$ again. Secondly reduce the the above boundary condition to this case, i.e. the corresponding function $\tilde{\lambda}(Z)=1$ on $\Gamma\cup\tilde{\Gamma}$ near $Z=t_l\,(0\le l\le 2N)$. In fact there exists an analytic function $\Phi_0(Z)$ in D'_Z satisfying the boundary condition

$$\text{Re}S(Z)=-\arg\tilde{\lambda}(Z) \text{ on } \Gamma\cup\tilde{\Gamma} \text{ near } t_l,\ \text{Im}S(t_l)=0,$$

and the estimate

$$C_\alpha[S(Z), D_l \cap D'_Z] \le M_6 = M_6(\delta, k, H, D, M_3) < \infty, \tag{3.44}$$

then the function $e^{jS(Z)}X(Z)W(Z)$ is satisfied the boundary condition

$$\mathrm{Re}[e^{iS(Z)}X(Z)W(Z)] = 0 \ \text{ on } \tilde{\Gamma} \text{ near } Z = t_l \, (1 \le j \le 2N).$$

Next we symmetrically extend the function $\Phi^*(Z)$ in D'_Z onto the symmetrical domain D^*_Z with respect to $\mathrm{Re}Z = t_l \, (1 \le j \le 2N)$, namely let

$$\hat{W}(Z) = \begin{cases} e^{iS(Z)}X(Z)W(Z) \ \text{in } D'_Z, \\ -\overline{e^{iS(Z^*)}X(Z^*)W(Z^*)} \ \text{in } D^*_Z, \end{cases}$$

where $Z^* = -\overline{(Z - t_l)} + t_l$, later on we shall omit the secondary part $e^{iS(Z)}$ and we can get

$$C_\delta[\tilde{\Phi}(Z), D_\varepsilon] \le M_7, C_\delta[X(Z)u_x, D_\varepsilon] \le M_7, C_\delta[X(Z)u_y, D_\varepsilon] \le M_7,$$

$$C_\delta[u_x, D'_\varepsilon] \le M_8, \ C_\delta[u_y, D'_\varepsilon] \le M_8, \tag{3.45}$$

in which $D_\varepsilon = \overline{D_Z} \cap \{\mathrm{dist}(Z, \tilde{L}) \ge \varepsilon\}, D'_\varepsilon = \overline{D_Z} \cap \{\mathrm{dist}(Z, \Gamma \cup \tilde{\Gamma} \cup T) \ge \varepsilon\}$, ε is arbitrary small positive constant, $M_7 = M_7(\delta, k, H, D_\varepsilon, M_3)$, $M_8 = M_8(\delta, k, H, D'_\varepsilon, M_3)$ are non-negative constants. In fact the first three estimates in (3.45) can be derived by the above integral representation of $\tilde{\Phi}(Z)$. Moreover from (3.22) and (3.42), denote $\tilde{g}(Z) = g(Z)$ in D_Z and $\tilde{g}(Z) = -\overline{g(\overline{Z})}$ in $\tilde{D}_Z$, $W(Z) = \hat{\Phi}(Z) + \hat{\Psi}(Z)$, $\hat{\Psi}(Z) = 2i\mathrm{Im}T\tilde{g}$ in D'_ε is bounded and Hölder continuous, $\hat{\Phi}(Z)$ in D'_Z is an analytic function, hence $\hat{\Phi}(Z) = H(y)\tilde{u}_y/2 = YF$ is a bounded harmonic function in D'_ε, and $u_x = O(Y^{2/(m+2)}F)$, F is is a continuous function. Thus similarly to the proof of Theorem 3.3, Chapter II, we can verify that the last two estimates in (3.45) are true.

After the above discussion, as stated in (3.22), the solution $X(Z)W(z)$ can be also expressed as $X(Z)W(Z) = \Phi(Z) + \Psi(Z) = \hat{\Phi}(Z) + \hat{\Psi}(Z)$, where $X(Z) = \tilde{X} + i\tilde{Y}$, $X(Z)$ is as stated in (3.38), $\Psi(Z), \hat{\Psi}(Z)$ in $\hat{D}_Z = \{D^*_Z \cup D'_Z\} \cap \{Y > 0\}$ are Hölder continuous, $\mathrm{Im}\Psi(Z) = 0$, $\mathrm{Re}\hat{\Psi}(Z) = 0$ in $\hat{D}_Z$, and $\Phi^*(Z) = \Phi(Z)$ or $\Phi^*(Z) = \hat{\Phi}(Z)$ is an analytic function in $\hat{D}_Z$ satisfying the boundary conditions in the form

$$\mathrm{Re}[\overline{\tilde{\lambda}(Z)}\Phi^*(Z)] = \tilde{R}(z) \ \text{on } \Gamma \cup \tilde{L},$$

$$u(a_1) = 0, \ \mathrm{Im}[\overline{\lambda(z'_l)}W(z'_l)] = 0, \, l = 2, ..., N,$$

because in the above case the index of $\tilde{\lambda}(Z)$ on ∂D_Z is $\tilde{K} = N/2 - 1$. Due to the function $\Phi^*(Z)$ in $D_l = \{|Z - t_l| < \varepsilon(> 0)\}$ is analytic, and $\Phi^*(t_l) = 0$, hence $\Phi^*(Z) = O(|Z - t_l|)$, $\Phi^{*\prime}(Z) = O(1)$ near $Z = t_l$, it is clear that $\text{Im}\Phi(Z) = \text{Im}X(Z)W(z)$ and $\text{Re}\hat{\Phi}(Z) = \text{Re}X(Z)W(z)$ extended are harmonic functions in $\hat{D}_Z$, and $\text{Re}\hat{\Phi}(Z)$, $\text{Im}\Phi(Z)$ can be expressed as

$$2\text{Re}\hat{\Phi}(Z) = \sum_{j,k=0}^{\infty} c_{jk}^{(1)} X^j Y^k,\ 2\text{Im}\Phi(Z) = \sum_{j,k=0}^{\infty} c_{jk}^{(2)} X^j Y^k$$

in D_l, herein $X = x - t_l$. Noting that $2\text{Re}\hat{\Phi}(Z) = \tilde{X}H(y)u_x + \tilde{Y}u_y = 0$, $2\text{Im}\Phi(Z) = \tilde{Y}H(y)u_x - \tilde{X}u_y = 0$ at $Z = t_l$ $(1 \le l \le 2)$, we have

$$\begin{aligned} 2\text{Re}\hat{\Phi}(Z) &= \tilde{X}Hu_x + \tilde{Y}u_y = Y\sum_{j,k=0}^{\infty} c_{jk+1}^{(1)} X^j Y^k = YF_1, \\ 2\text{Im}\Phi(Z) &= \tilde{Y}Hu_x - \tilde{X}u_y = \sum_{j,k=0}^{\infty} c_{jk}^{(2)} X^j Y^k = |Z - t_l|F_2 \end{aligned} \tag{3.46}$$

in $\tilde{D}_l$, where F_1, F_2 are continuous functions in D_l. From the system of algebraic equations, we can solve u_x, u_y as follows

$$\begin{aligned} u_x &= (\tilde{X}YF_1 + |Z - t_l|\tilde{Y}F_2)/H|X(Z)|^2, \\ u_y &= H(Y\tilde{Y}F_1 - |Z - t_l|\tilde{X}F_2)/H|X(Z)|^2, \text{ i.e.} \\ X(Z)Hu_x &= (\tilde{X}YF_1 + |Z - t_l|\tilde{Y}F_2)/\overline{X(Z)} = O(|Z - t_l|), \\ X(Z)u_y &= (Y\tilde{Y}F_1 - |Z - t_l|\tilde{X}F_2)/\overline{X(Z)} = O(|Z - t_l|), \text{ i.e.} \\ X(Z)u_x &= O(|Z - t_l|^{\delta}),\ X(Z)u_y = O(|Z - t_l|). \end{aligned}$$

Hence we have

$$C_\delta[X(Z)u_x, \tilde{D}_l] \le M_9,\ C_\delta[X(Z)u_y, \tilde{D}_l] \le M_9,,\ 1 \le l \le 2N, \tag{3.47}$$

where $M_9 = M_9(\delta, k, H, D, M_3)$ is a non-negative constant.

Hence by using the reduction to absurdity, we can prove that the assumption (3.42) is true. Thus the first estimate in (3.41) is derived. The second estimate in (3.41) is easily verified. The other cases can be similarly discussed. This completes the proof. Here we mention that the estimate of solutions of Problem B_1 in the set $\overline{D^+} \cap \{y > \delta(> 0)\}$ is not difficult to verify, and the estimate of solutions in $\overline{D^+} \cap \{0 \le y \le \delta\}$ can be locally used by the method of appropriate conformal mappings in N subsets, this way is seemly simpler than that as stated before.

Now we discuss Problem B_2 for equation (3.32).

Theorem 3.6 *If equation* (3.1) *satisfies Condition* C *and* (2.49) *below, then there exists a solution* $[w(z), v(z)]$ *of Problem* B_2 *for* (3.32), (3.33).

Proof Denote

$$D_0 = \overline{D^-} \cap \{a_l < \delta_0 \le x \le b_l, \delta \le y \le 0, l = 1, ..., N\}, \tag{3.48}$$

and the characteristics of families (3.23): s_1, s_2 emanate from any two points $(a_0, 0), (b_0, 0)(a_l + \delta_0 < a_0 < b_0 < b_l, l = 1, ..., N)$ respectively, which intersect at a point $(x, y) \in \overline{D^-}$, where δ, δ_0 are arbitrary small positive numbers.

We may only discuss the case of $K(y) = -|y|^m h(y)$. In order to find a solution of the system of integral equations (3.22), we need to add the condition

$$\frac{ay}{H(y)} = o(1), \text{ i.e. } \frac{|a|}{H(y)} = \frac{\varepsilon(y)}{|y|}, \; m \ge 2. \tag{3.49}$$

It is clear that for two characteristics $s_1 : x = x_1(y, z_0)$, $s_2 : x = x_2(y, z_0)$ passing through $P_0 = z_0 = x_0 + jy_0 \in D$, we have

$$\begin{gathered} |x_1 - x_2| \le 2\left|\int_0^{y_0} \sqrt{-K}dy\right| \le M|y_0|^{m/2+1}, \\ |y_0|^{1+m/2} \le \frac{k_0(m+2)}{2}|x_1 - x_2|, \end{gathered} \tag{3.50}$$

for any $z_1 = x_1 + jy \in s_1, z_2 = x_2 + jy \in s_2$, in which $M(> \max[2\sqrt{h(y)}, 1])$ is a positive constant. From (3.2), we can assume that the coefficients of (3.22) possess continuously differentiable with respect to $x \in \overline{D^-}$ and satisfy the condition

$$\begin{gathered} |\tilde{A}_l|, |\tilde{A}_{lx}|, |\tilde{B}_l|, |\tilde{B}_{lx}|, |\tilde{D}_l|, |\tilde{D}_{lx}|, |\tilde{E}_l|, |\tilde{E}_{lx}|, \\ |2\sqrt{h}|, |1/\sqrt{h}|, |h_y/h| \le k_0 \le M, z \in \bar{D}, \; l = 1, 2. \end{gathered} \tag{3.51}$$

Firstly, we choose $v_0 = 0, \xi_0 = 0, \eta_0 = 0$ and substitute them into the corresponding positions of v, ξ, η in the right-hand sides of (3.22). By the successive approximation, the sequences of functions $\{v_k\}, \{\xi_k\}, \{\eta_k\}$ are obtained, which satisfy the relations

$$v_{k+1}(z)=v_{k+1}(x)-2\int_0^y V_k(z)dy=v_{k+1}(x)+\int_0^y(\eta_k-\xi_k)dy,$$

$$\xi_{k+1}(z)=\zeta_{k+1}(z)+\int_0^y[\tilde{A}_1\xi_k+\tilde{B}_1\eta_k+\tilde{C}_1(\xi_k+\eta_k)+\tilde{D}_1u_k+\tilde{E}_1]dy,\ z\in s_1,$$

$$\eta_{k+1}(z)=\theta_{k+1}(z)+\int_0^y[\tilde{A}_2\xi_k+\tilde{B}_2\eta_k+\tilde{C}_2(\xi_k+\eta_k)+\tilde{D}_2u_k+\tilde{E}_1]dy,\ z\in s_2,$$

$$k=0,1,2,.... \tag{3.52}$$

Moreover we can prove that the systems of functions $\{v_k\}, \{\xi_k\}, \{\eta_k\}$ in D_0 satisfy the estimates

$$\begin{aligned}
&|v_k(z)-v_k(x)|, |\xi_k(z)-\zeta_k(z)|, |\eta_k(z)-\theta_k(z)| \le M'\gamma^{k-1}|y|^{1-\beta},\\
&|\xi_k(z_1)-\xi_k(z_2)|,|\eta_k(z_1)-\eta_k(z_2)|\le M'\gamma^{k-1}|x_1-x_2|^{\beta}|t|^{\beta'},\\
&|\tilde{\xi}_k(z_1)-\tilde{\xi}_k(z_2)|,|\tilde{\eta}_k(z_1)-\tilde{\eta}_k(z_2)|\le M'\gamma^{k-1}|x_1-x_2|^{\beta}|t|^{\beta'},\\
&|\xi_k(z)+\eta_k(z)-\zeta_k(z)-\theta_k(z)|\le M'\gamma^{k-1}|x_1-x_2|^{\beta}|y|^{\beta'},\\
&|\tilde{\xi}_k(z)+\tilde{\eta}_k(z)| \le M'\gamma^{k-1}|x_1-x_2|^{\beta}|y|^{\beta'}, 0\le|y|\le\delta,
\end{aligned} \tag{3.53}$$

where $z=x+jy$, $z=x+jt$ is the intersection point of s_1, s_2 passing through z_1, z_2, $\beta'=(1+m/2)(1-3\beta)$, δ, β are sufficiently small positive constants, $\gamma(<1)$ is a positive constant, and M' is a sufficiently large positive constant as stated in (2.44), Chapter V.

The formula (3.53) shows that these sequences of functions $\{v_k(z)\}$, $\{\xi_k(z)\}$, $\{\eta_k(z)\}$ in

$$\begin{aligned}
D_l=\overline{D^-}\cap\{|z-a_1|\ge1/l\}\cap\cdots\cap\{|z-a_N|\ge1/l\}\\
\cap\{|z-b_1|\ge1/l\}\cap\cdots\cap\{|z-b_N|\ge1/l\}\ (l>2)
\end{aligned}$$

are uniformly bounded and equicontinuous. In particular for $\delta_0=1/l$, l is a positive integer, from these sequences, we can choose the subsequences $\{v_k^l(z)\}$, $\{\xi_k^l(z)\}$, $\{\eta_k^l(z)\}$, which uniformly converge to $v_*(z), \xi_*(z)$, $\eta_*(z)$ in D_l respectively, and $v_*(z), \xi_*(z), \eta_*(z)$ satisfy the system of integral equations

$$
\begin{aligned}
&v_*(z)=v_*(x)-2\int_0^y V_*dy=u_*(x)+\int_0^y(\eta_*-\xi_*)dy,\\
&\xi_*(z)=\zeta_*(z)+\int_0^y[\tilde{A}_1\xi_*+\tilde{B}_1\eta_*+\tilde{C}_1(\xi_*+\eta_*)+\tilde{D}_1u_*+\tilde{E}_1]dy, z\in s_1,\\
&\eta_*(z)=\theta_*(z)+\int_0^y[\tilde{A}_2\xi_*+\tilde{B}_2\eta_*+\tilde{C}_2(\xi_*+\eta_*)+\tilde{D}_2u_*+\tilde{E}_2]dy, z\in s_2,
\end{aligned}\tag{3.54}
$$

and the function $v(z) = v_*(z)$ is a solution of Problem B_2 for (3.33), (3.34) in D_l. Moreover from $\{v_k^l(z)\}, \{\xi_k^l(z)\}, \{\eta_k^l(z)\}$ $(l = 1, 2...)$, we can select the diagonal sequence $\{v_l^l(z)\}, \{\xi_l^l(z)\}, \{\eta_l^l(z)\}$, which uniformly converge to $v_*(z), \xi_*(z), \eta_*(z)$ in D_0 respectively. Hence the function $u(z) = v(z)+u_0(z)$ is a solution of Problem T (or TB) for (3.1) in $D\cap\{-\delta < y \le 0\}$. Thus by the result in Section 1, the existence of solutions of Problem Q for equation (3.1) with $c = 0$ in D is proved. From the above discussion, we can see that the solution of Problem Q for (3.1) with $c = 0$ in D is unique.

From the above result, we have the following theorem.

Theorem 3.7 *Let equation* (3.1) *satisfy Condition C and* (3.49). *Then the Tricomi problem* (*Problem Q*) *or Tricomi-Bers Problem* (*Problem TB'*) *for* (3.1) *with $c = 0$ has a solution.*

Finally, we prove the following theorem.

Theorem 3.8 *Let equation* (3.1) *satisfy Condition C and* (3.49). *Then the Tricomi problem* (*Problem T*) *or Tricomi-Bers problem* (*Problem TB*) *for* (3.1) *has a solution. Especially Problem T* (*or TB*) *for the Chaplygin equation $K(y)u_{xx} + u_{yy} = 0$ is uniquely solvable.*

Proof From Theorem 3.7, we see that Problem Q for (3.1) has a solution $u^*(z)$ in D, if $u^*(a_l) = d_l, l = 1, ..., N$, then the solution $u^*(z)$ is just a solution of Problem P for (3.1). Otherwise, $[u^*(a_2), ..., u^*(a_N)] = [d_2^*, ..., d_N^*]$, we find $N-1$ solutions $u_2(z), ..., u_N(z)$ of Problem Q for the homogeneous linear equation

$$K(y)u_{xx} + u_{yy} + au_x + bu_y + cu = 0\ ,\ \text{in}\ D \tag{3.55}$$

with the boundary conditions

$$
\begin{aligned}
&\mathrm{Re}[\overline{\lambda(z)}u_{k\bar{z}}] = 0,\ z\in\Gamma\cup L',\ u(a_1) = 0,\\
&\mathrm{Im}[\overline{\lambda(z)}u_{k\bar{\bar{z}}}]|_{z=z_l}=0, \mathrm{Im}[\overline{\lambda(z)}u_{k\bar{\bar{z}}}]|_{z=z_l'}=\delta_{lk}, l, k=2, ..., N.
\end{aligned}\tag{3.56}
$$

It is obvious that $U(z) = \sum_{k=2}^N u_k(z) \not\equiv 0$ in $\overline{D}$, moreover we can verify that

$$J=\begin{vmatrix} u_1(a_2) & \dots & u_N(a_2) \\ \vdots & \ddots & \vdots \\ u_1(a_N) & \dots & u_N(a_N) \end{vmatrix} \neq 0,$$

where $a_2,...,a_N$ are as stated in (3.3), thus there exist $N-1$ real constants $C_2,...,C_N$, which are not equal to zero, such that

$$C_2u_2(a_k)+...+C_Nu_N(a_k)=d_k^*-d_k,\ k=2,...,N,$$

thus the function

$$u(z)=u^*(z)-\sum_{k=2}^{N}C_ku_k(z) \text{ in } \bar{D} \tag{3.57}$$

is just a solution of Problem T for the linear equation (3.1) with $c=0$. Moreover by using the method of parameter extension, the Schauder fixed-point theorem or the Leray-Schauder theorem, we can prove the solvability of Problem T for quasilinear equation (3.1).

In addition we can also generalize the above results to the general multiply connected domain similar to Theorems 2.5 and 2.6 in Section 2 by the similar method.

From the above discussion, we see that the open problem about Tricomi problem of Chaplygin equation in multiply connected domains posed by L. Bers in [9]1) is solved.

Remark 3.1 For the oblique derivative problem (Problem P or GTB) for equation (3.1), by the similar method, we can obtain the above corresponding results as stated in Theorems 3.1-3.8, but we require the index $K=N-1$ of Problem P in D^+, in this case there are $2N$ point conditions in boundary condition (3.3). As stated as in Section 5, Chapter II, there are in no harm in assuming that the elliptic domain D^+ is bounded the boundary $\Gamma\cup\tilde{L}$, $\tilde{L}=\cup_{j=1}^N\tilde{L}_l$, $\tilde{L}_l=(a_l,b_l)\,(l=1,...,N)$ are as stated in Subsection 3.1 and $\Gamma=\cup_{l=1}^N\Gamma_l^0$, herein $\Gamma_l^0\,(\in C_\mu^2,\ 0<\mu<1,\ l=1,...,N)$ are curves in the upper-half plane $\mathrm{Im}z>0$ with the end points at $a_l,b_l\,(l=1...,N)$ respectively, and the size of the inner angles of D^+ at $a_l,b_l\,(l=1,...,N)$ are equal to $\pi/2\,(l=1,..,2N)$.

4 The Oblique Derivative Problem for Equations of Mixed Type with Nonsmooth Degenerate Line

In [71]1) and [72], the authors discussed some boundary value problems of second order equations of mixed type with non-smooth degenerate line, but they only consider some special mixed equations. The present section deals with oblique derivative problem for general second order equations of mixed type with nonsmooth parabolic degenerate line. We first give the formulation and estimates of solutions of the problem for the equations, and then prove the existence of solutions for the above problem. Besides we also discuss the Tricomi problem for some nonlinear mixed equation of second order with parabolic degeneracy.

4.1 Formulation of oblique derivative problem for mixed equations with nonsmooth degeneracy

Let D be a simply connected bounded domain in the complex plane $\mathbf{C}$ with the boundary $\partial D = \Gamma\cup L$, in which $\Gamma\,(\subset \{y>0\}) \in C^2_\mu\,(0<\mu<1)$ is a curve with the end points $z=-1,1$, and $L_0=(-1,1)$ on $x-$axis, and $L=L_1\cup L_2\cup L_3\cup L_4$ are four characteristics with the slopes $-H_2(x)/H_1(y), H_2(x)/H_1(y), -H_2(x)/H_1(y), H_2(x)/H_1(y)$ passing through the points $z=x+iy=-1,0,0,1$ respectively as follows

$$
\begin{aligned}
&L_1=\{G_1(y)=\int_0^y H_1(t)dt=\int_x^{-1} H_2(t)dt=G_2(-1)-G_2(x), x\in(-1,x_1)\},\\
&L_2=\{G_1(y)=\int_0^y H_1(t)dt=\int_0^x H_2(t)dt = G_2(x), x\in(x_1,0)\},\\
&L_3=\{-G_1(y)=-\int_0^y H_1(t)dt=\int_0^x H_2(t)dt=G_2(x), x\in(0,x_2)\},\\
&L_4=\{-G_1(y)=-\int_0^y H_1(t)dt=\int_x^1 H_2(t)dt=G_2(1)-G_2(x), x\in(x_2,1)\},
\end{aligned}
$$

in which $H_1(y)=\sqrt{|K_1(y)|}$, $K_1(0)=0$, $H_2(x)=\sqrt{|K_2(x)|}$, $K_2(0)=0$, $K_1(y)=\text{sgn}y|y|_1^m h_1(y)$, $K_2(x)=\text{sgn}x|x|^{m_2}h_2(x)$, m_1,m_2 are positive constants, $h_1(y),h_2(x)$ in $\overline{D}$ are continuously differentiable positive functions. Similarly to Section 3, there is no harm in assuming that the boundary Γ of the domain D is a smooth curve, which possesses the form $\tilde{G}_2(x)-\tilde{G}_1(y)=\tilde{G}_2(-1)=h_1$ and $\tilde{G}_2(x)+\tilde{G}_1(y)=\tilde{G}_2(1)=h_2$ including the line segments $\text{Re}z=\pm1$ near the points $z=\pm1$ respectively and including a line segment $\mathrm{Im}z = y_0$ near the intersection point iy_0 of Γ and $\mathrm{Re}z = 0$, here $\tilde{G}_1(y), \tilde{G}_2(x)$ are the similar to that in Section 2, Chapter II, otherwise by using an appropriate conformal mapping, the requirement can be realized. We consider the quasilinear mixed equation with non-smooth degenerate line

$$K_1(y)u_{xx} + |K_2(x)|u_{yy} + au_x + bu_y + cu + d = 0 \text{ in } \overline{D}, \tag{4.1}$$

and its complex form is as in (4.16), (4.18) below, i.e.

$$W_{\overline{\overline{z}}}=A_1(z,u,W)W+A_2(z,u,W)\overline{W}+A_3(z,u,W)u+A_4(z,u,W) \text{ in } \overline{D}, \tag{4.2}$$

where $W(z) = [H_1(y)u_x - iH_2(x)u_y]/2 = U + iV$ in D^+, and $W(z) = [H_1(y)u_x - jH_2(x)u_y]/2 = U + jV = \xi e_1 + \eta e_2$ in D^-. Suppose that the coefficients of (4.2) are measurable in D^+ and continuous in $\overline{D^-}$ for any continuously differentiable function $u(z)$, and satisfy **Condition** C, namely

$$\begin{gathered} L_\infty[\eta, \overline{D^+}] \le k_0,\ \eta = a, b, c,\ L_\infty[d, \overline{D^+}] \le k_1,\ c \le 0 \text{ in } \overline{D^+}, \\ \tilde{C}[d, \overline{D^-}] = C[d, \overline{D^-}] + C[d_x, \overline{D^-}] \le k_1, \hat{C}[\eta, \overline{D^-}] \le k_0, \eta = a, b, c, \\ \eta|x|^{-m_2/2} = O(1) \text{ as } z = x + iy \to 0,\ \eta = a, b, c, d, \end{gathered} \tag{4.3}$$

and for any two continuously differentiable functions $u_1(z), u_2(z)$ in $D^* = \overline{D}\backslash(\{-1, 0, 1\}$, $F(z, u, u_z) = au_x + bu_y + cu + d$ satisfies the condition

$$F(z,u_1,u_{1z})-F(z,u_2,u_{2z})=\tilde{a}(u_1-u_2)_x+\tilde{b}(u_1-u_2)_y+\tilde{c}(u_1-u_2), \tag{4.4}$$

and the coefficients $\tilde{a}, \tilde{b}, \tilde{c}$ satisfy the same conditions as those of a, b, c in (4.3) respectively, where k_0, k_1 are non-negative constants. We can state the equivalent conditions about equation (4.2), i.e. $A_l(z, u, u_z)\,(l = 1, 2, 3)$ are measurable in $z \in D^+$ and continuous in $\overline{D^-}$ for all continuously differentiable functions $u(z)$ in D^* and satisfy

$$\begin{gathered} L_\infty[H_1\mathrm{Re}(A_1+A_2), D^+], L_\infty[x\mathrm{Re}(A_1-A_2), D^+], L_\infty[A_3, D^+], \\ L_\infty[y\mathrm{Im}(A_1+A_2), D^+], L_\infty[H_2\mathrm{Im}(A_1-A_2), D^+] \le k_0,\ A_3 \ge 0 \text{ in } D^+, \\ L_\infty[A_4, \overline{D^+}] \le k_1,\ \hat{C}[H_1(A_1+A_2), \overline{D^-}] = C[H_1(A_1+A_2), \overline{D^-}] \\ + C[H_1(A_1+A_2)_x, \overline{D^-}], \hat{C}[x\mathrm{Re}(A_1-A_2), \overline{D^-}], \hat{C}[y\mathrm{Im}(A_1+A_2), \overline{D^-}], \\ \hat{C}[H_2\mathrm{Im}(A_1-A_2), \overline{D^-}] \le k_0,\ \hat{C}[A_3, \overline{D^-}] \le k_0,\ \hat{C}[A_4, \overline{D^-}] \le k_1, \end{gathered} \tag{4.5}$$

and for any two continuously differentiable functions $u_1(z), u_2(z)$ in D^*, $G(z,u,u_{\tilde{z}})=A_1u_{\tilde{z}}+A_2u_{\overline{\tilde{z}}}+A_3u+A_4$ satisfies the condition

$$G(z,u_1,u_{1\tilde{z}})-G(z,u_2,u_{2\tilde{z}})=\tilde{A}_1(u_1-u_2)_{\tilde{z}}$$
$$+\tilde{A}_2(u_1-u_2)_{\overline{\tilde{z}}}+\tilde{A}_2(u_1-u_2) \text{ in } D, \tag{4.6}$$

in which $\tilde{A}_l=\tilde{A}_l(z,u_1,u_2)$ $(l=1,2,3)$ satisfy the conditions as those of $A_l(l=1,2,3,4)$ in (4.5) respectively, where $k_0(\geq \max_{l=1,2}[2\sqrt{h_l}, 1/\sqrt{h_l}])$, $k_1(\geq \max[1,12k_0])$ are positive constants.

If $H_1(y)=[|y|^{m_l}h_1(y)]^{1/2}$, $H_2(x)=[|x|^{m_2}h_2(x)]^{1/2}$, here m_1,m_2 are positive numbers, then

$$Y=G_1(y)=\int_0^y H_1(t)dt,\ |Y|\leq \frac{k_0}{m_1+2}|y|^{(m_1+2)/2},$$
$$X=G_2(x)=\int_0^x H_2(t)dt,|X|\leq \frac{k_0}{m_2+2}|x|^{(m_2+2)/2} \text{ in } \overline{D}, \tag{4.7}$$

and their inverse functions $y=\pm|G_1^{-1}(Y)|, y=\pm|G_2^{-1}(X)|$ satisfy the inequalities

$$|y|=|G_1^{-1}(Y)|\leq[k_0(m_1+2)|Y|]^{2/(m_1+2)}=J_1|Y|^{2/(m_1+2)},$$
$$|x|=|G_2^{-1}(X)|\leq[k_0(m_2+2)|X|]^{2/(m_2+2)}=J_2|X|^{2/(m_2+2)}. \tag{4.8}$$

The oblique derivative boundary value problem or general Tricomi-Rassias problem for equation (4.1) may be formulated as follows:

Problem *P* or *GTR* Find a continuous solution $u(z)$ of (4.1) in $\overline{D}$, where u_x, u_y are continuous in D^*, and satisfy the boundary conditions

$$\frac{1}{2}\frac{\partial u}{\partial\nu}=\frac{1}{H(x,y)}\text{Re}[\overline{\lambda(z)}u_{\tilde{z}}]=\text{Re}[\overline{\Lambda(z)}u_z]=r(z) \text{ on } \Gamma\cup L_1\cup L_4,$$
$$\frac{1}{H_1(y)}\text{Im}[\overline{\lambda(z)}u_{\tilde{z}}]|_{z=z_l}=\text{Im}[\overline{\Lambda(z)}u_z]|_{z=z_l}=b_l,\ l=1,2, \tag{4.9}$$
$$u(-1)=b_0,\ u(1)=b_3 \text{ or } \frac{1}{H_1(y)}\text{Im}[\overline{\lambda(z)}u_{\tilde{z}}]|_{z=z_3}=b_3,$$

in which ν is a given vector at every point $z\in\Gamma\cup L_1\cup L_4$, $u_{\tilde{z}}=[H_1(y)u_x-iH_2(x)u_y]/2$, $\Lambda(z)=\cos(\nu,x)-i\cos(\nu,y)$, $H(x,y)=H_1(y)$ or $H_2(x)$, and $\lambda(z)=\text{Re}\lambda(z)+i\text{Im}\lambda(z)$, if $z\in\Gamma\cup L_1\cup L_4$, $z_1=x_1-j|(-G_1)^{-1}(x_1)|$, $z_2=x_2-j|(-G_1)^{-1}(x_2)|$ are the intersection points of

L_1, L_2 and L_3, L_4 respectively, $z_3\,(\neq -1, 1)$ is a point of Γ, $b_l(l=0,1,2,3)$ are real constants, and $r(z), b_l(l=0,1,2,3)$ satisfy the conditions

$$C^1_\alpha[\lambda(z),\Gamma]\le k_0, C^1_\alpha[\lambda(x),L_1\cup L_4]\le k_0, C^1_\alpha[r(z),\Gamma]\le k_2,$$
$$C^1_\alpha[r(x),L_1\cup L_4]\le k_2,\ \cos(\nu,n)\ge 0 \text{ on } \Gamma\cup L_1\cup L_4,$$
$$|b_l|\le k_2, l=0,1,2,3, \max_{z\in L_1}\frac{1}{|a(z)-b(z)|}\le k_0, \max_{z\in L_4}\frac{1}{|a(z)+b(z)|}\le k_0, \tag{4.10}$$

in which $L_0 = L_0'\cup L_0''$, $L_0' = \{-1<x<0, y=0\}$, $L_0'' = \{0<x<1, y=0\}$, n is the outward normal vector at every point on Γ, $\alpha\,(0<\alpha<1), k_0, k_2$ are non-negative constants. For the last point condition in (4.9), we need to assume $c=0$ in (4.1). The number

$$K=\frac{1}{2}(K_1+K_2+K_3) \tag{4.11}$$

is called the index of Problem P, where

$$K_l=\left[\frac{\phi_l}{\pi}\right]+J_l, J_l=0 \text{ or } 1, e^{i\phi_l}=\frac{\lambda(t_l-0)}{\lambda(t_l+0)}, \gamma_l=\frac{\phi_l}{\pi}-K_l, l=1,2,3, \tag{4.12}$$

in which $t_1=-1, t_2=1, t_3=0$. Here $K=0$ on the boundary ∂D^+ of D^+ can be chosen, and we have the point condition $u(1)=b_3$.

It is clear that the Tricomi problem (Problem T) or Tricomi-Rassisa problem (Problem TR) with the boundary condition

$$u(z)=\phi(z) \text{ on } \Gamma,\ u_y(x)=r(x) \text{ on } L_0=L_0'\cup L_0'', \tag{4.13}$$

is a special case of Problem P or GTR, where

$$C^2_\alpha[\phi(z),\Gamma]\le k_2,\ C^1_\alpha[r(x),L_0]\le k_2,$$

where k_2 is a non-negative constant.

In [71]1), J. M. Rassian proposed and verified the uniqueness of solutions of the Tricomi problem for Chaplygin equation with non-smooth degenerate lines in several domains. Later on the solvability of the above problem for general equations of mixed type will be proved. We first find the derivative for (4.13) according to the parameter $s=\operatorname{Im} z=y$ on $\Gamma\cup L_0$, and obtain

$$
\begin{aligned}
&u_s = u_x x_y + u_y = \phi'(y), \text{ i.e. } \tilde{H}(y)H_1(y)u_x/H_1(y)\\
&+H_2(x)u_y/H_2(x) = \phi'(y) \text{ on } \Gamma \text{ near } x=-1,\\
&u_s = u_x x_y + u_y = \phi'(y), \text{ i.e. } \tilde{H}(y)H_1(y)u_x/H_1(y) \qquad (4.14)\\
&-H_2(x)u_y/H_2(x) = -\phi'(y) \text{ on } \Gamma \text{ near } x=1,\\
&H_2(x)u_y(x) = H_2(x)r(x) \text{ on } L_0 = L_0' \cup L_0'',
\end{aligned}
$$

in which $\tilde{H}(y) = \tilde{G}'(y) = [\tilde{G}_1(y)/\tilde{G}_2(x)]'$. It is not difficult to see the complex form of (4.14) as follows

$$
\begin{aligned}
\operatorname{Re}[\overline{\lambda(z)}(U+iV)] &= \operatorname{Re}[\overline{\lambda(z)}(H_1(y)u_x - iH_2(x)u_y)]/2\\
&= R(x) \text{ on } \Gamma \cup L_1 \cup L_4,
\end{aligned}
$$

where

$$
\lambda(z) = \begin{cases} -i, \\ i, \\ i, \end{cases} \quad R(x) = \begin{cases} H_2(x)\phi'(y)/2 \text{ on } \Gamma \text{ at } z=-1, \\ -H_2(x)\phi'(y)/2 \text{ on } \Gamma \text{ at } z=1, \\ -H_2(x)r(x)/2 \text{ on } L_0. \end{cases}
$$

Similarly to Section 2, We have

$$
\begin{aligned}
e^{i\phi_1} &= \frac{\lambda(t_1-0)}{\lambda(t_1+0)} = e^{-\pi i/2-\pi i/2} = e^{-\pi i}, \gamma_1 = -1-K_1 = 0,\ K_1 = -1,\\
e^{i\phi_2} &= \frac{\lambda(t_2-0)}{\lambda(t_2+0)} = e^{\pi i/2-\pi i/2} = e^{0\pi i},\ \gamma_2 = 0-K_2 = 0,\ K_2 = 0,\\
e^{i\phi_3} &= \frac{\lambda(t_3-0)}{\lambda(t_3+0)} = e^{\pi i/2-\pi i/2} = e^{0\pi i},\ \gamma_3 = 0-K_3 = 0, K_3 = 0,
\end{aligned}
$$

hence the index of $\lambda(z)$ on $\Gamma \cup L_0$ is

$$
K = \frac{1}{2}(K_1+K_2+K_3) = -\frac{1}{2},
$$

in this case we can add the point condition $u(0) = 0$. If we choose $\operatorname{Re}[\overline{\lambda(x)}W(x)] = 0$, $\lambda(x) = 1$ on L_0, then $\gamma_1 = \gamma_2 = -1/2, \gamma_3 = 0$, $K_1 = K_2 = K_3 = 0$, thus the index $K = 0$, in this case we can add the point condition $u(1) = b_3$ or $u(0) = 0$.

4.2 Representation of solutions of oblique derivative problem for mixed equations

In this section, we first write the complex form of equation (4.1). From (3.13), Chapter II, we have

$$\begin{aligned} W(z) &= U+iV=\frac{1}{2}[H_1(y)u_x-iH_2(x)u_y]\\ &=\frac{H_1(y)H_2(x)}{2}[u_X-iu_Y]=H_1(y)H_2(x)u_Z=u_{\tilde z},\\ &H_1(y)H_2(x)W_{\overline{Z}}=W_{\overline{\tilde z}}=\frac{H_1(y)H_2(x)}{2}[W_X+iW_Y]\\ &=A_1W+A_2\overline{W}+A_3u+A_4=g(Z)\ \text{in}\ D_Z^+, \end{aligned} \tag{4.15}$$

in which D_Z^+ is the image domain of D^+ with respect to the mapping $Z=Z(z)=X+iY$, and

$$\begin{aligned} A_1&=\frac{iH_2H_{1y}}{4H_1}+\frac{H_1H_{2x}}{4H_2}-\frac{a}{4H_1}-\frac{ib}{4H_2},\ A_3=\frac{-c}{4},\\ A_2&=\frac{iH_2H_{1y}}{4H_1}-\frac{H_1H_{2x}}{4H_2}-\frac{a}{4H_1}+\frac{ib}{4H_2},\ A_4=\frac{-d}{4}. \end{aligned} \tag{4.16}$$

Moreover we can obtain

$$\begin{aligned} W(z)&=U+jV=\frac{1}{2}[H_1(y)u_x-jH_2(x)u_y]\\ &=\frac{H_1(y)H_2(x)}{2}[u_X-ju_Y]=H_1(y)H_2(x)u_Z=u_{\tilde z},\\ &H_1(y)H_2(x)W_{\overline{Z}}=\frac{H_1(y)H_2(x)}{2}[W_X+jW_Y]\\ &=\frac{1}{2}[H_1(y)W_x+jH_2(x)W_y]=W_{\overline{\tilde z}}\ \text{in}\ \overline{D^-}, \end{aligned} \tag{4.17}$$

and

$$\begin{aligned} &-K_1(y)u_{xx}-|K_2(x)|u_{yy}=H_1(y)[H_1(y)u_x-jH_2(x)u_y]_x\\ &+jH_2(x)[H_1(y)u_x-jH_2(x)u_y]_y-jH_2(x)H_{1y}u_x+jH_1(y)H_{2x}u_y\\ &=2\{H_1[U+jV]_x+jH_2[U+jV]_y\}-j[H_2H_{1y}/H_1]H_1u_x\\ &+j[H_1H_{2x}/H_2]H_2u_y=4H_1(y)H_2(x)W_{\overline{Z}}-j[H_2H_{1y}/H_1]H_1u_x \end{aligned}$$

$$
\begin{aligned}
&+j[H_1H_{2x}/H_2]H_2u_y = au_x+bu_y+cu+d, \text{ i.e. } H_1(y)H_2(x)W_{\overline{Z}}\\
&= H_1H_2[W_X+jW_Y]/2 = H_1H_2\{(U+V)_\mu e_1+(U-V)_\nu e_2\}\\
&= \{2j[H_2H_{1y}/H_1]U + 2j[H_1H_{2x}/H_2]V + au_x+bu_y+cu+d\}/4\\
&= \{[jH_2H_{1y}/H_1+a/H_1](W+\overline{W})+[H_1H_{2x}/H_2-jb/H_2](W-\overline{W})\\
&+cu+d\}/4 = \{[jH_2H_{1y}/H_1+a/H_1+H_1H_{2x}/H_2-jb/H_2]W\\
&+[jH_2H_{1y}/H_1+a/H_1-H_1H_{2x}/H_2+jb/H_2]\overline{W}\\
&+cu+d\}/4 = \{[a/H_1+H_1H_{2x}/H_2+H_2H_{1y}/H_1-b/H_2](U+V)\\
&+[a/H_1-H_1H_{2x}/H_2+H_2H_{1y}/H_1+b/H_2](U+V)+cu+d\}e_1/4\\
&+\{[a/H_1-H_1H_{2x}/H_2-H_2H_{1y}/H_1-b/H_2](U+V)\\
&+[a/H_1+H_1H_{2x}/H_2-H_2H_{1y}/H_1+b/H_2](U-V)+cu+d\}e_2/4, \text{ i.e.}\\
&(U+V)_\mu e_1+(U-V)]_\nu e_2 = [\hat{A}_1(U+V)+\hat{B}_1(U-V)+\hat{C}_1u+\hat{D}_1]e_1\\
&+[\hat{A}_2(z)(U+V)+\hat{B}_2(U-V)+\hat{C}_2u+\hat{D}_2]e_2 \text{ in } D_\tau^-,
\end{aligned}
\tag{4.18}
$$

in which $e_1=(1+j)/2, e_2=(1-j)/2$, D_Z^-, D_τ^- are the image domains of D^- with respect to the mapping $Z = G_2(x)+jG_1(y) = Z(z)$, $\tau = \mu+j\nu = \tau(z)$ respectively, and the coefficients

$$
\begin{aligned}
\hat{A}_1 &= \frac{1}{4H_1H_2}[\frac{a}{H_1}+\frac{H_1H_{2x}}{H_2}+\frac{H_2H_{1y}}{H_1}-\frac{b}{H_2}],\\
\hat{B}_1 &= \frac{1}{4H_1H_2}[\frac{a}{H_1}-\frac{H_1H_{2x}}{H_2}+\frac{H_2H_{1y}}{H_1}+\frac{b}{H_2}],\\
\hat{A}_2 &= \frac{1}{4H_1H_2}[\frac{a}{H_1}-\frac{H_1H_{2x}}{H_2}-\frac{H_2H_{1y}}{H_1}-\frac{b}{H_2}],\\
\hat{B}_2 &= \frac{1}{4H_1H_2}[\frac{a}{H_1}+\frac{H_1H_{2x}}{H_2}-\frac{H_2H_{1y}}{H_1}+\frac{b}{H_2}],\\
\hat{C}_1 &= \hat{C}_2 = \frac{c}{4H_1H_2},\ \hat{D}_1 = \hat{D}_2 = \frac{d}{4H_1H_2}.
\end{aligned}
\tag{4.19}
$$

It is clear that a special case of equations (4.16), (4.18) is the complex equation

$$W_{\overline{Z}} = 0 \text{ in } \overline{D_Z}, \tag{4.20}$$

which can be rewritten in the form

$$\begin{aligned} &[(U+V)+i(U-V)]_{\mu-i\nu} = 0 \text{ in } D_\tau^+,\\ &(U+V)_\mu = 0,\ (U-V)_\nu = 0 \text{ in } \overline{D_\tau^-}, \end{aligned} \tag{4.21}$$

and the solution $(U+V)+i(U-V)$ of the first equation in (4.21) is an analytic function in the corresponding domain D_τ^+. The boundary value problem for equations (4.16), (4.18) with the boundary condition (4.9) and the relation: the first formula in (4.25) below will be called Problem A.

Similarly to the proof of Theorem 2.1, we can verify that there exists a solution of the Riemann-Hilbert problem (Problem A) for equation (4.20) in $\overline{D}$ with the boundary conditions

$$\begin{aligned} &\text{Re}[\overline{\lambda(z)}(U+jV)] = H_1(y)r(z) = R(z) \text{ on } \Gamma, u(-1) = b_0,\\ &\text{Re}[\overline{\lambda(z)}(U+iV] = H_1(y)r(x) = R(x) \text{ on } L_1 \cup L_4,\\ &\text{Im}[\overline{\lambda(z)}(U+iV]_{z=z_l} = H_1(y)|_{z=z_l} b_l = b_l',\ l=1,2,\\ &u(1) = b_3 \text{ or } \text{Im}[\overline{\lambda(z)}u_{\bar z}]|_{z=z_3} = H_1(\text{Im}z_3)b_3 = b_3', \end{aligned} \tag{4.22}$$

in which $\lambda(z) = a(z) + jb(z)$ on $\Gamma \cup L_0$.

Theorem 4.1 *Problem A of equation (4.20) or system (4.21) in $\overline{D}$ has a unique solution $W(z)$, which in D^- possesses the form*

$$W(z) = \begin{cases} \frac{1}{2}\{\frac{(1+j)(2R((X-Y+h_1)/2)-M(X,Y))}{a((X-Y+h_1)/2)-b((X-Y+h_1)/2)}+(1-j)R_0(X+Y)\},\\ M(X,Y)=[a((X-Y+h_1)/2)+b((X-Y+h_1)/2)]g(h_1) \text{ in } D_1^-,\\ \frac{1}{2}\{\frac{(1-j)(2R((X+Y+h_2)/2)-N(X,Y))}{a((X+Y+h_2)/2)+b((X+Y+h_2)/2)}+(1+j)R_0(X-Y)\},\\ N(X,Y)=[a((X+Y+h_2)/2)-b((X+Y+h_2)/2)]f(h_2) \text{ in } D_2^-,\\ [a(X_1)+b(X_1)]g(h_1) = H_1(y_1)r(Y_1) - b_1' \text{ or } 0,\\ [a(X_2)-b(X_2)]f(h_2) = H_1(y_2)r(Y_2) + b_2' \text{ or } 0, \end{cases} \tag{4.23}$$

where $h_1=G_2(-1), h_2=G_2(1), X=G_2(x), Y=G_1(y)$, $R_0(X)$ *on* $[h_1,h_2]$ *is an undetermined function, and the solution* $W(z)$ *satisfies the estimates*

$$C_\delta[u(z),\overline{D^-}]+C^1_\delta[u(z),\overline{D^-_\varepsilon}]\le M_1,$$

$$C_1[f(x),L_\varepsilon]+C_1[g(x),L_\varepsilon]\le M_2,$$

in which $\mu=X+G_1(y)$, $\nu=X-G_1(y)$, $f(\nu)=U(z)+V(z)$, $g(\mu)=U(z)-V(z), L'_0=(-1,0), L''_0=(0,1)$, $u(z)$ *is the corresponding function determined by the first formula in* (4.25) *below, where the function* $W(z)$ *is as stated in* (4.23),

$$D^-_\varepsilon=\overline{D^-}\cap\{|z|>\varepsilon\}\cap\{|z-1|>\varepsilon\}\cap\{|z-1|>\varepsilon(>0)\},$$

$$L_\varepsilon=\{L'_0\cup L''_0\}\cap D^-_\varepsilon,D^-_1=\overline{D^-}\cap\{x<0\},D^-_2=D^-\cap\{x>0\},$$

and $M_1=M_1(\delta,k_0,k_1,D^-_\varepsilon)$, $M_2=M_2(k_0,k_1,L_\varepsilon)$ *are non-negative constants.*

Proof From the boundary conditions in (4.22) on L_1 and L_4, denote by the functions $a(X)$, $b(X)$, $R(X)$ of X the functions $a(z)$, $b(z)$, $R(z)$ of z, we have

$$\begin{aligned}&a(X)U(x,y)-b(X)V(x,y)=R(X)\ \text{ on }\ L_1\cup L_4,\ \text{ i.e.}\\&[a(X)-b(X)]f(X-G_1(y))+[a(X)+b(X)]g(X+G_1(y))\\&=2R(X)\ \text{ on }\ L_1\cup L_4,\\&U(x)-V(x)=R_0(x)\ \text{ for }\ x=G_2^{-1}(X)\in(-1,0),\\&U(x)+V(x)=R_0(x)\ \text{ for }\ x=G_2^{-1}(X)\in(0,1),\end{aligned}$$

where $h_1=G_2(-1)$, $h_2=G_2(1)$, $G_2^{-1}(X)$ is the inverse function of $x=G_2(X)$, $R_0(X)$ on L_0 is an undetermined real function. Similarly to the proof of Theorem 2.1, we can obtain the above results.

In particular, when $G_1(y)=0$, the formula (4.23) becomes

$$W(x)=\begin{cases}\dfrac{1}{2}\{\dfrac{(1+j)(2R((X+h_1)/2)-M(X,0))}{a((X+h_1)/2)-b((X+h_1)/2)}+(1-j)R_0(X)\},\\M(X,0)=[a((X+h_1)/2)+b((X+h_1)/2)]g(h_1)\ \text{ on }(-1,0),\\g(h_1)=[R(X_1)-b'_1]/[a(X_1)+b(X_1)],\ \ X_1=G_2(x_1),\end{cases}$$

$$W(x)=\begin{cases}\dfrac{1}{2}\{(1+j)R_0(X)+\dfrac{(1-j)(2R((X+h_2)/2)-N(X,0))}{a((X+h_2)/2)+b((X+h_2)/2)}\},\\ N(X,0)=[a((X+h_2)/2)-b((X+h_2)/2)]f(h_2)\ \text{ on } (0,1),\\ f(h_2)=[R(X_2)+b_2']/[a(X_2)-b(X_2)],\ \ X_2=G_2(x_2),\end{cases}$$

and then

$$\begin{aligned}&\text{Re}W(x)+\text{Im}W(x)=U(x)+V(x)=f(X)\\ &=-\tilde{R}_0(X)=\frac{2R((X+h_1)/2)-M(X,0)}{a((X+h_1)/2)-b((x+h_1)/2)}\ \text{ on } (h_1,0),\\ &\text{Re}W(x)-\text{Im}W(x)=U(x)-V(x)=g(X)\\ &=\tilde{R}_0(X)=\frac{2R((X+h_2)/2)-N(x,0)}{a((X+h_2)/2)+b((X+h_2)/2)}\ \text{ on } (0,h_2).\end{aligned}\tag{4.24}$$

Now we state and verify the representation of solutions of Problem P for equation (4.1).

Theorem 4.2 *Under Condition C, any solution $u(z)$ of Problem P or GTR for equation* (4.1) *in $\overline{D}$ can be expressed as follows*

$$\begin{aligned}&u(z)=u(x)-2\int_0^y\frac{V(z)}{H_2(x)}dy\\ &=2\text{Re}\int_{-1}^{z}\left[\frac{\text{Re}w}{H_1(y)}+\begin{pmatrix}i\\ -j\end{pmatrix}\frac{\text{Im}w}{H_2(x)}\right]dz+b_0\ \text{ in }\begin{pmatrix}\overline{D^+}\\ \overline{D^-}\end{pmatrix},\\ &w(z)=\Phi(Z)+\Psi(Z)=\hat{\Phi}(Z)+\hat{\Psi}(Z),\\ &\Psi(Z)=2\text{Re}Tf,\ \hat{\Psi}(Z)=2i\text{Im}Tf\ \text{ in } \overline{D_Z},\\ &Tf=-\frac{2}{\pi}\iint_{D_t}\frac{f(t)}{t-Z}d\sigma_t\ \text{ in } \overline{D_Z},\\ &w(z)=\phi(z)+\psi(z)=\xi(z)e_1+\eta(z)e_2\ \text{ in } \overline{D_1^-}=D^-\cap\{x<0\},\\ &\xi(z)=\zeta(z)+\int_0^y g_1(z)dy=\zeta_0(z)+\int_{S_1}g_1(z)dy+\int_0^y g_1(z)dy,\end{aligned}$$

$$\eta(z) = \theta(z) + \int_0^y g_2(z)dy,\ z \in s_2,$$
$$g_l(z) = \tilde{A}_l(U+V) + \tilde{B}_l(U-V) + 2\tilde{C}_l U + \tilde{D}_l u + \tilde{E}_l,\ l = 1, 2, \tag{4.25}$$

in which $Z = X + iY = G_2(x) + iG_1(z)$, $f(Z) = g(Z)/H_1H_2$, $U = Hu_x/2, V = -H_2u_y/2$, $\phi[z(Z)] = \zeta(z)e_1 + \theta(z)e_2$ *is a solution of* (4.20) *in* D_Z^-, $\zeta_0(z) = \mathrm{Re}W(z) + \mathrm{Im}W(z)$, $\theta(z) = -\zeta(x + G(y))$ *in* $D_1^- = D^- \cap \{x < 0\}$, *and* $\theta_0(z) = \mathrm{Re}W(z) - \mathrm{Im}W(z)$, $\zeta(z) = -\theta(x - G(y))$ *in* $D_2^- = D_2^- \cap \{x > 0\}$, $W(z)$ *is as stated in Theorem* 4.1, s_1, s_2 *are two families of characteristics in* D^-:

$$s_1 : \frac{dx}{dy} = \frac{H_1(y)}{H_2(x)},\quad s_2 : \frac{dx}{dy} = -\frac{H_1(y)}{H_2(x)} \tag{4.26}$$

passing through $z = x + jy \in D^-$, S_1, S_2 *are the characteristic curves from the points on* L_1, L_4 *to the points on* L_0 *respectively, and*

$$w(z) = U(z) + jV(z) = \frac{1}{2}H_1u_x - \frac{j}{2}H_2u_y,$$
$$\xi(z) = \mathrm{Re}\psi(z) + \mathrm{Im}\psi(z),\ \eta(z) = \mathrm{Re}\psi(z) - \mathrm{Im}\psi(z),$$
$$\tilde{A}_1 = \frac{1}{4}\Big[\frac{h_{1y}}{h_1} + \frac{H_1h_{2x}}{H_2h_2} - \frac{2b}{H_2^2}\Big],\ \tilde{B}_1 = \frac{1}{4}\Big[\frac{h_{1y}}{h_1} - \frac{H_1h_{2x}}{H_2h_2} + \frac{2b}{H_2^2}\Big],$$
$$\tilde{A}_2 = \frac{1}{4}\Big[\frac{h_{1y}}{h_1} + \frac{H_1h_{2x}}{H_2h_2} + \frac{2b}{H_2^2}\Big],\ \tilde{B}_2 = \frac{1}{4}\Big[\frac{h_{1y}}{h_1} - \frac{H_1h_{2x}}{H_2h_2} - \frac{2b}{H_2^2}\Big], \tag{4.27}$$
$$\tilde{C}_1 = \frac{a}{2H_1H_2} + \frac{m_1}{4y},\ \tilde{C}_2 = -\frac{a}{2H_1H_2} + \frac{m_1}{4y},$$
$$\tilde{D}_1 = -\tilde{D}_2 = \frac{c}{2H_2},\ \tilde{E}_1 = -\tilde{E}_2 = \frac{d}{2H_2},$$

in which we choose $H_1(y) = [|y|^{m_1}h_1(y)]^{1/2}$, $H_2(x) = [|x|^{m_2}h_2(x)]^{1/2}$, $h_1(y)$, $h_2(x)$ *are continuously differentiable positive functions in* $\overline{D}$.

Proof From (4.18) we see that equation (4.1) in $\overline{D}$ can be reduced to the above system of integral equations.

Now we prove the uniqueness of solutions of Problem P for equation (4.1).

Theorem 4.3 *If equation* (4.1) *satisfies Condition* C *and* (2.24) *with* $m = m_1$, *then Problem* P *or GTR for* (4.1) *has at most one solution in* D.

Proof Let $u_1(z), u_2(z)$ be any two solutions of Problem P for (4.1). By Theorem 4.2, it is easy to see that $u(z) = u_1(z) - u_2(z)$ and $w(z) = H_1H_2u_Z$ satisfy the homogeneous equation and boundary conditions

$$w_{\overline{Z}} = A_1 w + A_2\overline{w} + A_3 u \text{ in } D, \tag{4.28}$$

$$\begin{gathered}\operatorname{Re}[\overline{\lambda(z)}w(z)] = R(z), z \in \Gamma \cup L_0, u(-1) = 0,\\ u(1) = 0 \text{ or } \operatorname{Im}[\overline{\lambda(z_3)}w(z_3)] = 0,\end{gathered} \tag{4.29}$$

where $R(z) = 0$ on Γ, $\lambda(x) = i$ on L_0, and $u(z) = 0$ in $\overline{D^-}$ is derived by the similar method in the proof of Theorem 3.4.

Now we verify that the above solution $u(z) \equiv 0$ in D^+. If the maximum $M = \max_{\overline{D^+}} u(z) > 0$, it is clear that the maximum point $z^* \notin D^+$. If the maximum M attains at a point $z^* \in \Gamma$ and $\cos(\nu, n) > 0$ at z^*, we get $\partial u/\partial\nu > 0$ at z^*, this contradicts the formula in (4.29) on Γ; if $\cos(\nu, n) = 0$ at z^*, denote by Γ' the longest curve of Γ including the point z^*, so that $\cos(\nu, n) = 0$ and $u(z) = M$ on Γ', then there exists a point $z' \in \Gamma\backslash\Gamma'$, such that

$$\cos(\nu, n) > 0, \partial u/\partial n > 0, \cos(\nu, s) > 0\,(<0), \partial u/\partial s \geq 0\,(\leq 0) \text{ at } z', \tag{4.30}$$

hence

$$\frac{\partial u}{\partial \nu} = \cos(\nu, n)\frac{\partial u}{\partial n} + \cos(\nu, s)\frac{\partial u}{\partial s} > 0 \text{ at } z' \tag{4.31}$$

holds, where s is the tangent vector of at $z' \in \Gamma$, it is impossible. This shows $z^* \notin \Gamma$. According to the proof of Theorem 3.2, Chapter II, we see that the maximum point $z^* \neq 0$ of $u(z)$. Moreover according to the proof of Theorems 3.2 and 5.2, Chapter II, wee can verify $z^* \notin L_0$. Hence $\max_{\overline{D^+}} u(z) = 0$. By the similar way, we can prove $\min_{\overline{D^+}} u(z) = 0$. Therefore $u(z) = 0$, $u_1(z) = u_2(z)$ in $\overline{D^+}$. Actually the proof of this theorem can be verified by the way as stated in the proof of Theorem 2.4, Chapter V.

4.3 Existence of solutions of oblique derivative problem for mixed equations

In this subsection, we prove the existence of solutions of Problem P or GTR for equation (4.1). Firstly we discuss the Riemann-Hilbert boundary value problem for the complex equation

$$w_{\bar{\bar{z}}} = A_1(z, u, w)w + A_2(z, u, w)\overline{w} + A_3(z, u, w)u + A_4(z, u, w) \text{ in } D, \tag{4.32}$$

with the relation

$$u(z)=2\mathrm{Re}\int_{-1}^{z}\left[\frac{\mathrm{Re}W(z)}{H_1(y)}+\binom{i}{-j}\frac{\mathrm{Im}W(z)}{H_2(x)}\right]dz+b_0 \text{ in } \binom{\overline{D^+}}{\overline{D^-}}, \tag{4.33}$$

where $H_1(y), H_2(x)$ are as stated in (4.7), and the coefficients in (4.1) satisfy the conditions as those in Condition C, and the boundary value problem (4.32), (4.33) with the boundary conditions

$$\begin{gathered}\mathrm{Re}[\overline{\lambda(z)}W(z)] = H_1(y)r(z) \text{ on } \Gamma\cup L_1\cup L_4,\\ u(-1)=b_0, \mathrm{Im}[\overline{\lambda(z_l)}W(z_l)]=b_l',\ l=1,2,3 \text{ or } u(1)=b_3,\end{gathered} \tag{4.34}$$

is called Problem A, where $\lambda(z), r(z), b_l\ (l=0,1,2,3)$ are as stated in (4.9), (4.10), similarly to Section 3, we can assume $R(z)=0$ on $\Gamma\cup L_1\cup L_4$, $b_0=b_1=b_2=b_3=0$. Moreover (4.32), (4.33) in D^+ with the boundary conditions

$$\begin{gathered}\mathrm{Re}[\overline{\lambda(z)}W(z)]=H_1(y)r(z)=0 \text{ on } \Gamma, u(-1)=0, u(1)=0,\\ \mathrm{Re}[iW(x)] = H_2(x)\hat{R}_0(x) \text{ on } L_0=L_0'\cup L_0'',\end{gathered} \tag{4.35}$$

is called Problem A^+, and (4.32),(4.33) in D^- with the boundary conditions

$$\begin{gathered}\mathrm{Re}[\overline{\lambda(z)}W(z)]=H_1(y)r(z)=0 \text{ on } L', \mathrm{Im}[\overline{\lambda(z_l)}W(z_l)]=b_l'=0, l=1,2,\\ \mathrm{Re}[(1-j)W(x)] = H_2(x)\tilde{R}_0(x) \text{ on } L_0,\end{gathered} \tag{4.36}$$

is called Problem A^-, where $\tilde{R}_0(x)$ on $L_0=L_0'\cup L_0''$ is similar to that in (4.24). The solvability of Problem A^+ can be obtained by the result in Theorem 3.4, Chapter II, and Problem A^- can be proved by the same method as stated in the proof of Theorems 2.3 and 2.4, hence we have

Theorem 4.4 *Under the same conditions as in Theorem 4.3, then Problem P or GTR for (4.1) in D has a solution.*

Finally we mention that the coefficients $K_1(y), K_2(x)$ in equation (4.1) can be replaced by functions $K_l(x,y)(l=1,2)$ with some conditions. In addition, we can discuss the uniqueness and existence of the Tricomi problem of the general mixed (elliptic and bi-hyperbolic) equations with bi-parabolic degeneracy, which includes the bi-parabolic, elliptic and bi-hyperbolic Tricomi problem (see [71]5)) as a special case. Besides in [72]2), the authors discussed the uniqueness and existence of solutions of Tricomi problem for equation (4.1) with the conditions $a=b=0$ in D and

$$c(x,y)=o(|x|^p|y|^q),\ d(x,y)=O(|x|^r|y|^s) \text{ as } |y|\to 0, \tag{4.37}$$

where p, q, r, s are some positive numbers. In [71]5), the author proved the uniqueness of solutions of Tricomi problem for the second order equation of mixed type

$$K_1(y)[M_2(x)u_x]_x + M_1(x)[K_2(y)u_y]_y + ru = f \text{ in } D, \tag{4.38}$$

which is a special case of equation (4.1). In fact, the above equation can be rewritten as

$$\begin{aligned} &K_1(y)M_2(x)u_{xx}+M_1(x)K_2(y)u_{yy}+K_1(y)M_{2x}u_x\\ &\quad +M_1(x)K_{2y}u_y + ru = f \text{ in } D, \end{aligned} \tag{4.39}$$

the above equation is divided by $K_2M_2(K_2 > 0, M_2 > 0)$ in D, we obtain

$$\begin{aligned} &Ku_{xx} + Mu_{yy} + K(\ln M_2)'u_x\\ &\quad +M(\ln K_2)'u_y + ru/M_2K_2 = f/M_2K_2, \end{aligned} \tag{4.40}$$

in which $K = K_1/K_2, M = M_1/M_2$. Noting that equation (4.40) is a special case of (4.1), hence similarly to Theorem 4.4, we can derive the solvability of the Tricomi problem for equation (4.40).

In addition, by using the method as in Theorems 2.5 and 2.6, we can also generalize the above results to the case of general domain D, which is called Problem *GFR*, because of the boundary value problem with general Frankl-Rassias boundary condition (see [71]2)).

Remark 4.1 If the domain D^- is replaced by $D^{-'}$, whose boundary is consists of the segment $L_0 = \{-1 < x < 1, y = 0\}$ and two characteristic lines $L = L_1' \cup L_2'$, where

$$L_1'=\{G_1(y)=\int_0^y H_1(t)dt=\int_x^{-1} H_2(t)dt=G_2(-1)-G_2(x), x\in(-1,0)\},$$

$$L_2'=\{-G_1(y)=-\int_0^y H_1(t)dt=\int_x^1 H_2(t)dt = G_2(1)-G_2(x), x\in(0,1)\},$$

in which $L_1 \cup L_4$ in the boundary conditions (4.9) is replaced by $L_1' \cup L_4'$, and the condition (4.10) is replaced by

$$L_\infty[\eta, D^+],\ \eta = a, b, c,\ L_\infty[d, \overline{D^+}] \le k_1,\ c \le 0 \text{ in } D^+,$$

$$\eta|x|^{-m_2/2} = O(1) \text{ as } z = x + iy \to 0,\ \eta = a, b, c, d,$$

$$|ay|/H_1=\varepsilon_1(y) \text{ in } \overline{D_1^-}, m_1\ge 2,\ |bx|/H_2=\varepsilon_2(x) \text{ in } \overline{D_2^-}, m_2\ge 2,$$

$$\tilde{C}[\eta, \overline{D^-}]=C[\eta, \overline{D^-}]+C[\eta_z, \overline{D^-}]\le k_0, \eta=a, b, c, \tilde{C}[d, \overline{D^-}]\le k_1,$$

in which $\varepsilon_1(y) \to 0$ as $y \to 0$, and $\varepsilon_2(x) \to 0$ as $x \to 0$. By using the similar method, we can prove the corresponding theorems 4.1-4.4.

5 The Oblique Derivative Problem for Second Order Equations of Mixed Type with Degenerate Rank 0

In Chapter V and above sections in this chapter, we discussed several boundary value problems for linear and quasilinear second order equations of mixed type with parabolic degeneracy, which possess the important application to gas dynamics. The present section deals with oblique derivative problem for general mixed equations with degenerate rank 0, which include the Tricomi problem as a special case. Firstly the formulation of the problem for the equations is posed, next the representations and estimates of solutions for the above problem are obtained, finally the existence of solutions for the problem is proved by the successive approximation and the method of parameter extension.

5.1 Formulation of oblique derivative problem for mixed equations with degenerate rank 0

Let D be a simply connected bounded domain in the complex plane $\mathbf{C}$ with the boundary $\partial D = \Gamma \cup L$, where $\Gamma(\subset \{y>0\}) \in C^2_\mu(0<\mu<1)$ is a curve with the end points $z=0,2$. Denote $D^+ = D \cap \{y>0\}$, $D^- = D \cap \{y<0\}$, and $H_l(y) = \sqrt{|K_l(y)|}\,(l=1,2)$, where $K_l(y) = \text{sgn}y|y|^{m_l}h_l(y)\,(l=1,2)$, $m_1, m_2\,(m_2<\min(1,m_1))$ are positive numbers, $h_l(y)(l=1,2)$ are continuously differentiable positive functions, and $H(y) = H_1(y)/H_2(y)$, $G(y) = \int_0^y H(t)dt$. There is no harm assuming that the boundary Γ of the domain D^+ is a smooth curve with the form $x-\tilde{G}(y)=0$ and $x+\tilde{G}(y)=2$ near the points $z=0$ and 2 respectively, which is the same as in Section 5, Chapter II, and $L = L_1 \cup L_2$, where

$$L_1 = \{x + \int_0^y H(t)dt = 0, x \in [0,1]\}, L_2 = \{x - \int_0^y H(t)dt = 2, x \in [1,2]\},$$

where $z_1 = x_1 + jy_1 = 1 + jy_1$ is the intersection point of L_1 and L_2. Consider second order quasilinear equation of mixed type with degenerate rank 0:

$$\begin{gathered} K_1(y)u_{xx} + |K_2(y)|u_{yy} + \hat{a}u_x + \hat{b}u_y + \hat{c}u + \hat{d} = 0, \text{ i.e.} \\ K(y)u_{xx} + u_{yy} + au_x + bu_y + cu + d = 0 \text{ in } D, \end{gathered} \tag{5.1}$$

where D is bounded by the segment $L_0=(0,2)$ and two characteristic lines

$$L_1:\frac{dx}{dy}=H(y)=\sqrt{|K(y)|},\,L_2:\frac{dx}{dy}=-H(y)=-\sqrt{|K(y)|},$$

emanating from 0 and 2 respectively, in which $K(y)=K_1(y)/|K_2(y)|$, $H(y)=H_1(y)/H_2(y)$, and $z_1=1+jy_1$ is the intersection point of L_1 and L_2, $a=\hat{a}/|K_2|$, $b=\hat{b}/|K_2|$, $c=\hat{c}/|K_2|, d=\hat{d}/|K_2|$, the coefficients $\hat{a},\hat{b},\hat{c},\hat{d}$ are real functions of $z(\in\overline{D}),u,u_x,u_y(\in\mathbf{R})$ satisfying **Condition** C: For any continuously differentiable function $u(z)$ in $\overline{D}\backslash\{0,2\}$, the functions $\hat{a},\hat{b},\hat{c},\hat{d}$ are continuous in D and satisfy

$$\begin{gathered}|\hat{a}y|/H_1H_2=\varepsilon(y),\ m_1+m_2\geq 2\ \text{ in }\ \overline{D^-}\\ \hat{C}[\eta,\overline{D}]=C[\eta,\overline{D}]+C[\eta_x,\overline{D}]\leq k_0,\eta=\hat{a},\hat{b},\hat{c},\ \hat{C}[\hat{d},\overline{D}]\leq k_1,\end{gathered}\tag{5.2}$$

where k_0, k_1 are positive constants, $\varepsilon(y)\to 0$ as $y\to 0$. Moreover for any continuously differentiable functions $u_1(z)$, $u_2(z)$ in D^*, $F(z,u,u_z)=\hat{a}u_x+\hat{b}u_y+\hat{c}u+\hat{d}$ satisfies the condition

$$\begin{gathered}F(z,u_1,u_{1z})-F(z,u_2,u_{2z})\\ =\tilde{a}(u_1-u_2)_x+\tilde{b}(u_1-u_2)_y+\tilde{c}(u_1-u_2)\text{ in }\overline{D},\end{gathered}\tag{5.3}$$

holds, where $\tilde{a},\tilde{b},\tilde{c}$ satisfy the same conditions as those of $\hat{a},\hat{b},\hat{c}$, in (5.2), (5.3), and k_0,k_1 are non-negative constants. If $H_l(y)=|y|^{m_l/2}$, $l=1,2$, $H(y),H_1(y),H_2(y),m_1,m_2$ are as stated before, then

$$\begin{gathered}G_l(y)=\int_0^y H_l(t)dy=-\frac{2}{m_l+2}|y|^{(m_l+2)/2},l=1,2,\\ Y=G(y)=\int_0^y H(t)dy=-\frac{2}{m+2}|y|^{(m+2)/2},m=m_1-m_2>-1\text{ in }\overline{D^-},\end{gathered}\tag{5.4}$$

and the inverse function of $Y=G(y)$ is

$$\begin{aligned}y&=-|G^{-1}(Y)|\\ &=-\left(\frac{m+2}{2}\right)^{2/(m+2)}|Y|^{2/(m+2)}=-J|Y|^{2/(m+2)}\ \text{ in }\overline{D^-}.\end{aligned}$$

The oblique derivative boundary value problem for equation (5.1) may be formulated as follows:

Problem P Find a continuous solution $u(z)$ of (5.1) in $\bar{D}$, where u_x, u_y are continuous in $D^* = \overline{D}\backslash\{0,2\}$, and satisfy the boundary conditions

$$\begin{gathered}\frac{1}{2}\frac{\partial u}{\partial \nu}=\frac{1}{H_1(y)}\text{Re}[\overline{\lambda(z)}u_{\tilde{z}}]=\text{Re}[\overline{\Lambda(z)}u_z]=r(z) \text{ on } \Gamma\cup L_1,\\ \frac{1}{H_1(y)}\text{Im}[\overline{\lambda(z)}u_{\tilde{z}}]|_{z=z_1}=b_1,\ u(0)=b_0,\\ u(2)=b_2 \text{ or } \text{Im}[\overline{\lambda(z)}u_{\tilde{z}}]|_{z=z_2}=b_2,\end{gathered}\tag{5.5}$$

in which ν is a given vector at every point $z\in\Gamma\cup L_1$, $u_{\tilde{z}}=[H_1(y)u_x - iH_2(y)u_y]/2$, $\Lambda(z)=\cos(\nu,x)-i\cos(\nu,y)$, $\lambda(z)=\text{Re}\lambda(z)+i\text{Im}\lambda(z)$, if $z\in\Gamma$, and $u_{\tilde{z}}=[H_1(y)u_x - jH_2(y)u_y]/2$, $\lambda(z)=\text{Re}\lambda(z)+j\text{Im}\lambda(z)$, if $z\in L_1$, $z_2(\neq 0,2)$ is a point on Γ, b_0,b_1,b_2 are real constants, and $r(z),b_0,b_1,b_2$ satisfy the conditions

$$\begin{gathered}C^1_\alpha[\lambda(z),\Gamma]\le k_0,\ C^1_\alpha[\lambda(x),L_1]\le k_0,\ C^1_\alpha[r(z),\Gamma]\le k_2,\ C^1_\alpha[r(x),L_1]\le k_2,\\ \cos(\nu,n)\ge 0 \text{ on } \Gamma\cup L_1,\ |b_0|,|b_1|,|b_2|\le k_2,\ \max_{z\in L_1}\frac{1}{|\text{Re}\lambda(z)-\text{Im}\lambda(z)|}\le k_0,\end{gathered}\tag{5.6}$$

in which n is the outward normal vector at every point on Γ, $\alpha(0<\alpha<1)$, k_0,k_2 are non-negative constants. For the last point condition in (5.5), we need to assume $c=0$ in equation (5.1). The number

$$K=\frac{1}{2}(K_1+K_2)\tag{5.7}$$

is called the index of Problem P, where

$$K_l=\left[\frac{\phi_l}{\pi}\right]+J_l,\ J_l=0 \text{ or } 1,\ e^{i\phi_l}=\frac{\lambda(t_l-0)}{\lambda(t_l+0)},\ \gamma_l=\frac{\phi_l}{\pi}-K_l,\ l=1,2,\tag{5.8}$$

in which $t_1=0$, $t_2=2$, $\lambda(t)=e^{i\pi/2}$ on $L_0=(0,2)$ and $\lambda(t_1+0)=\lambda(t_2-0)=\exp(i\pi/2)$. Here $K=0$ on the boundary ∂D^+ of D^+ can be chosen.

Problem P in the above case includes the Tricomi problem (Problem T, see [71]1)) as a special case. In fact, the boundary conditions of Problem T in D^+ are as follows

$$u(z)=\phi(x) \text{ on } \Gamma,\ u_y=r(x) \text{ on } L_1,\tag{5.9}$$

if the boundary Γ near $z=0,2$ possesses the form $x=\tilde{G}(y)$ or $x=2-\tilde{G}(y)$ respectively, we find the derivative for (5.9) according to the parameter

$s = \operatorname{Im} z = y$ on Γ, and obtain

$$u_s = u_x x_y + u_y = \phi'(y), \text{ i.e. } \tilde{H}(y)H_1(y)H_2(y)u_x/H_1(y)$$

$$+H_2(y)u_y = H_2(y)\phi'(y) \text{ on } \Gamma \text{ near } x = t_1 = 0,$$

$$u_s = u_x x_y + u_y = \phi'(y), \text{ i.e. } \tilde{H}(y)H_1(y)H_2(y)u_x/H_1(y)$$

$$-H_2(y)u_y = -H_2(y)\phi'(y) \text{ on } \Gamma \text{ near } x = t_2 = 2,$$

where $\tilde{H}(y) = \tilde{G}'(y)$, $H(y) = y^{m/2}$, it is clear that the complex form of above conditions is as follows

$$\operatorname{Re}[\overline{\lambda(z)}(U+iV)] = \operatorname{Re}[\overline{\lambda(z)}(H_1(y)u_x - iH_2(y)u_y)]/2 = R(z) \text{ on } \Gamma\cup L_0,$$

where

$$\lambda(z) = \begin{cases} -i, \\ i, \\ i, \end{cases} \quad R(z) = \begin{cases} R_1(x) \text{ on } \Gamma \text{ at } z=0, \\ R_2(x) \text{ on } \Gamma \text{ at } z=2, \\ -H_2(y)r(x)/2 = R_3(x) \text{ on } L_1, \end{cases}$$

here $R_1(x) = H_2(y)\phi'(y)/2$, $R_2(x) = -H_2(y)\phi'(y)/2$, and $R_3(x)$ is an undetermined real function. If we choose $\operatorname{Re}[\lambda(x)W(x)] = 0$, $\lambda(x) = 1$ on L_0, then

$$e^{i\phi_1} = \frac{\lambda(t_1-0)}{\lambda(t_1+0)} = e^{-\pi i/2 - 0\pi i} = e^{-\pi i/2}, \gamma_1 = \frac{-\pi/2}{\pi} - K_1 = -\frac{1}{2}, K_1 = 0,$$

$$e^{i\phi_2} = \frac{\lambda(t_2-0)}{\lambda(t_2+0)} = e^{0\pi i - \pi i/2} = e^{-\pi i/2}, \gamma_2 = \frac{-\pi/2}{\pi} - K_2 = -\frac{1}{2}, K_2 = 0, \tag{5.10}$$

this shows that Problem T is a special case of Problem P in the domain D^+ with the index $K = (K_1 + K_2)/2 = 0$. If we consider $\cos(\nu, n) = 0$ on L_0, and $\operatorname{Re}[\overline{\lambda(x)}W(x)] = R(x)$, $\lambda(x) = i$ on L_0, then $\gamma_1 = \gamma_2 = 0$, $K_1 = -1, K_2 = 0$, thus the index

$$K = \frac{1}{2}(K_1 + K_2) = -\frac{1}{2}.$$

For the index $K = -1/2$, the last point condition in the boundary condition (5.5) should be cancelled.

5.2 Representation of solutions of oblique derivative problem for degenerate mixed equations

In this section, we first write the equivalent complex form of equation (5.1). Denote

$$W(z)=U+jV=\frac{1}{2}[H_1(y)u_x-jH_2(y)u_y]=u_{\tilde z}=\frac{H_1(y)}{2}[u_x-ju_Y],$$

$$W_{\bar{\tilde z}}=\frac{1}{2}[H_1(y)W_x+jH_2(y)W_y]=\frac{H_1(y)}{2}[W_x+jW_Y]=H_1(y)W_{\overline{Z}}\text{ in }\overline{D^-},$$

$$W(z)=U+iV=\frac{1}{2}[H_1(y)u_x-iH_2(y)u_y]=u_{\tilde z}=\frac{H_1(y)}{2}[u_x-iu_Y],$$

$$W_{\bar{\tilde z}}=\frac{1}{2}[H_1(y)W_x+iH_2(y)W_y]=\frac{H_1(y)}{2}[W_x+iW_Y]=H_1(y)W_{\overline{Z}}\text{ in }\overline{D^+},$$

by using (5.11), Chapter II and (5.7), (5.8), Chapter III, the equation (5.1) can be rewritten in the complex form

$$W_{\bar{\tilde z}}=A_1(z,u,W)W+A_2(z,u,W)\overline{W}+A_3(z,u,W)u+A_4(z,u,W)\text{ in }D,$$

$$A_1=\begin{cases}\frac{1}{4}[-\frac{a}{H_1}+\frac{iH_2H_{1y}}{H_1}-i\frac{b}{H_2}+iH_{2y}],\\ \frac{1}{4}[\frac{a}{H_1}+\frac{jH_2H_{1y}}{H_1}-\frac{jb}{H_2}+jH_{2y}],\end{cases}\quad A_3=\begin{cases}-\frac{c}{4},\\ \frac{c}{4},\end{cases}$$

$$A_2=\begin{cases}\frac{1}{4}[-\frac{a}{H_1}+\frac{iH_2H_{1y}}{H_1}+i\frac{b}{H_2}-iH_{2y}],\\ \frac{1}{4}[\frac{a}{H_1}+\frac{jH_2H_{1y}}{H_1}+\frac{jb}{H_2}-jH_{2y}],\end{cases}\quad A_4=\begin{cases}-\frac{d}{4}\text{ in }\overline{D^+},\\ \frac{d}{4}\text{ in }\overline{D^-},\end{cases}\tag{5.11}$$

if $H_l(y)=\sqrt{|y|^{m_l}h_l(y)}$, $(l=1,2)$, in which $m_1,m_2(<1)$ are as stated before, $h_l(y)\,(l=1,2)$ are continuously differentiable positive functions, then the coefficients possess the form

$$A_1=\begin{cases}\frac{1}{4}\left[-\frac{a}{H_1}-\frac{ib}{H_2}+iH_2\sum_{l=1}^{2}\left(\frac{h_{ly}}{2h_l}+\frac{m_l}{2y}\right)\right],\\ \frac{1}{4}\left[\frac{a}{H_1}-\frac{jb}{H_2}+jH_2\sum_{l=1}^{2}\left(\frac{h_{ly}}{2h_l}+\frac{m_l}{2y}\right)\right],\end{cases}\quad A_3=\begin{cases}-\frac{c}{4},\\ \frac{c}{4},\end{cases}$$

$$A_2=\begin{cases}\dfrac{1}{4}\left[-\dfrac{a}{H_1}+\dfrac{ib}{H_2}+iH_2\displaystyle\sum_{l=1}^{2}(-1)^{l-1}\left(\dfrac{h_{ly}}{2h_l}+\dfrac{m_l}{2y}\right)\right],\\ \dfrac{1}{4}\left[\dfrac{a}{H_1}+\dfrac{jb}{H_2}+jH_2\displaystyle\sum_{l=1}^{2}(-1)^{l-1}\left(\dfrac{h_{ly}}{2h_l}+\dfrac{m_l}{2y}\right)\right],\end{cases}\quad A_4=\begin{cases}-\dfrac{d}{4}\text{ in }\overline{D^+},\\ \dfrac{d}{4}\text{ in }\overline{D^-},\end{cases}$$

and the relation of above functions $u(z)$ and $W(z)=U(z)+iV(z)$ in D^+, $W(z)=U(z)+jV(z)$ in D^- is as follows

$$u(z)=\begin{cases}2\text{Re}\displaystyle\int_0^z\left[\frac{U(z)}{H_1(y)}+\frac{iV(z)}{H_2(y)}\right]dz+b_0\text{ in }\overline{D^+}\\ 2\text{Re}\displaystyle\int_0^z\left[\frac{U(z)}{H_1(y)}-\frac{jV(z)}{H_2(y)}\right]dz+b_0\text{ in }\overline{D^-}\end{cases}\tag{5.12}$$

where b_0 is a real constant as stated in (5.5). For the second equation of (5.1), the complex form can be written as (2.6) in Section 2, where $H(y)=H_1(y)/H_2(y)$. It is clear that the complex equation

$$W_{\tilde{\bar{z}}}=0\text{ in }\overline{D}\tag{5.13}$$

can be rewritten in the system

$$\begin{aligned}&[(U+V)+i(U-V)]_{\mu-i\nu}=0\ \text{ in }\overline{D^+},\\&(U+V)_\mu=0,\ (U-V)_\nu=0\ \text{ in }\overline{D^-}.\end{aligned}\tag{5.14}$$

The boundary problem for equation (5.11) with the boundary condition (5.5) ($W(z)=u_{\tilde{z}}$) and the relation (5.12) will be called Problem A.

Now, we give the representation of solutions for the oblique derivative problem (Problem P) for system (5.14) in $\overline{D}$. For this, we first discuss the Riemann-Hilbert problem (Problem A) for the second system of (5.14) in $\overline{D^-}$ with the boundary conditions

$$\begin{aligned}&\text{Re}[\overline{\lambda(z)}(U+jV)]=\begin{cases}H_1(y)r(z)=R_1(z),\ z\in L_1,\\ R(x)=R_0(x),x\in L_0=[0,2],\end{cases}\\&\text{Im}[\overline{\lambda(z)}(U+jV)]|_{z=z_1}=H_1(\text{Im}z_1)b_1=b_1',\ u(0)=b_0,\end{aligned}\tag{5.15}$$

in which $H(y)=H_1(y)/H_2(y)$, $\lambda(z)=a(z)+jb(z)$ on L_1 and $\lambda(z)=1+j$ on L_0 and $R_0(x)$ is an undetermined real function. It is clear that the solution of Problem A for (5.14) in $\overline{D^-}$ can be expressed as

$$\begin{aligned}&\xi = U(z) + V(z) = f(\nu),\ \ \eta = U(z) - V(z) = g(\mu),\\&U(z)=[f(\nu) + g(\mu)]/2,\ V(z)=[f(\nu) - g(\mu)]/2,\ \text{i.e.}\\&W(z)=U(z)+jV(z) = [(1+j)f(\nu) + (1-j)g(\mu)]/2,\end{aligned}\tag{5.16}$$

where $f(t), g(t)$ are two arbitrary real continuous functions on $[0, 2]$. For convenience, denote by the functions $a(x), b(x), r(x)$ of x the functions $a(z), b(z), r(z)$ of z in (5.15), thus (5.15) can be rewritten as

$$\begin{aligned}&a(x)U(x,y)-b(x)V(x,y)=R_1(x) \text{ on } L_1,\\&U(x)\quad V(x) = R_0(x)\ \text{ on } L_0,\ \text{i.e.}\\&[a(x)-b(x)]f(x-G(y))+[a(x)+b(x)]g(x+G(y))=2R_1(x) \text{ on } L_1,\\&U(x) - V(x) = R_0(x)\ \text{ on } L_0,\ \text{i.e.}\\&[a(x) - b(x)]f(2x) + [a(x) + b(x)]g(0) = 2R_1(x), x\in[0, 1],\\&U(x)-V(x)=R_0(x), x\in[0, 2],\ \text{i.e.}\\&[a(t/2) - b(t/2)]f(t) + [a(t/2) + b(t/2)]g(0) = 2R_1(t/2)\\&= 2R_1(t/2), t \in [0, 2], U(t) - V(t) = R_0(t), t \in [0, 2],\end{aligned}$$

where

$$[a(1)+b(1)]g(0)=[a(1)+b(1)][U(z_1)-V(z_1)]=R_1(1)-b_1' \text{ or } 0.$$

Moreover we can derive

$$\begin{aligned}&f(\nu)=f(x-G(y))=\frac{2R_1(\nu/2)-(a(\nu/2)+b(\nu/2))g(0)}{a(\nu/2) - b(\nu/2)},\\&g(\mu) = g(x + G(y)) = R_0(\mu),\\&U(z) = \frac{1}{2}\{f(\nu) + R_0(\mu)\},\ V(z) = \frac{1}{2}\{f(\nu) - R_0(\mu)\},\end{aligned}\tag{5.17}$$

if $a(x) - b(x) \neq 0$ on $[0, 1]$. From the above formula, it follows that

$$\begin{aligned}&\mathrm{Re}[(1+j)W(x)]=U(x)+V(x)=\frac{2R_1(x/2) - (a(x/2) + b(x/2))g(0)}{a(x/2) - b(x/2)},\\&\mathrm{Re}[(1 - j)W(x)] = U(x) - V(x) = R_0(x),\ x \in [0, 2],\end{aligned}$$

if $a(x) - b(x) \neq 0$ on $[0,1]$. Thus we obtain

$$W(z)=\begin{cases}\frac{1}{2}\{(1+j)\frac{2R_1((x-G(y))/2)-M(x,y)}{a((x-G(y))/2)-b((x-G(y))/2)}\\ +(1-j)R_0(x+G(y))\},\\ M(x,y)=[a((x-G(y))/2)+b((x-G(y))/2)]g(0).\end{cases} \tag{5.18}$$

In particular, we have

$$\mathrm{Re}[\overline{(1+i)}(U(x)+iV(x))] = U(x)+V(x) = -\hat{R}_0(x)$$
$$= \frac{2R_1(x/2)[a(1)+b(1)]-[a(x/2)+b(x/2)][R_1(1)-b_1']}{[a(1)+b(1)][a(x/2)-b(x/2)]} \text{ on } L_0. \tag{5.19}$$

It is clear that if $r(z) = 0$, $R_1(z) = 0$ on L_1, $\hat{R}_0(x) = 0$ on L_0, and $b_0 = b_1 = b_2 = 0$, then $W(z) = U(z) + jV(z) = 0$ in $\overline{D^-}$. Next we find a solution of the Riemann-Hilbert boundary value problem for equation (5.13) in D^+ with the boundary conditions (5.19) and

$$\mathrm{Re}[\overline{\lambda(z)}(U(z)+iV(z))]=\mathrm{Re}\lambda(z)U(z)+\mathrm{Im}\lambda(z)V(z)=R_1(z) \text{ on } \Gamma. \tag{5.20}$$

Noting that the index of the above boundary condition is $K = 0$, by the method in [87]1), we know that the above Riemann-Hilbert problem has a unique solution $W(z)$ in D^+, and then

$$U(x) - V(x) = \mathrm{Re}[(1-j)(U(x)+jV(x)] = R_0(x) \text{ on } L_0 \tag{5.21}$$

is determined. This shows that Problem A for equation (5.13) is uniquely solvable, namely

Theorem 5.1 *Problem A of equation* (5.13) *or system* (5.14) *in $\overline{D}$ has a unique solution $W(z)$ as stated in* (5.18).

Now we state and can verify the representation of solutions of Problem P for equation (5.1).

Theorem 5.2 *Under Condition C, any solution $u(z)$ of Problem P for equation* (5.1) *in D^- can be expressed as follows*

$$u(z)=2\int_0^y V(z)dy=2\mathrm{Re}\int_0^z [\frac{\mathrm{Re}W}{H(y)}-j\mathrm{Im}W]dz \text{ in } \overline{D^-},$$

$$W(z)=\phi(z)+\psi(z)=\xi(z)e_1+\eta(z)e_2 \text{ in } \overline{D^-},$$

$$\xi(z)=\int_0^\mu \frac{g_1(z)}{H(y)}d\mu=\zeta(z)+\int_0^y g_1(z)dy=\int_{S_1} g_1(z)dy$$

$$+\int_0^y g_1(z)dy=\int_{y_1}^{|y|}\hat{g}_1(z)dy,\, z\in s_1,\ \eta(z)=\theta(z)+\int_0^y g_2(z)dy,\, z\in s_2,$$

$$g_l(z)=\tilde{A}_l(U+V)+\tilde{B}_l(U-V)+2\tilde{C}_lU+\tilde{D}_lu+\tilde{E}_l,\, l=1,2, \tag{5.22}$$

where $H(y)=H_1(y)/H_2(y), U=Hu_x/2, V=-u_y/2,\ Z(z)=x+jY=x+iY(y)$ *is a mapping from* $z(\in D^-)$ *to* Z, $\int_0^\mu[g_1(z)/2H_1(y)]d\mu$ *is the integral along characteristic curve* s_1 *from a point* $z_1=x_1+jy_1$ *on* L_1 *to the point* $z=x+jy\in\overline{D^-}$, $\theta(z)=-\zeta(x+G(y))$, $\zeta(z)e_1+\theta(z)e_2$ *is a solution of* (5.13) *in* D^-, *and* s_1,s_2 *are two families of characteristic curves in* D^- :

$$s_1:\ \frac{dx}{dy}=\sqrt{-K(y)}=H(y),\ s_2:\ \frac{dx}{dy}=-\sqrt{-K(y)}=-H(y) \tag{5.23}$$

passing through $z=x+jy\in D^-$, S_1 *is the characteristic curve from a point on* L_1 *to a point on* L_0, *and*

$$W(z)=U(z)+jV(z)=\frac{1}{2}[Hu_x-ju_y],$$

$$\xi(z)=\mathrm{Re}\psi(z)+\mathrm{Im}\psi(z),\eta(z)=\mathrm{Re}\psi(z)-\mathrm{Im}\psi(z),$$

$$\tilde{A}_1=\tilde{B}_2=\frac{1}{2}(\frac{h_y}{2h}-b),\ \tilde{A}_2=\tilde{B}_1=\frac{1}{2}(\frac{h_y}{2h}+b),$$

$$\tilde{C}_1=\frac{a}{2H}+\frac{m}{4y},\ \tilde{C}_2=-\frac{a}{2H}+\frac{m}{4y},$$

$$\tilde{D}_1=-\tilde{D}_2=\frac{c}{2},\ \tilde{E}_1=-\tilde{E}_2=\frac{d}{2},$$

in which we choose $H(y)=[|y|^mh(y)]^{1/2}$ *as stated in* (5.11), *and*

$$d\mu=d[x+G(y)]=2H(y)dy \text{ on } s_1, d\nu=d[x-G(y)]=-2H(y)dy \text{ on } s_2.$$

Proof Similarly to (2.5), equation (2.1) in $\overline{D^-}$ can be reduced to the system of integral equations: (5.22). Moreover we can extend the equation (5.11) onto the the symmetrical domain $\tilde{D}_Z$ of D_Z^- with respect to the real axis $\mathrm{Im}Z=0$, namely introduce the function $\hat{w}(Z)$ as follows:

$$\hat{w}(Z)=\begin{cases} w[z(Z)], \\ -\overline{w[z(\overline{Z})]}, \end{cases} \quad \hat{u}(z)=\begin{cases} u(Z) \text{ in } D_Z^-, \\ -u(\overline{Z}) \text{ in } \tilde{D}_Z, \end{cases} \tag{5.24}$$

and then the equation (5.11) is extended as

$$\hat{w}_{\bar{\bar{z}}} = \hat{A}_1\hat{w} + \hat{A}_2\overline{\hat{w}} + \hat{A}_3\hat{u} + \hat{A}_4 = \hat{g}(Z) \text{ in } \overline{D_Z^-} \cup \overline{\hat{D}_Z},$$

where

$$\hat{A}_l(Z) = \begin{cases} A_l(Z), \\ \overline{\tilde{A}_l(\overline{Z})}, \end{cases} l=1,2,3,\ \hat{A}_4(Z) = \begin{cases} A_4(Z), \\ -\overline{A_4(\overline{Z})}, \end{cases}$$

$$\hat{g}_l(Z) = \begin{cases} g_l(z) \text{ in } \overline{D_Z^-}, \\ -\overline{g_l(\overline{Z})} \text{ in } \overline{\tilde{D}_Z}, \end{cases} \quad l = 1, 2,$$

here $\tilde{A}_1(\overline{Z}) = A_2(\overline{Z}), \tilde{A}_2(\overline{Z}) = A_1(\overline{Z}), \tilde{A}_3(\overline{Z}) = A_3(\overline{Z})$. It is easy to see that the system of integral equations (5.22) can be written in the form

$$\begin{aligned} \xi(z) &= \zeta(z) + \int_0^{\tilde{y}} g_1(z)dy = \int_{y_1}^{\hat{y}} \hat{g}_1(z)dy, \\ \eta(z) &= \theta(z) - \int_0^{y} g_2(z)dy = \int_{y_1}^{\hat{y}} \hat{g}_2(z)dy, \hat{z} = x + j\hat{y} = x + j|y| \text{ in } \tilde{D}_Z, \end{aligned} \tag{5.25}$$

where $x_1 + jy_1$ is the intersection point of L_1 and the characteristic curve s_1 passing through $z = x + jy$, the function $\theta(z)$ is determined by $\zeta(z)$, the function $\theta(z)$ will be defined by $\theta(z) = -\zeta(z) = -\zeta(x + G(y))$, for the extended integral, later on the above form $\hat{g}_2(z)$ is written, and the numbers $\hat{y} - y_1, \hat{t} - y_1$ will be written by $\tilde{y}, \tilde{t}$ respectively.

5.3 Existence of solutions of oblique derivative problem for degenerate mixed equations

In this section, we prove the existence of solutions of Problem P for equation (5.1). Firstly we discuss the complex equation

$$w_{\bar{\bar{z}}} = A_1(z, u, w)w + A_2(z, u, w)\overline{w} + A_3(z, u, w)u + A_4(z, u, w) \text{ in } D, \tag{5.26}$$

with the relation

$$u(z) = \begin{cases} 2\text{Re} \displaystyle\int_0^z \left[\frac{\text{Re}w(z)}{H_1(y)} + \frac{i\text{Im}w(z)}{H_2(y)}\right] dz + b_0 \text{ in } \overline{D^+}, \\ 2\text{Re} \displaystyle\int_0^z \left[\frac{\text{Re}w(z)}{H_1(y)} - \frac{j\text{Im}w(z)}{H_2(y)}\right] dz + b_0 \text{ in } \overline{D^-}, \end{cases} \tag{5.27}$$

where $H_l(y)(l=1,2)$ and the coefficients in (5.26) are as stated in (5.11), and the boundary value problem (5.26), (5.27) with the boundary conditions

$$\begin{aligned}&\operatorname{Re}[\overline{\lambda(z)}w(z)]=H_1(y)r(z)=R(z) \ \text{ on } \Gamma\cup L_1,\\&\operatorname{Im}[\overline{\lambda(z_1)}w(z_l)]=b_l,\ l=1,2 \ \text{ or } u(2)=b_2\end{aligned}\tag{5.28}$$

is called Problem A, where $\lambda(z), r(z), z_l, b_l\ (l=1,2)$ are as stated in (5.5), (5.6). For the second equation in (5.1), its complex form is as stated in (2.6).

Noting that $\lambda(z), r(z)\in C^1_\alpha(\Gamma), \lambda(z), r(z)\in C^1(L_1)\,(0<\alpha<1)$, we can find two twice continuously differentiable functions $u_0^\pm(z)$ in $\overline{D}^\pm$, for instance, which are the solutions of oblique derivative problems with the boundary condition on $\Gamma\cup L_1$ in (5.5) and $b_0=b_1=b_2=0$, for harmonic equations in $D^\pm$, denote by $u(z)$ the solution of Problem P for (5.1), then the function $v=v^\pm(z)=u(z)-u_0^\pm(z)$ in $\overline{D}^\pm$ is a solution of the homogeneous boundary value problem P_0 for equation in the form

$$K_1(y)v_{xx}+\text{sgn}yK_2(y)v_{yy}+\tilde{a}v_x+\tilde{b}v_y+\tilde{c}u+\tilde{d}=0 \ \text{ in } D,$$

with the boundary conditions

$$\begin{aligned}&\operatorname{Re}[\overline{\lambda(z)}v_{\bar z}]=0 \ \text{ on } \Gamma\cup L_1,\ v(0)=0,\\&\operatorname{Im}[\overline{\lambda(z)}v_{\bar z}]|_{z=z_l}=b_l'\,(l=1,2) \ \text{ or } v(2)=0,\end{aligned}$$

where $b_1'=0$, the coefficients of the above equation satisfy the conditions similar to Condition C, $W(z)=U+iV=v_{\bar z}^+$ in D^+ and $W(z)=U+jV=v_{\bar z}^-$ in $\overline{D^-}$, hence later on we only discuss the case of $r(z)=0$ on $\Gamma\cup L_1$ and $b_0=b_1=0$ in (5.5) and the case of index $K=0$, the other case can be similarly discussed. From $v(z)=v^\pm(z)=u(z)-u_0^\pm(z)$ in $\overline{D^\pm}$, we have $u(z)=v^+(z)+u_0^+(z)$ in $\overline{D^+}$, $u(z)=v^-(z)+u_0^-(z)$ in $\overline{D^-}$, and

$$\begin{aligned}&v^+(z)=v^-(z)-u_0^+(z)+u_0^-(z),\ u_y=v_y^\pm+u_{0y}^\pm,\\&v_y^+=v_y^- -u_{0y}^+ +u_{0y}^-=2\hat{R}_0(x),\ v_y^-=2\tilde{R}_0(x) \ \text{ on } L_0=D\cap\{y=0\}.\end{aligned}$$

It is clear that the equation, the relation and the boundary condition of the corresponding Riemann-Hilbert problem (Problem $\tilde{A}$) are as follows

$$w_{\bar z}=A_1w+A_2(z)\overline{w}+A_3u+A_4 \ \text{ in } D,$$

$$w(z)=\begin{cases}2\mathrm{Re}\int_0^z\left[\frac{\mathrm{Re}w(z)}{H_1(y)}+i\frac{\mathrm{Im}w(z)}{H_2(y)}\right]dz+b_0 \text{ in } \overline{D^+},\\ 2\mathrm{Re}\int_0^z\left[\frac{\mathrm{Re}w(z)}{H_1(y)}-j\frac{\mathrm{Im}w(z)}{H_2(y)}\right]dz+b_0 \text{ in } \overline{D^-},\end{cases} \tag{5.29}$$

$$\begin{gathered}\mathrm{Re}[\overline{\lambda(z)}w(z)]=0 \text{ on } \Gamma\cup L_1,\ u(0)=0,\\ u(2)=0, \text{ or } \mathrm{Im}[\overline{\lambda(z_1)}w(z_1)]=b_2',\end{gathered} \tag{5.30}$$

hence we can only discuss Problem $\tilde{A}$ for (5.29) with the boundary condition (5.30). From Theorems 5.1 and 5.2, Problem $\tilde{A}$ can be divided into two problems, i.e. Problem A_1 of (5.29) in D^+ and Problem A_2 of (5.29) in D^-, the boundary conditions of Problems A_1 and A_2 are as follows:

$$\begin{gathered}\mathrm{Re}[\overline{\lambda(z)}w(z)]=R(z)=0 \text{ on } \Gamma,\ \mathrm{Re}[\overline{\lambda(z)}w(z)]=R(x) \text{ on } L_0,\\ u(0)=0, u(2)=0 \text{ or } \mathrm{Im}[\overline{\lambda(z_2)}w(z_2)]=b_2',\end{gathered} \tag{5.31}$$

and

$$\begin{gathered}\mathrm{Re}[\overline{\lambda(z)}w(z)]=R(z)=0 \text{ on } L_1, \mathrm{Re}[\overline{\lambda(z)}w(z)]=R(x) \text{ on } L_0,\\ u(0)=0,\ u(2)=0,\ \mathrm{Im}[\overline{\lambda(z_1)}w(z_1)]=0,\end{gathered} \tag{5.32}$$

respectively, in which $\lambda(z)=i$, $R(z)=-\hat{R}_0(x)$ and $\lambda(z)=j$, $R(z)=\tilde{R}_0(x)$ on L_0. For the second equation of (5.1), the corresponding boundary conditions are as those of Problem B_1, B_2 in Subsection 2.3. The solvability of Problem A_1 can be proved by the method in Chapter II. By using the result in Chapter III, the reduction to absurdity and the compactness principle, we can prove that there exists a solution $w(z)=u_{\tilde{z}}$ of Problem A_2 for the equation

$$\begin{gathered}K_1(y)u_{xx}+\mathrm{sgn}y K_2(y)u_{yy}+au_x+bu_y+cu+d=0, \text{ i.e.}\\ w_{\bar{\tilde{z}}}=A_1w+A_2\overline{w}+A_3u+A_4 \text{ in } D,\end{gathered}$$

thus Problem A_2 of equation (5.29), (5.32) in D^- is solvable, but in the following we can directly verify the solvability of Problem A_2 for (5.29), (5.30).

Theorem 5.3 *Let equation* (5.1) *satisfy Condition C and the last condition in* (5.33) *below. Then there exists a unique solution* $[w(z),v(z)]$ *of Problem A_2 for* (5.29) *in* D^-.

Proof Denote $D_0 = \overline{D^-} \cap \{a_0 \le x \le b_0, -\delta \le y \le 0\}$, and s_1, s_2 are the characteristics of families in Theorem 5.2 emanating from any two points $(a_0, 0), (b_0, 0)(0 < a_0 < b_0 = 2)$, which intersect at a point $z = x + jy \in \overline{D^-}$, and δ, δ_0 are sufficiently small positive constants.

We discuss the case of $K(y) = -|y|^m h(y)$, $m = m_1 - m_2, h(y)$ are as stated in Subsection 5.1. In order to find a solution of the system of integral equations (5.22), we need to add a condition for the coefficient $a = a(z, u, u_x, u_y)$ in equation (2.1), namely

$$\frac{ay}{H_1(y)H_2(y)} = o(1), \text{ i.e. } \frac{|a|}{H_1(y)H_2(y)} = \frac{\varepsilon(y)}{|y|}, \; m_1 + m_2 \ge 2, \tag{5.33}$$

where $\varepsilon(y) \to 0$ as $y \to 0$. It is clear that for two characteristics s_1, s_2 passing through a point $z = x + jy \in D$ and x_1, x_2 are the intersection points with the axis $y = 0$ respectively, for any two points $\tilde{z}_1 = \tilde{x}_1 + j\tilde{y} \in s_1, \tilde{z}_2 = \tilde{x}_2 + j\tilde{y} \in s_2$, we have

$$\begin{aligned} |\tilde{x}_1 - \tilde{x}_2| &\le |x_1 - x_2| = 2\left|\int_0^y \sqrt{-K(t)}dt\right| \\ &\le \frac{2k_0}{m+2}|y|^{1+m/2} \le \frac{k_1}{12}|y|^{1+m/2} \le M|y|^\beta, \end{aligned}$$

where $\beta = 1 + m/2$. From Condition C, we can assume that the coefficients in (5.29) are continuously differentiable with respect to $x \in L_0$ and satisfy the conditions

$$\begin{gathered} ||y|^{m_2}\tilde{A}_l|, ||y|^{m_2}\tilde{A}_{lx}|, ||y|^{m_2}\tilde{B}_l|, ||y|^{m_2}\tilde{B}_{lx}|, ||y|^{m_2}\tilde{D}_l|, \\ ||y|^{m_2}\tilde{D}_{lx}| \le k_0 \le k_1/12, ||y|^{m_2}\tilde{E}_l|, ||y|^{m_2}\tilde{E}_{lx}| \le k_1/2, \\ 2\sqrt{h},\ 1/\sqrt{h},\ |h_y/h| \le k_0 \le k_1/12 \text{ in } \overline{D^-},\ l = 1, 2, \end{gathered}$$

and later on we shall use the constants

$$\begin{gathered} M = 4\max[M_1, M_2, M_3], \\ M_1 = \max[8(k_1 d)^2/(1 - m'), M_3/k_1], \\ M_2 = (2 + m)k_0 d\delta^{-2-m}[4k_1\delta + 4\varepsilon_0 + m]/\delta, \\ M_3 = 2k_1^2|y_1'|^{-m'}[|y_1'|d + 1/2H(y_1')], \\ \gamma = \max[4k_1 d\delta^{\beta_1} + (4\varepsilon(y) + m)/2\beta'] < 1, \\ 0 \le |y| \le \delta, \end{gathered} \tag{5.34}$$

where $m' = m_2+\beta_1 < 1$, $\beta = 1-\beta_1$, $\beta' = (1+m/2)(1-3\beta_1)$, δ, β_1 are appropriate small positive constants, such that $(2+m)\beta_1 < 1$, and d is the diameter of D^-, $\varepsilon_0 = \max_{\overline{D^-}} \varepsilon(z)$, and $1/2H(y_1') \le k_0[(m+2)a_0/k_0]^{-m/(2+m)}$, and the positive number δ is small enough. We choose $v_0 = 0, \xi_0 = 0, \eta_0 = 0$ and substitute them into the corresponding positions of v, ξ, η in the right-hand sides of (5.22), and by the successive approximation, we find the sequences of functions $\{v_k\}, \{\xi_k\}, \{\eta_k\}$, which satisfy the relations

$$
\begin{aligned}
&v_{k+1}(z)=v_{k+1}(x)-2\int_0^y V_k(z)dy=v_{k+1}(x)+\int_0^y (\eta_k-\xi_k)dy,\\
&\xi_{k+1}(z)=\zeta_{k+1}(z)+\int_0^y g_{1k}(z)dy=\int_{y_1}^{\hat{y}} \hat{g}_{lk}dy,\\
&\eta_{k+1}(z)=\theta_{k+1}(z)+\int_0^y g_{2k}(z)dy=\int_{y_1}^{|y|} \hat{g}_{2k}(z)dy,\\
&g_{lk}(z)=\tilde{A}_l\xi_k+\tilde{B}_l\eta_k+\tilde{C}_l(\xi_k+\eta_k)+\tilde{D}_l v_k+\tilde{E}_l,\, l=1,2,\, k=0,1,2,...,
\end{aligned}
\tag{5.35}
$$

where $v(x) = u(x) - u_0(x)$ on L_0 as stated before, $z_1 = x_1 + jy_1$ is a point on L_1, which is the intersection of L_1 and the characteristic curve s_1 passing through the point $z = x + jy \in \overline{D^-}$. Setting that $\tilde{g}_{lk+1}(z) = g_{lk+1}(z) - g_{lk}(z)\,(l = 1, 2)$ and

$$
\begin{aligned}
&\tilde{y}=\hat{y}-y_1,\ \tilde{t}=\hat{t}-y_1,\ \tilde{v}_{k+1}(z)=v_{k+1}(z)-v_k(z),\\
&\tilde{\xi}_{k+1}(z)=\xi_{k+1}(z)-\xi_k(z),\ \tilde{\eta}_{k+1}(z)=\eta_{k+1}(z)-\eta_k(z),\\
&\tilde{\zeta}_{k+1}(z)=\zeta_{k+1}(z)-\zeta_k(z),\ \tilde{\theta}_{k+1}(z)=\theta_{k+1}(z)-\theta_k(z),
\end{aligned}
\tag{5.36}
$$

denote by $(k-m')!$ the product $(1-m')...(k-m')$, we can prove that $\{\tilde{v}_k\}$, $\{\tilde{\xi}_k\}$, $\{\tilde{\eta}_k\}$, $\{\tilde{\zeta}_k\}$, $\{\tilde{\theta}_k\}$ in D_0 satisfy the estimates

$$
\begin{aligned}
&|\tilde{v}_k(z)-\tilde{v}_k(x)|, |\tilde{\xi}_k(z)-\tilde{\zeta}_k(z)|, |\tilde{\eta}_k(z)-\tilde{\theta}_k(z)|\le M'\gamma^{k-1}|y|^{1-m'},\\
&|\tilde{\xi}_k(z)|, |\tilde{\eta}_k(z)| \le MM_0^{k-1}|\tilde{y}|^{k-m'}/(k-m')!,\text{ or } M'\gamma^{k-1},\\
&|\tilde{\xi}_k(z_1)-\tilde{\xi}_k(z_2)-\tilde{\zeta}_k(z_1)-\tilde{\zeta}_k(z_2)|, |\tilde{\eta}_k(z_1)-\tilde{\eta}_k(z_2)-\tilde{\theta}_k(z_1)-\tilde{\theta}_k(z_2)|\\
&\le M'\gamma^{k-1}[|x_1-x_2|^{1-m'}+|x_1-x_2|^{\beta}|t|^{\beta'}], 0\le|y|\le\delta, |\tilde{\xi}_k(z_1)-\tilde{\xi}_k(z_2)|,\\
&|\tilde{\eta}_k(z_1)-\tilde{\eta}_k(z_2)|\le MM_0^{k-1}|\tilde{t}|^{k-m'}|x_1-x_2|^{1-\beta}/(k-m')!,\\
&\text{or } M'\gamma^{k-1}|x_1-x_2|^{\beta}|t|^{\beta'}, |\tilde{\xi}_k(z)+\tilde{\eta}_k(z)-\tilde{\zeta}_k(z)-\tilde{\theta}_k(z)|\\
&\le M'\gamma^{k-1}|x_1-x_2|^{\beta}|y|^{\beta'}, |\tilde{\xi}_k(z)+\tilde{\eta}_k(z)| \le MM_0^{k-1}|\tilde{y}|^{k-m'}
\end{aligned}
$$

$$\times|x_1-x_2|^{1-\beta}/(k-m')!, \text{ or } M'\gamma^{k-1}|x_1-x_2|^\beta|y|^{\beta'}, \tag{5.37}$$

where $z = x + jy$, $z = x + jt$ is the intersection point of s_1, s_2 in (5.23) passing through z_1, z_2, γ is as stated in (5.34), d is the diameter of D^-, and M, M' is a sufficiently large positive constants as in (2.44), Chapter V and (5.34).

On the basis of the estimate (5.37) and the convergence of sequences $\{\tilde{v}_n(z)\}$, $\{\tilde{\xi}_n(z)\}$, $\{\tilde{\eta}_n(z)\}$, and the comparison test, we can derive that $\{v_n(z)\}$, $\{\xi_n(z)\}$, $\{\eta_n(z)\}$ in D_0 uniformly converge to $v_*(z)$, $\xi_*(z)$, $\eta_*(z)$ satisfying the system of integral equations

$$v_*(z)=v_*(x)-2\int_0^y V_*(z)dy=v_*(x)+\int_0^y(\eta_*(z)-\xi_*(z))dy,$$

$$\xi_*(z)=\zeta_*(z)+\int_0^y[\tilde{A}_1\xi_*+\tilde{B}_1\eta_*+\tilde{C}_1(\xi_*+\eta_*)+\tilde{D}_1u_*+\tilde{E}_1]dy,\ z\in s_1,$$

$$\eta_*(z)=\theta_*(z)+\int_0^y[\tilde{A}_2\xi_*+\tilde{B}_2\eta_*+\tilde{C}_2(\xi_*+\eta_*)+\tilde{D}_2u_*+\tilde{E}_2]dy,\ z\in s_2.$$

Noting that the arbitrariness of δ_0, hence there exists a solution of Problem A_2 for equation (5.29) in $D^-\cap\{-\delta\le y\le 0\}$. Next we apply the similar method in Section 1, the corresponding result in $D^-\cap\{y\le -\delta\}$ can be obtained. Hence $u(z) = v(z) + u_0(z)$ is a solution of Problem P for (5.1) in D^-. Thus the existence of solutions of Problem P for equation (5.1) is proved.

From the above result, we can obtain the following theorem.

Theorem 5.4 *If equation* (5.1) *satisfy Condition C and the condition* (5.33), *then the oblique derivative problem* (*Problem P*) *for* (5.1) *has a unique solution.*

Proof Now we prove the uniqueness of solutions of Problem P for equation (5.1), it suffices to verify that the corresponding homogeneous problem (Problem P_0) only has the trivial solution. The homogeneous equation (5.1) can be rewritten as

$$\begin{gathered}K_1(y)u_{xx}+\text{sgn}yK_2(y)u_{yy}+au_x+bu_y+cu=0, \text{ i.e.}\\ w_{\bar{\bar{z}}}=A_1w+A_2\overline{w}+A_3u \text{ in } D,\end{gathered} \tag{5.38}$$

where $u_{\bar{\bar{z}}}=w(z)$. Similarly to the proof of Theorem 4.3, we can prove that the solution $u(z)$ cannot attain the positive maximum in $\overline{D^+}\backslash L_0$, and its positive maximum M attains at a point on $L_0 = (0,2)$. However by means of the way as in Theorems 3.3 and 3.4, we can prove the solution $u(z) \equiv 0$ in $\overline{D^-}$ of Problem P for equation (5.38), obviously $u(z) = 0$ on L_0. Hence $u(z) \equiv 0$ in $\overline{D}$. This proves the uniqueness of the solution of Problem P For equation (5.1).

In [74], M. M. Smirnov investigated the unique solvability of the Tricomi problem for equation (5.1) with $K_1(y) = 1$, $K_2(y) = |y|^{m_2}\,(0 < m_2 < 1)$, $a = b = c = d = 0$ in D, by using the method of integral equations. In particular, when $m_2 = 1/2$, the Tricomi problem is a model of the plane-parallel symmetric Laval nozzle of given shape (the direct problem of Laval nozzle theory) (see [28]4)).

Remark 5.1 In the above discussion, we assume $m_2 < 1$, but if $|K_l| = |y|^{m_l}h_l(y)$, where $h_l(y)\,(l = 1,2)$ are continuously differentiable positive functions in $\overline{D}$, and the coefficients a, b, c, d of equations (5.1) satisfy the conditions, namely $a/|y|^{[m_2]}, b/|y|^{[m_2]}, c/|y|^{[m_2]}, d/|y|^{[m_2]}$ in D^+ are bounded and in $\overline{D}^-$ are continuous, where $[m_2]$ is the integer part of m_2, then the positive integers m_1, m_2 can be arbitrary, provided that $m_1 > [m_2]$. Moreover the coefficients $K_l(y)(j = 1,2)$ in equation (5.1) can be replaced by functions $K_l(x,y)(l = 1,2)$ with some conditions. Besides if the boundary condition (5.5) is replaced by

$$\frac{1}{2}\frac{\partial u}{\partial \nu} = \frac{1}{H_1(y)}\mathrm{Re}[\overline{\lambda(z)}u_{\tilde{z}}] = \mathrm{Re}[\overline{\Lambda(z)}u_z] = r(z) \text{ on } \Gamma\cup L_2,$$

$$\mathrm{Im}[\overline{\lambda(z)}u_{\tilde{z}}]|_{z=z_1} = b_1,\ u(0) = b_0,\ u(2) = b_2 \text{ or } \mathrm{Im}[\overline{\lambda(z)}u_{\tilde{z}}]|_{z=z_2} = b_2',$$

then similarly we can also obtain the above corresponding results.

Finally we mention that there are some results about the Tricomi problem for second order equations of mixed type with degenerate plane in higher dimensional domains (see [13], [36], [48], [71]2),6) and so on), but a lot of open problems remain to be continuously investigated, for instance see [71]2). We conjecture that it is possible the above problems can be discussed by the method of Clifford analysis (see [43]). Besides the Tricomi problem and oblique derivative problem for second order systems of nonlinear equations of mixed type with parabolic degeneracy are worth considering.

References

[1] R. Adams. *Sobolev spaces.* Academic Press, New York, 1975.

[2] S. Agmon, L. Nirenberg and M. H. Protter. *A maximum principle for a class of hyperbolic equations and applications to equations of mixed elliptic-hyperbolic type.* Comm. Pure Appl. Math., 6 (1953), 455-470.

[3] H. Begehr. 1) *Boundary value problems for mixed kind systems of first order partial differential equations.* 3. Roumanian-Finnish Seminar on Complex Analysis, Bucharest 1976, Lecture Notes in Math. 743, Springer-Verlag, Berlin, 1979, 600-614.
2) *Complex analytic methods for partial differential equations.* World Scientific, Singapore, 1994.

[4] H. Begehr and A. Dzhureav. *An introduction to several complex variables and partial differential equations.* Addison Wesley Longman, Harlow, 1997.

[5] H. Begehr and R. P. Gilbert. 1) *Pseudohyperanalytic functions.* Complex Variables, Theory Appl., 9 (1988), 343-357.
2) *Transformations, transmutations, and kernel functions* I, II. Longman, Harlow, 1992, 1993.

[6] H. Begehr and A. Jeffrey. 1) *Partial differential equations with complex analysis.* Longman, Harlow, 1992.
2) *Partial differential equations with real analysis.* Longman, Harlow, 1992.

[7] H. Begehr and Wei Lin. *A mixed-contact boundary problem in orthotropic elasticity.* Partial Differential Equations with Real Analysis, Longman, Harlow, 1992, 219-239.

[8] H. Begehr and Guo-chun Wen. 1) *The discontinuous oblique derivative problem for nonlinear elliptic systems of first order.* Rev. Roumanie Math. Pures Appl., 33 (1988), 7-19.
2) *A priori estimates for the discontinuous oblique derivative problem for elliptic systems.* Math. Nachr., 142 (1989), 307-336.
3) *Nonlinear elliptic boundary value problems and their applications.* Addison Wesley Longman, Harlow, 1996.

[9] L. Bers. 1) *Theory of pseudoanalytic functions.* New York University, New York, 1953.
2) *Mathematical aspects of subsonic and transonic gas dynamics.* Wiley, New York, 1958.

[10] L. Bers, F. John and M. Schechter. *Partial differential equations.* Interscience Publ., New York, 1964.

[11] L. Bers and L. Nirenberg. 1) *On a representation theorem for linear elliptic systems with discontinuous coefficients and its application.* Conv. Intern. Eq. Lin. Derivate Partiali, Trieste, Cremonense, Roma, 1954, 111-140.
2) *On linear and nonlinear elliptic boundary value problems in the plane.* Conv. Intern. Eq. Lin. Derivate Partiali, Trieste, Cremonense, Roma, 1954, 141-167.

[12] A. V. Bitsadze. 1) *Differential equations of mixed type.* Mac Millan Co., New York, 1964.
2) *Boundary value problems for elliptic equations of second order.* Nauka, Moscow, 1966 (Russian); Engl. Transl. North Holland Publ. Co., Amsterdam, 1968.
3) *Some classes of partial differential equations.* Gordon and Breach, New York, 1988.
4) *Partial differential equations.* World Scientific, Singapore, 1994.

[13] A. V. Bitsadze, and A. N. Nakhushev. *Theory of degenerating hyperbolic equations.* Dokl. Akad. Nauk SSSR, 204 (1972), 1289-1291 (Russian).

[14] B. Bojarski. 1) *Generalized solutions of a system of differential equations of first order and elliptic type with discontinuous coefficients.* Mat. Sb. N. S., 43(85) (1957), 451-563 (Russian).
2) *Subsonic flow of compressible fluid.* Arch. Mech. Stos., 18 (1966), 497-520; Mathematical Problems in Fluid Mechanics, Polon. Acad. Sci., Warsaw, 1967, 9-32.

[15] B. Bojarski and T. Iwaniec. *Quasiconformal mappings and non-linear elliptic equations in two variables* I, II. Bull. Acad. Polon. Sci., ér. Sci. Math. Astr. Phys., 22 (1974), 473-478; 479-484.

[16] F. Brackx, R. Delanghe and F. Sommen. *Clifford analysis.* Pitman, London, 1982.

[17] S. A. Chaplygin. *Gas jets, complete works.* Moscow-Leningrad, Vol. 2, 1933.

[18] D. Colton. 1) *Partial differential equations in the complex domain.* Pitman, London, 1976.
2) *Analytic theory of partial differential equations.* Pitman, London, 1980.

[19] R. Courant and H. Hilbert. *Methods of mathematical physics* II. Interscience Publ., New York, 1962.

[20] I. I. Daniljuk. 1) *Nonregular boundary value problems in the plane.* Izdat. Nauka, Moscow, 1975 (Russian).
2) *Selected works.* Naukova Dumka, Kiev, 1995 (Russian etc.).

[21] Ju. V. Devingtal′. *Existence and uniqueness of the solution of the Frankl problem.* Uspehi Mat. Nauk, 14 (1959), no.1 (85), 177-182 (Russian).

[22] Zheng-zhong Ding. *The general boundary value problem for a class of semilinear second order degenerate elliptic equations.* Acta Math. Sinica, 27 (1984), 177-191 (Chinese).

[23] Guang-chang Dong. 1) *Boundary value problems for degenerate elliptic partial differential equations.* Sci. Sinica, 13 (1964), 697-708.
2) *Nonlinear second order partial differential equations.* Amer. Math. Soc., Providence, RI, 1991.

[24] Guang-chang Dong and Min-you Chi. *Influence of Tricomi's mathematical work in China.* Mixed Type Equations, BSB Teubner, Leipzig, 90, 1986, 105-111.

[25] A. Douglis and L. Nirenberg. *Interior estimates for elliptic systems of partial differential equations.* Comm. Pure Appl. Math., 8 (1953), 503-538.

[26] A. Dzhuraev. 1) *Methods of singular integral equations.* Nauk SSSR, Moscow, 1987 (Russian); Engl. Transl. Longman, Harlow, 1992.
2) *Degenerate and other problems.* Longman, Harlow, 1992.

[27] Ai-nong Fang. *Quasiconformal mappings and the theory of functions for systems of nonlinear elliptic partial differential equations of first order.* Acta Math. Sinica, 23 (1980), 280-292 (Chinese).

[28] F. I. Frankl. 1) *On the problems of Chaplygin for mixed subsonic and supersonic flows.* Izv. Akad. Nuak SSSR Ser. Mat., 9 (1945), 121-143.
2) *Two gas-dynamical applications of the Lavrent′ev-Bitsadze boundary value problem.* Vest. Leningrad Univ. Ser. Mat. Meh. Astronom., 6 (1951), 3-7 (Russian).
3) *Gas flows past profiles with a local supersonic zone ending in a direct shock wave.* Prikl. Mat. Meh., 20 (1956), 196-202 (Russian).
4) *A generalization of the Tricomi problem and its application in solving the direct problem of the theory of the Laval nozzle.* Mat. Sb., 54(96) (1961), 225-236 (Russian). 5) *Selected works on gas dynamics.* Nauka, Moscow, 1973 (Russian).

[29] F. D. Gakhov. *Boundary value problems.* Fizmatgiz, Moscow, 1963 (Russian); Pergamon, Oxford, 1966.

[30] D. Gilbarg and N. S. Trudinger. *Elliptic partial differential equations of second order.* Springer-Verlag, Berlin, 1977.

[31] R. P. Gilbert. 1) *Function theoretic methods in partial differential equations.* Academic Press, New York, 1969.
2) *Plane ellipticity and related problems.* Amer. Math. Soc., Providence, RI, 1981.

[32] R. P. Gilbert and J. L. Buchanan. *First order elliptic systems: A function theoretic approach.* Academic Press, New York, 1983.

[33] R. P. Gilbert and G. N. Hile. 1) *Generalized hypercomplex function theory.* Trans. Amer. Math. Soc., 195 (1974), 1-29.
2) *Degenerate elliptic systems whose coefficients matrix has a group inverse.* Complex Variables, Theory Appl., 1 (1982), 61-88.

[34] R. P. Gilbert and Wei Lin. 1) *Algorithms for generalized Cauchy kernels.* Complex Variables, Theory Appl., 2 (1983), 103-124.
2) *Function theoretic solutions to problems of orthotropic elasticity.* J. Elasticity, 15 (1985), 143-154.

[35] R. P. Gilbert and Guo-chun Wen. 1) *Free boundary problems occurring in planar fluid dynamics.* Nonlinear Analysis, 13 (1989), 285-303.
2) *Two free boundary problems occurring in planar filtrations.* Nonlinear Analysis, 21 (1993), 859-868.

[36] Chao-hao Gu and Jia-xing Hong. *Some developments of the theory of partial differential equations of mixed type.* Mixed Type Equations, BSB Teubner, Leipzig, 90, 1986, 120-135.

[37] W. Haack and W. Wendland. *Lectures on Pfaffian and partial differential equations.* Pergamon Press, Oxford, 1972.

[38] G. N. Hile and M. H. Protter. 1) *Maximum principles for a class of first order elliptical systems.* J. Diff. Eq., 24(1) (1977), 136-151.
2) *Properties of overdetermined first order elliptic systems.* Arch. Rat. Mech. Anal., 66 (1977), 267-293.

[39] L. Hörmander. *Linear partial differential operators.* Springer-Verlag, Berlin, 1964.

[40] G. C. Hsiao and W. Wendland. *A finite element method for some integral equations of the first kind.* J. Math. Anal. Appl., 58 (1977), 449-481.

[41] Loo-Keng Hua. 1) *On Lavrent'ev's partial differential equation of the mixed type.* Sci. Sinica, 13 (1964), 1755-1762 (Chinese).
2) *Starting with the unit circle.* Springer-Verlag, Berlin, 1981.

[42] Loo-keng Hua, Wei Lin and Tzu-chien Wu. *Second order systems of partial differential equations in the plane.* Pitman, London, 1985.

[43] Sha Huang, Yu-ying Qiao and Guo-chun Wen. *Real and Complex Clifford Analysis.* Springer Verlag, Heidelberg, 2006.

[44] T. Iwaniec. *Quasiconformal mapping problem for general nonlinear systems of partial differential equations.* Symposia Math. 18, Acad. Press, London, 1976, 501-517.

[45] T. Iwaniec and A. Mamourian. *On the first order nonlinear differential systems with degeneration of ellipticity.* Proc. Second Finnish-Polish Summer School in Complex Analysis, (Jyväskylä 1983), Univ. Jyväskylä, 28 (1984), 41-52.

[46] Xin-hua Ji and De-quan Chen. 1) *The equations of mixed type of elliptic and hyperbolic in the n-dimensional real projective space.* Acta Math. Sinica, 23 (1980), 908-921.
2) *The non-homogeneous equations of mixed type in the real projective planer.* Mixed Type Equations, BSB Teubner, Leipzig, 90, 1986, 280-300.

[47] M. V. Keldych. *On certain cases of degeneration of equations of elliptic type on the boundary of a domain.* Dokl. Akad. Nauk SSSR (N.S.), 77 (1951), 181-183 (Russian).

[48] A. G. Kuz′min. *Equations of mixed type with non-classical behavior of characteristics.* Mixed Type Equations, BSB Teubner, Leipzig, 90, 1986, 180-194.

[49] O. A. Ladyzhenskaja and N. N. Uraltseva. *Linear and quasilinear elliptic equations.* Academic Press, New York, 1968.

[50] E. Lanckau and W. Tutschke. *Complex analysis: Methods, trends, and applications.* Akademie-Verlag, Berlin, 1983.

[51] M. A. Lavrent′ev and A. V. Bitsadze. *The problem of equations of mixed type.* Dokl. AN SSSR, 1950, Vol. 70, 373-376.

[52] M. A. Lavrent′ev and B. V. Shabat. *Methods of function theory of a complex variable.* GITTL, Moscow, 1958 (Russian).

[53] J. Leray. *Hyperbolic differential equations.* Princeton Univ. Press, 1954.

[54] J. Leray and J. Schauder. *Topologie et équations fonczionelles.* Ann. Sci. École Norm. Sup., 51 (1934), 45-78; YMH, 1 (1946), 71-95 (Russian).

[55] Ming-de Li and Yu-chun Qin. *On boundary value problems for singular elliptic equations.* Hangzhoudaxue Xuebao (Ziran Kexue), 1980., no.2, 1-8 (Chinese).

[56] Jian-bing Lin. *On some problems of Frankl.* Vestnik Leningrad Univ., 16 (1961), no.13, 28-34 (Russian).

[57] Jian-ke Lu. 1) *Boundary value problems for analytic functions.* World Scientific, Singapore, 1993.
2) *Complex variable methods in plane elasticity.* World Scientific, Singapore, 1995.

[58] Jian-ke Lu and Guo-chun Wen. *Boundary value problems, integral equations and related problems.* World Scientific, Singapore, 2000.

[59] L. G. Mikhailov. *A new class of singular integral equations and its applications to differential equations with singular coefficients.* Wolters-Noordhoff, Groningen, 1970.

[60] C. Miranda. *Partial differential equations of elliptic type.* Springer-Verlag, Berlin, 1970.

[61] V. N. Monakhov. 1) *Transformations of multiply connected domains by the solutions of nonlinear L-elliptic systems of equations.* Dokl. Akad. Nauk SSSR, 220 (1975), 520-523 (Russian).
2) *Boundary value problems with free boundaries for elliptic systems.* Amer. Math. Soc., Providence, RI, 1983.

[62] N. I. Mushelishvili. 1) *Singular integral equations.* Noordhoff, Groningen, 1953.
2) *Some basic problems of the mathematical theory of elasticity.* Nauka, Moscow, 1946 (Russian); Noordhoff, Groningen, 1953.

[63] J. Naas and W. Tutschke. *Some probabilistic aspects in partial complex differential equations.* Complex Analysis and its Applications. Akad. Nauk SSSR, Izd. Nauka, Moscow, 1978, 409-412.

[64] L. Nirenberg. 1) *On nonlinear elliptic partial differential equations and Hölder continuity.* Comm. Pure Appl. Math., 6 (1953), 103-156.
2) *An application of generalized degree to a class of nonlinear problems.* Coll. Analyse Fonct. Liége, 1970, Vander, Louvain, 1971, 57-74.

[65] O. A. Oleinik. *On equations of elliptic type degenerating on the boundary of a region.* Dokl. Akad. Nauk SSSR (N.S.), 87 (1952), 885-888.

[66] N. I. Popivanov and M. Schneider. *The Darboux problem in* $\mathbf{R}^3$ *for a class of degenerating hyperbolic equations.* J. Math. Anal. and Appl., 175(1993), 537-578.

[67] M. H. Protter. 1) *The Cauchy problem for a hyperbolic second order equation with data on the parabolic line.* Can. J. Math., 6 (1954), 542-553.
2) *An existence theorem for the generalized Tricomi problem.* Duke Math. J., 21 (1954), 1-7.

[68] M. H. Protter and H. F. Weinberger. *Maximum principles in differential equations.* Prentice-Hall, Englewood Cliffs, N. J., 1967.

[69] Deqian Pu. 1) *Function-theoretic process of hyperbolic equations.* Integral Equations and Boundary Value Problems, World Scientific, Singapore, 1991, 161-169.
2) *Monogenic functions of a complex variable.* Complex Variables and Elliptic Equations, 51 (2006), 999-1019.
3) *Monogenic functions of a complex variable.* Complex Variables and Elliptic Equations, 51(2007), 999-1019.

[70] Yu-ying Qiao. *The Cauchy problem for a hyperbolic equation of second order.* Boundary Value problems, Integral Equations and Related Problems, World Scientific, Singapore, 2000, 143-149.

[71] J. M. Rassias. 1) *Mixed type equations.* BSB Teubner, Leipzig, 90, 1986.
2) *Lecture notes on mixed type partial differential equations.* World Scientific, Singapore, 1990.
3) *The well-posed Tricomi problem in Euclidean plane.* Geometry Analysis ans Mechanics, World Sci. Publishing, 1994, 189-195.
4) *Uniqueness of quasi-regular solutions for a parabolic elliptic-hyperbolic Tricomi problem.* Bull. Inst. Math. Acad. Sinica, 25 (1997), no.4, 277-287.
5) *Uniqueness of quasi-regular solutions for a bi-parabolic elliptic bi-hyperbolic Tricomi problem.* Complex Variables, 47 (2002), 707-718.
6) *Mixed type partial differential equations in* **R**. Ph. D. dissertation, University of California, Berkeley, 1977.

[72] M. S. Salakhitdinov and B. Islomov. 1) *The Tricomi problem for the general linear equation of mixed type with a nonsmooth line of degeneracy.* Soviet Math. Dokl, 34 (1987), 133-136.
2) *Boundary value problems for an equation of mixed type with two interior lines of degeneracy.* Soviet Math. Dokl, 43 (1991), 235-238.

[73] M. Schneider. *The existence of generalized solutions for a class of quasilinear equations of mixed type.* J. Math. Anal. Appl., 107 (1985), 425-445.

[74] M. M. Smirnov. *Equations of mixed type.* Amer. Math. Soc., Providence, RI, 1978.

[75] S. L. Sobolev. *Applications of functional analysis in mathematical physics.* Amer. Math. Soc., Providence, RI, 1963.

[76] He-sheng Sun. 1) *The problems of the rigidity of the surfaces with mixed Gauss curvature and boundary value problems for the equations of mixed type.* Proc. 1980 Beijing Sym. Diff. Geom. Diff. Eq., Beijing, 1982, 1441-1450.
2) *Generalized Tricomi problem for the equation of mixed type of second order.* Kexue Tongbao, 30 (1985), 1441-1446.
3) *Tricomi problem for nonlinear equation of mixed type.* Sci. in China (Series A), 35 (1992), 14-20.
4) *First order system of equations of mixed type and second order equation of mixed type.* J. Partial Diff. Eqs., 6 (1993), 155-163.
5) *Nonlinear system of equations of mixed type and nonlinear complex equation of mixed type.* Nonlinear Evolutionary Partial Differential Equations, AMS/IP Stud. Adv. Math., 3, 1997, 393-401.

[77] Mao-ying Tian. *The mixed boundary value problem for nonlinear degenerate elliptic equations of second order in the plane.* Acta Sci. Natur. Univ. Peki., 1982, 5:1-13 (Chinese).

[78] F. G. Tricomi. 1) *Sulle equazioni lineari alle derivate parziali di 2° ordine.* di tipo misto, atti Accad. Naz. Lincei Mem. Cl. Sci. Fis. Nat., (5) 14 (1923), 133-247 (1924).
2) *Repertorium der Theorie der Differentialgleichungen.* Springer-Verlag, Berlin, 1968.

[79] N. S. Trudinger. *Nonlinear oblique boundary value problems for nonlinear elliptic equations.* Trans. Amer. Math. Soc., 295 (1986), 509-546.

[80] W. Tutschke. 1) *The Riemann-Hilbert problem for nonlinear systems of differential equations in the plane.* Complex Analysis and its Applications. Akad. Nauk SSSR, Izd. Nauka, Moscow, 1978, 537-542.
2) *Boundary value problems for generalized analytic functions for several complex variables.* Ann. Pol. Math., 39 (1981), 227-238.

[81] I. N. Vekua. 1) *Generalized analytic functions.* Pergamon, Oxford, 1962.
2) *New methods for solving elliptic equations.* North-Holland Publ., Amsterdam, 1967.

[82] N. D. Vvedenskaya. *On a boundary problem for equations of elliptic type degenerating on the boundary of a region.* Dokl. Akad. Nauk SSSR (N.S.), 91 (1953), 711-714 (Russian).

[83] Chung-fang Wang. *Dirichlet problems for singular elliptic equations.* Hangzhoudaxue Xuebao (Ziran Kexue), 1978, no.2, 19-32 (Chinese).

[84] W. Warschawski. *On differentiability at the boundary in conformal mapping.* Proc. Amer. Math. Soc., 12 (1961), 614-620.

[85] E. Wegert. *Nonlinear boundary value problems for homeomorphic functions and singular integral equations.* Akademie-Verlag, Berlin, 1992.

[86] Guo-chun Wen. 1) *Modified Dirichlet problem and quasiconformal mappings for nonlinear elliptic systems of first order.* Kexue Tongbao (A monthly J. Sci.), 25 (1980), 449-453.
2) *The Riemann-Hilbert boundary value problem for nonlinear elliptic systems of first order in the plane.* Acta. Math. Sinica, 23 (1980), 244-255 (Chinese).
3) *The singular case of the Riemann-Hilbert boundary value problem.* Beijingdaxue Xuebao (Acta Sci. Natur. Univ. Peki.), 1981, no.4, 1-14 (Chinese).
4) *The mixed boundary value problem for nonlinear elliptic equations of second order in the plane.* Proc. 1980 Beijing Sym. Diff. Geom. Diff. Eq., Beijing, 1982, 1543-1557.
5) *On a representation theorem of solutions and mixed boundary value problems for second order nonlinear elliptic equations with unbounded measurable coefficients.* Acta Math. Sinica, 26 (1983), 533-537 (Chinese).
6) *Oblique derivative boundary value problems for nonlinear elliptic systems of second order.* Scientia Sinica, Ser. A, 26 (1983), 113-124.
7) *Some nonlinear boundary value problems for nonlinear elliptic equations of second order in the plane.* Complex Variables, Theory Appl., 4 (1985), 189-210.
8) *Applications of complex analysis to nonlinear elliptic systems of partial differential equations.* Analytic Functions of one Complex Variable, Amer. Math. Soc., Providence, RI, 1985, 217-234.
9) *Linear and nonlinear elliptic complex equations.* Shanghai Science Techn. Publ., Shanghai, 1986 (Chinese).
10) *Some boundary value problems for nonlinear degenerate elliptic complex equations.* Lectures on Complex Analysis, World Scientific, Singapore, 1988, 265-281.
11) *Conformal mappings and boundary value problems.* Amer. Math. Soc., Providence, RI, 1992.
12) *Function theoretic methods for partial differential equations and their applications and computations.* Advan. Math., 22 (1993), 391-402.

13) *Nonlinear irregular oblique derivative problems for fully nonlinear elliptic equations.* Acta Math. Sci., 15 (1995), 82-90.
14) *Some boundary value problems for linear mixed complex equations of first and second order.* J. Yantai Univ. (Natur. Sci. and Engin.), 1997, no.3, 1-8.
15) *A free boundary problem in axisymmetric filtration with two fluids.* Acta Math. Appl. Sinica, 1998, 139-143.
16) *Complex forms and boundary value problems of second order quasi-linear equations of mixed type.* J. Yantai Univ. (Natur. Sci. and Eigin.), 11(1998), 235-242.
17) *Oblique derivative problems for linear second order equations of mixed type.* Science in China, 41(1998), 346-356.
18) *Applications of complex analysis to second order equations of mixed type.* Ann. Polon. Math., LXX(1998), 221-231.
19) *A free boundary problem in axisymmetric filtration with two fluids.* Acta Math. App. Sinica, 21(1998), 139-143 (Chinese).
20) *A new approach to quasilinear equations of mixed type in general domains.* Progress in Natural Science, 9(1999), 580-586.
21) *The Riemann-Hilbert problem for first order complex equations of mixed type.* Partial Differential and Integral Equations, Kluwer, 1999, 97-105.
22) *The Riemann-Hilbert boundary value problem for first order quasi-linear complex equations of mixed type.* Sys. Sci. and Math. Sci., 12 (1999), 357-365.
23) *Oblique derivative problems for second order nonlinear equations of mixed type in multiply connected domains.* Comm. in Nonlinear Sci. & Numer. Simu., 1999, 281-284.
24) *Complex analytic method for hyperbolic equations of second order.* Complex Methods for Partial Differential Equations, Kluwer Academic Publishers, 1999, 271-288.
25) *Approximate methods and numerical analysis for elliptic complex equations.* Gordon and Breach Science Publishers, Amsterdam, 1999.
26) *Linear and nonlinear parabolic complex equations.* World Scientific, Singapore, 1999.
27) *Nonlinear partial differential complex equations.* Science Press, Beijing, 1999 (Chinese).
28) *The Riemann-Hilbert problem for nonlinear elliptic complex equations of first order in multiply connected infinite domains.* Partial Differential and Integral Equations, Kluwer, 1999, 85-95.
29) *Discontinuous boundary value problems for first order systems of mixed type.* Boundary Value problems, Integral Equations and Related Problems, World Scientific, Singapore, 2000, 209-216.

30) *Discontinuous oblique derivative problems for quasilinear mixed equations of second order.* Advan. in Math., 29 (2000), 554-562.
31) *Oblique derivative problems for second order equations of mixed type in multiply connected domains.* Proc. of the Second ISAAC Congress, 2000, 1539-1546.
32) *The Frankl problem for second order quasilinear equations of mixed type.* Acta Math. Appl. Sinica, 24 (2001), 204–211 (Chinese).
33) *Linear and quasilinear equations of hyperbolic and mixed types.* Taylor and Francis, London, 2002.
34) *Boundary value problems for degenerate hyperbolic systems of first order equations.* Complex Variables, Theory Appl., 47 (2002), 839-851.
35) *Uniqueness of oblique derivative problems of second order equations of mixed type with nonsmooth degenerate line.* Complex Variables, Theory and Application, 49 (2004), 257-270.
36) *The mixed problem for degenerate elliptic equations of second order.* Analysis in Theory and Applications, 21 (2005), 118-131.
37) *The Tricomi problem for second order linear mixed equations with parabolic degeneracy.* Applicable Analysis, 84 (2005), 1267-1286.
38) *The Riemann-Hilbert problem for first order elliptic complex equations with parabolic degeneracy.* Acta Sci. Natur., Univ. Bekin. 41 (2005), 47-54.
39) *Mixed boundary value problems for nonlinear elliptic equations of second order equations with measurable coefficients in high dimensional domains.* Complex Variables and Elliptic Equations, 51 (2006), 267-276.
40) *The mixed problem for a nonlinear degenerate elliptic equation of second order.* Comm. in Nonlinear Sci. & Numer. Simu., 11 (2006), 709-720.
41) *The Cauchy problem for general quasilinear hyperbolic equations with degenerate rank 0.* J. Sichuan Normal Univ.(Natur. Sci.), 29 (2006), 127-134.
42) *The oblique derivative problem for degenerate elliptic equations of second order.* Acta Sci. Natur., Univ. Bekin., 42 (2006), 29-34.
43) *The exterior Tricomi problem for generalized mixed equations with parabolic degeneracy.* Acta Mathematica Sinica, 22 (2006), 1358-1398.
44) *Oblique derivative problems for second order quasilinear equations of mixed type with parabolic degeneracy.* Acta Sci. Natur., Univ. Bekin., 43 (2006), 285-293.
45) *The mixed boundary value problem for second order elliptic equations with degenerate curve on the sides of angle.* Math. Nachr. 279 (2006), 1602-1613.

[87] Guo-chun Wen and H. Begehr. 1) *Boundary value problems for elliptic equations and systems.* Longman, Harlow, 1990.
2) *Existence of solutions of Frankl problem for general Lavrent′ev-Bitsadze equations.* Rev. Roumaine Math. Pure Appl., 45 (2000), 141-160.

[88] Guo-chun Wen and De-chang Chen. 1) *Oblique derivative problems for nonlinear elliptic equations of second order equations with measurable coefficients in high dimensional domains.* Comm. in Nonlinear Sci. & Numer. Simu. 10 (2005), 559-570.
2) *Some boundary value problems for nonlinear elliptic equations of second order in high dimensional domains.* Applicable Analysis, 85(2006), 1011-1023.
3) *Discontinuous Riemann-Hilbert problems for quasilinear degenerate elliptic complex equations of first order.* Complex Variables, Theory Appl., 50(2005), 707-718.

[89] Guo-chun Wen and Ping-qian Li. 1) *The weak solution of Riemann-Hilbert problems for elliptic complex equations of first order.* Appl. Anal., 45 (1992), 209-227.
2) *Oblique derivative problems for nonlinear systems of second order equations.* Boundary Value problems, Integral Equations and Related Problems, World Scientific, Singapore, 2000, 98-104.

[90] Guo-chun Wen and Zhao-fu Luo. *Hyperbolic complex functions and hyperbolic pseudoregular functions.* J. Ningxia Univ. (Natur. Sci.), 19 (1998), no.1, 12-18.

[91] Guo-chun Wen, Chung-wei Tai and Mao-ying Tian. *Function theoretic methods of free boundary problems and their applications to mechanics.* Higher Education Press, Beijing, 1995 (Chinese).

[92] Guo-chun Wen and Mao-ying Tian. 1) *Solutions for elliptic equations of second order in the whole plane.* J. Math., 2 (1982), no.1, 23-36 (Chinese).
2) *Oblique derivative problems for nonlinear equations of mixed type in general domains.* Comm. in Nonlinear Sci. & Numer. Simu., 34 (1998), 148-151.
3) *Generalizations of Keldych-Sedov formula and their applications.* Acta Sci. Natur., Univ. Bekin., 35(1999), 47-55.
4) *Oblique derivative problems for quasilinear equations of mixed type in general domains* I. Progress in Natural Science, 4 (1999), no.1, 88-95.
5) *The oblique derivative problem for general second order quasilinear equations of mixed type.* Applicable Analysis, 73(1999), 255-270.

6) *Discontinuous boundary value problems for elliptic complex equations.* Boundary Value Problems, Integral Equations and Related Problems, 2000, 183-190.

[93] Guo-chun Wen and Wen-sui Wu. *The complex form of hyperbolic systems of first order equations in the plane.* J. Sichuan Normal University (Natur. Sci.), 1994, no.2, 92-94 (Chinese).

[94] Guo-chun Wen and C. C. Yang. 1) *Some discontinuous boundary value problems for nonlinear elliptic systems of first order in the plane.* Complex Variables, Theory Appl., 25 (1994), 217-226.
2) *On general boundary value problems for nonlinear elliptic equations of second order in a multiply connected domain.* Acta Applicandae Mathematicae, 43 (1996), 169-189.

[95] Guo-chun Wen, Guang-wu Yang and Sha Huang. *Generalized analytic functions and their generalizations.* Hebei Education Press, Hebei, 1989 (Chinese).

[96] Guo-chun Wen and Zhen Zhao. *Integral equations and boundary value problems.* World Scientific, Singapore, 1991.

[97] W. Wendland. 1) *On the imbedding method for semilinear first order elliptic systems and related finite element methods.* Continuation methods, Academic Press, New York, 1978, 277-336.
2) *Elliptic systems in the plane.* Pitman, London, 1979.

[98] Xi-jin Xiang. *Pan-complex functions and its applications in mathematics and physics.* Northeast Normal University Press, Changchun, 1988 (Chinese).

[99] C. C. Yang, G. C. Wen, K. Y. Li and Y. M. Chiang. *Complex analysis and its applications.* Longman, Harlow, 1993.

Index